Lecture Notes in Physics

Springer
Berlin
Heidelberg
New York
Barcelona
Hong Kong
London
Milan
Paris
Singapore
Tokyo

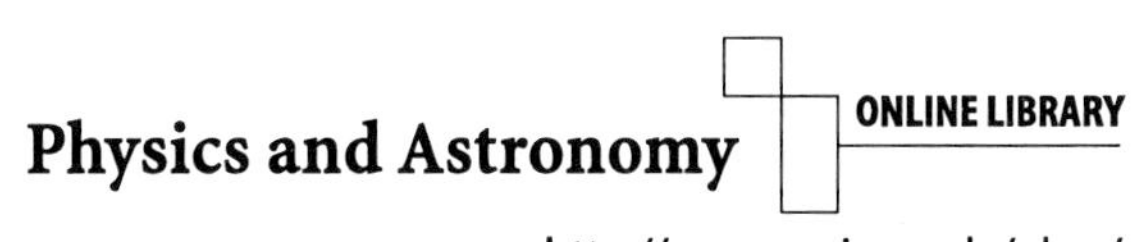

The Editorial Policy for Proceedings

The series Lecture Notes in Physics reports new developments in physical research and teaching – quickly, informally, and at a high level. The proceedings to be considered for publication in this series should be limited to only a few areas of research, and these should be closely related to each other. The contributions should be of a high standard and should avoid lengthy redraftings of papers already published or about to be published elsewhere. As a whole, the proceedings should aim for a balanced presentation of the theme of the conference including a description of the techniques used and enough motivation for a broad readership. It should not be assumed that the published proceedings must reflect the conference in its entirety. (A listing or abstracts of papers presented at the meeting but not included in the proceedings could be added as an appendix.)
When applying for publication in the series Lecture Notes in Physics the volume's editor(s) should submit sufficient material to enable the series editors and their referees to make a fairly accurate evaluation (e.g. a complete list of speakers and titles of papers to be presented and abstracts). If, based on this information, the proceedings are (tentatively) accepted, the volume's editor(s), whose name(s) will appear on the title pages, should select the papers suitable for publication and have them refereed (as for a journal) when appropriate. As a rule discussions will not be accepted. The series editors and Springer-Verlag will normally not interfere with the detailed editing except in fairly obvious cases or on technical matters.
Final acceptance is expressed by the series editor in charge, in consultation with Springer-Verlag only after receiving the complete manuscript. It might help to send a copy of the authors' manuscripts in advance to the editor in charge to discuss possible revisions with him. As a general rule, the series editor will confirm his tentative acceptance if the final manuscript corresponds to the original concept discussed, if the quality of the contribution meets the requirements of the series, and if the final size of the manuscript does not greatly exceed the number of pages originally agreed upon. The manuscript should be forwarded to Springer-Verlag shortly after the meeting. In cases of extreme delay (more than six months after the conference) the series editors will check once more the timeliness of the papers. Therefore, the volume's editor(s) should establish strict deadlines, or collect the articles during the conference and have them revised on the spot. If a delay is unavoidable, one should encourage the authors to update their contributions if appropriate. The editors of proceedings are strongly advised to inform contributors about these points at an early stage.
The final manuscript should contain a table of contents and an informative introduction accessible also to readers not particularly familiar with the topic of the conference. The contributions should be in English. The volume's editor(s) should check the contributions for the correct use of language. At Springer-Verlag only the prefaces will be checked by a copy-editor for language and style. Grave linguistic or technical shortcomings may lead to the rejection of contributions by the series editors. A conference report should not exceed a total of 500 pages. Keeping the size within this bound should be achieved by a stricter selection of articles and not by imposing an upper limit to the length of the individual papers. Editors receive jointly 30 complimentary copies of their book. They are entitled to purchase further copies of their book at a reduced rate. As a rule no reprints of individual contributions can be supplied. No royalty is paid on Lecture Notes in Physics volumes. Commitment to publish is made by letter of interest rather than by signing a formal contract. Springer-Verlag secures the copyright for each volume.

The Production Process

The books are hardbound, and the publisher will select quality paper appropriate to the needs of the author(s). Publication time is about ten weeks. More than twenty years of experience guarantee authors the best possible service. To reach the goal of rapid publication at a low price the technique of photographic reproduction from a camera-ready manuscript was chosen. This process shifts the main responsibility for the technical quality considerably from the publisher to the authors. We therefore urge all authors and editors of proceedings to observe very carefully the essentials for the preparation of camera-ready manuscripts, which we will supply on request. This applies especially to the quality of figures and halftones submitted for publication. In addition, it might be useful to look at some of the volumes already published. As a special service, we offer free of charge LaTeX and TeX macro packages to format the text according to Springer-Verlag's quality requirements. We strongly recommend that you make use of this offer, since the result will be a book of considerably improved technical quality. To avoid mistakes and time-consuming correspondence during the production period the conference editors should request special instructions from the publisher well before the beginning of the conference. Manuscripts not meeting the technical standard of the series will have to be returned for improvement.
For further information please contact Springer-Verlag, Physics Editorial Department II, Tiergartenstrasse 17, D-69121 Heidelberg, Germany

Series homepage – http://www.springer.de/phys/books/lnpp

Bernd G. Schmidt (Ed.)

Einstein's Field Equations and Their Physical Implications

Selected Essays in Honour of Jürgen Ehlers

Springer

Editor

Bernd G. Schmidt
Max-Planck-Institut für Gravitationsphysik
Albert-Einstein-Institut
Am Mühlenberg 1
14476 Golm, Germany

Library of Congress Cataloging-in-Publication Data applied for.

Die Deutsche Bibliothek - CIP-Einheitsaufnahme

Einstein's field equations and their physical implications : selected essays
in honour of Jürgen Ehlers / Bernd G. Schmidt (ed.). - Berlin ;
Heidelberg ; New York ; Barcelona ; Hong Kong ; London ; Milan ;
Paris ; Singapore ; Tokyo : Springer, 2000
(Lecture notes in physics ; Vol. 540)
ISBN 3-540-67073-4

ISSN 0075-8450
ISBN 3-540-67073-4 Springer-Verlag Berlin Heidelberg New York

Typesetting: Camera-ready by the authors/editor
Cover design: *design & production*, Heidelberg

Printed on acid-free paper
SPIN: 10720694 55/3144/du - 5 4 3 2 1 0

Jürgen Ehlers

In the 1950s the mathematical department of Hamburg University, with its stars Artin, Blaschke, Collatz, Kähler, Peterson, Sperner and Witt had a strong drawing power for Jürgen Ehlers, student of mathematics and physics. Since he had impressed his teachers he could well have embarked on a distinguished career in mathematics had it not been for Pascual Jordan and – I suspect – Hermann Weyl's *Space–Time–Matter*.

Jordan had just published his book *"Schwerkraft und Weltall"* which was a text on Einstein's theory of gravitation, developing his theory of a variable gravitational "constant". Only the rudiments of this theory had been formulated and Jordan, overburdened with countless extraneous commitments, was eager to find collaborators to develop his theory. This opportunity to break new ground in physics enticed Jürgen Ehlers and Wolfgang Kundt to help Jordan with his problems, and their work was acknowledged in the 1955 second edition of Jordan's book.

It didn't take Jürgen, who always was a systematic thinker, long to realize that not only Jordan's generalization but also Einstein's theory itself needed a lot more work. This impression was well described by Kurt Goedel in 1955 in a letter to Carl Seelig: "My own work in relativity theory refers to the pure gravitational theory of 1916 of which I believe that it was left by Einstein himself and the whole contemporary generation of physicists as a torso – and in every respect, physically, mathematically, and its applications to cosmology".

When asked by Seelig to elaborate , Goedel added: "Concerning the completion of gravitational theory of which I wrote in my last letter I do not mean a completion in the sense that the theory would cover a larger domain of phenomena (Tatsachenbereich), but a mathematical analysis of the equations that would make it possible to attempt their solution systematically and to find their general properties. Until now one does not even know the analogs of the fundamental integral theorems of Newtonian theory which, in my opinion, have to exist without fail. Since such integral theorems and other mathematical lemmas would have a physical meaning, the physical understanding of the theory would be enhanced. On the other hand, a closer analysis of the physical content of the theory could lead to such mathematical theorems".

Such a view of Einstein's theory was also reflected in the talks and discussions of the "Jordan Seminar". This was a weekly meeting of Jordan's coworkers in the Physics Department of Hamburg University to discuss Jordan's theory of a variable gravitational scalar. However, under Jürgen's leadership, the structure and interpretation of Einstein's original theory became the principal theme of nearly all talks. Jordan, who found little time to contribute actively to his theory, reluctantly went along with this change of topic. Through grants from the US Air Force and other sources he provided the logistic support for his research group. For publication of the lengthy research papers on Einstein's theory of gravitation by Ehlers, Kundt, Ozsvath, Sachs and Trümper, he made the proceedings of the Akademie der Wissenschaften und der Literatur in Mainz available. Jordan appeared often as coauthor, but I doubt whether he contributed much more than suggestions in style, like never to start a sentence with a formula. Some results were also written up as reports for the Air Force and became known as the Hamburg Bible.

It was a principal concern in Jürgen's contributions to Einstein's theory to clarify the mathematics, separate proof from conjecture and insist on invariance as well as elegance. This clear and terse style, which always kept physical interpretation in mind, appeared already in his Hamburg papers. His work in relativity resulted not only in books, published papers, supervised theses, critical remarks in discussions and suggestions for future work. By establishing the "Albert–Einstein–Institut" Jürgen designed a unique international center for research in relativity. As the founding director of this "Max–Planck–Institut für Gravitationsphysik" in Brandenburg, he has led it to instant success. Through his leadership, research on Einstein's theory in Germany is flourishing again and his work and style has set a standard for a whole generation of researchers.

Engelbert Schücking

Preface

The contributions in this book are dedicated to Jürgen Ehlers on the occasion of his 70th birthday. I have tried to find topics which were and are near to Jürgen's interests and scientific activities. I hope that the book – even in the era of electronic publishing – will serve for some time as a review of the themes treated; a source from which, for example, a PhD student could learn certain things thoroughly. In initiating the project of the book, the model I had in mind was the "Witten book".

Early in his career Jürgen Ehlers worked on exact solutions, and demonstrated how one goes about characterizing exact solutions invariantly and searching for their intrinsic geometrical properties. So, it seems appropriate to begin the book with the article by J. Bičák: "Selected Solutions of Einstein's Field Equations: Their Role in General Relativity and Astrophysics." Certainly not all of the large number of known exact solutions are of equal weight; this article describes the most important ones and explains their role for the development and understanding of Einstein's theory of gravity.

The second contribution is the article by H. Friedrich and A. Rendall: "The Cauchy Problem for the Einstein Equations". It contains a careful exposition of the local theory, including the delicate gauge questions and a discussion of various ways of writing the equations as hyperbolic systems. Furthermore, it becomes clear that an understanding of the Cauchy problem really gives new insight into properties of the equations and the solutions and not just "uniqueness and existence".

"Post-Newtonian Gravitational Radiation" is the title of the article by L. Blanchet. It deals with a topic Jürgen has contributed to and thought about deeply. However, these matters have developed in such a way that presently only a small number of experts understand all the technical details and subtleties. Hopefully, this present contribution will help us gain some understanding of certain aspects of post-Newtonian approximations.

The fourth contribution, "Duality and Hidden Symmetries in Gravitational Theories", by D. Maison, outlines how far one of Jürgen's creations, the "Ehlers transformation" has evolved. From a "trick" to produce new solutions from known ones, the presence of such transformations in the space of solutions is now seen as a structural property of various gravitational theories, which at present attract a lot of attention.

The contribution, by R. Beig and B. Schmidt, "Time-Independent Gravitational Fields" collects and describes what is known about global properties of time-independent spacetimes. It contains, in particular, a fairly self-contained description of the multipole expansion at infinity.

V. Perlick has written on "Gravitational Lensing from a Geometric Viewpoint". In the last ten years, lensing has become a fascinating new part of observational astrophysics. However, there are still important and interesting conceptual and mathematical questions when one tries to compare practical astrophysical applications with their mathematical modelling in Einstein's theory of gravity. Some of those issues are treated in this contribution.

Obviously, there are some subjects missing, for which I was not able to find a contribution. What I regret most is that there is no article on cosmology, a field in which Jürgen has always been very interested.

An intruiging thought about the book is that Juergen would have read all these contributions before publication and no doubt improved them by his constructive criticism. For a short while I had in mind to ask Jürgen to do just this, but finally I decided that this would be too much of a burden for a birthday present.

Finally, I would like to thank the authors, friends and colleagues who have helped me and have given valuable advice.

Bernd Schmidt

Contents

Selected Solutions of Einstein's Field Equations: Their Role in General Relativity and Astrophysics

Jiří Bičák

Institute of Theoretical Physics,
Charles University, Prague

1 Introduction and a Few Excursions

The primary purpose of all physical theory is rooted in reality, and most relativists pretend to be physicists. We may often be members of departments of mathematics and our work oriented towards the mathematical aspects of Einstein's theory, but even those of us who hold a permanent position on "scri", are primarily looking there for gravitational waves. Of course, the builder of this theory and its field equations was *the* physicist. Jürgen Ehlers has always been very much interested in the conceptual and axiomatic foundations of physical theories and their rigorous, mathematically elegant formulation; but he has also developed and emphasized the importance of such areas of relativity as kinetic theory, the mechanics of continuous media, thermodynamics and, more recently, gravitational lensing. Feynman expressed his view on the relation of physics to mathematics as follows [1]:

"The physicist is always interested in the special case; he is never interested in the general case. He is talking about something; he is not talking abstractly about anything. He wants to discuss the gravity law in three dimensions; he never wants the arbitrary force case in n dimensions. So a certain amount of reducing is necessary, because the mathematicians have prepared these things for a wide range of problems. This is very useful, and later on it always turns out that the poor physicist has to come back and say, 'Excuse me, when you wanted to tell me about four dimensions...' " Of course, this is Feynman, and from 1965...

However, physicists are still rightly impressed by special explicit formulae. Explicit solutions enable us to discriminate more easily between a "physical" and "pathological" feature. Where are there singularities? What is their character? How do test particles and fields behave in given background spacetimes? What are their global structures? Is a solution stable and, in some sense, generic? Clearly, such questions have been asked not only within general relativity.

By studying a *special* explicit solution one acquires an intuition which, in turn, stimulates further questions relevant to more *general* situations. Consider, for example, charged black holes as described by the Reissner–Nordström solution. We have learned that in their interior a Cauchy horizon exists and that the singularities are timelike. We shall discuss this in greater detail in Sect. 3.1. The singularities can be seen by, and thus exert an influence on, an observer travelling in their neighborhood. However, will this violation of the (strong) cosmic censorship persist when the black hole is perturbed by weak ("linear") or even strong ("nonlinear") perturbations? We shall see that, remarkably, this question can also be studied by explicit exact special model solutions. Still more surprisingly, perhaps, a similar question can be addressed and analyzed by means of explicit solutions describing completely diverse situations – the collisions of plane waves. As we shall note in Sect. 8.3, such collisions may develop Cauchy horizons and subsequent timelike singularities. The theory of black holes and the theory of colliding waves have intriguing structural similarities which, first of all, stem from the circumstance that in both cases there exist two symmetries, i.e. two Killing fields. What, however, about more general situations? This is a natural question inspired by the explicit solutions. Then "the poor physicists have to come back" to a mathematician, or today alternatively, to a numerical relativist, and hope that somehow they will firmly learn whether the cosmic censorship is the "truth", or that it has been a very inspirational, but in general false conjecture. However, even *after* the formulation of a conjecture about a general situation inspired by particular exact solutions, *newly* discovered exact solutions can play an important role in verifying, clarifying, modifying, or ruling out the conjecture. And also "old" solutions may turn out to act as asymptotic states of general classes of models, and so become still more significant.

Exact explicit solutions have played a crucial role in the development of many areas of physics and astrophysics. Later on in this Introduction we will take note of some general features which are specific to the solutions of Einstein's equations. Before that, however, for illustration and comparison we shall indicate briefly with a few examples what influence exact explicit solutions have had in other physical theories. Our next introductory excursion, in Sect. 1.2, describes in some detail the (especially early) history of Einstein's route to the gravitational field equations for which his short stay in Prague was of great significance. The role of Ernst Mach (who spent 28 years in Prague before Einstein) in the construction of the first modern cosmological model, the Einstein static universe, is also touched upon. Section 1.3 is devoted to a few remarks on some old and new impacts of the other simplest "cosmological" solutions of Einstein's equations – the Minkowski, the de Sitter, and the anti de Sitter spacetimes. Some specific features of solutions in Einstein's theory, such as the observability and interpretation of metrics, the role of general covariance, the problem of the equivalence of two metrics, and of geometrical characterization of solutions are mentioned in Sect. 1.4. Finally, in the last (sub)sections of the "Introduction" we give some reasons why we consider our choice of solutions to be "a natural selection", and we briefly outline the main body of the article.

1.1 A Word on the Role of Explicit Solutions in Other Parts of Physics and Astrophysics

Even in a linear theory like Maxwell's electrodynamics one needs a good sample, a useful kit, of exact fields like the homogeneous field, the Coulomb field, the dipole, the quadrupole and other simple solutions, in order to gain a physical intuition and understanding of the theory. Similarly, of course, with the linearized theory of gravity. Going over to the Schrödinger equation of standard quantum mechanics, again a linear theory, consider what we have learned from simple, explicitly soluble problems like the linear and the three-dimensional harmonic oscillator, or particles in potential wells of various shapes. We have acquired, for example, a transparent insight into such basic quantum phenomena as the existence of minimum energy states whose energy is not zero, and their associated wave functions which have a certain spatial extent, in contrast to classical mechanics. The three-dimensional problems have taught us, among other things, about the degeneracy of the energy levels. The case of the harmonic oscillator is, of course, very exceptional since Hamiltonians of the same type appear in all problems involving quantized oscillations. One encounters them in quantum electrodynamics, quantum field theory, and likewise in the theory of molecular and crystalline vibrations. It is thus perhaps not so surprising that the Hamiltonian and the wave functions of the harmonic oscillator arise even in the minisuperspace models associated with the Hartle–Hawking no-boundary proposal for the wave function of the universe [2], and in the minisuperspace model of homogeneous spherically symmetric dust filled universes [3].

In *nonlinear* problems explicit solutions play still a greater role since to gain an intuition of nonlinear phenomena is hard. Landau and Lifshitz in their Fluid Mechanics (Volume 6 of their course) devote a whole section to the exact solutions of the nonlinear Navier–Stokes equations for a viscous fluid (including Landau's own solution for a jet emerging from the end of a narrow tube into an infinite space filled with fluid).

Although Poisson's equation for the gravitational potential in the classical theory of gravity is linear, the combined system of equations describing both the field and its *fluid* sources (not rigid bodies, these are simple!) characterized by Euler's equations and an equation of state are nonlinear. In classical astrophysical fluid dynamics perhaps the most distinct and fortunate example of the role of explicit solutions is given by the exact descriptions of ellipsoidal, uniform density masses of self-gravitating fluids. These "ellipsoidal figures of equilibrium" [4] include the familiar Maclaurin spheroids and triaxial Jacobi ellipsoids, which are characterized by rigid rotation, and a wider class discovered by Dedekind and Riemann, in which a motion of uniform vorticity exists, even in a frame in which the ellipsoidal surface is at rest. The solutions representing the rotating ellipsoids did not only play an inspirational role in developing basic concepts of the theory of rigidly rotating stars, but quite unexpectedly in the study of inviscid, differentially rotating polytropes. These closely resemble Maclaurin spheroids, although they do not maintain rigid rotation. As noted in the well-known monograph on rotating stars [5], "the classical work on uniformly rotating, homogeneous spheroids has a range of validity much greater than was usually anticipated". It also influenced galactic dynamics [6]: the existence of Jacobi ellipsoids suggested that a rapidly rotating galaxy may not remain axisymmetric, and the Riemann ellipsoids demonstrated that there is a distinction between the rate at which the matter in a triaxial rotating body streams and the rate at which the figure of the body rotates. Since rotating incompressible ellipsoids adequately illustrate the general feature of rotating axisymmetric bodies, they are also used in the studies of double stars whose components are close to each other. The disturbances caused by a neighbouring component are treated as first order perturbations. Relativistic effects on the rotating incompressible ellipsoids have been investigated in the post-Newtonian approximation by various authors, recently with a motivation to understand the coalescence of binary neutron stars near their innermost stable circular orbit (see [7] for the latest work and a number of references).

As for the last subject, which has a more direct connection with exact explicit solutions of Einstein's equations, we want to say a few words about integrable systems and their soliton solutions. Soliton theory has been one of the most interesting developments in the past decades both in physics and mathematics, and gravity has played a role both in its birth and recent developments. It has been known from the end of the last century that the celebrated Korteweg–de Vries nonlinear evolution equation, which governs one dimensional surface gravity waves propagating in a shallow channel of water, admits solitary wave solutions. However, it was not until Zabusky and Kruskal (the Kruskal of Sect. 2.4 below) did extensive *numerical* studies of this equation in 1965 that the remarkable properties of the solitary waves were discovered: the nonlinear solitary waves, named solitons by Zabusky and Kruskal, can interact and then continue, preserving their shapes and velocities. This discovery has stimulated extensive studies of other nonlinear equations, the inverse scattering methods of their solution, the proof of the existence of an infinite number of conservation laws associated with such equations, and the construction of explicit solutions (see [8] for a recent comprehensive treatise). Various other nonlinear equations, similar to the sine-Gordon equation or the nonlinear Schrödinger equation, arising for example in plasma physics, solid state physics, and nonlinear optics, have also been successfully tackled by these methods. At the end of the 1970s several authors discovered that Einstein's vacuum equations for axisymmetric stationary systems can be solved by means of the inverse scattering methods, and it soon became clear that one can employ them also in situations when both Killing vectors are spacelike (producing, for example, soliton-type cosmological gravitational waves). Dieter Maison, one of the pioneers in applying these techniques in general relativity, describes the subject thoroughly in this volume. We shall briefly meet the soliton methods when we discuss the uniformly rotating disk solution of Neugebauer and Meinel (Sect. 6.3), colliding plane waves (Sect. 8.3), and inhomogeneous cosmological models (Sect. 12.2). Our aim, however, is to understand the meaning of solutions, rather than generation techniques of finding them. From this viewpoint it is perhaps first worth noting the interplay between numerical and analytic studies of the soliton solutions – hopefully, a good example of an interaction for numerical and mathematical relativists. However, the explicit solutions of integrable models have played important roles in various other contexts. The most interesting *multi*-dimensional integrable equations are the four-dimensional self-dual Yang–Mills equations arising in field theory. Their solutions, discovered by R. Ward using twistor theory, on one hand stimulated Donaldson's most remarkable work on inequivalent differential structures on four-manifolds. On the other hand, Ward indicated that many of the known integrable systems can be obtained by dimensional reduction from the self-dual Yang–Mills equations. Very recently this view has been substantiated in the monograph by Mason and Woodhouse [9]. The words by which these authors finely express the significance of exact solutions in integrable systems can be equally well used for solutions of Einstein's equations: "they combine tractability with nonlinearity, so they make it possible to explore nonlinear phenomena while working with explicit solutions".[1]

1.2 Einstein's Field Equations

Since Jürgen Ehlers has always been, among other things, interested in the history of science, he will hopefully tolerate a few remarks on the early history of Einstein's equations to which not much attention has been paid in the literature. It was during his stay in Prague in 1911 and 1912 that Einstein's intensive interest in quantum theory diminished, and his systematic effort in constructing a relativistic theory of gravitation began. In his first "Prague theory of gravity" he assumed that gravity can be described by a single function – the local velocity of light. This assumption led to insurmountable difficulties. However, Einstein learned much in Prague on his way to general relativity [11]: he understood the local significance of the principle of equivalence; he realized that the equations describing the gravitational field must be nonlinear and have a form invariant with respect to a larger group

[1] In 1998, in the discussion after his Prague lecture on the present role of physics in mathematics, Prof. Michael Atiyah expressed a similar view that even with more powerful supercomputers and with a growing body of general mathematical results on the existence and uniqueness of solutions of differential equations, the exact, explicit solutions of nonlinear equations will not cease to play a significant role. (As it is well known, Sir Michael Atiyah has made fundamental contributions to various branches of mathematics and mathematical physics, among others, to the theory of solitons, instantons, and to the twistor theory of Sir Roger Penrose, with whom he has been interacting "under the same roof" in Oxford for 17 years [10].)

of transformations than the Lorentz group; and he found that "*spacetime coordinates lose their simple physical meaning*", i.e. they do not determine directly the distances between spacetime points.[2] In his "Autobiographical Notes" Einstein says: "Why were seven years ... required for the construction of general relativity? The main reason lies in the fact that it is not easy to free oneself from the idea that coordinates must have an immediate metrical meaning"... Either from Georg Pick while still in Prague, or from Marcel Grossmann during the autumn of 1912 after his return to Zurich (cf. [11]), Einstein learned that an appropriate mathematical formalism for his new theory of gravity was available in the work of Riemann, Ricci, and Levi–Civita. Several months after his departure from Prague and his collaboration with Grossmann, Einstein had general relativity almost in hand. Their work [13] was already based on the generally invariant line element

$$ds^2 = g_{\mu\nu} dx^\mu dx^\nu \tag{I}$$

in which the spacetime metric tensor $g_{\mu\nu}(x^\rho), \mu, \nu, \rho = 0, 1, 2, 3$, plays a dual role: on the one hand it determines the spacetime geometry, on the other it represents the (ten components of the) gravitational potential and is thus a dynamical variable. The disparity between geometry and physics, criticized notably by Ernst Mach,[3] had thus been removed. When searching for the field equations for the metric tensor, Einstein and Grossmann *had* already realized that a natural candidate for generally covariant field equations would be the equations relating – in present-day terminology – the Ricci tensor and the energy-momentum tensor of matter. However, they erroneously concluded that such equations would not yield the Poisson equation of Newton's theory of gravitation as a first approximation for weak gravitational fields (see both §5 in the "Physical part" in [13] written by Einstein and §4, below equation (46), in the "Mathematical part" by M. Grossmann). Einstein then rejected the general covariance. In a subsequent paper with Grossmann [14], they supported this mis-step by a well-known "hole" meta-argument and obtained (in today's terminology) four gauge conditions such that the field equations were covariant only with respect to transformations of coordinates permitted by the gauge conditions. We refer to, for example, [15] for more detailed information on the further developments leading to the final version of the field equations. Let us only summarize that in late 1915 Einstein first readopted the generally covariant field equations from 1913, in which the Ricci tensor $R_{\mu\nu}$ was, up to the gravitational coupling constant, equal to the energy-momentum tensor $T_{\mu\nu}$ (paper submitted to the Prussian Academy on

[2] At that time Einstein's view on the future theory of gravity are best summarized in his reply to M. Abraham [12], written just before departure from Prague.

[3] Mach spent 28 years as Professor of Experimental Physics in Prague, until 1895, when he took the History and Theory of Inductive Natural Sciences chair in Vienna.

November 4). From his *vacuum field equations*

$$R_{\mu\nu} = 0, \tag{II}$$

where $R_{\mu\nu}$ depends nonlinearly on $g_{\alpha\beta}$ and its first derivatives, and linearly on its second derivatives, he was able to explain the anomalous part of the perihelion precession of Mercury – in the note presented to the Academy on November 18. And finally, in the paper [16] submitted on November 25 (published on December 2, 1915), the final version of the *gravitational field equations*, or *Einstein's field equations* appeared:[4]

$$R_{\mu\nu} - \frac{1}{2} g_{\mu\nu} R = \frac{8\pi G}{c^4} T_{\mu\nu}, \tag{III}$$

where the scalar curvature $R = g^{\mu\nu} R_{\mu\nu}$. Newton's gravitational constant G and the velocity of light c are the (only) fundamental constants appearing in the theory. If not stated otherwise, in this article we use the geometrized units in which $G = c = 1$, and the same conventions as in [18] and [19].

Now it is well known that Einstein further generalized his field equations by adding a cosmological term $+\Lambda g_{\mu\nu}$ on the left side of the field equations (III). The cosmological constant Λ appeared first in Einstein's work "Cosmological considerations in the General Theory of Relativity" [20] submitted on February 8, 1917 and published on February 15, 1917, which contained the *closed static model of the Universe* (the Einstein static universe) – an exact solution of equations (III) with $\Lambda > 0$ and an energy-momentum tensor of incoherent matter ("dust"). This solution marked the birth of modern cosmology.

We do not wish to embark upon the question of the role that Mach's principle played in Einstein's thinking when constructing general relativity, or upon the intriguing issues relating to aspects of Mach's principle in present-day relativity and cosmology[5] – a problem which in any event would far exceed the scope of this article. Although it would not be inappropriate to

---

[4] David Hilbert submitted his paper on these field equations five days before Einstein, though it was published only on March 31, 1916. Recent analysis [17] of archival materials has revealed that Hilbert made significant changes in the proofs. The originally submitted version of his paper contained the theory which is not generally covariant, and the paper did not include equations (III).

[5] It was primarily Einstein's recognition of the role of Mach's ideas in his route towards general relativity, and in his christening them by the name "Mach's principle" (though Schlick used this term in a vague sense three years before Einstein), that makes Mach's Principle influential even today. After the 1988 Prague conference on Ernst Mach and his influence on the development of physics [21], the 1993 conference devoted exclusively to Mach's principle was held in Tübingen, from which a remarkably thorough volume was prepared [22], covering all aspects of Mach's principle and recording carefully all discussion. The clarity of ideas and insights of Jürgen Ehlers contributed much to both conferences and their proceedings. For a brief more recent survey of various aspects of Mach's

include it here since exact solutions (such as Gödel's universe or Ozsváth's and Schücking's closed model) have played a prominent role in this context. However, it should be at least stated that Einstein originally invented the idea of a closed space in order to eliminate boundary conditions at spatial infinity. The boundary conditions "flat at infinity" bring with them an inertial frame unrelated to the mass-energy content of the space, and Einstein, in accordance with Mach's views, believed that merely mass-energy can influence inertia. Field equations (III) are not inconsistent with this idea, but they admit as the simplest solution an empty flat *Minkowski space* ($T_{\mu\nu} = 0,\ g_{\mu\nu} = \eta_{\mu\nu} = \text{diag}\,(-1, +1, +1, +1)$), so some restrictive boundary conditions are essential if the idea is to be maintained. Hence, Einstein introduced the cosmological constant Λ, hoping that with this space will always be closed, and the boundary conditions eliminated. But it was also in 1917 when *de Sitter* discovered the *solution* [25] of the vacuum field equations (II) with added cosmological term ($\Lambda > 0$) which demonstrated that a nonvanishing Λ does not necessarily imply a nonvanishing mass-energy content of the universe.

1.3 "Just So" Notes on the Simplest Solutions: The Minkowski, de Sitter, and Anti-de Sitter Spacetimes

Our brief intermezzo on the cosmological constant brought up three explicit simple exact solutions of Einstein's field equations – the Minkowski, Einstein, and de Sitter spacetimes. To these also belongs the anti de Sitter spacetime, corresponding to a negative Λ. The de Sitter spacetime has the topology $R^1 \times S^3$ (with R^1 corresponding to the time) and is best represented geometrically as the 4-dimensional hyperboloid $-v^2 + w^2 + x^2 + y^2 + z^2 = (3/\Lambda)$ in 5-dimensional flat space with metric $ds^2 = -dv^2 + dw^2 + dx^2 + dy^2 + dz^2$. The anti de Sitter spacetime has the topology $S^1 \times R^3$, and can be visualized as the 4-dimensional hyperboloid $-U^2 - V^2 + X^2 + Y^2 + Z^2 = (-3/\Lambda)$, $\Lambda < 0$, in flat 5-dimensional space with metric $ds^2 = -dU^2 - dV^2 + dX^2 + dY^2 + dZ^2$. As is usual (cf. e.g. [26,27]), we mean by "anti de Sitter spacetime" the universal covering space which contains no closed timelike lines; this is obtained by unwrapping the circle S^1.

These spacetimes will not be discussed in the following sections. Ocassionally, for instance, in Sects. 5 and 10, we shall consider spacetimes which become asymptotically de Sitter. However, since these solutions have played a crucial role in many issues in general relativity and cosmology, and most recently, they have become important prerequisites on the stage of the theoretical physics of the "new age", including string theory and string cosmology, we shall make a few comments on these solutions here, and give some references to recent literature.

---

principle in general relativity, see the introductory section in the work [23], in which Mach's principle is analyzed in the context of perturbed Robertson–Walker universes. Most recently, Mach's principle seems to enter even into M theory [24].

The basic geometrical properties of these spaces are analyzed in the Battelle Recontres lectures by Penrose [27], and in the monograph by Hawking and Ellis [26], where also references to older literature can be found. The important role of the de Sitter solution in the theory of the expanding universe is finely described in the book by Peebles [28], and in much greater detail in the proceedings of the Bologna 1988 meeting on the history of modern cosmology [29].

The Minkowski, de Sitter and anti de Sitter spacetimes are the simplest solutions in the sense that their *metrics* are *of constant* (zero, positive, and negative) *curvature*. They admit the same number (ten) of independent Killing vectors, but the interpretations of corresponding symmetries differ for each spacetime. Together with the Einstein static universe, they all are conformally flat, and can be represented as portions of the Einstein static universe [26,27]. However, their *conformal structure* is globally different. In Minkowski spacetime one can go to infinity along timelike geodesics and arrive to the future (or past) *timelike infinity* i^+ (or i^-); along null geodesics one reaches the future (past) *null infinity* $\mathcal{J}^+(\mathcal{J}^-)$; and spacelike geodesics lead to *spatial infinity* i^0. Minkowski spacetime can be compactified and mapped onto a finite region by an appropriate conformal rescaling of the metric. One thus obtains the well-known Penrose diagram in which the three types of infinities are mapped onto the boundaries of the compactified spacetime – see for example the boundaries on the "right side" in the Penrose diagram of the Schwarzschild—Kruskal spacetime in Fig. 3, Sect. 2.4, or the Penrose compactified diagram of boost-rotation symmetric spacetimes in Fig. 13, Sect. 11. (The details of the conformal rescaling of the metric and resulting diagrams are given in [26,27] and in standard textbooks, for example [18,19,30].) In the de Sitter spacetime there are only past and future conformal infinities $\mathcal{J}^-, \mathcal{J}^+$, both being *spacelike* (cf. the Penrose diagram of the "cosmological" Robinson–Trautman solutions in Fig. 11, Sect. 10); the conformal infinity in anti de Sitter spacetime is *timelike*.

These three spacetimes of constant curvature offer many basic insights which have played a most important role elsewhere in relativity. To give just a few examples (see e.g. [26,27]): both the particle (cosmological) horizons and the event horizons for geodesic observers are well illustrated in the de Sitter spacetime; the Cauchy horizons in the anti de Sitter space; and the simplest acceleration horizons in Minkowski space (hypersurfaces $t^2 = z^2$ in Fig. 12, Sect. 11). With the de Sitter spacetime one learns (by considering different cuts through the 4-dimensional hyperboloid) that the concept of an "open" or "closed" universe depends upon the choice of a spacelike slice through the spacetime. There is perhaps no simpler way to understand that Einstein's field equations are of local nature, and that the spacetime topology is thus not given a priori, than by considering the following construction in Minkowski spacetime. Take the region given in the usual coordinates by $|x| \leq 1, |y| \leq 1, |z| \leq 1$, remove the rest and identify pairs of boundary points of the form $(t, 1, y, z)$ and $(t, -1, y, z)$, and similarly for y and z. In this way the spatial sections are identified to obtain a 3-torus – a flat but closed manifold.[6]

The *spacetimes of constant curvature* have been resurrected as basic *arenas of new physical theories* since their first appearance. After the role of the de Sitter universe decreased with the refutation of the steady-state cosmology, it has inflated again enormously in connection with the theory of early quasi-exponential phase of expansion of the universe, due to the false-vacuum state of a hypothetical scalar (inflaton) field(s) (see e.g. [28]). We shall mention the de Sitter space as the asymptotic state of cosmological models with a nonvanishing Λ (so verifying the "cosmic no-hair conjecture") in Sect. 10 on Robinson–Trautman spacetimes. Motivated by its importance in inflationary cosmologies, several new useful papers reviewing the properties of de Sitter spacetime have appeared [33,34]; they also contain many references to older literature. For the most recent work on the quantum structure of de Sitter space, see [35].

In the last two years, anti de Sitter spacetime has come to the fore in light of Maldacena's conjecture [36] relating string theory in (asymptotically) anti de Sitter space to a non-gravitational conformal field theory on the boundary at spatial infinity, which is timelike as mentioned above (see, e.g. [37], where among others, in the Appendix various coordinate systems describing anti de Sitter spaces in arbitrary dimensions are discussed).

Amazingly, the Minkowski spacetime has recently entered the active new area of so called pre-big bang string cosmology [38]. String theory is here applied to the problem of the big bang. The idea is to start from a simple Minkowski space (as an "asymptotic past triviality") and to show that it is in an unstable false-vacuum state, which leads to a long *pre*-big bangian inflationary phase. This, at later times, should provide a hot big bang. Although such a scenario has been criticized on various grounds, it has attractive features, and most importantly, can be probed through its observable relics [38].

Since it is hard to forecast how the roles of these three spacetimes of constant curvature will develop in new and exciting theories in the next millennium, let us better conclude our "just so" notes by stating three "stable" results of complicated, rigorous mathematical analyses of (the classical) Einstein's equations.

In their recent treatise [39], Christodoulou and Klainerman prove that any smooth, asymptotically flat initial data set which is "near flat, Minkowski data" leads to a unique, smooth and geodesically complete solution of Ein-

[6] This very simple point was apparently unknown to Einstein in 1917, although soon after the publication of his cosmological paper, E. Freundlich and F. Klein pointed out to him that an elliptical topology (arising from the identification of antipodal points) could have been chosen instead of the spherical one considered by Einstein. Although topological questions have been followed with a great interest in recent decades, the chapter by Geroch and Horowitz in "An Einstein Centenary Survey" [31] remains the classic; for more recent texts, see for example [32] and references therein.

stein's vacuum equations with vanishing cosmological constant. This demonstrates the *stability of the Minkowski space* with respect to nonlinear (vacuum) perturbations, and the existence of singularity-free, asymptotically flat radiative vacuum spacetimes. Christodoulou and Klainerman, however, are able to show only a somewhat weaker decay of the field at null infinity than is expected from the usual assumption of a sufficient smoothness at null infinity in the framework of Penrose (see e.g. [40] for a brief account).

Curiously enough, in the case of the vacuum Einstein equations with a *nonvanishing cosmological constant*, a more complete picture has been known for some time. By using his regular conformal field equations, Friedrich [41] demonstrated that initial data sufficiently close to de Sitter data develop into solutions of Einstein's equations with a positive cosmological constant, which are "asymptotically simple" (with a smooth conformal infinity), as required in Penrose's framework. More recently, Friedrich [42] has shown the existence of asymptotically simple solutions to the Einstein vacuum equations with a negative cosmological constant. For the latest review of Friedrich's thorough work on asymptotics, see [43].

Summarizing, thanks to these profound mathematical achievements we know that the *Minkowski, de Sitter*, and *anti de Sitter spacetimes* are the solutions of Einstein's field equations which are *stable with respect to general, nonlinear* (though "weak" in a functional sense) *vacuum perturbations.* A result of this type is not known for any other solution of Einstein's equations.

1.4 On the Interpretation and Characterization of Metrics

Suppose that a metric satisfying Einstein's field equations is known in some region of spacetime and in a given coordinate (reference) system x^μ. A fundamental question, frequently "forgotten" to be addressed in modern theories which extend upon general relativity, is whether the *metric tensor* $g_{\alpha\beta}(x^\mu)$ *is a measurable quantity.* Classical general relativity offers (at least) three ways of giving a positive answer, depending on what objects are considered as "primitive tools" to perform the measurements. The first, elaborated and emphasized primarily by Møller [44], employs standard rigid rods in the measurements. However, a "rigid rod" is not really a simple primitive concept. The second procedure, due to Synge [45], accepts as the basic concepts a "particle" and a "standard clock". If x^μ and $x^\mu + dx^\mu$ are two nearby events contained in the worldline of a clock, then the separation (the spacetime interval) between the events is equal to the interval measured by the clock. The main drawback of this approach appears to lie in the fact that it does not explain why the same functions $g_{\alpha\beta}(x^\mu)$ describe the behavior of the clock as well as paths of free particles, as explained in more detail by Ehlers, Pirani and Schild [46], in the motivation for their own axiomatic but constructive procedure for setting up the spacetime geometry. Their method, inspired by the work of Weyl and others, uses neither rods nor clocks, but instead, light rays and freely falling test particles, which are considered as basic tools for measuring the metric and determining the spacetime geometry. (For a simple description of how this can be performed, see exercise 13.7 in [18]; for some new developments which build upon, among others, the Ehlers-Pirani-Sachs approach, see [47].) After indicating that the metric tensor is a measurable quantity let us briefly turn to the *role of spacetime coordinates.*

In special relativity there are infinitely many global inertial coordinate systems labelling events in the Minkowski manifold $\mathbb{R}^4$; they are related by elements of the Poincaré group. The inertial coordinates labels X^0, X^1, X^2, X^3 of a given event do not thus have intrinsic meaning. However, the spacetime interval between two events, determined by the Minkowski metric $\eta_{\mu\nu}$, represents an intrinsic property of spacetime. Since the Minkowski metric is so simple, the differences between inertial coordinates can have a metrical meaning (recall Einstein's reply to Abraham mentioned in Sect. 1.2). In principle, however, both in special and general relativity, it is the metric, the line element, which exhibits intrinsically the geometry, and gives all relevant information. As Misner [48] puts it, if you write down for someone the Schwarzschild metric in the "canonical" form (equation (2) in Sect. 2.2) and receive the reaction "that [it] tells me the $g_{\mu\nu}$ gravitational potentials, now tell me in which (t, r, θ, φ) coordinate system they have these values?", then there are two valid responses: (a) indicate that it is an indelicate and unnecessary question, or (b) ignore it. Clearly, the Schwarzschild metric describes the geometrical properties of the coordinates used in (2). For example, it implies that worldlines with fixed r, θ, φ are timelike at $r > 2M$, orthogonal to the lines with $t =$ constant. It determines local null cones (given by $ds^2 = 0$), i.e. the *causal structure* of the spacetime. In addition, in Schwarzschild coordinates the metric (2) indicates how to *measure* the radial coordinate of a given event, because the proper area of the sphere going through the event is given just by the Euclidean expression $4\pi r^2$ (r is thus often called "the curvature coordinate"). On each sphere the angular coordinates θ, φ have the same meaning as on a sphere in Euclidean space. The Schwarzschild coordinate time t, geometrically preferred by the timelike (for $r > 2M$) Killing vector, which is just equal to $\partial/\partial t$, can be measured by radar signals sent out from spatial infinity ($r \gg 2M$) where t is the proper time (see e.g. [18]). The coordinates used in (2) are in fact "more unique" than the inertial coordinates in Minkowski spacetime, because the only possible continuous transformations preserving the form (2) are rigid rotations of a sphere, and $t \to t +$ constant. Such a simple interpretation of coordinates is exceptional. However, the simple case of the Schwarzschild metric clearly demonstrates that all intrinsic information is contained in the line element.

It is interesting, and for some purposes useful, to consider not just one Schwarzschild metric with a given mass M but the *family* of such metrics for all possible M. In order to cover also the future event horizon let us describe the metrics by using Eddington–Finkelstein ingoing coordinates as in equation (4), Sect. 2.3. This equation can be interpreted as a family of metrics with various values of M given on a *fixed background manifold* $\bar{\mathcal{M}}_1$, with $v \in \mathbb{R}, r \in (0,\infty)$, and $\theta \in [0,\pi], \varphi \in [0,2\pi)$. Alternatively, however, we may use, for example, the Kruskal null coordinates $\tilde{U}, \tilde{V}$ in which the metric is given by equation (6), Sect. 2.3, with $\tilde{U} = V - U, \tilde{V} = V + U$. We may then consider metrics on a background manifold $\bar{\mathcal{M}}_2$ given by $\tilde{U} \in \mathbb{R}, \tilde{V} \in (0,\infty)$, $\theta \in [0,\pi]$, and $\varphi \in [0,2\pi)$, which corresponds to $\bar{\mathcal{M}}_1$. However, these two background manifolds are *not* the same: the transformation between the Eddington–Finkelstein coordinates and the Kruskal coordinates is not a map from $\bar{\mathcal{M}}_1$ to $\bar{\mathcal{M}}_2$ because it depends on the value of mass M. Therefore, the "background manifold" used frequently in general relativity, for example in problems of conservation of energy, or in quantum gravity, is not defined in a natural, unique manner. The above simple pedagogical observation has recently been made in connection with gauge fixing in quantum gravity by Hájíček [49] in order to explain the old insight by Bergmann and Komar, that the gauge group of general relativity is much larger than the diffeomorphism group of one manifold. To identify points when working with backgrounds, one usually fixes coordinates in all solution manifolds by some gauge condition, and identifies those points of all these manifolds which have the same value of the coordinates.

Returning back to a single solution $(\mathcal{M}, g_{\alpha\beta})$, described by a manifold $\mathcal{M}$ and a metric $g_{\alpha\beta}$ in some coordinates, a notorious (local) *"equivalence problem"* often arises. A given (not necessarily global) solution has the variety of representations which equals the variety of choices of a 4-dimensional coordinate system. Transitions from one choice to another are isomorphic with the group of 4-dimensional diffeomorphisms which expresses the general covariance of the theory.[7] Given another set of functions $g'_{\alpha\beta}(x'^{\gamma})$ which satisfy Einstein's equations, how do we learn that they are not just transformed components of the metric $g_{\alpha\beta}(x^{\gamma})$? In 1869 E. B. Christoffel raised a more general question: under which conditions is it possible to transform a quadratic form $g_{\alpha\beta}(x^{\gamma})dx^{\alpha}dx^{\beta}$ in n-dimensions into another such form $g'_{\alpha\beta}(x'^{\gamma})dx'^{\alpha}dx'^{\beta}$ by means of smooth transformation $x^{\gamma}(x'^{\kappa})$? As Ehlers emphasized in his paper

[7] As pointed out by Kretschmann soon after the birth of general relativity, one can always make a theory generally covariant by taking more variables and inserting them as new dynamical variables into the (enlarged) theory. Thus, standard Yang–Mills theory is covariant with respect to the transformations of Yang–Mills potentials, corresponding to a particular group, say $SU(2)$. However, the theory is usually formulated on a fixed background spacetime with a given metric. The evolution of a dynamical Yang–Mills solution is thus "painted" on a given spacetime. When the metric – the gravitational field – is incorporated as a dynamical variable in the Einstein–Yang–Mills theory, the whole spacetime metric and Yang–Mills field are "built-up" from given data (cf. the article by Friedrich and Rendall in this volume). The resulting theory is covariant with respect to a much larger group. The dual role of the metric, determined only up to 4-dimensional diffeomorphisms, makes the character of the solutions of Einstein's equations unique among solutions of other field theories, which do not consider spacetime as being dynamical.

[50] on the meaning of Christoffel's equivalence problem in modern field theories, Christoffel's results apply to metrics of arbitrary signature, and can be thus used directly in general relativity. Without going into details let us say that today the solution to the equivalence problem as presented by Cartan is most commonly used. For both metrics $g_{\alpha\beta}$ and $g'_{\alpha\beta}$ one has to find a frame (four 1-forms) in which the frame metric is constant, and find the frame components of the Riemann tensor and its covariant derivatives up to – possibly – the 10th order. The two metrics $g_{\alpha\beta}$ and $g'_{\alpha\beta}$ are then equivalent if and only if there exist coordinate and Lorentz transformations under which one whole set of frame components goes into the other. In a practical algorithm given by Karlhede [51], recently summarized and used in [52], the number of derivations required is reduced.

A natural first idea of how to solve the equivalence problem is to employ the scalar invariants from the Riemann tensor and its covariant derivatives. This, however, does not work. For example, in all Petrov type *N* and *III* nonexpanding and nontwisting solutions all these invariants vanish as shown recently (see Sect. 8.2), as they do in Minkowski spacetime.

However, even without regarding invariants, at present much can be learnt about an exact solution (at least locally) in geometrical terms, without reference to special coordinates. This is thanks to the progress started in the late 1950s, in which the group of Pascual Jordan in Hamburg has played the leading role, with Jürgen Ehlers as one of its most active members. Ehlers' dissertation[8] [54] from 1957 is devoted to the characterization of exact solutions.

The problem of exact solutions also forms the content of his contribution to the Royaumont GR-conference [55], as well as his plenary talk in the London GR-conference [56]. A detailed description of the results of the Hamburg group on invariant geometrical characterization of exact solutions by using and developing the Petrov classification of Weyl's tensors, groups of isometries, and conformal transformations are contained in the first paper [57] in the (today "golden oldies") series of articles published in the "Abhandlungen der Akademie der Wissenschaften in Mainz". An English version, in a somewhat shorter form, was published by Ehlers and Kundt [53] in the "classic" 1962 book "Gravitation: An Introduction to Current Research" compiled by L. Witten. (We shall meet these references in the following sections.) In the second paper of the "Abhandlungen" [58], among others, algebraically special vacuum solutions are studied, using the formalism of the 2-component spinors, and in particular, geometrical properties of the congruences of null rays are analyzed in terms of their expansion, twist, and shear.

---

[8] The English translation of the title of the dissertation reads: "The construction and characterization of the solutions of Einstein's gravitational field equations". In [53] the original German title is quoted, as in our citation [54], but "of the solutions" is erroneously omitted. This error then reemerges in the references in [19].

These tools became essential for the discovery by Roy Kerr in 1963 of the solution which, when compared with all other solutions of Einstein's equations found from the beginning of the renaissance of general relativity in the late 1950s until today, has played the most important role. As Chandrasekhar [59] eloquently expresses his wonder about the remarkable fact that all stationary and isolated black holes are *exactly* described by the Kerr solution: "This is the only instance we have of an exact description of a macroscopic object. Macroscopic objects, as we see them all around us, are governed by a variety of forces, derived from a variety of approximations to a variety of physical theories. In contrast, the only elements in the construction of black holes are our basic concepts of space and time ..." The Kerr solution can also serve as one of finest examples in general relativity of "the incredible fact that a discovery motivated by a search after the beautiful in mathematics should find its exact replica in Nature..." [60].

The technology developed in the classical works [53,57], and in a number of subsequent contributions, is mostly concerned with the local geometrical characterization of exact spacetime solutions. A well-known feature of the solutions of Einstein's equations, not shared by solutions in other physical theories, is that it is often very complicated to analyze their global properties, such as their extensions, completeness, or topology. If analyzed globally, almost any solution can tell us something about the basic issues in general relativity, like the nature of singularities, or cosmic censorship.

1.5 The Choice of Solutions

Since most solutions, when properly analyzed, can be of potential interest, we are confronted with a richness of material which puts us in danger of mentioning many of them, but remaining on a general level, and just enumerating rather than enlightening. In fact, because of lack of space (and of our understanding) we shall have to adopt this attitude in many places. However, we have selected some solutions, hopefully the fittest ones, and when discussing their role, we have chosen particular topics to be analyzed in some detail, and left other issues to brief remarks and references.

Firstly, however, let us ask what do we understand by the term "exact solution". In the much used "exact-solution-book" [61], the authors "do not intend to provide a definition", or, rather, they have decided that what they "chose to include was, by definition, an exact solution". A mathematical relativist-purist would perhaps consider solutions, the existence of which has been demonstrated in the works of Friedrich or Christodoulou and Klainerman, mentioned at the end of Sect. 1.3, as "good" as the Schwarzschild metric. Most recently, Penrose [62] presented a strong conjecture which may lead to a general vacuum solution described in the complicated (complex) formalism of his twistor theory. Although in this article we do not mean by exact solutions those just mentioned, we also do not consider as exact solutions only those explicit solutions which can be written in terms of elementary functions on half of a page. We prefer, recalling Feynman, simple "special cases", but we also discuss, for example, the late-time behaviour of the Robinson–Trautman solutions for which rigorously convergent series expansions can be obtained, which provide sufficiently rich "special information".

Concerning the selection of the solutions, the builder of general relativity and the gravitational field equations (III) himself indicates which solutions should be preferred [63]: "The theory avoids all internal discrepancies which we have charged against the basis of classical mechanics... But, it is similar to a building, one wing of which is made of fine marble (left part of the equation), but the other wing of which is built of low grade wood (right side of equation). The phenomenological representation of matter is, in fact, only a crude substitute for a representation which would correspond to all known properties of matter. There is no difficulty in connecting Maxwell's theory... so long as one restricts himself to space, free of ponderable matter and free of electric density..."

Of course, Einstein was not aware when he was writing this of Yang–Mills–Higgs fields, or of the dilaton field, etc. However, remaining on the level of field theories with a clear classical meaning, his view has its strength and motivates us to prefer (electro)vacuum solutions. A physical interpretation of the vacuum solutions of Einstein's equations have been reviewed in papers by Bonnor [64], and Bonnor, Griffiths and MacCallum [65] five years ago. Our article, in particular in emphasizing and describing the role of solutions in giving rise to various concepts, conjectures, and methods of solving problems in general relativity, and in the astrophysical impacts of the solutions, is oriented quite differently, and gives more detail. However, up to some exceptions, like, for example, metrics for an infinite line-mass or plane, which are discussed in [64], and new solutions which have been discovered after the reviews [64,65] appeared as, for example, the solution describing a rigidly rotating thin disk of dust, our choice of solutions is similar to that of [64,65].

In selecting particular topics for a more detailed discussion we will be led primarily by following overlapping aspects: (i) the "commonly acknowledged" significance of a solution – we will concentrate in particular on the Schwarzschild, the Kerr, the Taub-NUT, and plane wave solutions, and (ii) the solutions and their properties that I (and my colleagues) have been directly interested in, such as the Reissner–Nordström metric, vacuum solutions outside rotating disks, or radiative solutions such as cylindrical waves, Robinson–Trautman solutions, and the boost-rotation symmetric solutions. Some of these have also been connected with the interests of Jürgen Ehlers, and we shall indicate whenever we are aware of this fact.

Vacuum cosmological solutions are discussed in less detail than they deserve. A possible excuse – from the point of view of being a relativist, a rather unfair one – could be that a special recent issue of Reviews of Modern Physics (Volume 71, 1999), marking the Centennial of the American Physical Society, contains discussion of the Schwarzschild, the Reissner–Nordström and other black hole solutions, and even remarks on the work of Bondi et al. [66] on radiative solutions, but among the cosmological solutions only the standard models are mentioned. A real reason is the author's lack of space, time, and energy. In the concluding remarks we will try to list at least the most important solutions (not only the Friedmann models!) which have not been "selected" and give references to the literature in which more information can be found.

1.6 The Outline

Since the titles of the following sections characterize the contents rather specifically, we restrict ourselves to only a few explanatory remarks. In our discussion of the Schwarschild metric, after mentioning its role in the solar system, we indicate how the Schwarzschild solution gave rise to such concepts as the event horizon, the trapped surface, and the apparent horizon. We pay more attention to the concept of a bifurcate Killing horizon, because this is usually not treated in textbooks, and in addition, Jürgen Ehlers played a role in its first description in the literature. Another point which has not received much attention is Penrose's nice presentation of evidence against Lorentz-covariant field theoretical approaches to gravity, based on analysis of the causal structure of the Schwarzschild spacetime. Among various astrophysical implications of the Schwarzschild solution we especially note recent suggestions which indicate that we may have evidence of the existence of event horizons, and of a black hole in the centre of our Galaxy.

The main focus in our treatment of the Reissner–Nordström metric is directed to the instability of the Cauchy horizon and its relation to the cosmic censorship conjecture. We also briefly discuss extreme black holes and their role in string theory.

About the same amount of space as that given to the Schwarzschild solution is devoted to the Kerr metric. After explaining a few new concepts the metric inspired, such as locally nonrotating frames and ergoregions, we mention a number of physical processes which can take place in the Kerr background, including the Penrose energy extraction process, and the Blandford–Znajek mechanism. In the section on the astrophysical evidence for a Kerr metric, the main attention is paid to the broad iron line, the character of which, as most recent observations indicate, is best explained by assuming that it originates very close to a maximally rotating black hole. The discussion of recent results on black hole uniqueness and on multi-black hole solutions concludes our exposition of spacetimes representing black holes. In the section on axisymmetric fields and relativistic disks a brief survey of various static solutions is first given, then we concentrate on relativistic disks as sources of the Kerr metric and other stationary fields; in particular, we summarize briefly the recent work on uniformly rotating disks.

An intriguing case of Taub-NUT space is introduced by a new constructive derivation of the solution. Various pathological features of this space are then briefly listed.

Going over to radiative spacetimes, we analyze in some detail plane waves – also in the light of the thorough study by Ehlers and Kundt [53]. Some new developments are then noted, in particular, impulsive waves generated by boosting various "particles", their symmetries, and recent use of the Colombeau algebra of generalized functions in the analyses of impulsive waves. A fairly detailed discussion is devoted to various effects connected with colliding plane waves.

In our treatment of cylindrical waves we concentrate in particular on two issues: on the proof that these waves provide explicitly given spacetimes, which admit a smooth global null infinity, even for strong initial data within a $(2+1)$-dimensional framework; and on the role that cylindrical waves have played in the first construction of a midisuperspace model in quantum gravity. Various other developments concerning cylindrical waves are then summarized only telegraphically.

A short section on Robinson–Trautman solutions points out how these solutions with a nonvanishing cosmological constant can be used to give an exact demonstration of the cosmic no-hair conjecture under the presence of gravitational radiation, and also of the existence of an event horizon which is smooth but not analytic.

As the last class of radiative spacetimes we analyze the boost-rotation symmetric solutions representing uniformly accelerated objects. They play a unique role among radiative spacetimes since they are asymptotically flat, in the sense that they admit global smooth sections of null infinity. And as the only known radiative solutions describing finite sources they can provide expressions for the Bondi mass, the news function, or the radiation patterns in explicit forms. They have also been used as test-beds in numerical relativity, and as the model spacetimes describing the production of black hole pairs in strong fields.

Vacuum cosmological solutions such as the vacuum Bianchi models and Gowdy solutions are mentioned, and their significance in the development of general relativity is indicated in the last section. Special attention is paid to their role in understanding the behaviour of a general model near an initial singularity.

In the concluding remarks, several important, in particular *non*-vacuum solutions, which have not been included in the main body of the paper, are at least listed, together with some relevant references. A few remarks on the possible future role of exact solutions ends the article.

Although we give over 360 references in the bibliography, we do not at all pretend to give all relevant citations. When discussing more basic facts and concepts, we quote primarily textbooks and monographs. Only when mentioning more recent developments do we refer to journals. The complete titles of all listed references will hopefully offer the reader a more complete idea of the role the explicit solutions have played on the relativistic stage and in the astrophysical sky.

2 The Schwarzschild Solution

In his thorough "Survey of General Relativity Theory" [67], Jürgen Ehlers begins with an empirical motivation of the theory, goes in depth and detail through his favourite topics such as the axiomatic approach, kinetic theory, geometrical optics, approximation methods, and only in the last section turns to spherically symmetric spacetimes. As T. S. Eliot says, "to make an end is to make a beginning – the end is where we start from", and so here we start with a few remarks on spherical symmetry.

2.1 Spherically Symmetric Spacetimes

In the early days of general relativity spherical symmetry was introduced in an intuitive manner. It is because of the existence of exact solutions which are singular at their centres (such as the Schwarzschild or the Reissner–Nordström solutions), and a realization that spherically symmetric, topologically non-trivial smooth spacetimes without any centre may exist [68], that today the group-theoretical definition of spherical symmetry is preferred (for a detailed analysis, see e.g. [19,26,67]).

Following Ehlers [67], we define a spacetime $(\mathcal{M}, g_{\alpha\beta})$ to be spherically symmetric if the rotation group SO_3 acts on $(\mathcal{M}, g_{\alpha\beta})$ as an isometry group with simply connected, complete, spacelike, 2-dimensional orbits. One can then prove the theorem [67,69] that a spherically symmetric spacetime is the direct product $\mathcal{M} = S^2 \times N$, where S^2 is the 2-sphere manifold with the standard metric g_S on the unit sphere; and N is a 2-dimensional manifold with a Lorentzian (indefinite) metric g_N, and with a scalar r such that the complete spacetime metric $g_{\alpha\beta}$ is "conformally decomposable", i.e. $r^{-2}g_{\alpha\beta}$ is the direct sum of the 2-dimensional parts g_N and g_S. Leaving further technicalities aside (see e.g. [26,67,69]) we write down the final spherically symmetric line element in the form

$$ds^2 = -e^{2\phi}dt^2 + e^{2\lambda}dr^2 + r^2(d\theta^2 + \sin^2\theta\, d\varphi^2), \tag{1}$$

where (following [67]) we permit $\phi(r,t)$ and $\lambda(r,t)$ to have an imaginary part $i\pi/2$ so that the signs of dt^2 and dr^2 in (1), and thus the role of r and t as space- and time- coordinates may interchange (a lesson learned from the vacuum Schwarzschild solutions – see below). The "curvature coordinate" r is defined invariantly by the area, $4\pi r^2$, of the 2-spheres r = constant, t = constant. There is no a priori relation between r and the proper distance from the centre (if there is one) to the spherical surface.

2.2 The Schwarzschild Metric and Its Role in the Solar System

Starting from the line element (1) and imposing Einstein's *vacuum* field equations, but allowing spacetime to be in general dynamical, we are led uniquely (cf. Birkhoff's theorem discussed e.g. in [18,26]) to the Schwarzschild metric

$$ds^2 = -\left(1 - \frac{2M}{r}\right) dt^2 + \left(1 - \frac{2M}{r}\right)^{-1} dr^2 + r^2 \left(d\theta^2 + \sin^2\theta \, d\varphi^2\right), \quad (2)$$

where $M =$ constant has to be interpreted as a mass, as test particle orbits show. The resulting spacetime is static at $r > 2M$ (no spherically symmetric gravitational waves exist), and asymptotically flat at $r \to \infty$.

Undoubtedly, the Schwarzschild solution, describing the exterior gravitational field of an arbitrary – static, oscillating, collapsing or expanding – spherically symmetric body of (Schwarzschild) mass M, is among the most influential solutions of the gravitational field equations, if not of any type of field equations invented in the 20th century. It is the first exact solution of Einstein's equations obtained – by K. Schwarzschild in December 1915, still before Einstein's theory reached its definitive form and, independently, in May 1916, by J. Droste, a Dutch student of H. A. Lorentz (see [70] for comprehensive survey).

However, in its exact form (involving regions near $r \approx 2M$) the metric (2) has not yet been experimentally tested (a more optimistic recent suggestion will be mentioned in Sect. 2.6). When in 1915 Einstein explained the perihelion advance of Mercury, he found and used only an approximate (to second order in the gravitational potential) spherically symmetric solution. In order to find the value of the deflection of light passing close to the surface of the Sun, in his famous 1911 Prague paper, Einstein used just the equivalence principle within his "Prague gravity theory", based on the variable velocity of light. Then, in 1915, he obtained this value to be twice as big in general relativity, when, in addition to the equivalence principle, the curvature of space (determined from (2) to first order in M/r) was taken into account.

Despite the fact that for the purpose of solar-system observations the Schwarzschild metric in the form (2) is, quoting [18], "too accurate", it has played an important role in experimental relativity. Eddington, Robertson and others introduced the method of expanding the Schwarzschild metric at the order beyond Newtonian theory, and then multiplying each post-Newtonian term by a dimensionless parameter which should be determined by experiment. These methods inspired the much more powerful *PPN ("Parametrized post-Newtonian") formalism* which was developed at the end of the 1960s and the beginning of the 1970s for testing general relativity and alternative theories of gravity. It has been very effectively used to compare general relativity with observations (see e.g. [18,71,72] and references therein). In order to gain at least some concrete idea, let us just write down the simplest generalization of (2), namely the metric

$$ds^2 = -\left[1 - \frac{2M}{r} + 2\,(\beta - \gamma)\,\frac{M^2}{r^2}\right] dt^2 + \left(1 + 2\gamma\frac{M}{r}\right) dr^2$$
$$+r^2\left(d\theta^2 + \sin^2\theta\; d\varphi^2\right)\ , \qquad (3)$$

which is obtained by expanding the metric (2) in M/r up to one order beyond the Newtonian approximation, and multiplying each post-Newtonian term by dimensionless parameters which distinguish the post-Newtonian limits of different metric theories of gravity, and should be determined experimentally. (In general, one needs not just two but ten PPN parameters [18,71,72].) In Einstein's theory: $\beta = \gamma = 1$. Calculating from metric (3) the advance of the pericentre of a test particle orbiting a central mass M on an ellipse with semi-major axis a and eccentricity e, one finds $\Delta\phi = \frac{1}{3}(2+2\gamma-\beta)6\pi M/[a(1-e^2)]$, whereas the total deflection angle of electromagnetic waves passing close to the surface of the body is $\Delta\psi = 2(1+\gamma)M/r_0$, where r_0 is the radius of closest approach of photons to the central body.

Measurements of the deflection of radio waves and microwaves by the Sun (recently also of radio waves by Jupiter) at present restrict γ to $\frac{1}{2}(1+\gamma) = 1.0001 \pm 0.001$ [71,72]. Planetary radar rangings, mainly to Mercury, give from the perihelion shift measurements the result $(2\gamma+2-\beta)/3 = 1.00 \pm 0.002$, so that $\beta = 1.000 \pm 0.003$, whereas the measurements of periastron advance for the binary pulsar systems such as PSR 1913+16 implied agreement with Einstein's theory to better than about 1% (see e.g. [71,72] for reviews). There are other solar-system experiments verifying the leading orders of the Schwarzschild solution to a high accuracy, such as gravitational redshift, signal retardation, or lunar geodesic precession. A number of advanced space missions have been proposed which could lead to significant improvements in values of the PPN parameters, and even to the measurements of post-post-Newtonian effects [72].

Hence, though in an approximate form, the Schwarzschild solution has had a great impact on *experimental relativity.* In addition, the observational effects of gravity on light propagation in the solar system, and also today routine observations of gravitational lenses in cosmological contexts [73], have significantly increased our confidence in taking seriously similar predictions of general relativity in more extreme conditions.

2.3 Schwarzschild Metric Outside a Collapsing Star

I recall how Roger Penrose, at the beginning of his lecture at the 1974 Erice Summer School on gravitational collapse, placed two figures side by side. The first illustrated schematically the bending of light rays by the Sun (surprisingly, Penrose did not write "Prague 1911" below the figure). I do not remember exactly his second figure but it was similar to Fig. 1 below: the spacetime

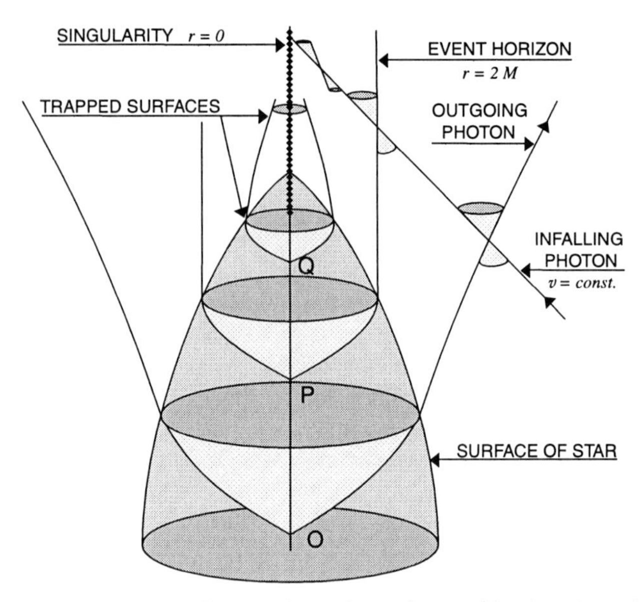

Fig. 1. The gravitational collapse of a spherical star (the interior of the star is shaded). The light cones of the three events, O, P, Q, at the centre of the star, and of the three events outside the star are illustrated. The event horizon, the trapped surfaces, and the singularity formed during the collapse are also shown. Although the singularity appears to lie in a "time direction", from the character of the light cone outside the star but inside the event horizon it is seen that it has a spacelike character.

diagram showing spherical gravitational collapse through the Schwarzschild radius into a spherical black hole.

It is in all modern books on general relativity that the Schwarzschild radius at $R_s = 2M$ is the place where Schwarzschild coordinates t, r are unsuitable, and that metric (2) has a coordinate singularity but not a physical one. One has to introduce other coordinates to extend the Schwarzschild metric through R_s. In order to describe all spacetime outside a collapsing spherical body it is advantageous to use ingoing Eddington–Finkelstein coordinates (v, r, θ, φ) where $v = t + r + 2M \log(r/2M - 1)$. Metric (2) takes the form

$$ds^2 = -\left(1 - \frac{2M}{r}\right) dv^2 + 2dvdr + r^2\left(d\theta^2 + \sin^2\theta\, d\varphi^2\right), \tag{4}$$

(v, θ, φ) = constant are ingoing radial null geodesics. Figure 1, plotted in these coordinates, demonstrates well several basic concepts and facts which were introduced and learned after the end of 1950s when a more complete understanding of the Schwarzschild solution was gradually achieved. The metric (4) holds only outside the star, there will be another metric in its interior, for example the Oppenheimer–Snyder collapsing dust solution (i.e. a portion of a collapsing Friedmann universe), but the precise form of the interior solution is not important at the moment. Consider a series of flashes of light emitted from the centre of the star at events O, P, Q (see Fig. 1) and assume that the stellar material is transparent. As the Sun has a focusing effect on the light rays, so does matter during collapse. As the matter density becomes higher and higher, the focusing effect increases. At event P a special wavefront will start to propagate, the rays of which will emerge from the surface of the star with zero divergence, i.e. the null vector $k^\alpha = dx^\alpha/dw$, w being an affine parameter, tangent to null geodesics, satisfies $k^\alpha_{;\alpha} = 0$. The wavefront then "stays" at the hypersurface $r = 2M$ in metric (4), and the area of its 2-dimensional cross-section remains constant. The null hypersurface representing the history of this critical wavefront is the (future) *event horizon.* Note that the light cones turn more and more inwards as the event horizon is approached. They become tangential to the horizon in such a way that radial outgoing photons stay at $r = 2M$ whereas ingoing photons fall inwards, and will eventually reach the curvature singularity at $r = 0$. As Fig. 1 indicates, wavefronts emitted still later than the critical one, as for example that emitted from event Q, will be focused so strongly that their rays will start to converge, and will form (closed) *trapped surfaces.* The light cones at trapped surfaces are so turned inwards that both ingoing and outgoing radial rays converge, and their area decreases.

Consider a family of spacelike hypersurfaces $\Sigma(\tau)$ foliating spacetime (τ is a time coordinate, e.g. $v - r$). The boundary of the region of $\Sigma(\tau)$ which contains trapped surfaces lying in $\Sigma(\tau)$ is called the *apparent horizon* in $\Sigma(\tau)$.

In general, the apparent horizon is different from the intersection of the event horizon with $\Sigma(\tau)$, as a nice simple example (based again on an exact solution) due to Hawking [74] shows. Assume that after the spherical collapse of a star a spherical thin shell of mass m surrounding the star collapses and eventually crashes at the singularity at $r = 0$ (Fig. 2). In the vacuum region inside the shell there is the Schwarzschild metric (4) with mass M, and outside the shell with mass $M + m$. Hence the apparent horizon on $\Sigma(\tau_1)$ will be at $r = 2M$ and will remain there until $\Sigma(\tau_2)$ when it discontinuously jumps to $r = 2(M + m)$. One can determine the apparent horizon on a given hypersurface. In order to find the event horizon one has to know the whole spacetime solution. The future event horizon separates events which are visible from future infinity, from those which are not, and thus forms the boundary of a *black hole.*

From the above example of a shell collapsing onto a Schwarzschild black hole we can also learn about the "teleological" nature of the horizon: the motion of the horizon depends on what will happen to the horizon in the future (whether a collapsing shell will cross it or not). This *teleological behaviour of the horizon* has later been discovered in a variety of astrophysically realistic situations such as the behaviour of a horizon perturbed by a mass orbiting a black hole (see [75] for enlightening discussions of such effects).

By studying the Schwarzschild solution and spherical collapse it became evident that one has to turn to *global methods* to gain a full understanding of general relativity. The intuition acquired from analyzing the Schwarzschild metric helped crucially in defining and understanding such concepts as the trapped surface, the event horizon, or the apparent horizon in general situations without symmetry. Nowadays these concepts are explained in several advanced textbooks and monographs (e.g. [18,19,26,32,76]).

Following from the example of spherical collapse one is led to ask whether generic gravitational collapses lead to spacetime singularities and whether these are always surrounded by an event horizon. The Penrose-Hawking singularity theorems [19,26] show that singularities do arise under quite generic circumstances (the occurrence of a closed trapped surface is most significant for the appearance of a singularity). The second question is the essence of the cosmic censorship hypothesis. Various exact solutions have played a role in attempts to "prove" or "disprove" this "one of the most important issues" of classical relativity. We shall meet it in several other places later on, in particular in Sect. 3.1. There a more detailed formulation is given.

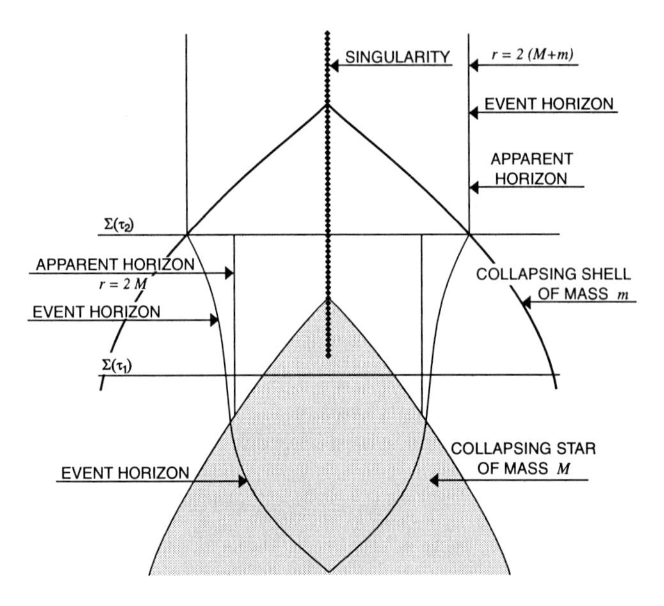

Fig. 2. The "teleological" behaviour of the event horizon during the gravitational collapse of a star, followed by the collapse of a shell. The event horizon moves outwards because it will be crossed by the shell. The apparent horizon moves outwards discontinuously (adapted from [74]).

2.4 The Schwarzschild–Kruskal Spacetime

In the remarks above we considered the Schwarzschild solution outside a static (possibly oscillating, or expanding from $r > 2M$) star, and outside a star collapsing into a black hole. It is not excluded that just these situations will turn out to be physically relevant. Nevertheless, in connection with the Schwarzschild metric it would be heretical not to mention the enormous impact which its maximal vacuum analytic extension into the Schwarzschild–Kruskal spacetime has had. This is today described in detail in many places (see e.g. [18,19,26,76]). We need two sets of the Schwarzschild coordinates to cover the complete spacetime, and we obtain two asymptotically flat spaces, i.e. the spacetime with two ("right" and "left") infinities. The metric in Kruskal coordinates U, V, related to the Schwarzschild r, t (in the regions with $r > 2M$) by

$$\begin{aligned} U &= \pm(r/2M - 1)^{1/2} e^{r/4M} \cosh(t/4M)\,, \\ V &= \pm(r/2M - 1)^{1/2} e^{r/4M} \sinh(t/4M)\,, \end{aligned} \tag{5}$$

takes the form

$$ds^2 = \frac{32M^3}{r} e^{-r/2M} \left(-dV^2 + dU^2\right) + r^2 \left(d\theta^2 + \sin^2\theta\, d\varphi^2\right). \tag{6}$$

The introduction of the Kruskal coordinates which remove the singularity of the Schwarzschild metric (2) at the horizon $r = 2M$ and cover the complete spacetime manifold (every geodesic either hits the singularity or can be continued to the infinite values of its affine parameter), was the most influential example which showed that one has to distinguish carefully between just a coordinate singularity and the real, physical singularity. It also helped us to realize that the definition of a singularity itself is a subtle issue in which the concept of geodesic completeness plays a significant role (see [77] for a recent analysis of spacetime singularities).

The character of the Schwarzschild–Kruskal spacetime is best seen in the Penrose diagram given in Fig. 3, in which the spacetime is compactified by a suitable conformal rescaling of the metric. Both right and left infinities are represented, and the causal structure is well illustrated because worldlines of radial light signals (radial null geodesics) are 45-degree lines in the diagram. In particular the black hole region *II* and a "newly emerged" (as a consequence of the analytical continuation) *white hole* region *IV* (with the white-hole singularity at $r = 0$) are exhibited. For more detailed analyses of the Penrose diagram of the Schwarzschild–Kruskal spacetime the reader is referred to e.g. [18,19,26,76]. Here we wish to turn in some detail to two very important concepts in black hole theory which were first understood by the analytic extension of the Schwarzschild solution, and which are not often treated in standard textbooks. These are the concepts of the *bifurcate horizon* and of the *horizon surface gravity*. Jürgen Ehlers played a somewhat indirect, but important and noble part in their introduction into literature.

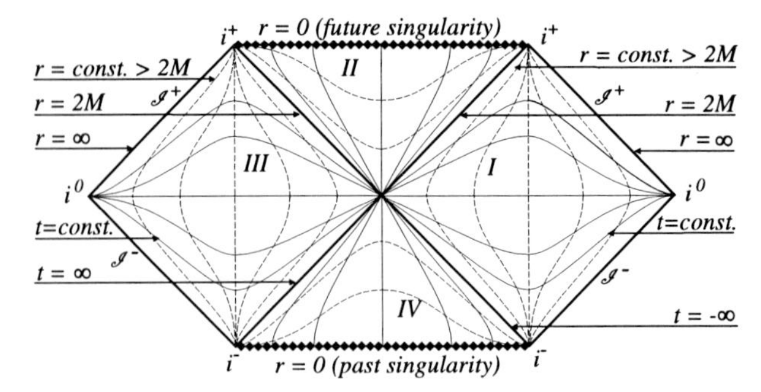

Fig. 3. The Penrose diagram of the compactified Schwarzschild–Kruskal spacetime. Radial null geodesics are 45-degrees lines. Timelike geodesics reach the future (or past) timelike infinities i^+ (or i^-), null geodesics reach the future (or past) null infinities $\mathcal{J}^+$ (or $\mathcal{J}^-$) and spacelike geodesics lead to spatial infinities i^0. (Notice that at i^0 the lines $t =$ constant are tangent to each other – this is often not taken into account in the literature – see e.g. [26,30].)

These concepts were the main subject of the last work of Robert Boyer who became one of the victims of a mass murder on August 1, 1966, in Austin, Texas. Jürgen Ehlers was authorized by Mrs. Boyer to look through the scientific papers of her husband, and together with John Stachel, prepared posthumously the paper [78] from R. Boyer's notes. Ehlers inserted his own discussions, generalized the main theorem on bifurcate horizons, but the paper [78] was published with R. Boyer as the only author.

In the Schwarzschild spacetime there exists the timelike Killing vector, $\partial/\partial t$, which when analytically extended into all Schwarzschild–Kruskal manifold, becomes null at the event horizon $r = 2M$, and is spacelike in the regions II and IV with $r < 2M$. In Kruskal coordinates it is given by

$$k^\alpha = \left(k^V = U/4M,\ k^U = V/4M,\ k^\theta = 0,\ k^\varphi = 0\right). \tag{7}$$

Hence it vanishes at all points with $U = V = 0$, $\theta \in [0,\pi], \varphi \in [0, 2\pi)$. These points, forming a spacelike 2-sphere which we denote B (in Schwarzschild coordinates given by $r = 2M$, $t =$ constant), are fixed points of the 1-dimensional group G of isometries generated by k^α (see Fig. 3). At the event horizon the corresponding 1-dimensional orbits are null geodesics, with k^α being a tangent vector. However, since k^α vanishes at B, these orbits are incomplete.

This (and similar observations for other black hole solutions) motivated a general analysis of the bifurcate Killing horizons given in [78]. There it is proven for spacetimes admitting a general Killing vector field ξ^α, which generates a 1-dimensional group of isometries, that (i) a 1-dimensional orbit is a complete geodesic if the gradient of the square ξ^2 vanishes on the orbit, (ii) if a geodesic orbit is incomplete, then it is null and $(\xi^2)_{,\alpha} \neq 0$. In addition, if $\xi^\alpha = dx^\alpha/dv$ (v being the group parameter), the affine parameter along the geodesic is $w = e^{\kappa v}$, where $\kappa = $ constant satisfies

$$(-\xi^2)^{,\alpha} = 2\kappa\xi^\alpha. \tag{8}$$

In the Schwarzschild case, with $\xi^\alpha = k^\alpha = (\partial/\partial t)^\alpha$, and considering the part $V = U$ of the horizon, we get $\kappa = 1/4M$. The relation $w = e^{\kappa v}$ is just the familiar equation $\tilde{V} = e^{v/4M}$, where $\tilde{V} = V + U$ is the Kruskal null coordinate and v is the Eddington–Finkelstein ingoing null coordinate used in (4). (Notice that $\tilde{V}$ is indeed the affine parameter along the null geodesics at the horizon $V = U$.) The quantity κ, first introduced in [78], has become fundamental in modern black hole theory, and also in its generalizations in string theory. It is the well-known *surface gravity* of the black hole horizon.

With $\kappa \neq 0$, the limit points corresponding to $v \to -\infty, w = 0$ are fixed points of G. (Unless the spacetime is incomplete, there exists a continuation of each null geodesic beyond these fixed points to $w < 0$.) One can show that the fixed points form a spacelike 2-dimensional manifold B, given by $U = V = 0$ in the Schwarzschild case; this "bifurcation surface" is a totally geodesic submanifold. By the original definition [79], a Killing horizon is a G invariant null hypersurface N on which $\xi^2 = 0$. (A recent definition [80,81] specifies a Killing horizon to be any union of such hypersurfaces.) If $\kappa \neq 0$, at each point of B there is one null direction orthogonal to B which is not tangent to $\bar{N} = N \cup B$. The null geodesics intersecting B in these directions form another null hypersurface, $\tilde{N}$, which is also a Killing horizon. The union $N \cup \tilde{N}$ is called a *bifurcate Killing horizon* (Fig. 4).

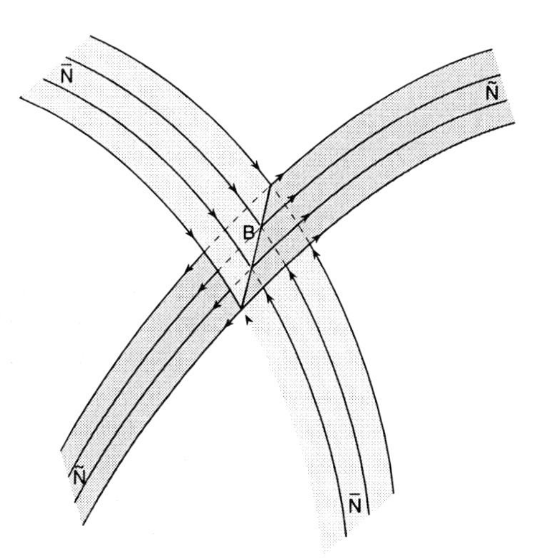

Fig. 4. The bifurcate Killing horizon consisting of two null hypersurfaces $\tilde{N}$ and $\bar{N}$ which intersect in the spacelike 2-dimensional "bifurcation surface" B.

Bifurcate Killing horizons exist also in flat and other curved spacetimes. For example, in the boost-rotation symmetric spacetimes (Sect. 11), null hypersurfaces $z = \pm t$ form the bifurcate Killing horizon corresponding to the boost Killing vector; B, given by $z = t = 0$, is then not compact. (As in the Schwarzschild–Kruskal spacetime, a bifurcate Killing horizon locally divides the spacetime into four wedges.) However, the first motivation for analyzing Killing horizons came from the black hole solutions.

Both Killing horizons and surface gravity play an important role in *black hole thermodynamics* and *quantum field theory on curved backgrounds* [82], in particular in their two principal results: the *Hawking effect* of particle creation by black holes; and the *Unruh effect* showing that a thermal bath of particles will be seen also by a uniformly accelerated observer in flat spacetime when the quantum field is in its vacuum state with respect to inertial observers. Recently, new results were obtained [83] which support the view that a spacetime representing the final state of a black hole formed by collapse has indeed a bifurcate Killing horizon, or the Killing horizon is degenerate ($\kappa = 0$).

2.5 The Schwarzschild Metric as a Case Against Lorentz-Covariant Approaches

There are many other issues on which the Schwarzschild solution has made an impact. Some of astrophysical applications will be very briefly mentioned later on. As the last theoretical point in this section I would like to discuss in some detail the causal structure of the Schwarzschild spacetime including infinity. By analyzing this structure, Penrose [84] presented evidence against various Lorentz (Poincaré)-covariant field theoretical approaches, which regard the physical metric tensor g to be not much different from any other tensor in Minkowski spacetime with flat metric η (see e.g. [85,86]). I thought it appropriate to mention this point here, since Jürgen Ehlers, among others, certainly does not share a field theoretical viewpoint.

The normal procedure of calculating the metric g in these approaches is from a power series expansion of Lorentz-covariant terms (in quantum theory this corresponds to an infinite summation of Feynman diagrams). The derived field propagation has to follow the true null cones of the curved metric g instead of those of η. However, as Penrose shows, in a satisfactory theory the null cones defined by g should not extend outside the null cones defined by η, or "the causality defined by g should not violate the background η-causality". Following [84], let us write this condition as $g < \eta$. Now at first sight we may believe that $g < \eta$ is satisfied in the Schwarzschild field since its effect is to "slow down" the velocity of light (cf. "signal retardation" mentioned in 2.2). However, in the field-theoretical approaches one of the main emphasis is in a consistent formulation of scattering theory. This requires a good behaviour at infinity. But with the Schwarzschild metric, null geodesics with respect to metric g "infinitely deviate" from those with respect to η: for example, the radial outgoing g null geodesics $\theta, \varphi =$ constant, and $u = t - r - 2M \log(r/2M - 1) =$ constant at $r \to \infty$ go "indefinitely far" into the retarded time $t - r$ of η, and hence, do not correspond to outgoing η-null geodesics $t - r =$ constant. One can try to use a different flat metric associated with the Schwarzschild metric g which does not lead to pathological behaviour at infinity, but then it turns out that $g < \eta$ is violated locally. In fact, Penrose [84] proves the theorem, showing that there is an essential incompatibility between the causal structures in the Schwarzschild and Minkowski spacetimes which appears either asymptotically or locally.

This incompatibility is easily understood with the exact Schwarzschild solution, but it is generic, since one is concerned only with the behaviour of the space at large distances from a positive-mass source, i.e. with the causal properties in the neighbourhood of spacelike infinity i^0.

In the present post-Minkowskian approximation methods for the generation of gravitational waves by relativistic sources, a suitable (Bondi-type) coordinate system [66] is constructed at all orders in the far wave zone, which in particular corrects for the logarithmic deviation of the true light cones with respect to the coordinate flat light cones (cf. contribution by L. Blanchet in this volume).

2.6 The Schwarzschild Metric and Astrophysics

In his introductory chapter "General Relativity as a Tool for Astrophysics" for the Seminar in Bad Honnef in 1996 [87], Jürgen Ehlers remarks that "The interest of black holes for astrophysics is obvious... The challenge here is to find observable features that are truly relativistic, related, for example, to horizons, ergoregions... Indications exist, but – as far as I am aware – no firm evidence."

There are many excellent recent reviews on the astrophysical evidence for black holes (see e.g. [88–90]). It is true, that the evidence points towards the presence of dark massive objects – stellar-mass objects in binaries, and supermassive objects in the centres of galaxies – which are associated with deep gravitational potential wells where Newtonian gravity cannot be used, but it does not offer a clear diagnostic of general relativity.

Many investigations of test particle orbits in the strong-gravity regions ($r \leq 10M$) have shown basic differences between the motion in the Schwarzschild metric and the motion in the central field in Newton's theory (e.g. [18,76,91]). For example for $3M < r < 6M$ unstable circular particle orbits exist which are energetically unbound, and thus perturbed particles may escape to infinity; at $r = 3M$ circular photon orbits occur and there are no circular orbits for $r < 3M$. Particles are trapped by a Schwarzschild black hole if they reach the region $r < 3M$.

About ten years ago, the study of the *behaviour of particles and gyroscopes in the Schwarzschild field* revived interest in the "classical" problem of the definition of gravitational, centrifugal, and other inertial "forces" acting on particles and gyros moving on the Schwarzschild or on a more general curved backgrounds, usually axisymmetric and stationary (see e.g. [92,93], and many references therein). One would like to have a split of a covariantly defined quantity (like an acceleration) into non-covariant parts, the physical meaning of which would increase our intuition of relativistic effects in astrophysical problems. If, for example, we adopt the view that the "gravitational force" is velocity-independent, then we find that at the orbits outside the circular photon orbit ($r > 3M$), the centrifugal force is as in classical physics, repulsive, while it becomes attractive inside this orbit, being zero exactly at the orbit.[9]

Relativistic effects will, of course, play a role in many astrophysical situations involving spherical accretion, the structure of accretion disks around compact stars and black holes, their optical appearance etc. They have become an important part of the arsenal of astrophysicists, and they have entered standard literature (see e.g. [95,96]). Though this whole field of science lies beyond the scope of this article, I would like to mention three recent issues which provide us with hope that we may perhaps soon meet the challenge noted in Jürgen Ehlers' remarks made in Bad Honnef in 1996.

The first concerns our Galactic centre. Thanks to new observations of stars in the near infrared band it was possible to detect the transverse motions of stars (for which the radial velocities are also observed) within 0.1 pc in our Galactic centre. The stellar velocities up to 2000 km/sec and their dependence on the radial distance from the centre are consistent with a black hole of mass $2.5 \times 10^6 M_\odot$. In the opinion of some leading astrophysicists, our Galactic centre now provides "the most convincing case for a supermassive hole, with the single exception of NGC 4258" [88]. (In NGC 4258 a disk is observed whose inner edge is orbiting at 1080 km/sec, implying a black hole – "or something more exotic" [88] – with a mass of $3.6 \times 10^7 M_\odot$.) Perhaps we shall be able to observe relativistic effects on the proper motions of stars in our Galactic centre in the not too distant future.

The second issue concerns the fundamental question of whether observations can bring convincing proof of the existence of black hole event horizons. Very recently some astrophysicists [89] claimed that new observations, in particular of X-ray binaries, imply such evidence. The idea is that thin disk accretion cannot explain the spectra of some of X-ray binaries. One has to use a different accretion model, a so called advection-dominated accretion flow model (ADAF) in which most of the gravitational energy released in the infalling gas is carried (advected) with the flow as thermal energy, which falls on the central object. (In thin disks most of this energy is radiated out from the disk.) If the central compact object (for example a neutron star) has a hard surface, the thermal energy stored in the flow is re-radiated after the flow hits the surface. However, some of the X-ray binaries show such low luminosities that a very large fraction of the energy in the flow must be ad-

---

[9] Curiously enough, Feynman in his 1962-63 lectures on gravitation [94] writes that "inside $r = 2M$ [not $3M$!]... the 'centrifugal force' apparently acts as an attraction rather than a repulsion".

vected through an event horizon into a black hole [89]. Although Rees [88], for example, considers this evidence "gratifyingly consistent with the high-mass objects in binaries being black holes", he believes that it "would still not convince an intelligent sceptic, who could postulate a different theory of strong-field gravity or else that the high-mass compact objects were (for instance) self-gravitating clusters of weakly interacting particles...".

For a sceptical optimistic relativist, the most challenging observational issue related to black holes probably is to find astrophysical evidence for a Kerr metric. We shall come to this point in Sect. 4.3.

The last (but certainly not the least) issue lies more in the future, but eventually should turn out to be most promising. It is connected with both the Numerical Relativity Great Challenge Alliance and the "great challenge" of experimental relativity: to calculate reliable gravitational wave-forms and to detect them. When gravitational waves from stars captured by a supermassive black hole, or from a newly forming supermassive black hole, or, most importantly, from coalescing supermassive holes will be detected and compared with the predictions of the theory, we should learn significant facts about black holes [88,97]. Are these so general remarks entirely inappropriate in the section on the Schwarzschild solution?

One of the most important roles of the Schwarzschild solution in the development of mathematical relativity and especially of relativistic astrophysics stems from its simplicity, in particular from its spherical symmetry. This has enabled us to develop the mathematically beautiful theory of linear perturbations of the Schwarzschild background and employ it in various astrophysically realistic situations (see e.g. [75,76,91], and many references therein). Surprisingly enough, this theory does not only give reliable results in such problems as the calculation of waves emitted by pulsating neutron stars, or waves radiated out from stars falling into a supermassive black hole. Very recently we have learned that one can use perturbation theory of a single Schwarzschild black hole as a "close approximation" to black hole collisions. Towards the end of the collision of two black holes, they will not in fact be two black holes, but will merge into a highly distorted single black hole [98]. When compared with the numerical results on a head-on collision it has been found that this approximation gives predictions for separations Δ as large as $\Delta/M \sim 7$.

3 The Reissner–Nordström Solution

This spherically symmetric solution of the Einstein–Maxwell equations was derived independently[10] by H. Reissner in 1916, H. Weyl in 1917, and G.

[10] In the literature one finds the solution to be repeatedly connected only with the names of Reissner and Nordström, except for the "exact-solutions-book" [61]: there in four places the solution is called as everywhere else, but in one place (p. 257) it is referred to as the "Reissner–Weyl solutions". An enlightening

Nordström in 1918. It represents a spacetime with no matter sources except for a radial electric field, the energy of which has to be included on the right-hand side of the Einstein equations. Since Birkhoff's theorem, mentioned in connection with the Schwarzschild solution in Sect. 2.2, can be generalized to the electrovacuum case, the Reissner–Nordström solution is the unique spherical electrovacuum solution. Similarly to the Schwarzschild solution, it thus describes the exterior gravitational and electromagnetic fields of an arbitrary – static, oscillating, collapsing or expanding – spherically symmetric, charged body of mass M and charge Q. The metric reads

$$ds^2 = -\left(1 - \frac{2M}{r} + \frac{Q^2}{r^2}\right) dt^2 + \left(1 - \frac{2M}{r} + \frac{Q^2}{r^2}\right)^{-1} dr^2 + r^2 \left(d\theta^2 + \sin^2\theta \, d\varphi^2\right), \tag{9}$$

the electromagnetic field in these spherical coordinates is described by the "classical" expressions for the time component of the electromagnetic potential and the (only non-zero) component of the electromagnetic field tensor:

$$A_t = -\frac{Q}{r}, \quad F_{tr} = -F_{rt} = -\frac{Q}{r^2}. \tag{10}$$

A number of authors have discussed spherically symmetric, static charged dust configurations producing a Reissner–Nordström metric outside, some of them with a hope to construct a "classical model" of a charged elementary particle (see [61] for references). The main influence the metric has exerted on the developments of general relativity, and more recently in supersymmetric and superstring theories (see Sect. 3.2), is however in its analytically extended electrovacuum form when it represents charged, spherical black holes.

3.1 Reissner–Nordström Black Holes and the Question of Cosmic Censorship

The analytic extensions have qualitatively different character in three cases, depending on the relationship between the mass M and the charge Q. In the case $Q^2 > M^2$ (corresponding, for example, to the field outside an electron), the complete electrovacuum spacetime is covered by the coordinates (t, r, θ, φ), $0 < r < \infty$. There is a *naked singularity* (visible from infinity) at $r = 0$ in which the curvature invariants diverge. If $Q^2 < M^2$, the metric (9)

---

discussion on p. 209 in [61] shows that the solution belongs to a more general "Weyl's electrovacuum class" of electrostatic solutions discovered by Weyl (in 1917) which follow from an Ansatz that there is a functional relationship between the gravitational and electrostatic potentials. As will be noticed also in the case of cylindrical waves in Sect. 9, if "too many" solutions are given in one paper, the name of the author is not likely to survive in the name of an important subclass...

describes a (generic) *Reissner–Nordström black hole*; it becomes singular at two radii:

$$r = r_{\pm} = M \pm (M^2 - Q^2)^{\frac{1}{2}}. \quad (11)$$

Similarly to the Schwarzschild case, these are only coordinate singularities. Graves and Brill [99] discovered, however, that the analytic extension and the causal structure of the Reissner–Nordström spacetime with $M^2 > Q^2$ is fundamentally different from that of the Schwarzschild spacetime. There are two null hypersurfaces, at $r = r_+$ and $r = r_-$, which are known as the *outer (event) horizon* and the *inner horizon*; the Killing vector $\partial/\partial t$ is null at the horizons, timelike at $r > r_+$ and $r < r_-$, but spacelike at $r_- < r < r_+$. The character of the extended manifold is best seen in the Penrose diagram in Fig. 5, in which the spacetime is compactified by a suitable conformal rescaling of the metric (see, e.g. [18,26,30]). As in the compactified Kruskal-Schwarzschild diagram in Fig. 3, the causal structure is well illustrated because worldlines of radial light signals are 45-degree lines. There are again two infinities illustrated - the right and left - in regions *I* and *III*. However, the maximally extended Reissner–Nordström geometry consists of an *infinite chain of asymptotic regions* connected by "wormholes" between the real singularities (with divergent curvature invariants) at $r = 0$. In Fig. 5, the right and left (past null) infinities in regions I' and III' are still seen - the others are obtained by extending the diagram vertically in both directions.

An important lesson one has learned is that the character of the singularity need not be spacelike as it is in the Schwarzschild case, or with the big bang singularities in standard cosmological models. Indeed, the *singularities* in the Reissner–Nordström geometry are *timelike*: they do not block the way to the future. By solving the geodesic equation one can show that there are test particles which start in "our universe" (region *I*), cross the outer horizon at $r = r_+$ and the inner horizon at $r = r_-$, avoid the singularity and through a "white hole" (the outer horizon between regions IV' and I') emerge into "another universe" I' with its own asymptotically flat region. Such a *gravitational bounce* can occur not only with test particles. The studies of the gravitational collapse of charged spherical shells ([100] and references therein) and of charged dust spheres ([101] and references therein) have shown that a bounce can take place also in fully dynamical cases.[11] The part of Fig. 5 which is "left" from the worldline of the surface of the sphere or the shell is "covered" by the interior of the sphere or flat space inside the shell. As observed in [100], the outcome of the bounce of a shell can be

[11] An intuitive explanation [101] of this bounce is that as the sphere (the shell) contracts, the volume of the exterior region increases, and hence also the total energy in the electric field, which eventually exceeds the energy in the sphere. However, the external plus internal energy does not change during collapse (there are no waves), and so in the neighborhood of a highly contracted charged object, the gravitational field must have a repulsive character corresponding to a negative mass-energy.

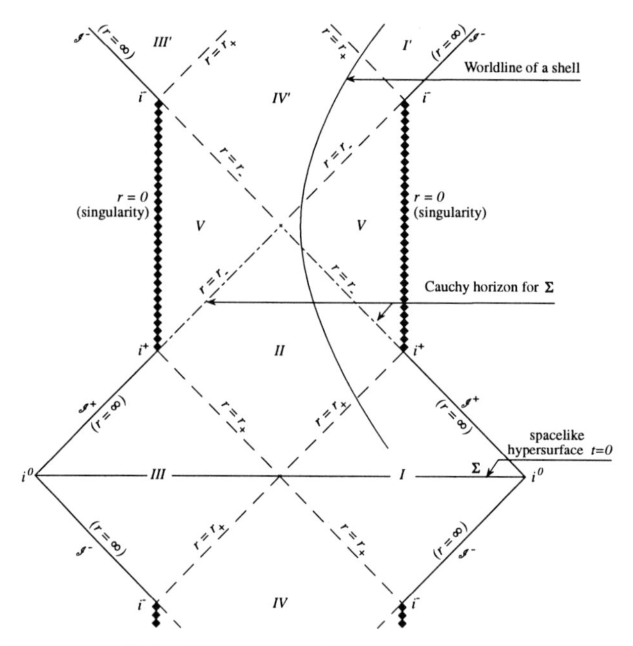

Fig. 5. The compactified Reissner–Nordström spacetime representing a non-extreme black hole consists of an indefinite chain of asymptotic regions ("universes") connected by "wormholes" between timelike singularities. The worldline of a shell collapsing from "universe" I and re-emerging in "universe" I' is indicated. The inner horizon at $r = r_-$ is the Cauchy horizon for spacelike hypersurface Σ. It is unstable and will thus very likely prevent such a process occuring.

different, depending on the value of the shell's total mass, charge and rest mass. The shell may crash into the "right" singularity or it may continue to expand and emerge in region I'. If the rest mass of the shell is negative the collapse may even lead to a naked singularity.

Now even if the shell collapses into a black hole, and after a bounce, emerges in region I', a *locally* naked singularity is present: the timelike singularity at $r = 0$, to the "right" from the wordline of the shell. An observer travelling into the future "between" the shell and the singularity can be surprised by a signal coming from the singularity (see Fig. 5). Penrose's *strong cosmic censorship conjecture* (see e.g. [102]) suggests that this should not happen. In its physical formulation, as given by Wald (see [19] also for the precise formulation), it says that all physically reasonable spacetimes are globally hyperbolic, i.e. apart from a possible initial (big bang-type) singularity, *no singularity is visible to any observer.* It was just the example of the Reissner–Nordström solution (and a similar property of the Kerr solution) which inspired Penrose to formulate the strong cosmic censorship conjecture, in addition to its *weak version* which only requires that from generic nonsingular initial data on a Cauchy hypersurface *no spacetime singularity* develops which is *visible from infinity.* As Penrose [102] puts it, "it seems to be comparatively unimportant whether the observer himself can escape to infinity".

It is evident from Fig. 5 that the Reissner–Nordström spacetime is not globally hyperbolic, i.e. it does not possess a Cauchy hypersurface Σ, the initial data on which (for a test field, say) would determine the development of data in the entire future. If data are given on the spacelike hypersurface Σ "connecting" left and right infinities of regions *III* and *I*, the Cauchy development will predict what happens only in regions *III* and *I* above Σ, and in region *II*, i.e. not beyond the null hypersurfaces (inner horizons) $r = r_-$ between region *II* and regions *V* in the figure. The inner horizons $r = r_-$ represent the *Cauchy horizon* for a typical initial hypersurface like Σ. As noticed above, what is happening at an event in regions *V* is in general influenced not only by data on Σ but also by what is happening at the (locally) naked singularities (which cannot be predicted since the physics at a singularity cannot be controlled).

Penrose was also the first who predicted that the *inner (Cauchy) horizon is unstable* [27]. If this is true, a null singularity, or possibly even a spacelike singularity may arise during a general collapse, so preventing a violation of the strong cosmic censorship conjecture. The instability of the Cauchy horizon can in fact be expected by using first the following simple geometrical-optics argument.

Introduce the ingoing Eddington–Finkelstein null coordinate v by

$$v = t + r^* = t + \int f(r) dr,$$
$$f = 1 - 2M/r + Q^2/r^2, \tag{12}$$

which brings the metric (9) into the form as of equation (4) in the Schwarzschild case. Consider a freely falling observer who is approaching the inner horizon given by $r = r_-, v = \infty$. Denoting the observer's constant specific energy parameter (see e.g. [18]) by $\tilde{E} = -\boldsymbol{U\xi}$, where the Killing vector $\boldsymbol{\xi} = \partial/\partial v$, observer's four-velocity $\boldsymbol{U} = d/d\tau$ (τ - observer's proper time), the geodesic equations imply

$$\dot{r}^2 + f = \tilde{E}^2, \quad \dot{v} = f^{-1}[\tilde{E} - (\tilde{E}^2 - f)^{\frac{1}{2}}], \tag{13}$$

where a dot denotes $d/d\tau$. Between the horizons, $\boldsymbol{\xi}$ is spacelike and $\tilde{E}$ can be negative. Geodesic equations (13) imply $dr/dv \cong \frac{1}{2}f$ for an observer with $\tilde{E} < 0$, approaching $r = r_-$ from region *II*. Expanding f near $r = r_-$, $f \cong f'(r_-)(r - r_-) = -2\kappa(r - r_-)$, where

$$\kappa = (M^2 - Q^2)^{\frac{1}{2}}/r_-^2 \tag{14}$$

is the surface gravity of the inner horizon (as it follows from definition (8)), and integrating, we get f near r_-. From the second equation in (13) we then obtain the "asymptotic formula"

$$\dot{v} \simeq \text{constant} \mid \tilde{E} \mid e^{\kappa\tau} \cong \frac{1}{\kappa \Delta\tau}, \; v \to \infty, \Delta\tau \to 0, \tag{15}$$

where $\Delta\tau$ is the amount of proper time the observer needs to reach the inner (Cauchy) horizon.

Imagine now two nearby events A_{out} and B_{out} in the outside world I, for example the emission of two photons from a given fixed $r > r_+$, which are connected by the ingoing null geodesics $v = \text{constant}$ and $v + dv = \text{constant}$ with events A_{in} and B_{in} on the worldline of the observer approaching r_-. The interval of proper time between A_{out} and B_{out} is $d\tau_{out} \sim dv$, whereas (15) implies that the interval of proper time between A_{in} and B_{in}, as measured by the observer approaching r_-, is $d\tau_{in} \sim e^{-\kappa v} dv$. Therefore,

$$\frac{d\tau_{in}}{d\tau_{out}} \sim e^{-\kappa v}, \tag{16}$$

so that as $v \to +\infty$ the events (clocks) in the outside world are measured to proceed increasingly fast by the inside observer approaching the inner horizon. In the limit, when the observer crosses the Cauchy horizon, he sees the whole future history (from some event as A_{out}) of the external universe to "proceed in one flash": $d\tau_{in} \to 0$ with $d\tau_{out} \to \infty$. An intuitive explanation is given by the fact that the observers in region I need infinite proper time to reach $v = \infty$, whereas inside observers only finite proper time. The infalling radiation will thus be unboundedly blue-shifted at the inner horizon which in general will lead to a divergence of the energy density there. This *infinite blueshift at the inner (Cauchy) horizon makes it generally unstable to perturbations ("the blue-sheet instability").*

There exists an extensive literature analyzing the exciting questions of the *black hole interiors* (see for example [76], the introductory review [103] in the proceedings of the recent workshop devoted entirely to these issues, and other contributions in the proceedings which give also many further references). Some crucial questions are still the subject of much debate. One of the following two approaches to the problem is usually chosen: (i) a linear perturbation analysis of the behaviour of fields at the Cauchy horizon, (ii) the simplified nonlinear, spherically symmetric models of black hole interiors.

In the first approach one considers the evolution of linear perturbations, representing scalar, electromagnetic, or gravitational fields, on the Reissner–Nordström background. Since there is a nonvanishing background electric field, the *electromagnetic and gravitational* perturbations are *coupled.*[12] It is

[12] This leads to various interesting phenomena. For example, the scattering of incident electromagnetic and gravitational waves by the Reissner–Nordström black hole allows for the partial conversion of electromagnetic waves into gravitational waves and vice versa [91]. When studying stationary electromagnetic fields due to sources located outside the Reissner–Nordström black hole, one discovers that

a remarkable fact that "wave equations" for certain gauge-invariant combinations of perturbations can be derived from which all perturbations can eventually be constructed [91,105,106]. In the simplest case of the scalar field Φ on the Reissner–Nordström background, after resolving the field into spherical harmonics and putting $\Phi = r\Psi$, the wave equation has the form [107]

$$\Psi_{,tt} - \Psi_{,r^*r^*} + F_l(r^*)\Psi = 0, \tag{17}$$

where the curvature-induced potential barrier is given by

$$F_l(r^*) = \left(1 - \frac{2M}{r} + \frac{Q^2}{r^2}\right)\left[\frac{2}{r^3}\left(M - \frac{Q^2}{r^2}\right) + \frac{l(l+1)}{r^2}\right], \tag{18}$$

where r is considered to be a function of r^* (cf. (12)). In order to determine the evolution of the field below the outer horizon in a real gravitational collapse, one first concentrates on the evolution of the field outside a collapsing body (star). A nonspherically symmetric scalar test field (generated by a nonspherical distribution of "scalar charge" in the star) serves as a prototype for (small) asymmetries in the external gravitational and electromagnetic fields, which are generated by asymmetries in matter and charge distributions inside the star. Now when a slightly nonspherical star starts to collapse, the perturbations become dynamical and propagate as waves. Their evolution can be determined by solving the wave equation (17). Because of the potential barrier (18) the waves get backscattered and produce slowly decaying radiative tails, as shown in the classical papers by Price [108,109], and generalized to the Reissner–Nordström case in [107,110]. The tails decay in the vicinity of the outer event horizon r_+ (i.e. between regions *I* and *II* in Fig. 5) as $\Psi \sim v^{-2(l+1)}$ for l-pole perturbations.[13] The decaying tails provide the initial data for the "internal problem" – the behaviour of the field near the Cauchy horizon. Calculations show (see [103] and references therein) that near the Cauchy horizon the behaviour of the field remains qualitatively the same: $\Psi(u, v \to \infty) \sim v^{-2(l+1)} + \{\text{slowly varying function of } u\}$, where $u = 2r^* - v$ is constant along outgoing radial null geodesics in region *II*. However, as a consequence of the "exponentially growing blueshift", given by formula (15), the *rate of change of the field diverges as the observer approaches the Cauchy horizon*: $d\Psi/d\tau = (\partial\Psi/\partial x^\alpha)U^\alpha \simeq \Psi_{,v}\,\dot{v} \sim v^{-2l-3}e^{\kappa v}$. Therefore, the measured energy density in the field would also diverge, causing an instability of the Cauchy horizon, which would be expected to create a curvature singularity. More detailed considerations [103] show that the singularity, at least for

closed magnetic field lines not linking any current source may exist, since gravitational perturbations constitute, via the background Maxwell tensor, an effective source [104].

[13] This result is true for a general charged Reissner–Nordström black hole with $Q^2 < M^2$. In the extremal case, $Q^2 = M^2$, the field decays only as $\Psi \sim v^{-(l+2)}$ [107].

large $|u|$, is null and weak (the metric is well-defined, only the Riemann tensor is singular). Any definitive picture of the Cauchy horizon instability can come however only from a fully nonlinear analysis which takes into account the backreaction of spacetime geometry to the growing perturbations.

The second approach to the study of the Cauchy horizon instabilities employs a simplified, *spherically symmetric model which treats the nonlinearities exactly* [111]. The ingoing radiation is modelled by a stream of ingoing charged null dust [112] which is infinitely blueshifted at the inner horizon. There is, however, also an outgoing stream of charged null dust considered to propagate into region *II* towards the inner horizon. The outgoing flux may represent radiation coming from the stellar surface below the outer horizon, as well as a portion of the ingoing radiation which is backscattered in region *II*, and irradiates thus the inner horizon. A detailed analysis based on exact spherically symmetric solutions revealed a remarkable effect: an effective *internal* gravitational-mass parameter of the hole unboundedly increases at the inner (Cauchy) horizon (though the external mass of the hole remains finite). This *"mass inflation phenomenon"* causes the divergence of some curvature scalars at the Cauchy horizon [111]. In reality, the classical laws of general relativity will break down when the curvature reaches Planckian values.

It is outside the scope of this review to discuss further the fascinating issues of black hole interiors. They involve deep questions of classical relativity, of quantum field theory on curved background (as, for example, in discussions of electromagnetic pair production and vacuum polarization effects inside black holes), and they lead us eventually to quantum gravity. We refer again especially to [76] and [103] for more information. Let us only add three further remarks. We mentioned above the work on the inner structure of Reissner–Nordström black holes because this is the most explored (though not closed) area. However, Kerr black holes (Sect. 4) possess also inner horizons and there are many papers concerned with the instabilities of the Kerr Cauchy horizons (see [76,103] for references). Secondly, at the beginning of 1990s, it was shown that the inner horizons of the Reissner–Nordström-de Sitter and Kerr–de Sitter black holes are classically stable in the case when the surface gravity at the inner horizons is smaller than the surface gravity at the cosmological horizon ([103] and references therein, in particular, the review [113]). Penrose [114] even suggested that "it may well be that cosmic censorship requires a zero (or at least a nonpositive) cosmological constant". Very recently, however, three experts in the field [115] have claimed that outgoing modes near to the black hole (outer) event horizon lead to instability for all values of the parameters of Reissner–Nordström-de Sitter black holes. Let me borrow again a statement from Penrose [114]: "My own feelings are left somewhat uncertain by all these considerations".

Finally, a new contribution [116] to the old problem of testing the *weak* cosmic censorship by employing a Reissner–Nordström black hole indicates that one can overcharge a near extreme ($Q^2 \to M^2$) black hole by throwing in a charged particle appropriately. However, the backreaction effects remain to be explored more thoroughly. The question of cosmic censorship thus remains as interesting as ever.

3.2 On Extreme Black Holes, *d*-Dimensional Black Holes, String Theory and "All That"

In the previous section we considered generic Reissner–Nordström black holes with $M^2 > Q^2$. They have outer and inner horizons given by (11), with nonvanishing surface gravities (cf. (14) for the inner horizon). For $M^2 = Q^2$ the two horizons coincide at $r_+ = r_- = M$. Defining the ingoing null coordinate v as in (12), we obtain the ingoing extension of the Reissner–Nordström metric (9) in the form

$$ds^2 = -\left(1 - \frac{M}{r}\right)^2 dr^2 + 2dvdr + r^2(d\theta^2 + \sin^2\theta d\varphi^2). \qquad (19)$$

This is the metric of *extreme Reissner–Nordström black holes.* Frequently, these holes are called "degenerate". At the horizon $r = M$, the Killing vector field $\boldsymbol{k} = \partial/\partial v$ obeys the equation $(k^\alpha k_\alpha)_{,\beta} = 0$, so that regarding the general relation (8), the surface gravity $\kappa = 0$, i.e. the *Killing horizon is degenerate.* Using $(k^2)_{,\beta} = 0$ and the Killing equation, we easily deduce that the horizon null generators with tangent $k^\alpha = dx^\alpha/dv$ satisfy the geodesic equation with affine parameter v. The generators have infinite affine length to the past given by $v \to -\infty$ (in contrast to the generators of a bifurcate Killing horizon – cf. Sect. 2.4). This part of the extreme Reissner–Nordström spacetime, given by $r = M, v \to -\infty$, is called an *"internal infinity"*. That there is no "wormhole" joining two asymptotically flat regions and containing a minimal surface 2-sphere like in the non-extreme case can also be seen from the metric in the original Schwarzschild-type coordinates. Considering an embedding diagram $t = \text{constant}, \theta = \pi/2$ in flat Euclidean space one finds that an infinite "tube", or an asymptotically cylindrical region on each t = constant hypersurface develops. The boundary of the cylindrical region is the internal infinity. It is a compact 2-dimensional spacelike surface. The hypersurfaces t = constant do not intersect the horizon but only approach such an intersection at the internal infinity. (See [79] for the conformal diagram and a detailed discussion, including analysis of the electrovacuum Robinson–Bertotti universe as the asymptotic limit of the extreme Reissner–Nordström geometry at the internal infinity.)

There has been much interest in the extreme Reissner–Nordström black holes within standard Einstein–Maxwell theory. They admit surprisingly simple solutions of the perturbation equations [117]. Some of them appear to be stable with respect to both classical and quantum processes, and there are attempts to interpret them as solitons [118]. Also, they admit supersymmetry [119].

The quotation marks in the title of this section play a double role: the last two words are just "quoting" from the end of the title of a general review on string theory and supersymmetry prepared for the special March 1999 issue of the Reviews of Modern Physics in honor of the centenary of the American Physical Society by Schwarz and Seiberg [120], but they also should "self-ironically" indicate my ignorance in these issues. In addition, unified theories of the type of string theory appear to be somewhat outside the direct interest of Jürgen Ehlers, who has always emphasized the depth and economy of general relativity because it is a "background-independent" theory: string theories still suffer from the lack of a background-independent formulation. Nevertheless, they are beautiful, consistent, and very challenging constructions, representing one of the most active areas of theoretical physics. Recently, string theory provided an explanation of the Bekenstein-Hawking prediction of the entropy of extreme and nearly extreme black holes. From the point of view of this review we should emphasize that many of the techniques that have been used to obtain exact solutions – mostly exact black hole solutions – in generalized theories like string theory were motivated by classical general relativity. There are also results in classical general relativity which are finding interesting generalizations to string theories, as we shall see with one example below.

Before making a few amateurish comments on new results concerning extreme black holes in string theories, let us point out that in many papers from the last 20 years, *black hole solutions* were studied in spacetimes with the number of *dimensions* either *lower* or *higher* than four. The lower-dimensional cases are usually analyzed as "toy models" for understanding the complicated problems of quantum gravity. The higher-dimensional models are motivated by efforts to find a theory which unifies gravity with the other forces. The most surprising and popular (2+1)-dimensional black hole is the BTZ (Bañados-Teitelboim-Zanelli) black hole in the Einstein theory with a negative cosmological constant. Locally it is isometric to anti de Sitter space but its topology is different. In [121] the properties of (2+1)-dimensional black holes are reviewed. In (1+1)-dimensions one obtains black holes only if one includes at least a simple dilaton scalar field; the motivation for how to do this comes from string theory. In higher dimensions one can find generalizations of all basic black hole solutions in four dimensions [122]. Interesting observations concerning higher-dimensional black holes have been given a few years ago [123]. Perhaps one does not need to quantize gravity in order to remove the singularities of classical relativity. It may well be true that some new classical physics intervenes below Planckian energies. In [123] it is demonstrated that certain singularities of the four-dimensional extreme dilaton black holes can be resolved by passing to a higher-dimensional theory of gravity in which usual spacetime is obtained only below some compactification scale. A useful, brief pedagogical introduction to black holes in unified theories is contained in [76].

One of the most admirable recent results of string theory, which undoubtedly converted some relativists and stimulated many string theorists, has been the derivation of the *exact value of the entropy of extreme and nearly extreme black holes.* I shall just paraphrase a few statements from the March 1999 review for the centenary of the American Physical Society by Horowitz and Teukolsky [124]. There are very special states in string theory called BPS (Bogomol'ny–Prasad–Sommerfield) states which saturate an equality $M \geq c|Q|$, with M being the mass, Q the charge, and c is a fixed constant. The mass of these special states does not get any quantum corrections. The strength of the interactions in string theory is determined by a coupling constant g. One can count BPS states at large Q and small g. By increasing g one increases gravity, and then all of these states become black holes. (The BPS states are supersymmetric and one can thus follow the states from weak to strong coupling.) But they all become identical extreme Reissner–Nordström black holes, because there is only one black hole for given $M = |Q|$. When one counts the number N of BPS states in which an extreme hole can exist, and compares this with the entropy $S_{bh} = \frac{1}{4}A$ of the hole as obtained in black hole thermodynamics [82,125], where A is the area of the event horizon ($A = 4\pi M^2$ for the extreme Reissner–Nordström black hole), one finds exactly the "classical" result: $S_{bh} = \log N!$. The entropy of the classical black hole configuration is given in terms of the number of quantum microstates associated with that configuration, by the basic formula of statistical physics. For more detailed recent reviews, see [126,127], and references therein. Remarkably, the results for the black hole entropy have been obtained also within the canonical quantization of gravity [128]. A comprehensive review [129] of black holes and solitons in string theory appeared very recently.

Allow me to finish this "all that" section with a personal remark. In 1980 L. Dvořák and I found that in the Einstein–Maxwell theory, external magnetic flux lines are expelled from the black hole horizon as the hole becomes an extreme Reissner–Nordström black hole [104]. Hence, extreme black holes exhibit some sort of *"Meissner effect"* known from superconductivity. Last year it was demonstrated by Chamblin, Emparan and Gibbons [130] that this effect occurs also for black hole solutions in string theory and Kaluza–Klein theory. Other extremal solitonic objects in string theory (like p-branes) can also have superconducting properties. Within the Einstein–Maxwell theory this effect was first studied to linear order in magnetic field – we analyzed Reissner–Nordström black holes in the presence of magnetic fields induced by current loops. However, we also used an exact solution due to Ernst [131], describing a charged black hole in a background magnetic field, which asymptotically goes over to a Melvin universe, and found the same effect (see also [132] for the case of the magnetized Kerr–Newman black hole). In [130] the techniques of finding exact solutions of Einstein's field equations are employed within string theory and Kaluza–Klein theory to demonstrate the "Meissner effect" in these theories.

4 The Kerr Metric

The discovery of the Kerr metric in 1963 and the proof of its unique role in the physics of black holes have made an immense impact on the development of general relativity and astrophysics. This can hardly be more eloquently demonstrated than by an emotional text from Chandrasekhar [60]: "In my entire scientific life, extending over forty-five years, the most shattering experience has been the realization that an exact solution of Einstein's equations of general relativity, discovered by the New Zealand mathematician Roy Kerr, provides the absolutely exact representation of untold numbers of massive black holes that populate the Universe..."

In Boyer–Lindquist coordinates the Kerr metric [133] looks as follows (see e.g. [18,30]):

$$\begin{aligned} ds^2 = & -\left(1 - \frac{2Mr}{\Sigma}\right) dt^2 - 2\frac{2aMr\sin^2\theta}{\Sigma} dt\, d\varphi + \\ & + \frac{\Sigma}{\Delta} dr^2 + \Sigma d\theta^2 + \frac{\mathcal{A}}{\Sigma}\sin^2\theta\, d\varphi^2, \end{aligned} \tag{20}$$

where

$$\begin{aligned} \Sigma = r^2 + a^2\cos^2\theta, \quad & \Delta = r^2 - 2Mr + a^2, \\ \mathcal{A} = \Sigma(r^2 + a^2) & + 2Mra^2\sin^2\theta, \end{aligned} \tag{21}$$

where M and a are constants.

4.1 Basic Features

The Boyer–Lindquist coordinates follow naturally from the symmetries of the Kerr spacetime. The scalars t and φ can be fixed uniquely (up to additive constants) as parameters varying along the integral curves of (unique) stationary and axial Killing vector fields $\boldsymbol{k}$ and $\boldsymbol{m}$; and the scalars r and θ can be fixed (up to constant factors) as parameters related as closely as possible to the (geometrically preferred) principal null congruences, which in the Kerr spacetime exist (see e.g. [18,30]), and their projections on to the two-dimensional spacelike submanifolds orthogonal to both $\boldsymbol{k}$ and $\boldsymbol{m}$ (see [134] for details). The Boyer–Lindquist coordinates represent the natural generalization of Schwarzschild coordinates. With $a = 0$ the metric (20) reduces to the Schwarzschild metric.

By examining the Kerr metric in the asymptotic region $r \to \infty$, one finds that M represents the mass and $J = Ma$ the angular momentum pointing in the z-direction, so that a is the angular momentum per unit mass. One can arrive at these results by considering, for example, the weak field and slow motion limit, $M/r \ll 1$ and $a/r \ll 1$. The Kerr metric (20) can then be written in the form

$$ds^2 = -\left(1 - \frac{2M}{r}\right) dt^2 + \left(1 + \frac{2M}{r}\right) dr^2$$
$$+r^2\left(d\theta^2 + \sin^2\theta \, d\varphi^2\right) - \frac{4aM}{r}\sin^2\theta \, d\varphi \, dt, \quad (22)$$

which is the weak field metric generated by a central body with mass M and angular momentum $J = Ma$. A general, rigorous way of interpreting the parameters entering the Kerr metric starts from the *definition of multipole moments* of asymptotically flat, stationary vacuum spacetimes. This is given in physical space by Thorne [135], using his "asymptotically Cartesian and mass centered" coordinate systems, and by Hansen [136], who, generalizing the definition of Geroch for the static case, gives the coordinate independent definition based on the conformal completion of the 3-dimensional manifold of trajectories of a timelike Killing vector $\boldsymbol{k}$. The exact Kerr solution has served as a convenient "test-bed" for such definitions.[14] The mass monopole moment – the mass – is M, the mass dipole moment vanishes in the "mass-centered" coordinates, the quadrupole moment components are $\frac{1}{3}Ma^2$ and $-\frac{2}{3}Ma^2$. The current dipole moment – the angular momentum – is nonvanishing only along the axis of symmetry and is equal to $J = Ma$, while the current quadrupole moment vanishes. All other nonvanishing l-pole moments are proportional to Ma^l [135,136]. Because these specific values of the multipole moments depend on only two parameters, the Kerr solution clearly cannot represent the gravitational field outside a general rotating body. In Sect. 6.2 we indicate how the Kerr metric with general values of M and a can be produced by special disk sources. The fundamental significance of the Kerr spacetime, however, lies in its role as the *only vacuum rotating black hole solution.*

Many texts give excellent and thorough discussions of properties of Kerr black holes from various viewpoints [18,19,26,30,70,75,76,79,91,95,138]. The Kerr metric entered the new edition of "Landau and Lifshitz" [139]. A few years ago, a book devoted entirely to the Kerr geometry appeared [140]. Here we can list only a few basic points.

As with the Reissner–Nordström spacetime, one can make the maximal analytic extension of the Kerr geometry. This, in fact, has much in common with the Reissner–Nordström case. Loosely speaking, the "repulsive" characters of both charge and rotation have somewhat similar manifestations. When $a^2 < M^2$, the metric (20) has coordinate singularities at $\Delta = 0$, i.e. at (cf. (21))

$$r = r_\pm = M \pm \left(M^2 - a^2\right)^{\frac{1}{2}}. \quad (23)$$

[14] For the most complete, rigorous treatment of the asymptotic structure of stationary spacetimes characterized uniquely by multipole moments defined at spatial infinity, see the work by Beig and Simon [137], the article by Beig and Schmidt in this volume, and references therein.

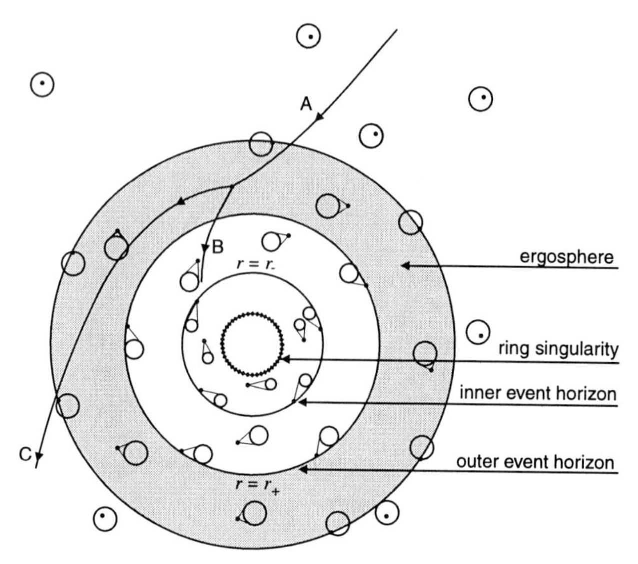

Fig. 6. A schematic picture of a Kerr black hole with two horizons, ergosphere, and the ring singularity; local light wavefronts are also indicated. Particle A, entering the ergosphere from infinity, can split inside the ergosphere into particles B and C in such a manner that C arrives at infinity with a higher energy than particle A came in.

The intrinsic three-dimensional geometry at $r = r_{\pm}$ reveals that these are null hypersurfaces – the (outer) event horizon and the inner horizon (Fig. 6). As with the Reissner–Nordström metric, the inner horizon – the Cauchy hypersurface – is unstable (see a more detailed discussion in Sect. 3.2 and references there). And as in the Penrose diagram in Fig. 5, one finds infinitely many asymptotically flat regions in the analogous Penrose diagram for the Kerr black hole spacetime.

A crucial difference between the Reissner–Nordström and Kerr geometry is the existence of the *ergosphere* (or, more precisely, ergoregion) in the Kerr case. This is caused by the *dragging of inertial frames* due to a non-vanishing angular momentum. The timelike Killing vector $\boldsymbol{k}$, given in the Boyer–Lindquist coordinates by $\partial/\partial t$, becomes null "sooner", at $r = r_0$, than at the event horizon, $r_0 > r_+$ at $\theta \neq 0, \pi$, as a consequence of this dragging:

$$\begin{aligned} k^\alpha k_\alpha &= -g_{tt} = 1 - 2Mr/\Sigma = 0, \\ r &= r_0 = M + (M^2 - a^2 \cos^2\theta)^{\frac{1}{2}}. \end{aligned} \tag{24}$$

This is the location of the *ergosurface*, the ergoregion being between this surface and the (outer) horizon. In the ergosphere, schematically illustrated in Fig. 6, the "rotating geometry" drags the particles and light (with wavefronts indicated in the figure) so strongly that all physical observers must corotate with the hole, and so rotate with respect to distant observers – "fixed stars" –

at rest in the Boyer–Lindquist coordinates.[15] Static observers, whose worldlines $(r, \theta, \varphi) =$ constant would have $\boldsymbol{k}$ as tangent vectors, cannot exist since $\boldsymbol{k}$ is spacelike in the ergosphere. Indeed, a non-spacelike worldline with r, θ fixed must satisfy the condition

$$g_{tt}dt^2 + g_{\varphi\varphi}d\varphi^2 + 2g_{\varphi t}d\varphi dt \leq 0, \tag{25}$$

in which $g_{tt} = -k^\alpha k_\alpha$, $g_{\varphi\varphi} = m^\alpha m_\alpha$, $g_{\varphi t} = k^\alpha m_\alpha$ are invariants. In the ergosphere, the metric (20), (21) yields $g_{tt} > 0$, $g_{\varphi\varphi} > 0, g_{\varphi t} < 0$, so that $d\varphi/dt > 0$ – an observer moving along a non-spacelike worldline must corotate with the hole. The effect of dragging on the forms of photon escape cones in a general Kerr field (without restriction $a^2 < M^2$) has been numerically studied and carefully illustrated in a number of figures only recently [142].

In order to "compensate" the dragging, the congruence of *"locally nonrotating frames"* (LNRFs), or *"zero-angular momentum observers"* (ZAMOS), has been introduced. These frames have also commonly been used outside relativistic, rapidly rotating stars constructed numerically, but the Kerr metric played an inspiring role (as, after all, in several other issues, such as in understanding the ergoregions, etc.). The four-velocity of these (not freely falling!) observers, given in Boyer–Lindquist coordinates by

$$e^\alpha_{(t)} = \left[(\mathcal{A}/\Sigma\Delta)^{\frac{1}{2}}, 0, 0, 2aMr/(\mathcal{A}\Sigma\Delta)^{\frac{1}{2}}\right], \tag{26}$$

is orthogonal to the hypersurfaces $t =$ constant. The particles falling from rest at infinity with zero total angular momentum fall exactly in the radial direction in the locally nonrotating frames with an orthogonal triad tied to the r, θ, φ coordinate directions (see [143] for the study of the shell of such particles falling on to a Kerr black hole).

Now going down from the ergosphere to the outer horizon, we find that both Killing vectors $\boldsymbol{k}$ and $\boldsymbol{m}$ are tangent to the horizon, and are spacelike there (with $\boldsymbol{k}$ "rotating" with respect to infinity). The null geodesic generators of the horizon are tangent to the null vectors $\boldsymbol{l} = \boldsymbol{k} + \Omega\boldsymbol{m}$, where $\Omega = a/2Mr_+ =$ constant is called the *angular velocity of the hole.* Ω is constant over the horizon so that the *horizon rotates rigidly.* Since $\boldsymbol{l}$ is a Killing vector, the horizon is a Killing horizon (cf. Sect. 2.4).

Another notable difference from the Reissner–Nordström metric is the character of the singularity at $\Sigma = 0$, i.e. at $r = 0$, $\theta = \pi/2$. It is timelike, as in the Reissner–Nordström case, but it is a *ring* singularity (see Fig. 6). In the maximal analytic extension of the Kerr metric one can go through the ring to negative values of the coordinate r and discover *closed timelike*

[15] It is instructive to analyze the somewhat "inverse problem" of gravitational collapse of a slowly rotating dust shell which produces the Kerr metric, linearized in a/M, outside (cf. Eq.(22)), and has flat space inside. Fixed distant stars seen from the centre of such shell appear to rotate due to the dragging of inertial frames, as was discussed in detail recently [141].

lines since $g_{\varphi\varphi} < 0$ there. If the Kerr parameters are such that $a^2 > M^2$, the Kerr metric does not represent a black hole. It describes the gravitational field with a *naked ring singularity.* The Kerr ring singularity has a repulsive character near the rotation axis. It gives particles outward accelerations and collimates them along the rotation axis [144], which might be relevant in the context of the formation and precollimation of cosmic jets. However, the cosmic censorship conjecture is a very plausible, though difficult "to prove" hypothesis, and Kerr naked singularities are unlikely to form in nature. However, the Kerr geometry with $a^2 > M^2$, with a region containing the ring singularity "cut out", can be produced by thin disks; though if they should be composed of physical matter, they cannot be very relativistic (see Sect. 6.2 and references therein).

If $a^2 = M^2$, the Kerr solution represents an *extreme Kerr black hole*, as is the analogous Reissner–Nordström case with $Q^2 = M^2$. The inner and outer horizons then coincide at $r = M$. The horizon is degenerate with infinite affine length. Almost extreme Kerr black holes probably play the most important role in astrophysics (see below). In realistic astrophysical situations accreting matter will very likely have a sufficient amount of angular momentum to turn a Kerr hole to an almost extreme state.

There exists a charged, electrovacuum generalization of the Kerr family found by Newman et al. [145]. The *Kerr–Newman metric* in the Boyer–Lindquist coordinates can be obtained from the Kerr metric (20) if all of the terms $2Mr$ explicitly appearing in (20), (21) are replaced by $2Mr - Q^2$, with Q being the charge. The metric describes *charged, rotating black holes* if $M^2 > a^2 + Q^2$, with two horizons located at $r_\pm = M \pm (M^2 - a^2 - Q^2)^{\frac{1}{2}}$. These become extreme when $M^2 = a^2 + Q^2$, and with $M^2 < a^2 + Q^2$ one obtains naked (ring) singularities. The analytic extension, the presence of ergoregions and the structure of the singularity is similar to the Kerr case.

In addition to the gravitational field, there exists a stationary electromagnetic field which is completely determined by the charge Q and rotation parameter a. The vector potential of this field is given by the 1-form

$$A_\alpha dx^\alpha = -(Qr/\Sigma)(dt - a\sin^2\theta\, d\varphi), \tag{27}$$

so that if $a \neq 0$ the electric field is supplemented by a magnetic field. At large distances ($r \to \infty$) the field corresponds to a *monopole electric field* with charge Q and a *dipole magnetic field* with magnetic moment $\mu = Qa$. Since the gyromagnetic ratio of a charged system with angular momentum J is defined by $\gamma = \mu/J$, one finds the charged-rotating-black hole *gyromagnetic ratio* to satisfy the relation $\gamma = Q/M$, i.e. it is twice as large as that of classical matter, and the same as that of an electron. By examining a black hole with a loop of rotating charged matter around it, the radius of the loop changing from large values to the size of the horizon, it is possible to gain some understanding of this result [146].

4.2 The Physics and Astrophysics Around Rotating Black Holes

In the introduction to their new 770 page monograph on black hole physics, Frolov and Novikov [76] write: "... there are a lot of questions connected with black hole physics and its applications. It is now virtually impossible to write a book where all these problems and questions are discussed in detail. Every month new issues of Physical Review D, Astrophysical Journal, and other physical and astrophysical journals add scores of new publications on the subject of black holes..." Although Frolov and Novikov have also black hole-like solutions in superstring and other theories on their minds, we would not probably be much in error, in particular in the context of astrophysics, if we would claim the same just about Kerr black holes. Hence, first of all, we must refer to the same literature as in the previous Sect. 4.1. A few more references will be given below.

A remarkable fact which stands at the roots of these developments is that the wave equation is separable, and the geodesic equations are integrable in the Kerr geometry. Carter [79], who explicitly demonstrated the separability of the Hamilton-Jacobi equation governing the geodesic motion, has emphasized that one can in fact *derive* the Kerr metric as the simplest nonstatic generalization of the Schwarzschild solution, by requiring the separability of the covariant Klein-Gordon wave equation.[16]

A thorough and comprehensive analysis of the behaviour of freely falling particles in the Kerr field would produce material for a book. We refer to e.g. [18,76,91,134,138,144,147] for fairly detailed accounts and a number of further references. From the point of view of astrophysical applications the following items appear to be most essential: in contrast to the Schwarzschild case, where the stable circular orbits exist only up to $r = 3r_+ = 6M$, in the field of rotating black holes, the stable direct (i.e. with a positive angular momentum) circular orbits in the equatorial plane can reach regions of "deeper potential well". With an extreme Kerr black hole the last stable direct circular orbit occurs at $r = r_+ = M$. (See [147] for a clear discussion of the positions of the innermost stable, innermost bound, and photon orbits as the hole becomes extreme and a long cylindrical throat at the horizon develops.) A *"spin-orbit-coupling"* effect *increases the binding energy of the direct orbits* and decreases the binding energy of the retrograde (with a negative angular momentum) orbits relative to the Schwarzschild values. The binding energy of the last stable direct circular orbit is $\Delta E = 0.0572\mu$ (μ is the particle's proper mass) in the Schwarzschild case, whereas $\Delta E = 0.4235\mu$ for an extreme Kerr hole. A particle slowly spiralling inward due the emission of gravitational waves

[16] Although this is still not a "constructive, analytic derivation of the Kerr metric which would fit its physical meaning", as required by Landau and Lifshitz [139], it is certainly more intuitive than the original derivation by Kerr. On the other hand, despite various hints like the existence of the Killing tensor field (in addition to the Killing vectors) in the Kerr geometry (see e.g. [134]), it does not seem to be clear *why* the Kerr geometry makes it possible to separate these equations.

would radiate the total energy equal to this binding energy; hence much more – 42% of its rest energy – in the Kerr case. The second significant effect is the dragging of the particles moving on orbits outside of the equatorial plane. The *dragging*[17] will make the orbit of a star around a supermassive black hole to precess with angular velocity $\sim 2J/r^3$. The star may go through a disk around the hole, subsequently crossing it at different places [150]. One can also show that as a result of the joint action of the gravomagnetic effect and the viscous forces in an accretion disk, the disk tends to be oriented in the equatorial plane of the central rotating black hole (the "Bardeen–Petterson effect").

The above examples demonstrate specific effects in the Kerr background which very likely play a significant role in astrophysics (see also Sect. 4.3 below). The best known process in the field of a rotating black hole is probably astrophysically unimportant, but is of principal significance in the black hole physics. This is the *Penrose process for extracting energy from rotating black holes.* It is illustrated schematically in Fig. 6: particle A comes from infinity into the ergosphere, splits into two particles, B and C. Whereas C is ejected back to infinity, B falls inside the black hole. The process can be arranged in such a way that particle C comes back to infinity with higher energy than with which particle A was coming in. The gain in the energy is caused by the decrease of rotational energy of the hole. Such process is possible because the Killing vector $\boldsymbol{k}$ becomes spacelike in the ergosphere, so that the (conserved) energy of particle B "as measured at infinity" (see e.g. [18]), $E_B = -k_\alpha p^\alpha_B$, can be negative. Unfortunately, the "explosion" of particle B requires such a big internal energy that the process is not realistic astrophysically.

More general considerations of the interaction of black holes with matter outside have led to the formulation of the four laws of *black hole thermodynamics* [19,76,82,125]. These issues, in particular after the discovery of the *Hawking effect* that black holes emit particles thermally with temperature $T = \kappa\hbar/2\pi kc$ (κ-surface gravity, k-Boltzmann's constant), have been an inspiration in various areas of theoretical physics, going from general relativity and statistical physics, to quantum gravity and string theory (see [125,126] and some remarks and references in Sect. 3.2). The Kerr solution played indeed the most crucial role in these developments. I recall how during my visits to Moscow in the middle of the 1970s Zel'dovich and his colleagues were somewhat regretfully admitting that they were on the edge of discovering the Hawking effect. They realized that an analogue of the Penrose process occurs with the waves (in so called *superradiant scattering*) which get amplified if their energy per unit angular momentum is smaller than the angular velocity

[17] Relativists often consider the effects produced by moving mass currents as "the dragging of inertial frames", but the concept of the gravomagnetic field, or *gravomagnetism* has some advantages, as has been stressed recently [148]. The gravomagnetic viewpoint, however, has also been used in many works in the past – see, e.g. [71,75], and in particular [149] and references therein.

Ω of the horizon. Zel'dovich then suggested that there should be spontaneous emission of particles in the corresponding modes but did not study quantum fields on a nonrotating background (cf. an account of these developments, including the visit of Hawking to Moscow in 1973, by Israel [70]).

Returning back to Earth or, rather, up to heavens, it is not so well known that an astrophysically more realistic example of the Penrose process exists: this is the *Blandford–Znajek mechanism* – see [75,88,151] – in which a magnetic field threading a rotating hole (the field being maintained by external currents in an accretion disk, for example) can extract the hole's rotational energy and convert it into a Poynting flux and electron-positron pairs. A Kerr black hole with angular momentum parallel to an external magnetic field acts (by "unipolar induction") like a rotating conductor in an external field. There will be an induced electric field and a potential difference between the pole and the equator. If these are connected, an electric current will flow and power will be dissipated. In fact, this appears to be until now the most plausible process to explain gigantic relativistic jets emanating from the centres of some of the most active galaxies. The BZ-mechanism has its problems: extremely rotating black holes expel magnetic flux [75,152] – there probably exists a value of the angular momentum $J_0 < J_{max}$ for which the power extracted will be greatest. It is not clear whether the process can be efficiently maintained [153]; and perhaps more importantly, new astrophysical estimates of seed magnetic fields seem to be too low to make the mechanism efficient [154]. The BZ-mechanism will probably attract more attention in the coming years, in particular in view of the recent discovery of two "microquasars" in our own Galaxy, which generate double radio structures and apparent superluminal jets similar to extragalactic strong radio sources [155].

A remarkable achievement of pure mathematical physics, with a great impact on astrophysics, has not only been the discovery of the Kerr solution itself but also the development of *the theory of Kerr metric perturbations* [75,76,91,156]. By employing the Newman–Penrose null tetrad formalism, invented and extensively used in mathematical relativity, in particular in gravitational radiation theory, it has been possible to separate completely all perturbation equations for non-zero spin fields. In particular, a *single* "master equation" – called the *Teukolsky equation* – governs scalar, electromagnetic and gravitational perturbations of a Kerr black hole.[18] If no sources are present on the right-hand side, the equation looks as follows:

$$\left[\frac{(r^2+a^2)^2}{\Delta} - a^2\sin^2\theta\right]\frac{\partial^2\psi}{\partial t^2} + \frac{4Mar}{\Delta}\frac{\partial^2\psi}{\partial t\partial\phi} + \left[\frac{a^2}{\Delta} - \frac{1}{\sin^2\theta}\right]\frac{\partial^2\psi}{\partial\phi^2}$$
$$-\Delta^{-s}\frac{\partial}{\partial r}\left(\Delta^{s+1}\frac{\partial\psi}{\partial r}\right) - \frac{1}{\sin\theta}\frac{\partial}{\partial\theta}\left(\sin\theta\frac{\partial\psi}{\partial\theta}\right) - 2s\left[\frac{a(r-M)}{\Delta} + \frac{i\cos\theta}{\sin^2\theta}\right]\frac{\partial\psi}{\partial\phi}$$

[18] In the case of a Kerr–Newman black hole, the electromagnetic and gravitational perturbations necessarily couple. Until now, in contrast to the spherical Reissner–Nordström case, a way of how to decouple them has not been discovered.

$$-2s\left[\frac{M(r^2-a^2)}{\Delta}-r-ia\cos\theta\right]\frac{\partial\psi}{\partial t}+(s^2\cot^2\theta-s)\psi=0. \tag{28}$$

The coordinates are the Boyer–Lindquist coordinates used in (20), Δ is defined in (21), and s is the spin weight of the perturbing field; $s=0,\pm1,\pm2$. The variables in the Teukolsky equation can be separated by decomposing ψ according to

$${}_s\psi_{lm}=(1/\sqrt{2\pi})\,{}_sR_{lm}(r,\omega)\,{}_sS_{lm}(\theta)e^{im\varphi}e^{-i\omega t}, \tag{29}$$

where ${}_sS_{lm}$ are so called spin weighted spheroidal harmonics. By solving the radial Teukolsky equation for ${}_sR_{lm}$ with appropriate boundary conditions one can find answers to a number of (astro)physical problems of interest like the structure of stationary electromagnetic or gravitational fields due to external sources around a Kerr black hole (e.g. [75,157]), the emission of gravitational waves from particles plunging into the hole (e.g. [76,97,158]), or the scattering of the waves from a rotating black hole (e.g. [76,156] and references therein). At present, the Teukolsky equation is being used to study the formation of a rotating black hole from a head-on collision of two holes of equal mass and spin, initially with small separation, to find the wave forms of gravitational radiation produced in this process [159]. The first studies of second-order perturbations of a Kerr black hole are also appearing [160].

To find all gravitational (metric) perturbations by solving the complete system of equations in the Newman–Penrose formalism is in general a formidable task. As Chandrasekhar's "last observation" at the end of his chapter on gravitational perturbations of the Kerr black hole reads [91]: "The treatment of the perturbations of the Kerr spacetime in this chapter has been prolixious in its complexity. Perhaps, at a later time, the complexity will be unravelled by deeper insights. But mean time, the analysis has led us into a realm of the rococo: splendorous, joyful, and immensely ornate."

4.3 Astrophysical Evidence for a Kerr Metric

Very recently new observations seem to have opened up real possibilities of testing gravity in the strong-field regime. In particular, it appears feasible to distinguish the Kerr metric from the Schwarzschild, i.e. to measure a/M. Our following remarks on these developments are based on the review by M. Rees [88], and in particular, on the very recent survey by A. Fabian [161], an authority on diagnosing relativistic rotation from the character of the emission lines of accretion disks around black holes.

The interest here is not in optical lines since the optical band comes from a volume much larger than the hole. However, the X-rays are produced in the innermost parts of an accretion flow, and should thus display substantial gravitational redshifts as well as Doppler shifts. This only became possible to observe quite recently, when the ASCA X-ray satellite started to operate, and the energy resolution and sensitivity became sufficient to analyze line shapes.

Typically the profile of a line emitted by a disk from gas orbiting around a compact object has a double-horned shape. The disk can be imagined to be composed of thin annuli of orbiting matter – the total line is then the sum of contributions from each annulus. If the disk is not perpendicular to our line of sight its approaching sides will – due to classical Doppler shifts – produce blue peaks, receding sides red peaks. The broadest parts of the total line come from the innermost annuli because the motion there is fastest. In addition, there are relativistic effects: they imply that the emission is beamed in the direction of motion, transverse Doppler shifted, and gravitationally redshifted. As a result, the total line is broad and skewed in a characteristic manner. Such lines are seen in the X-ray spectra of most Seyfert 1 galaxies. In the Seyfert galaxy MCG-6-30-15 the fluorescent iron line was observed to be (red)shifted further to lower energies.[19]

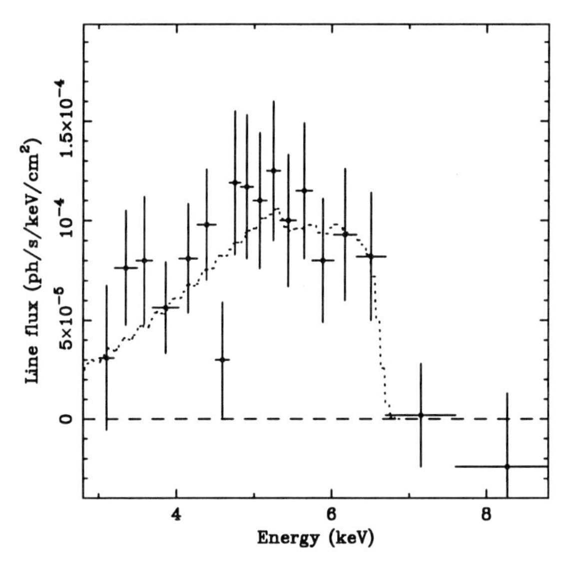

Fig. 7. The broad iron line from MCG-6-30-15. The best-fitting, maximally spinning Kerr black hole model is shown (from Iwasawa, K. et al. (1996), Mon. Not. Roy. Astron. Soc. **282**, 1038).

This suggests that the emission took place below $3R_s = 6M$, i.e. below the innermost stable orbit for a Schwarzschild black hole. In 1996, the line shape was well fitted by the assumption that the line is produced in a close orbit

[19] In Seyfert 1 galaxies hard flares occur which irradiate the accretion disk, and produce a reflection component of continuum peaking at ~ 30 keV and the fluorescent iron line at about 6.5 keV.

around maximally rotating (extreme) Kerr black hole. In 1997, the parameter a/M was quantified as exceeding 0.95. Hence, it has been "tentatively concluded that the line was the first spectroscopic evidence for a Kerr hole" [161].

There are alternative models for a broad skew iron line, including Comptonization by cold electrons, or the emission from irradiated matter falling from the inner edge of the disk around a nonrotating Schwarzschild black hole. It appears, however, that the data speak against these possibilities [161]. In any case, with future X-ray detectors, which will yield count rates orders of magnitude higher than ASCA, the line shapes should reveal in a much greater detail specific features of the Kerr metric.

In addition, other possibilities to determine a/M exist. These include:
(i) Observations of stars in relativistic orbits going through a disk around a supermassive rotating black hole [88,150].
(ii) Characteristic frequencies of the vibrational modes in disks or tori around rotating black holes [88,162].
(iii) The precession of a disk which is tilted with respect to the hole's spin axis. This precession arises because of frame dragging and produces a periodic modulation of the X-ray luminosity.
(iv) Astrophysically most important would be a discovery showing that the properties of cosmic jets depend on the value of a/M. This could indicate that the Blandford–Znajek mechanism (see Sect. 4.2) is really going on. Its likelihood would increase if jets were found with Lorentz factors γ significantly exceeding 10 (see [88] for more details).
(v) Last but not least, future observations of gravitational waves from black hole collisions [97] offer great hopes of a clean observation of a black hole geometry, without astrophysical complications.

It is hard to point out any other exact solution of Einstein's field equations (or of any kind of field equations?) discovered in the second half of the 20th century which has had so many impacts on so many diverse areas of physics, astrophysics, astronomy, and even space science as has had the Kerr metric.

5 Black Hole Uniqueness and Multi-black Hole Solutions

Since black holes can be formed from the collapse of various matter configurations, it is natural to expect that there will be many solutions of Einstein's equations describing black holes. It is expected that the asymptotic final state of a collapse can be represented by a *stationary* spacetime, i.e. one which admits a 1-dimensional group of isometries whose orbits are timelike near infinity. Strong arguments show [26] that the event horizon of a stationary black hole must be a Killing horizon. One of the most remarkable and surprising results of black hole theory are the sequence of theorems showing rigorously that the only stationary solution of the Einstein electrovacuum equations that is asymptotically flat and has a regular event horizon is the Kerr–Newman solution. There is a number of papers on this issue – recent detailed reviews are given in [80,81]. The intuition gained from exact black hole solutions in proving the theorems has been essential.

Roughly speaking, the uniqueness proof consists of the following three parts. First, one demonstrates the "rigidity theorem", which claims that nondegenerate ($\kappa \neq 0$) stationary electrovacuum analytic black holes are either static or axially symmetric. One then establishes that the Reissner–Nordström nondegenerate electrovacuum black holes are all static (nonrotating) nondegenerate black holes in electrovacuum. Finally, one separately proves that the nondegenerate Kerr–Newman black holes represent all nondegenerate axially symmetric stationary electrovacuum black holes.

Although such results were proved more than 10 years ago, recently there has been new progress in the understanding of the global structure of stationary black holes. Again, exact solutions have been inspiring: by gluing together two copies of the Kerr spacetime in a certain way, Chruściel [80] constructed a black hole spacetime which is stationary but not axisymmetric, demonstrating thus that the standard formulation and proof of the rigidity theorem [26] is not correct. (The reason being essentially that when one extends the isometries from a neighbourhood of the horizon by analytic continuation one has no guarantee that the maximal analytic extension is unique.) Chruściel proved "a corrected version of the black hole rigidity theorem"; in the connected case one can prove a uniqueness theorem for static electrovacuum black holes with *degenerate* horizons. The uniqueness theorem for static degenerate black holes which demonstrates that the extreme Reissner–Nordström black hole is the only case, is of importance also in string theory. The most unsatisfactory feature of the rigidity theorem is the assumption of analyticity of the metric in a neighborhood of the event horizon. In this context, Chruściel [80] mentions the case of Robinson–Trautman exact analytic metrics, which can be smoothly but not analytically extended through an event horizon [163]. We shall discuss this issue in somewhat greater detail in Sect. 10.

The black hole uniqueness theorems indicated above are concerned with only single black holes. (Corresponding spacetimes contain an asymptotically flat hypersurface Σ with compact interior and compact connected boundary $\partial\Sigma$ which is located on the event horizon.) Consequently a question naturally arises as to whether one can generalize the theorems to some multi-black hole solutions. In classical physics a solution exists in which a system of n arbitrarily located charged mass points with charges q_i and masses m_i, such that $|q_i| = \sqrt{G}m_i$, is in static equilibrium. In relativity the metric

$$ds^2 = -V^{-2}dt^2 + V^2\left(dx^2 + dy^2 + dz^2\right) , \tag{30}$$

with time-independent V satisfying Laplace's equation

$$\nabla^2 V = \frac{\partial^2 V}{\partial x^2} + \frac{\partial^2 V}{\partial y^2} + \frac{\partial^2 V}{\partial z^2} = 0 , \tag{31}$$

is a solution of the Einstein–Maxwell equations with the electric field

$$E = \nabla V^{-1}, \tag{32}$$

where $\nabla = (\partial/\partial x, \partial/\partial y, \partial/\partial z)$. (In standard units $E = \sqrt{G}\nabla V^{-1}$.) The simplest solution of this form is the Majumdar–Papapetrou metric, corresponding to a linear combination of n "monopole sources" with masses $m_i > 0$ and charges $q_i = m_i$, located at arbitrary points $\boldsymbol{x}_i$:

$$V = 1 + \sum_{i=1}^{n} \frac{m_i}{\mid \boldsymbol{x} - \boldsymbol{x}_i \mid}. \tag{33}$$

Hartle and Hawking [164] have shown that every such spacetime can be analytically extended to a spacetime representing n degenerate charged black holes in static equilibrium. The points $\boldsymbol{x} = \boldsymbol{x}_i$ are actually event horizons of area $4\pi m_i^2$. For the case of one black hole, the metric (30) is just the extreme Reissner–Nordström black hole in isotropic coordinates.

A uniqueness theorem for the Majumdar–Papapetrou metrics is not available, although some partial answers are known (see [165,166] for more details). It is believed that these are the only asymptotically flat, regular multi-black hole solutions of the Einstein–Maxwell equations. In fact, such a result would exclude an interesting possibility that *a repulsive gravitational spin-spin interaction* between two (or more) rotating, possibly charged, black holes can overcome their gravitational attraction and thus that there exists in Einstein's theory of gravitation – in contrast to Newton's theory – a stationary solution of the two-body problem.

Among new solutions discovered by modern generating techniques, there are the solutions of Kramer and Neugebauer which represent a nonlinear superposition of Kerr black holes (see [167] for a review). These solutions have been the subject of a number of investigations which have shown that spin-spin repulsion is not strong enough to overcome attraction. In particular, two symmetrically arranged equal black holes cannot be in stationary equilibrium. The situation might change if one considers two Kerr–Newman black holes [168]. Here one has four forces to reckon with: gravitational and electromagnetic Coulomb-type interactions, and gravitational and electromagnetic spin-spin interactions. One can then satisfy the conditions which render the system of two Kerr–Newman black hole free of singularities on the axis, and make the total mass of the system positive. However, there persists a singularity in the plane of symmetry away from the axis [168]. In view of this result we have conjectured that even with electromagnetic forces included one cannot achieve balance for two black holes, except for the exceptional case of two nonrotating extreme (degenerate) Reissner–Nordström black holes. Recently, some new rigorous results concerning the (non)existence of multi-black hole stationary, axisymmetric electrovacuum spacetimes have been obtained [169] (see also [80]), but the "decisive theorem" is still missing.

In connection with the problem of the balance of gravity by a gravitational spin-spin interaction, we should mention that there exists the solution of Dietz and Hoenselaers [170] in which balance of the two rotating "particles" is achieved. However, the "sources" are complicated naked singularities which become Curzon–Chazy "particles" (see Sect. 6.1) if the rotation goes to zero, and it is far from clear whether appropriate physical interior solutions can be constructed.

In 1993, Kastor and Traschen [171] found an interesting family of solutions to the Einstein–Maxwell equations with a non-zero cosmological constant Λ. They describe an arbitrary number of charged black holes in a "background" de Sitter universe. In the limit of $\Lambda = 0$ these solutions become Majumdar–Papapetrou static metrics. In contrast to these metrics, the *cosmological multi-black hole solutions with* $\Lambda > 0$ *are dynamical.* Remarkably, one can construct solutions which describe coalescing black holes. In some cases cosmic censorship is violated – a naked singularity is formed as a result of the collision [172]. Although these solutions do not have smooth horizons, the singularities are mild, and geodesics can be extended through them. The metric is always at least C^2. Since the solutions are dynamical, one may interpret the non-smoothness of the horizons as a consequence of gravitational and electromagnetic radiation. In this sense, the situation is analogous to the case of the Robinson–Trautman spacetimes discussed in Sect. 10. In five or more dimensions, however, one can construct *static* multi-black hole solutions with $\Lambda = 0$, which do not have smooth horizons [173]. The solutions of Kastor and Traschen also inspired a new and careful analysis [174] of the global structure of the Reissner–Nordström–de Sitter spacetimes characterized by mass, charge, and cosmological constant. The structure is considerably richer than that with $\Lambda = 0$. Most recently, the hoop conjecture (giving the criterion as to whether a black hole forms from a collapsing system) was discussed [175] by analyzing the solution of Kastor and Traschen.

6 On Stationary Axisymmetric Fields and Relativistic Disks

6.1 Static Weyl Metrics

The static axisymmetric vacuum metrics in Weyl's canonical coordinates $\rho \in [0, \infty), z, t \in I\!R, \varphi \in [0, 2\pi)$ have the form

$$ds^2 = e^{-2U} \left[e^{2k} \left(d\rho^2 + dz^2 \right) + \rho^2 d\varphi^2 \right] - e^{2U} dt^2. \tag{34}$$

The function $U(\rho, z)$ satisfies flat-space Laplace's equation

$$\frac{\partial^2 U}{\partial \rho^2} + \frac{1}{\rho} \frac{\partial U}{\partial \rho} + \frac{\partial^2 U}{\partial z^2} = 0. \tag{35}$$

The function $k(\rho, z)$ is determined from U by quadrature up to an additive constant. The axis $\rho = 0$ is free of conical singularities at places where $\lim_{\rho \to 0} k = 0$.

The mathematically simplest example is the Curzon–Chazy solution in which $U = -m/\sqrt{\rho^2 + z^2}$ is the Newtonian potential of a spherical point particle. The spacetime, however, is not spherically symmetric. In fact, one of the lessons which one has learned from this solution is the *directional character of the singularity* at $\rho^2 + z^2 = 0$. For example, the limit of the invariant $R_{\alpha\beta\gamma\delta}R^{\alpha\beta\gamma\delta}$ depends on the direction of approach to the singularity. The singularity has a character of a ring through which some timelike geodesics may pass to a Minkowski region [176].

Various studies of the Weyl metrics indicated explicitly how important it is always to check whether a result is not just a consequence of the choice of coordinates. There is the subclass of Weyl metrics generated by the Newtonian potential of a constant density line mass ("rod") with total mass M and (coordinate) length l, which is located along the z-axis with the middle point at the origin. These are Darmois-Zipoy-Vorhees metrics, called also the γ-metrics [64]. The Schwarzschild solution (a spherically symmetric metric!) is a special case in this subclass: it is given by the potential of the rod with $l = 2M$. Clearly, in general there is no correspondence between the geometry of the physical source and the geometry of the Newtonian "source" from the potential of which a Weyl metric is generated.

A survey of the best known Weyl metrics, including some specific solutions describing fields due to circular disks is contained in [64]. More recently, Bičák, Lynden-Bell and Katz [177] have shown that most vacuum static Weyl solutions, including the Curzon and the Darmois-Vorhees-Zipoy solutions, can arise as the metrics of counterrotating relativistic disks (see [177] also for other references on relativistic disks). The simple idea which inspired their work is commonly used in Newtonian *galactic dynamics* [6]: imagine a point mass placed at a distance b below the centre $\rho = 0$ of a plane $z = 0$. This gives a solution of Laplace's equation above the plane. Then consider the potential obtained by reflecting this $z \geq 0$ potential in $z = 0$ so that a symmetrical solution both above and below the plane is obtained. It is continuous but has a discontinuous normal derivative on $z = 0$, the jump in which gives a positive surface density on the plane. In galactic dynamics one considers general line distributions of mass along the negative z-axis and, employing the device described above, one finds the potential-density pairs for general axially symmetric disks. In [178], an infinite number of new static solutions of Einstein's equations were found starting from realistic potentials used to describe flat galaxies, as given recently by Evans and de Zeeuw [179].

Although these disks are Newtonian at large distances, in their central regions interesting relativistic features arise, such as velocities close to the velocity of light, and large redshifts. In a more mathematical context, some particular cases are so far the only explicit examples of spacetimes with a "polyhomogeneous" null infinity (cf. [180] and Sect. 9), and spacetimes with a meaningful, but infinite ADM mass [178]. New Weyl vacuum solutions generated by Newtonian potentials of flat galaxies correspond to both finite and semi-infinite rods, with the line mass densities decreasing according to general power laws. It is an open question what kinds of singularities rods with different density profiles represent.

Very recently, new interesting examples of the static solutions describing self-gravitating disks or rings, and disks or rings around static black holes have been constructed [181–183] and the effects of the fields on freely moving test particles studied [181]. Exact disks with electric currents and magnetic fields have also been considered [184].

Employing the Weyl formalism, one can describe nonrotating *black holes strongly distorted* by the surrounding matter. The influence of the matter can be so strong that it may even cause the horizon topology to be changed from spherical to toroidal (see [75] and references therein).

Finally, we have to mention two solutions in the Weyl class, which were found soon after the birth of general relativity, and have not lost their influence even today. The first, discovered by Bach and Weyl, is assigned by Bonnor [64] as "probably the most perspicacious of all exact solutions in GR". It refers to two Curzon–Chazy "monopoles" on the axis of symmetry. One finds that the metric function k has the property that $\lim_{\rho \to 0} k \neq 0$, so that there is a stress described by a conical singularity between the particles, which holds particles apart. A similar solution can be constructed for the Schwarzschild "particles" (black holes) held apart by a stress. These cases can serve as one of the simplest demonstrations of the difference between the Einstein theory and field theories like the Maxwell theory: it is only in general relativity in which field equations involve also *equations of motion.*

The second "old" solution which has played a very significant role is the metric discovered by Levi–Civita. It belongs to the class of degenerate (type D) static vacuum solutions which form a subclass of the Weyl solutions. In the invariant classification of the degenerate solutions by Ehlers and Kundt [53], this solution is contained in the last, third subclass. That is why Ehlers and Kundt called it the C-metric, and it is so well known today. We shall discuss the C-metric later (Sect. 11) in greater detail since, as it has been learned in the 1970s, it is actually a radiative solution representing uniformly accelerated black holes. What Levi–Civita found and Ehlers and Kundt analyzed is only a portion of spacetime in which the boost Killing vector is timelike, and the coordinates can thus be found there (analogous to the coordinates in a uniformly accelerated frame in special relativity) in which the metric is time-independent.

6.2 Relativistic Disks as Sources of the Kerr Metric and Other Stationary Spacetimes

Thanks to the black hole uniqueness theorems (Sect. 5), the Kerr metric represents the unique solution describing all rotating vacuum black holes. Nevertheless, although the cosmic censorship conjecture, on which the physical relevance of the Kerr metric rests, is a very plausible hypothesis, it remains, as was noted in several places above, one of the central unresolved issues in relativity. It would thus support the significance of the Kerr metric if a physical source were found which produces the Kerr field. The situation would then resemble the case of the spherically symmetric Schwarzschild metric which can represent both a black hole and the external field due to matter.

This has been realized by many workers. The review on the "Sources for the Kerr Metric" [185] written in 1978, contains 71 references, and concludes with: "Destructive statements denying the existence of a material source for the Kerr metric should be rejected until (if ever) they are reasonably justified." The work from 1991 gives "a toroidal source", consisting of "a toroidal shell ..., a disk ... and an annulus of matter interior to the torus" [186]. The masses of the disk and annulus are negative. To summarize in Hermann Bondi's way, the sources suggested for the Kerr metric have not been the easiest materials to buy in the shops ...

The situation is somewhat different in the special case of the *extreme* Kerr metric, where there is a definite relationship between mass and angular momentum. The numerical study [187] of uniformly rotating disks indicated how the extreme Kerr geometry forms around disks in the "ultrarelativistic" limit. These numerical results have been supported by important analytical work (see Sect. 6.3). However, in the case of a general Kerr metric physical sources had not been found before 1993.

A method similar to that of constructing disk sources of static Weyl spacetimes (described in Sect. 5.1) has been shown to work also for axisymmetric, reflection symmetric, and *stationary* spacetimes [188,189]. It is important to realize that although now no metric function solves Laplace's equation as in the static case, we may view the procedure described in Sect. 5.1 as the *identification* of the surface $z = b$ with the surface $z = -b$. The field then remains continuous, but the jump of its normal derivatives induces a matter distribution in the disk which arises due to the identification of the surfaces. What remains to be seen, is whether the material can be "bought in the shops". This idea can be employed for all known asymptotically flat stationary vacuum spacetimes, for example for the Tomimatsu-Sato solutions, for the "rotating" Curzon solution, or for other metrics (cf. [61] for references).

Any stationary axisymmetric vacuum metric can be written in canonical coordinates (t, φ, ρ, z) in the form [61]

$$ds^2 = e^{-2U}\left[e^{2k}\left(d\rho^2 + dz^2\right) + \rho^2 d\varphi^2\right)] - e^{2U}(dt + Ad\varphi)^2 , \qquad (36)$$

where U, k, and A are functions of ρ, z. For the Kerr solution (mass M, specific angular momentum $a \geq 0$), the functions U, k, A are ratios of polynomials when expressed in spheroidal coordinates [61].

Now, identify the "planes" $z = b =$ constant > 0 and $z = -b$ (this identification leads to disks with zero radial pressure). With the Kerr geometry the matching is more complicated than in the static cases, and therefore, one has to turn to Israel's covariant formalism (see [190] for its recent exposition). Using this formalism one is able to link the surface stress-energy tensor of the disk arising from this identification, to the jump of normal extrinsic curvature across the timelike hypersurface given by $z = b$ (with the jump being determined by the discontinuities in the normal derivatives in functions U, k, A).

The procedure leads to physically plausible disks made of two streams of collisionless particles, that circulate in opposite directions with differential velocities [188,189]. Although extending to infinity, the disks have finite mass and exhibit interesting relativistic properties such as high velocities, large redshifts, and dragging effects, including ergoregions. Physical disk sources of Kerr spacetimes with $a^2 > M^2$ can be constructed (though these are "less relativistic"). And the procedure works also for electrovacuum stationary spacetimes. The disks with electric current producing Kerr–Newman spacetimes are described in [191], where the conditions for the existence of (electro)geodesic streams are also discussed.

The power and beauty of the Einstein field equations is again illustrated: the character of exact vacuum fields determines fully the physical characteristics of their sources. In a more sophisticated way, this is seen in the problem of relativistic rigidly (uniformly) rotating disks of dust.

6.3 Uniformly Rotating Disks

The structure of an infinitesimally thin, finite relativistic disk of dust particles which rotate uniformly around a common centre was first explored by J. Bardeen and R. Wagoner 25 years ago [187]. By developing an efficient expansion technique in the quantity $\delta = z_c/(1 + z_c)$, z_c denoting the central redshift, they obtained numerically a fairly complete picture of the behaviour of the disk, even in the ultrarelativistic regime ($\delta \to 1$). In their first letter from 1969 they noted that "there may be some hope of finding an analytic solution". Today such a hope has been substantiated, thanks to the work of G. Neugebauer and R. Meinel (see [192,193] and references therein). The solution had, in fact, to wait until the "soliton-type-solution generating techniques" for nonlinear partial differential equations had been brought over from applied mathematics and other branches of physics to general relativity, starting from the end of the 1970s.

These techniques have been mainly applied only in the vacuum cases so far, but this is precisely what is in this case needed: the structure of the thin disk enters the field equations only through the boundary conditions at $z = 0$, $0 \leq \rho \leq a$ (a is the radius of the disk). The specific procedures which enabled Neugebauer and Meinel to tackle the problem are sophisticated and lengthy. Nevertheless, we wish to mention them telegraphically at least, since they represent the first example of solving *the boundary value problem* for a *rotating* object in Einstein's theory by analytic methods.

In the stationary axisymmetric case, Einstein's vacuum field equations for the metric (18) imply the well-known Ernst equation (see e.g. [61]) – a nonlinear partial differential equation for a complex function f of ρ and z:

$$(\mathrm{Re} f)\left[f_{,\rho\rho} + f_{,zz} + \frac{1}{\rho} f_{,\rho}\right] = f_{,\rho}^2 + f_{,z}^2, \tag{37}$$

where the Ernst potential

$$f(\rho, z) = e^{2U} + ib, \tag{38}$$

with $U(\rho, z)$ being the function entering the metric (36), function $b(\rho, z)$ is a "potential" for $A(\rho, z)$ in (37),

$$A_{,\rho} = \rho\, e^{-4U} b_{,z}, \qquad A_{,z} = -\rho\, e^{-4U} b_{,\rho}, \tag{39}$$

and the last function $k(\rho, z)$ in (37) can be determined from U and b by quadratures.

The Ernst equation can be regarded *as the integrability condition* of a system of *linear* equations for a complex matrix Φ, which is a function of $\rho + iz, \rho - iz$, and of a (new) complex parameter λ. Knowing Φ, one can determine f from Φ at $\lambda = 1$. Now the problem of solving the linear system can be reformulated as the so called *Riemann–Hilbert problem* in complex function theory. (This, very roughly, means the following: let K be a closed curve in the complex plane and $F(K)$ a matrix function given on K; find a matrix function Φ_{in} which is analytic inside L, and Φ_{out} analytic outside K such that $\Phi_{in}\Phi_{out} = F$ on K.) The Riemann–Hilbert problem can be formulated as an integral equation. The hardest problem with which Neugebauer and Meinel were faced was in connecting the specific physical boundary values of f on the disk with the functions entering the Riemann–Hilbert problem (with contour K being determined by the position of the disk in the ρ, z plane), and with the corresponding integral equation. The fact that they succeeded and found the solution of their integral equation is a remarkable achievement in mathematical physics. The gravitational field and various physical characteristics of the disk (e.g. the surface density) are given up to quadratures in terms of ultraelliptic functions [192], which can be numerically evaluated without difficulties. This result, however, may appear as a "lucky case": it does not imply that one will be able to tackle similarly more complicated situations as, for example, thin disks with pressure, with non-uniform rotation, or 3-dimensional rotating bodies such as neutron stars.

Many physical characteristics of uniformly rotating relativistic disks such as their surprisingly high binding energies, the high redshifts of photons emitted from the disks, or the dragging of inertial frames in the vicinity of the disks, were already obtained with remarkable accuracy in [187], as the exact solution now verifies. Here we only wish to demonstrate the fundamental difference between the Newtonian and relativistic case, as it is illustrated

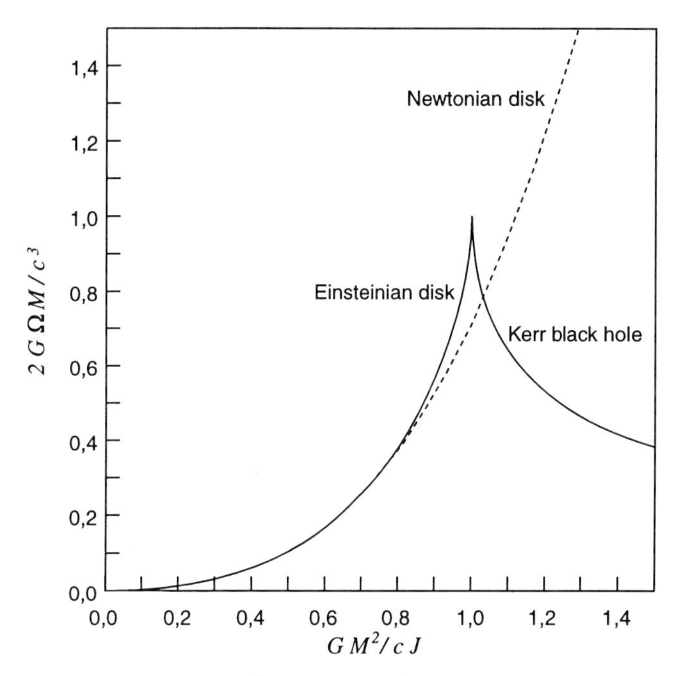

Fig. 8. The general relativistic ("Einsteinian") thin disk of rigidly rotating dust constructed by Neugebauer and Meinel, compared with the analogous disk in Newtonian theory. If the angular momentum is too low, the disk forms a rotating (Kerr) black hole. (From [193].)

in Fig. 8. The rigidly rotating disk of dust of Neugebauer and Meinel represents the relativistic analogue of a classical Maclaurin disk. For the Maclaurin disk, it is easy to show that the (dimensionless) quantities $y = 2G\Omega M/c^3$ and $x = GM^2/cJ$ (M and J are the total mass and angular momentum respectively, and Ω is the angular velocity) are related by $y = (9\pi^2/125)x^3$. For a fixed M the angular velocity $\Omega \sim y$ can be increased arbitrarily, with J being correspondingly decreased. For relativistic disks, however, there is an upper bound on Ω given by $\Omega_{max} = c^3/2GM$, whereas J is restricted by the lower bound $J_{min} = GM^2/c$. With an angular momentum too low, a *rigidly* rotating disk cannot exist. If we "prepare" such a disk, it immediately begins to collapse and forms – assuming the cosmic censorship – a rotating Kerr black hole with $x = GM^2/cJ > 1$. (Notice that the assumption of rigid rotation is here crucial: the differentially rotating disks considered in the preceding section can have an arbitrary value of x.) Since one can define the angular velocity $\Omega(M, J)$ of the horizon of a Kerr hole, one may consider $y(x)$ for black hole states with $x > 1$ (cf. Fig 8). The rigidly rotating disk states and the black hole states just "meet" at $y = x = 1$. In the ultrarelativistic limit the gravitational field outside the disk starts to be unaffected by the detailed structure of the disk – it approaches the field of an extremely rotating Kerr black hole with $x = 1$. Such a result had already been obtained by Bardeen and Wagoner. However, it is only now, with the exact solution available, that it can be investigated with full rigor. It gives indirect evidence that Kerr black holes are really formed in the gravitational collapse of rotating bodies.

As noticed also by Bardeen and Wagoner, in the ultrarelativistic limit the disk itself "becomes buried in the horizon of the extreme Kerr metric, surrounded by its own infinite, non asymptotically flat universe" (see [194] for a recent detailed analysis of the ultrarelativistic limit). Similar phenomena arise also in the case of some spherical solutions of Einstein–Yang–Mills–Higgs equations (cf. [195] and Sect. 13).

7 Taub-NUT Space

The name of this solution of vacuum Einstein's equations fits both to the names of its discoverers (Taub–Newman–Unti–Tamburino) and to its curious properties. Owing to these properties (which induced Misner [196] to consider the solution "as a counterexample to almost anything"), this spacetime has played a significant role in exhibiting the type of effects that can arise in strong gravitational fields.

Taub [197] discovered an empty universe with four global Killing vectors almost half a century ago, during his pioneering study of metrics with several symmetries. By continuing the Taub universe through its horizon one arrives in NUT space. NUT space, however, was only discovered in 1963 by a different method [198]. In fact, it could have been obtained earlier by applying the transformation given in Jürgen Ehlers' dissertation [54] and his talk at the GR2 conference in Royaumont in 1959 [55]. This transformation gives the recipe for obtaining stationary solutions from static ones. How the NUT space can be obtained by applying this transformation to the Schwarzschild metric was demonstrated explicitly by Ehlers at GR4 in London in 1965 [56].

7.1 A New Way to the NUT Metric

Here we shall briefly mention a simple, physically appealing new derivation of the NUT metric given recently by Lynden-Bell and Nouri-Zonoz (LBNZ) [199]. Their work also shows how even uncomplicated solutions may still be of interest in unexpected contexts. LBNZ's inspiration to study the NUT space has in fact come from Newton's Principia! In one of his scholia Newton discusses motion under the standard central force plus a force which is normal to the surface swept out by the radius vector to the body which is describing the non-coplanar path. A simple interesting case is the motion of mass m_0 satisfying the equation

$$m_0 \, \frac{d^2 \boldsymbol{r}}{dt^2} = -V'(r) \, \hat{\mathbf{r}} + \frac{m_0}{c} \, \boldsymbol{v} \times \boldsymbol{B}_g, \tag{40}$$

where $\hat{\mathbf{r}} = \boldsymbol{r}/r$,

$$\boldsymbol{B}_g = -Q\,\hat{\mathbf{r}}/r^2,\ Q = \tilde{Q}\,c/m_0, \tag{41}$$

Q and $\tilde{Q}$ are constants, and c is the velocity of light. Here we write c explicitly though $c = 1$, to make the analogy with magnetism. Indeed, $\boldsymbol{B}_g$ is the field of a "gravomagnetic" monopole of strength Q. The classical orbits of particles lie on cones which, if the monopole is absent, flatten into a plane [199].

It was known that NUT space corresponds to the mass with a gravomagnetic monopole, but this was never used in such a physical way as by LBNZ for its derivation. The main point is to start from the well-known split of the stationary metrics as described in Landau and Lifshitz [139] (see [200] for a covariant approach, and the contribution of Beig and Schmidt in the present volume)

$$ds^2 = -e^{-2\nu}(dt - A_i dx^i)^2 + \gamma_{ij}dx^i dx^j, \tag{42}$$

where ν, A_i, γ_{ij} are independent of t. This form is unique up to the choice of time zero: $t^{'} = t + \chi(x^i)$ implies again the metric in the form (42) in $(t^{'}, x^i)$, with the "vector potential" undergoing a gauge transformation $A_i' = A_i + \nabla_i\chi$. Writing down the equation of motion of a test particle in metric (42), in analogy with the equation of motion of a charged particle in an electromagnetic field, one is naturally led to define the "gravoelectric" and "gravomagnetic" fields by

$$\boldsymbol{E}_g = \nabla\nu,\ \ \boldsymbol{B}_g = \nabla \times \boldsymbol{A}, \tag{43}$$

where "$\nabla\times$" is with respect to γ_{ij}. Following the problem of §95 in [139] one then rewrites all Einstein's equations in terms of the fields (43), the metric γ_{ij}, and their derivatives. To find the vacuum *spherically symmetric* spatial γ-metric one takes $\gamma_{ij}dx^i dx^j = e^{2\lambda}dr^2 + r^2(d\theta^2 + \sin^2\theta\, d\varphi^2)$, and one assumes $\nu = \nu(r)$, and $B_g^r = -Qe^{-\lambda}/r^2$. The Einstein equations then imply the spacetime metric, which is *not* spherically symmetric, in the form

$$\begin{aligned} ds^2 = -e^{-2\nu}\Big(dt - 2q\,(1+\cos\theta)\,d\varphi\Big)^2 + \left(1 - q^2/r^2\right)^{-1} e^{2\nu} dr^2 \\ + \ r^2\left(d\theta^2 + \sin^2\theta\ d\varphi^2\right), \end{aligned} \tag{44}$$

where $q = Q/2 =$ constant,

$$e^{-2\nu} = 1 - 2r^{-2}\left(q^2 + m\sqrt{r^2 - q^2}\right), \tag{45}$$

and the vector potential $A_\varphi = 2q(1 + \cos\theta)$ satisfies (43). (The factor $(1 - q^2/r^2)$ should be raised to the power -1 in equation (3.22) in [199], as it is clear from (3.20).) Equation (44) is the NUT metric, with r being the curvature coordinate of spheres $r =$ constant. With $q = 0$ the metric (44) becomes the Schwarzschild metric in the standard Schwarzschild coordinates.

More commonly the metric (44) is written in the form

$$ds^2 = -V\left(d\tilde{t} + 4q\sin^2\frac{\theta}{2}\,d\varphi\right)^2 + V^{-1}d\tilde{r}^2 + \left(\tilde{r}^2 + q^2\right)\left(d\theta^2 + \sin^2\theta\,d\varphi^2\right), \tag{46}$$

$$V = 1 - 2\,\frac{m\tilde{r} + q^2}{\tilde{r}^2 + q^2}, \tag{47}$$

which can be obtained from (44) by putting

$$\tilde{r} = \sqrt{r^2 - q^2}, \qquad \tilde{t} = t - 4q\varphi. \tag{48}$$

Recently, Ehlers [201] considered a Newtonian limit of NUT space within his frame theory which encompasses general relativity and Newton–Cartan theory (a slight generalization of Newton's theory). The main purpose of Ehlers' frame theory is to define rigorously what is meant by the statement that a one parameter family of relativistic spacetime models converges to a Newton–Cartan model or, in particular, to a strictly Newtonian model.

The strictly Newtonian limit occurs when the Coriolis angular velocity field $\boldsymbol{\omega}$, related to the connection coefficients Γ^i_{tj} in the Newton–Cartan theory, depends on time only. NUT spacetimes approach a truly Newton–Cartan limit with spatially *non*-constant radial Coriolis field $\omega^{\tilde{r}} = -q/\tilde{r}^2$, which in this limit coincides with the Newtonian gravomagnetic field. As in the analogous classical problem with the equation of motion (40), the geodesics in NUT space lie on cones. This result has been used to study gravitational lensing by gravomagnetic monopoles [199]: they twist the rays that pass them in a characteristic manner, different from that due to rotating objects.

The metrics (44) and (46) appear to have a preferred axis of fixed points of symmetry. This is a false impression since we can switch the axis into any direction by a gauge transformation. For example, the metric (44) has a conical singularity at $\theta = 0$ but is regular at $\theta = \pi$, whereas the metric (46) has a conical singularity at $\theta = \pi$ but is regular at $\theta = 0$. The metrics are connected by the simple gauge transformation, i.e. $t \to \tilde{t} = t - 4q\varphi$. A mass endowed with a gravomagnetic monopole appears as a spherically symmteric object but the spacetime is *not* spherically symmetric according to the definition given in Sect. 2.1. Nevertheless, there exist equivalent coordinate systems in which the axis can be made to point in any direction – just as the axis of the vector potential of a magnetic monopole can be chosen arbitrarily. For further references on interpreting the NUT metric as a gravomagnetic monopole, see [199] and the review by Bonnor [64].

7.2 Taub-NUT Pathologies and Applications

By introducing two coordinate patches, namely the coordinates of metric (44) to cover the south pole ($\theta = \pi$) and those of (46) the north pole ($\theta = 0$), the rotation axis can be made regular. However since φ is identified with period

2π, equation (48) implies that t and $\tilde{t}$ have to be identified with the period $8\pi q$. Then observers with $(\tilde{r}, \theta, \varphi) =$ constant follow *closed timelike lines* if V in (47) is positive, i.e. if $\tilde{r} > \tilde{r}_0 = m + (m^2 + q^2)^{\frac{1}{2}}$. The hypersurface $\tilde{r} = \tilde{r}_0$ is the null hypersurface – horizon, below which lines $(t, \theta, \varphi) =$ constant become spacelike. Because of the periodic identification of t and $\tilde{t}$, the hypersurfaces of constant $\tilde{r}$ change the topology from $S^2 \times R^1$ to S^3, on which $t/2q$, θ, φ become Euler angle coordinates.

The region with $V < 0$ is the Taub universe: it has homogeneous but non-isotropic space sections $\tilde{r} =$ constant. The coordinate $\tilde{r}$, allowed to run from $-\infty$ to $+\infty$, is a timelike coordinate, and is naturally denoted by t in the Taub region.

In addition to the closed timelike lines in the NUT region there are further intriguing pathologies exhibited by the Taub-NUT solutions. Here we just list some of them and refer to the relevant literature [26,27,196]. The Taub region is globally hyperbolic: its entire future and past history can be determined from conditions given on a spacelike Cauchy hypersurface. However, this is not the case with the whole Taub-NUT spacetime. As in the Reissner–Nordström spacetimes (Sect. 3), there are Cauchy horizons $H^{\pm}(\Sigma)$ of a particular spacelike section Σ of maximal proper volume lying between the globally hyperbolic Taub regions and the causality violating NUT regions. $H^{\pm}(\Sigma)$ are smooth, compact null hypersurfaces diffeomorphic to S^3 – the generators of such null surfaces are *closed null geodesics*. The Taub region is limited between $t_- \leq t \leq t_+$, where $t_{\pm}$ are roots of V in equation (47) (with the interchange $t \leftrightarrow \tilde{r}$). This region is compact but there are timelike and null geodesics which remain within it and are not complete. (See [27] for a nice picture of these geodesics spiralling around and approaching $H_+(\Sigma)$ asymptotically.) This pathological behaviour of "the incomplete geodesics imprisoned in a compact neighbourhood of the horizon" was inspirational in the definition of singularities [77] – one meets here the example in which the geodesic incompleteness is not necessarily connected with strong gravitational fields. It can be shown, however, that after an addition of even the slightest amount of matter this pathological behaviour will not take place – true singularities arise.

This enables one to consider the time between t_- and t_+ as the lifetime of the Taub universe. Wheeler [202] constructed a specific case of the Taub universe which will live as long as a typical Friedmann closed dust model ($\sim 10^{10}$ years) but will have a volume at maximum expansion smaller by a factor of 5×10^{10}. This example thus appears to be a difficulty for the anthropic principle.

Taub space seems also to be the only known example giving the possibility of making inequivalent NUT-like extensions which lead to a non-Hausdorff spacetime manifold [26,200,203].

The Taub-NUT solution plays an important role in cosmology and quantum gravity. Here we wish to note yet two other recent applications of this space. About ten years ago, interest was revived in closed timelike lines, time machines, and wormholes. One of the leaders in this activity, Kip Thorne, explains in [204] in a pedagogical way the main recent results on closed timelike curves and wormholes by using "Misner space" – Minkowski spacetime with identification under a boost, which Misner introduced as a simplified version of Taub-NUT space.

The second application of the Taub-NUT space is still more remarkable – it plays an important role outside general relativity. The asymptotic motion of monopoles in (super-)Yang–Mills theories corresponds to the geodesic motion in *Euclidean* Taub-NUT space [205]. Euclidean Taub-NUT spaces have been discussed in many further works on monopoles in gauge theories. One of the latest of these works [206], on the exact T-duality (which relates string theories compactified on large and small tori) between Taub-NUT spaces and so called "calorons" (instantons at finite temperature defined on $R^3 \times S^1$), gives also references to previous contributions.

8 Plane Waves and Their Collisions

8.1 Plane-Fronted Waves

The history of gravitational plane waves had began already by 1923 with the paper on spaces conformal to flat space by Brinkmann. Interest in these waves was revived in 1937 by Rosen, and in the late 1950s by Bondi, Pirani and Robinson, Holy, and Peres (see [53,61] for references). A comprehensive geometrical approach to these spacetimes soon followed in the classical treatise by Jordan, Ehlers and Kundt [57], and in the subsequent well-known chapter by Ehlers and Kundt [53]. As an application of various newly developed methods to analyze gravitational radiation, and as a simple background to test various physical theories, plane waves have proved to be a useful and stimulating arena which offers interesting contests even today, as we shall indicate by a few examples in Sect. 8.2.

Consider a congruence of null geodesics (rays) $x^\alpha(v)$ such that $dx^\alpha/dv = k^\alpha, k_\alpha k^\alpha = 0, k_{\alpha;\beta}k^\beta = 0$, v being an affine parameter. In general a geodesic congruence is characterized by its expansion θ, shear $|\sigma|$ and twist ω given by (see e.g. [19])

$$\theta = \frac{1}{2}k^\alpha_{;\alpha}, \qquad |\sigma| = \sqrt{\frac{1}{2}k_{(\alpha;\beta)}k^{\alpha;\beta} - \theta^2}, \tag{49}$$

$$\omega = \sqrt{\frac{1}{2}k_{[\alpha;\beta]}k^{\alpha;\beta}}. \tag{50}$$

According to the definition given by Ehlers and Kundt [53] a vacuum spacetime is a *"plane-fronted gravitational wave"* if it contains a shearfree $|\sigma| = 0$ geodesic null congruence, and if it admits "plane wave surfaces" (spacelike 2-surfaces orthogonal to k^α). This definition is inspired by plane electromagnetic waves in Maxwell's theory. Electromagnetic plane waves are *null fields* ("pure radiation fields"): there exists a null vector k^α, tangent to the rays, which is transverse to the electromagnetic field $F_{\alpha\beta}$, i.e. $F_{\alpha\beta}k^\beta = 0$, $F^*_{\alpha\beta}k^\beta = 0$, and the quadratic invariants of which vanish, $F_{\alpha\beta}F^{\alpha\beta} = 0 = F_{\alpha\beta}F^{*\alpha\beta}$, where $F^*_{\alpha\beta}$ is dual to $F_{\alpha\beta}$. Analogously, Petrov type N gravitational fields (see [61]) are null fields with rays tangent to k^α (the "quadruple Debever-Penrose null vector"), and with the Riemann tensor satisfying $R_{\alpha\beta\gamma\delta}k^\delta = 0$, $R_{\alpha\beta\gamma\delta}R^{\alpha\beta\gamma\delta} = 0$, and $R^{\alpha\beta\gamma\delta}R^*_{\alpha\beta\gamma\delta} = 0$.[20] Then the Bianchi identities and the Kundt-Thompson theorem for type N solutions in vacuum spacetimes (also more generally, under the presence of a nonvanishing cosmological constant) imply that the shear of k^α must necessarily vanish (see [61,207]). Because of the existence of plane wave surfaces, the expansion (49) and twist (50) must vanish as well, $\theta = \omega = 0$. In this way we arrive at the Kundt class of nonexpanding, shearfree and twistfree gravitational waves [61]. The best known subclass of these waves are *"plane-fronted gravitational waves with parallel rays" (pp-waves)* which are defined by the condition that the null vector k^α is covariantly constant, $k_{\alpha;\beta} = 0$. Thus, automatically k^α is the Killing vector, and $\theta = |\sigma| = \omega = 0$.

Ehlers and Kundt [53] give several equivalent characterizations of the pp-waves and show, following their previous work [57], that in suitable null coordinates with a null coordinate u such that $k_\alpha = u_{,\alpha}$ and $k^\alpha = (\partial/\partial v)^\alpha$, the metric has the form

$$ds^2 = 2d\zeta d\bar{\zeta} - 2dudv - 2H(u,\zeta,\bar{\zeta})du^2, \tag{51}$$

where H is a real function dependent on u, and on the complex coordinate ζ which spans the wave 2-surfaces $u =$ constant, $v =$ constant. These 2-surfaces with Euclidean geometry are thus contained in the wave hypersurfaces $u =$ constant and cut the rays given by $(u,\zeta) =$ constant, v changing. The vacuum field equations imply 2-dimensional Laplace's equation

$$H_{,\zeta\bar{\zeta}} = 0, \tag{52}$$

so that we can write

$$2H = f(u,\zeta) + \bar{f}(u,\bar{\zeta}), \tag{53}$$

where $f(u,\zeta)$ is an arbitrary function of u, analytic in ζ. To characterize the curvature in the waves and their effect on test particles it is convenient to introduce the null complex tetrad, such that at each spacetime point, together with the preferred null vector k^α, we have a null vector $l^\alpha, l^\alpha k_\alpha = -1$, and complex spacelike vector m^α satisfying $m_\alpha \bar{m}^\alpha = 1, m_\alpha k^\alpha = m_\alpha l^\alpha = 0$. For

[20] This algebraic (local) analogy between null fields exists also between electromagnetic and gravitational shocks (possible discontinuities across null hypersurfaces), and in the asymptotic behaviour of fields at large distances from sources (the "peeling property" – see e.g. [19,27]).

the metric (51) the only nonvanishing projection of the Weyl (in the vacuum case, the Riemann) tensor onto this tetrad is the (Newman–Penrose) scalar

$$\Psi_4 = C_{\alpha\beta\gamma\delta} l^\alpha \bar{m}^\beta l^\gamma m^\delta = H_{,\bar{\zeta}\bar{\zeta}} , \tag{54}$$

which denotes a *transverse* component of the wave propagating in the k^α direction. As shown by Ehlers and Kundt [53] (see also e.g. [61]), though in a somewhat different notation, we can use again an analogy with the electromagnetic field – described for an analogous plane wave by the transverse component $\phi_2 = F_{\alpha\beta}\bar{m}^\alpha l^\beta$ – and write $\Psi_4 = A\, e^{i\Theta}$, where real $A > 0$ is considered as the *amplitude* of the wave, and at each spacetime point associate Θ with the plane of polarization. Vacuum pp-waves with Θ = constant are called *linearly polarized.*

Consider a free test particle (observer) with 4-velocity $\mathbf{u}$ and a neighbouring free test particle displaced by a "connecting" vector $Z^\alpha(\tau)$. Introducing then the physical frame $\mathbf{e}_{(i)}$ which is connected with the observer such that $\mathbf{e}_{(0)} = \mathbf{u}$ and $\mathbf{e}_{(i)}$ are connected with the null tetrad vectors by

$$\begin{aligned} \mathbf{m} &= \tfrac{1}{\sqrt{2}}\left(\mathbf{e}_{(1)} + i\mathbf{e}_{(2)}\right), & \bar{\mathbf{m}} &= \tfrac{1}{\sqrt{2}}\left(\mathbf{e}_{(1)} - i\mathbf{e}_{(2)}\right), \\ \mathbf{l} &= \tfrac{1}{\sqrt{2}}\left(\mathbf{u} - \mathbf{e}_{(3)}\right), & \mathbf{k} &= \tfrac{1}{\sqrt{2}}\left(\mathbf{u} + \mathbf{e}_{(3)}\right), \end{aligned} \tag{55}$$

we find that the equation of geodesic deviation in spacetime with only $\Psi_4 \neq 0$ implies (see [207])

$$\ddot{Z}^{(1)} = -A_+ Z^{(1)} + A_\times Z^{(2)},\ \ddot{Z}^{(2)} = A_+ Z^{(2)} + A_\times Z^{(1)},\ \ddot{Z}^{(3)} = 0, \tag{56}$$

where $A_+ = \frac{1}{2}$ Re $\Psi_4, A_\times = \frac{1}{2}$ Im Ψ_4 are amplitudes of "+" and "×" polarization modes, and $Z^{(i)}$ are the frame components of the connecting vector $\mathbf{Z}$. Since the frame vector $\mathbf{e}_{(3)}$ is chosen in the longitudinal direction (the direction of the rays), equation (56) clearly exhibits the transverse character of the wave. If particles, initially at rest, lie in the $(\mathbf{e}_{(1)}, \mathbf{e}_{(2)})$ plane, there is no motion in the longitudinal direction of $\mathbf{e}_{(3)}$. The ring of particles is deformed into an ellipse, the axes of different polarizations are shifted one with respect to the other by $\frac{\pi}{4}$ (such behaviour is typical for linearized gravitational waves – cf. e.g. [18]). Making a rotation in the transverse plane by an angle ϑ,

$$\mathbf{e}'_{(1)} = \cos\vartheta\, \mathbf{e}_{(1)} + \sin\vartheta\, \mathbf{e}_{(2)} , \quad \mathbf{e}'_{(2)} = -\sin\vartheta\, \mathbf{e}_{(1)} + \cos\vartheta\, \mathbf{e}_{(2)} , \tag{57}$$

and taking $\vartheta = \vartheta_+(\tau) = -\frac{1}{2}$ Arg $\Psi_4 = -\frac{1}{2}\Theta$, then $A'_+ = \frac{1}{2}|\Psi|$, $A'_\times = 0$ – the wave is purely "+" polarized. If Θ = constant, the rotation angle is independent of time – the wave is rightly considered as linearly polarized.

Hence, with the discovery of pp-waves, the understanding of the properties of gravitational radiation has become deeper and closer to physics. In addition, the pp-waves can easily be "linearized" by taking the function H in the metric (51) to be so small that the spacetime can be considered as a perturbation of Minkowski space within the linearized theory. Such an "easy way" from the linear to fully nonlinear spacetimes is of course paid by their simplicity.

In general, in fact, the pp-waves have only the single isometry generated by the Killing vector $k^\alpha = (\partial/\partial v)^\alpha$. However, a much larger *group of symmetries* may exist for various particular choices of the function $H(u, \zeta, \bar{\zeta})$. Jordan, Ehlers and Kundt [57] (see also [53,61]) gave a complete classification of the pp-waves in terms of their symmetries and corresponding special forms of H. For example, in the best known case of plane waves to which we shall turn in greater detail below, Ψ_4 is independent of ζ, so that after removing linear terms in ζ by a coordinate transformation, we have

$$H(u, \zeta, \bar{\zeta}) = A(u)\zeta^2 + \bar{A}(u)\bar{\zeta}^2, \tag{58}$$

with $A(u)$ being an arbitrary function of u. This spacetime admits five Killing vectors.

Recently, Aichelburg and Balasin [208,209] generalized the classification given in [57] by admitting distribution-valued profile functions and allowing for non-vacuum spacetimes with metric (51), but with H which in general does not satisfy (52). They have shown that with H in the form of delta-like pulses,

$$H(u, \zeta, \bar{\zeta}) = f(\zeta, \bar{\zeta})\delta(u), \tag{59}$$

new symmetry classes arise even in the vacuum case.

The main motivation to consider impulsive pp-waves stems from the metrics describing a black hole or a "particle" boosted to the speed of light. The simplest metric of this type, given by Aichelburg and Sexl [210], is a Schwarzschild black hole with mass m boosted in such a way that $\mu = m/\sqrt{1 - w^2}$ is held constant as $w \to 1$. It reads

$$ds^2 = 2d\zeta d\bar{\zeta} - 2dudv - 4\mu \log(\zeta\bar{\zeta})\delta(u)du^2, \tag{60}$$

with H clearly in the form (59). This is not a vacuum metric: the energy-momentum tensor $T_{\alpha\beta} = \mu\delta(u)\delta(\zeta)k_\alpha k_\beta$ indicates that there is a "point-like particle" moving with the speed of light along $u = 0$. The Aichelburg-Sexl metric and its more recent generalizations have found interesting applications even outside of general relativity. Some of them will be briefly mentioned in Sect. 8.2.

Let us now turn to the simplest class of pp-waves, which comprises of the best known and illuminating examples of exact gravitational waves. These are the *plane waves*. They are defined as homogeneous pp-waves in the sense that the curvature component Ψ_4 (see (54)) is constant along the wave surfaces so that function H is in the form (58). One can write H as in (53) where

$$f(u, \zeta) = \frac{1}{2}\mathcal{A}(u)e^{i\Theta(u)}\zeta^2, \tag{61}$$

with linear terms being removed by a coordinate transformation. Just as a plane electromagnetic wave, a plane gravitational wave is thus completely represented by its amplitude $\mathcal{A}(u)$ and polarization angle $\Theta(u)$ as functions of the phase u.

The plane waves, including their generalization into the Einstein–Maxwell theory (an additional term $B(u)\zeta\bar{\zeta}$ then appears in H, both Ψ_4 and the electromagnetic quantity Φ_2 being independent of ζ), were already studied in 1926 (see [61]). A real understanding however came only in the late 1950s. Ehlers and Kundt [53] give various characterizations of this class. For example, they prove that a non-flat vacuum field is a pp-wave if and only if the curvature tensor is complex recurrent, i.e. if $P_{\alpha\beta\gamma\delta,\mu} = P_{\alpha\beta\gamma\delta}q_\mu$, where $P_{\alpha\beta\gamma\delta} = R_{\alpha\beta\gamma\delta} + i\,{}^{*}\!R_{\alpha\beta\gamma\delta}$; and it is a plane wave if and only if the recurrence vector q_μ is collinear with a real null vector. They also state a nice theorem showing that the plane wave spacetimes defined by the metric (51), H and f given by (53), (69), $\zeta = x + iy$, and with coordinate ranges $-\infty < x, y, u, v < \infty$, are geodesically complete if functions $\mathcal{A}(u)$ and $\Theta(u)$ are C^1-functions. Quoting directly from [53], *"there exist ... complete solutions free of sources (singularities), proving to think of a graviton field independent of any matter by which it be generated. This corresponds to the existence of source-free photon fields in electrodynamics"*. Ehlers and Kundt [53] also state an open problem which, as far as I am aware, has not yet been solved: to prove that plane waves are the only geodesically complete pp-waves.

The most telling examples of plane waves are *sandwich waves*. The amplitude $\mathcal{A}(u)$ in (61) need not be smooth: either it can only be continuous and nonvanishing on a finite interval of u (sandwich), or a step function (shock), or a delta function (impulse). A physical interpretation of such waves is better achieved in other coordinate systems, in which the metric "before" and "after" the wave is not Minkowskian but has a higher degree of smoothness. For linearly polarized waves (Θ equal to zero), a convenient coordinate system can be introduced by setting (see e.g. [211]) $\zeta = (1/\sqrt{2})(px + iqy)$, $v = (1/2)(t + z + pp'x^2 + qq'y^2)$, $u = t - z$, where $' = d/du$, and functions $p = p(u)$ and $q = q(u)$ solve equations $p'' + \mathcal{A}(u)p = 0$ and $q'' - \mathcal{A}(u)q = 0$. In these coordinates the metric turns out to be

$$ds^2 = -dt^2 + p^2dx^2 + q^2dy^2 + dz^2. \tag{62}$$

In double-null coordinates $\tilde{u}$, $\tilde{v}$, with $\tilde{u} = u = t - z$, $\tilde{v} = t + z$, and with a general polarization, the metric can be cast into the form (see e.g. [65,212])

$$ds^2 = -d\tilde{u}d\tilde{v} + e^{-U}(e^V \cosh W dx^2 + e^{-V} \cosh W dy^2 - 2\sinh W dx dy), \tag{63}$$

where U, V, W depend on $\tilde{u}$ only. This so called Rosen form was used in the classical paper on exact plane waves by Bondi, Pirani and Robinson [213].

A simple, textbook example [214] of a sandwich wave is the wave with a "square profile": $\mathcal{A}(u) = 0$ for $u < 0$ and $u > a^2, \mathcal{A}(u) = a^{-2} =$ constant for $0 \leq u \leq a^2$. The functions p and q which enter (62) are then $p = q = 1$ at $u \leq 0, p = \cos(u/a), q = \cosh(u/a)$ at $0 \leq u \leq a^2$, and $p = -(u/a)\sin a +$

constant, $q = (u/a)\sinh a +$ constant at $a^2 \leq u$. This example can be used to demonstrate explicitly various typical features of plane sandwich gravitational waves within the exact theory: (i) the wave fronts travel with the speed of light; (ii) the discontinuities of the second derivatives of the metric tensor are permitted along a null hypersurface, but must have a special structure; (iii) the waves have a transverse character and produce relative accelerations in test particles; (iv) the waves focus astigmatically initially parallel null congruences (rays) that are pointing in other directions than the waves themselves; (v) as a consequence of the focusing, Rosen-type line elements contain coordinate singularities on a hypersurface behind the waves, and in general caustics will develop there [214].

The focusing effects imply a remarkable property of plane wave spacetimes: no spacelike global hypersurface exists on which initial data can be specified, i.e. *plane wave spacetimes contain no global Cauchy hypersurface.* This can be understood from Fig. 9. Considering a point Q in flat space in front of the wave, Penrose [215] has shown that its future null cone is distorted as it passes through the wave in such a manner that it is refocused to either a point R or a line passing through R parallel to the wave front. Any possible Cauchy hypersurface going through Q must lie below the future null cone through Q, i.e. below the past null cone of R. Hence, it cannot extend as a spacelike hypersurface to spatial infinity.

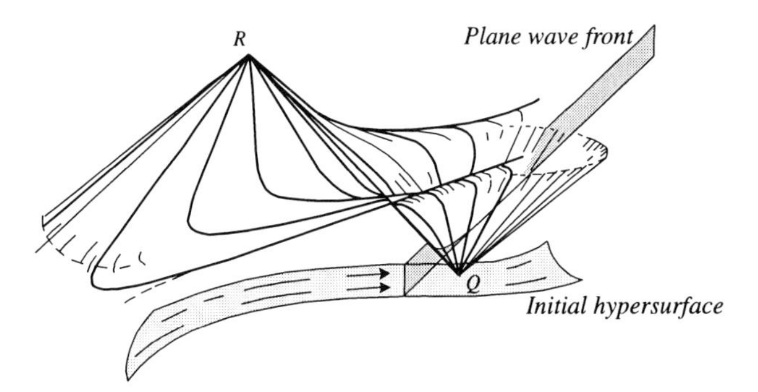

Fig. 9. The future null cone of the event Q is distorted as it passes through the plane wave, and refocused at the event R in such a manner that no Cauchy initial hypersurface going through Q exists. (From [215].)

8.2 Plane-Fronted Waves: New Developments and Applications

The interest in impulsive waves generated by boosting a "particle" at rest to the velocity of light by means of an appropriate limiting procedure persists up to the present. The ultrarelativistic limits of Kerr and Kerr–Newman black holes were obtained in [216–218], and recently, boosted static multipole

(Weyl) particles were studied [219]. Impulsive gravitational waves were also generated by boosting the Schwarzschild–de Sitter and Schwarzschild–anti de Sitter metrics to the ultrarelativistic limit [220,221].

These types of spacetimes, especially the simple Aichelburg-Sexl metrics, have been employed in current problems of the generation of gravitational radiation from axisymmetric black hole collisions and black hole encounters. The recent monograph by d'Eath [222] gives a comprehensive survey, including the author's new results. There is good reason to believe that spacetime metrics produced in high speed collisions will be simpler than those corresponding to (more realistic) situations in which black holes start to collide with low relative velocities. The spacetimes corresponding to the collisions at exactly the speed of light is an interesting limit which can be treated most easily. Aichelburg-Sexl metrics are used to describe limiting "incoming states" of two black holes, moving one against the other with the speed of light. An approximation method has been developed in which a large Lorentz boost is applied so that one has a weak shock propagating on a strong shock. One finds an estimate of 16.8 % for the efficiency of gravitational wave generation in a head-on speed-of-light collision [222].

Great interest has been stimulated by 't Hooft's [223] work on the quantum scattering of two pointlike particles at centre-of-mass energies higher or equal to the Planck energy. This quantum process has been shown to have close connection with classical black hole collisions at the speed of light (see [222,224] and references therein).

Recently, the Colombeau algebra of generalized functions, which enables one to deal with singular products of distributions, has been brought to general relativity and used in the description of impulsive pp-waves in various coordinate systems [225], and also for a rigorous solution of the geodesic and geodesic deviation equations for impulsive waves [226]. The investigation of the equations of geodesics in non-homogeneous pp-waves (with $f \sim \zeta^3$) has shown that the motion of test particles in these spacetimes is described by the Hénon-Heiles Hamiltonian which implies that the motion is chaotic [227].

Plane-fronted waves have been used as simple metrics in various other contexts, for example, in quantum field theory on a given background (see [228] for recent work), and in string theory [229]. As emphasized very recently by Gibbons [230], since for pp-waves and type *N* Kundt's class (see the beginning of Sect. 8.1) all possible invariants formed from the Weyl tensor and its covariant derivatives vanish [231], these metrics suffer no quantum corrections to all loop orders. Thus they may offer insights into the behaviour of a full quantum theory. The invariants vanish also in type *III* spacetimes with nonexpanding and nontwisting rays [232].

8.3 Colliding Plane Waves

As with a number of other issues in gravitational (radiation) theory, the pioneering ideas on colliding plane gravitational waves are connected with

Roger Penrose. It does not seem to be generally recognized that the basic idea appeared six years before the well-known paper by Khan and Penrose [233] in which the metric describing the general spacetime representing a collision of two parallel-polarized impulsive gravitational waves was obtained. Having demonstrated the surprising fact that general relativistic plane wave spacetimes admit no Cauchy hypersurface due to the focusing effect the waves exert on null cones, Penrose [215] (in footnote 12) remarks: "This fact has relevance to the question of two colliding weak plane sandwich waves. Each wave warps the other until singularities in the wave fronts ultimately appear. This, in fact, causes the spacetime to acquire genuine physical singularities in this case. The warping also produces a scattering of each wave after collision so that they cease to be sandwich waves when they separate (and they are no longer plane – although they have a two-parameter symmetry group)."

The first detailed study of colliding plane waves, independently of Khan and Penrose, was also undertaken by Szekeres (see [234,235]). He formulated the problem as a characteristic initial value problem for a system of hyperbolic equations in two variables (null coordinates) u, v with data specified on the pair of null hypersurfaces $u = 0, v = 0$ intersecting in a spacelike 2-surface (Fig. 10). In the particular case of spacetimes representing plane waves propagating before the collision in a flat background, Szekeres has shown that coordinates (of the "Rosen type", as known from the case of one wave – see Eq. (63)) exist in which the metric reads

$$\begin{aligned} ds^2 = & - e^{-M} du\, dv + \\ & + e^{-U} \left[e^{V} \cosh W dx^2 + e^{-V} \cosh W dy^2 - 2 \sinh W dx\, dy \right] , \end{aligned} \tag{64}$$

where M, U, V and W are functions of u and v. Coordinates x and y are aligned along the two commuting Killing vectors $\partial/\partial x$ and $\partial/\partial y$, which are assumed to exist in the whole spacetime representing the colliding waves (cf. the note by Penrose above). In almost all recent work on colliding waves, region *IV* in Fig. 10, where $u < 0$, $v < 0$, is assumed to be flat. The null lines $u = 0, v < 0$ and $v = 0, u < 0$ are wavefronts, and in regions *II* $(u < 0, v > 0)$ and *III* $(u > 0, v < 0)$ one has the standard plane wave metric corresponding to two approaching plane waves from opposite directions. In region *II*, functions M, U, V, W depend on v only, and in region *III* only on u. The waves collide at the 2-surface $u = v = 0$, in region *I* they interact. The spacetime here can be determined by the initial data posed on the $v \geq 0$ portion of the hypersurface $u = 0$ (which in Fig. 10 are "supplied" by the wave propagating to the right) and by the data on the $u \geq 0$ portion of the hypersurface $v = 0$ (given by the wave propagating to the left). Unfortunately, the integration of such an initial value problem does not seem to be possible for general incoming wave forms and polarizations. If, however, the approaching waves have constant and aligned (parallel) polarizations, one may set the function $W = 0$ globally. The solution of the initial value problem then reduces to a one dimensional integral for the function V, and two quadratures for the function

M. (The function $\exp(-U)$ must have the form $f(u)+g(v)$ as a consequence of the field equations everywhere, and it can be determined easily from the initial data.) Despite these simplifications it is very difficult to obtain exact solutions in closed analytic form. Szekeres [235] found a solution (as he puts it "more or less by trial and error") which, as special cases, includes the solution given by himself earlier [234] and the solution obtained independently and simultaneously by Khan and Penrose [233]. Although Szekeres' formulation of a general solution for the problem of colliding parallel-polarized waves is difficult to use for constructing other specific explicit examples, it has been employed in a general analysis of the structure of the singularities produced by the collision [236], which will be discussed in the following.

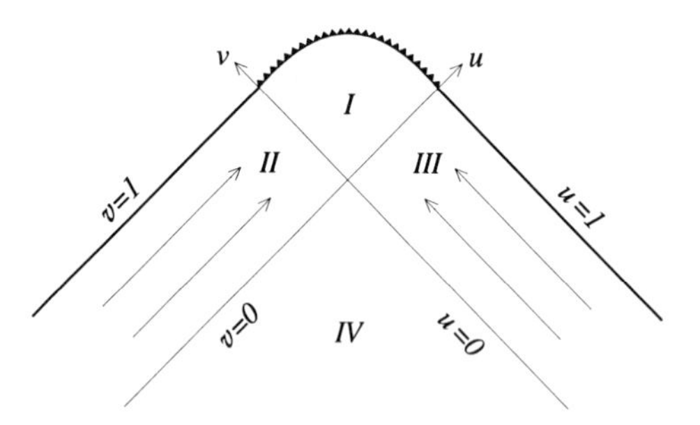

Fig. 10. The spacetime diagram indicating the collision of two plane-fronted gravitational waves which come from regions *II* and *III*, collide in region *I*, and produce a spacelike singularity. Region *IV* is flat.

It has also inspired an important, difficult piece of mathematical physics which was developed at the beginning of the 1990s in the series of papers by Hauser and Ernst [237]. Their new method of analyzing the initial value problem can be used also for the case when the polarization of the approaching waves is not aligned. They formulated the initial value problem in terms of the equivalent matrix Riemann–Hilbert problem in the complex plane. Their techniques are related to those used by Neugebauer and Meinel to analyze and construct the rotating disk solution as a boundary value problem (Sect. 6.3). No analogous solution for colliding waves in the noncollinear case is available at present, but investigations in this direction are still in progress. Most recently, Hauser and Ernst prepared an extensive treatise [238] in which they give a general description and detailed mathematical proofs of their study of the solutions of the hyperbolic Ernst equation.

The approach of Khan and Penrose for obtaining exact solutions describing colliding plane waves starts in the region *I* where the waves interact: (i) find a solution with two commuting spacelike Killing vectors $\partial/\partial x$ and $\partial/\partial y$, transform to null coordinates, and look back in time whether this solution can be extended across the null hypersurface $u=0, v=0$ so that it describes a plane wave propagating in the u-direction in region *II* and another plane wave propagating in the v-direction in region *III* ; (ii) satisfy boundary conditions not only across boundaries between regions *I* and *II*, and regions *I* and *III*, but also across the boundaries between *II* and *IV*, and *III* and *IV* in such a manner that *IV* is flat. The original prescription of Khan and Penrose for extending the solution from region *I* to regions *II* and *III* consists in the substitutions $uH(u)$ and $vH(v)$ in place of u and v everywhere in the metric coefficients; here $H(u) = 1$ for $u \geq 0$, $H = 0$ for $u < 0$ is the usual Heaviside function. We then get the metric as a function of v (respectively u) in region *II* (respectively *III*) corresponding to the wave propagating to the right (respectively to the left) in Fig. 10. Finally, it remains to investigate carefully the structure of discontinuities and possible singularities on the null boundaries between these regions. In the original Khan and Penrose solutions the Riemann tensor has a δ-function character on the boundaries between *II* and *IV*, and *III* and *IV*; but inside regions *II* and *III* themselves the spacetime is flat (the collision of impulsive plane waves). In the solution obtained by Szekeres [235], regions *II* and *III* are not flat, and the Riemann tensor at the boundaries between *II* (respectively *III*) and *IV* is just discontinuous (the collision of shock waves).

Nutku and Halil [239] constructed an exact solution describing the collision of two impulsive plane waves with non-aligned polarizations. In the limit of collinear polarizations their solution reduces to the solution of Khan and Penrose. All of these solutions reveal that the spacelike singularity always develops in region *I* (given by $u^2 + v^2 = 1$ in Fig. 10) – in agreement with the original suggestion of Penrose. Moreover, the singularity "propagates backward" and so called *fold singularities*, analyzed in detail in 1984 by Matzner and Tipler [240], appear also at $v = 1$ and $u = 1$ in regions *II* and *III*. This new type of singularity provides evidence of how even relatively recent studies of explicit exact solutions may reveal unexpected global features of relativistic spacetimes.

The remarkable growth of interest in colliding plane waves owes much to the systematic (and symptomatic) effort of S. Chandrasekhar who, since 1984, together with V. Ferrari, and with B. Xanthopoulos, published a number of papers on colliding plane vacuum gravitational waves [241,242], and on gravitational waves coupled with electromagnetic waves, with null dust, and with perfect fluid (see [212] for references). The basic strategy of their approach follows that of Khan and Penrose: first a careful analysis of the possible solution is done in the interaction region *I*, and then one works backward in time, extending the solutions to regions *II*, *III* and *IV*.

The main new input consists in carrying over the techniques known from stationary, axisymmetric spacetimes with one timelike and one spacelike Killing vector to the case of two spacelike Killing vectors, $\partial/\partial x, \partial/\partial y$, and exploring new features.

Taking a simple linear solution of the Ernst equation, $E = P\eta + iQ\mu$, where P and Q are real constants which satisfy $P^2 + Q^2 = 1$, and η, μ are suitable time and space coordinates, Chandrasekhar and Ferrari [241] show that one arrives at the Nutku-Halil solution. In particular, if $Q = 0$, the Khan-Penrose solution emerges. Since by starting from the same simplest form of the Ernst function in the axisymmetric stationary case one arrives at the Kerr solution (or at the Schwarzschild solution for the real Ernst function), we may conclude that in region I the solutions of Khan and Penrose and of Nutku and Halil are, for spacetimes with two spacelike Killing vectors, the analogues of the Schwarzschild and Kerr solutions. This mathematical analogy can be generalized to colliding electromagnetic and gravitational waves within the Einstein–Maxwell theory – Chandrasekhar and Xanthopoulos [243] found the analogue of the charged Kerr–Newman solution. Such a generalization is also of interest from a conceptual viewpoint: the δ-function singularity in the Weyl tensor of an impulsive gravitational wave might imply a similar singularity in the Maxwell stress tensor, which would seem to suggest that the field itself would contain "square roots of the δ-function".

In the most important paper [242] of the series, Chandrasekhar and Xanthopoulos, starting from the simplest linear solution for the Ernst conjugate function $E^+ = P\eta + iQ\mu, P^2 + Q^2 = 1$, obtained a new exact solution for colliding plane impulsive gravitational waves accompanied by shock waves. This solution results in the development of a nonsingular Killing–Cauchy horizon instead of a spacelike curvature singularity. The metric can be analytically extended across this horizon to produce a maximal spacetime which contains timelike singularities. (The spacelike singularity in region I in Fig. 10 is changed into the horizon, to the future of which timelike singularities occur.) In the region of interaction of the colliding waves, the spacetime is isometric to the spacetime in a region interior to the ergosphere.

Many new interesting solutions were discovered by using the Khan and Penrose approach. In addition, inverse scattering (soliton) methods and other tools from the solution generation techniques were applied. They are reviewed in detail in [65,212,244].

Although very attractive mathematical methods are contained in these works, one feels that physical interpretation has receded into the background – as seemed to be the case when the new solution generating techniques were exploited in all possible directions for stationary axisymmetric spacetimes. It is therefore encouraging that a more physical and original approach to the problem has been initiated by Yurtsever. In a couple of papers he discusses Killing–Cauchy horizons [245] and the structure of the singularities produced by colliding plane waves [236]. Similar to the Cauchy horizons in black hole physics, one finds that the Killing–Cauchy horizons are unstable. We thus expect that the horizon will be converted to a spacelike singularity. By using the approach of Szekeres described at the beginning of this section, it is possible to relate the asymptotic form of the metric near the singularity – which approaches an inhomogeneous Kasner solution (see Sect. 12.1) – to the initial data given along the wavefronts of the incoming waves. For specific choices of initial data the singularity degenerates into a coordinate singularity and a Killing–Cauchy horizon arises. However, Yurtsever's analysis [236] shows that such horizons are unstable (within *full nonlinear* theory) against small but generic perturbations of the initial data. These results are stronger than those on the instability of the inner horizons of the Reissner–Nordström or Kerr black holes. In particular, Yurtsever constructs an interesting (though unstable) solution which, when analytically extended across its Killing–Cauchy horizon, represents a Schwarzschild black hole created out of the collision between two plane sandwich waves propagating in a cylindrical universe [236].

Yurtsever also introduced "almost plane wave spacetimes" and analyzed collisions of almost plane waves [246]. These waves have a finite but very large transverse sizes. Some general results can be proved (for example, that almost plane waves cannot have a sandwich character, but always leave tails behind), and an order-of-magnitude analysis can be used in the discussion of the outcome of the collision of two almost plane waves; i.e. whether they will focus to a finite minimum size and then disperse, or whether a black hole will be created. Although in the case of almost plane waves one can hardly hope to find an exact spacetime in an explicit form, this is a field which was inspired by exact explicit solutions, and may play a significant role in other parts of general relativity.

9 Cylindrical Waves

In 1913, before the final formulation of general relativity, Einstein remarked in a discussion with Max Born that, in the weak-field limit, gravitational waves exist and propagate with the velocity of light (Poincaré pioneered the idea of gravitational waves propagating with the velocity of light in 1905 – see [15]). Yet, in 1936 Einstein wrote to Born [247]: "... gravitational waves do not exist, though they had been assumed a certainty to the first approximation. This shows the nonlinear general relativistic field equations can tell us more, or, rather, limit us more than we have believed up to now. If only it were not so damnably difficult to find rigorous solutions". However, after finding a mistake in his argumentation (with the help of H. Robertson) and discovering with Nathan Rosen cylindrical gravitational waves [248] as the first exact radiative solutions to his vacuum field equations, Einstein changed his mind. In fact, cylindrical waves were found more than 10 years before Einstein and Rosen by Guido Beck in Vienna [249]. Beck was mainly interested in time-independent axisymmetric Weyl fields, but he realized that through a complex transformation of coordinates ($z \to it, t \to iz$) one obtains cylindrically symmetric time-dependent fields which represent cylindrical gravitational waves, and wrote down equations (71) and (72) below. The work of Einstein and Rosen is devoted explicitly to gravitational waves. It investigates conditions for the existence of standing and progressive waves, and even notices that the waves carry away energy from the mass located at the axis of symmetry. We shall thus not modify the tradition and will call this type of waves Einstein–Rosen waves (which some readers may wish to shorten to EROS-waves).

This type of waves, symmetric with respect to the transformation $z \to -z$ (z – the axis of symmetry), contains one degree of freedom of the radiation field and corresponds to a fixed state of polarization. The metric can be written in the form

$$ds^2 = e^{2(\gamma-\psi)}(-dt^2 + d\rho^2) + e^{2\psi}dz^2 + \rho^2 e^{-2\psi}d\varphi^2, \tag{65}$$

where ρ and t are invariants ("Weyl-type canonical coordinates"), and $\psi = \psi(t,\rho)$, $\gamma = \gamma(t,\rho)$. The Killing vectors $\partial/\partial\varphi$ and $\partial/\partial z$ are both spacelike and hypersurface orthogonal.

The metric containing a second degree of freedom was discovered by Jürgen Ehlers (working in the group of Pascual Jordan), who used a trick similar to Beck's on the generalized (stationary) Weyl metrics, and independently by Kompaneets (see the discussion in [250]). In the literature (e.g. [251,252]) one refers to the Jordan-Ehlers-Kompaneets form of the metric:

$$ds^2 = e^{2(\gamma-\psi)}\left(-dt^2 + d\rho^2\right) + e^{2\psi}\left(dz + \omega d\varphi\right)^2 + \rho^2 e^{-2\psi}d\varphi^2. \tag{66}$$

Here, the additional function $\omega(t,\rho)$ represents the second polarization.

Despite the fact that cylindrically symmetric waves cannot describe exactly the radiation from bounded sources, both the Einstein–Rosen waves and their generalization (66) have played an important role in clarifying a number of complicated issues, such as the energy loss due to gravitational waves [253], the interaction of waves with cosmic strings [254,255], the asymptotic structure of radiative spacetimes [250], the dispersion of waves [256], testing the quasilocal mass-energy [257], testing codes in numerical relativity [251], investigation of the cosmic censorship [258], and quantum gravity in a simplified but field theoretically interesting context of midisuperspaces [259–261].

In the following we shall discuss in some detail the asymptotic structure and midisuperspace quantization since in these two issues cylindrical waves have played the pioneering role. Some other applications of cylindrical waves will be briefly mentioned at the end of the section.

9.1 Cylindrical Waves and the Asymptotic Structure of 3-Dimensional General Relativity

In recent work with Ashtekar and Schmidt [262,263], which started thanks to the hospitality of Jürgen Ehlers' group, we considered gravitational waves with a space-translation Killing field ("generalized Einstein–Rosen waves"). In (2+1)-dimensional framework the Einstein–Rosen subclass forms a simple instructive example of *explicitly given spacetimes which admit a smooth global null (and timelike) infinity even for strong initial data.* Because of the symmetry, the 4-dimensional Einstein vacuum equations are equivalent to the 3-dimensional Einstein equations with certain matter sources. This result has roots in the classical paper by Jordan, Ehlers and Kundt [57] which includes "reduction formulas" for the calculation of the Riemann tensor of spaces which admit an Abelian isometry group.

Vacuum spacetimes which admit a spacelike, hypersurface orthogonal Killing vector $\partial/\partial z$ can be described conveniently in coordinates adapted to the symmetry:

$$ds^2 = V^2(x)dz^2 + \bar{g}_{ab}(x)dx^a dx^b, \quad a, b, \ldots = 0, 1, 2, \tag{67}$$

where $x \equiv x^a$ and $\bar{g}_{ab}$ is a metric with Lorentz signature. The field equations can be simplified if one uses a metric in the 3-space which is rescaled by the norm of the Killing vector, and writes the norm of the Killing vector as an exponential. Then (67) becomes

$$ds^2 = e^{2\psi(x)}dz^2 + e^{-2\psi(x)}g_{ab}(x)dx^a dx^b, \tag{68}$$

and the field equations,

$$R_{ab} - 2\nabla_a\psi\nabla_b\psi = 0, \qquad g^{ab}\nabla_a\nabla_b\psi = 0, \tag{69}$$

where ∇ denotes the derivative with respect to the metric g_{ab}, can be reinterpreted as Einstein's equations in 3 dimensions with a scalar field $\Phi = \sqrt{2}\psi$ as source. Thus, 4-dimensional vacuum gravity is equivalent to 3-dimensional gravity coupled to a scalar field. In 3 dimensions, there is no gravitational radiation. Hence, the local degrees of freedom are all contained in the scalar field. One therefore expects that Cauchy data for the scalar field will suffice to determine the solution. For data which fall off appropriately, we thus expect the 3-dimensional Lorentzian geometry to be asymptotically flat in the sense of Penrose [27,264], i.e. that there should exist a 2-dimensional boundary representing null infinity.

In general cases, this is analyzed in [262]. Here we shall restrict ourselves to the Einstein–Rosen waves by assuming that there is a further spacelike, hypersurface orthogonal Killing vector $\partial/\partial\varphi$ which commutes with $\partial/\partial z$. Then, as is well known, the equations simplify drastically. The 3-metric is given by

$$d\sigma^2 = g_{ab}dx^a dx^b = e^{2\gamma}(-dt^2 + d\rho^2) + \rho^2 d\varphi^2, \tag{70}$$

the field equations (69) become

$$\gamma' = \rho(\dot{\psi}^2 + \psi'^2), \qquad \dot{\gamma} = 2\rho\dot{\psi}\psi', \tag{71}$$

and

$$-\ddot{\psi} + \psi'' + \rho^{-1}\psi' = 0, \tag{72}$$

where the dot and the prime denote derivatives with respect to t and ρ respectively. The last equation is the wave equation for the non-flat 3-metric (70) as well as for the flat metric obtained by setting $\gamma = 0$.

Thus, we can first solve the axisymmetric wave equation (72) for ψ on Minkowski space and then solve (71) for γ – the only unknown metric coefficient – by quadratures. The "method of descent" from the Kirchhoff formula in 4 dimensions gives the representation of the solution of the wave equation in 3 dimensions in terms of Cauchy data $\Psi_0 = \psi(t=0,x,y), \Psi_1 = \psi_{,t}(t=0,x,y)$ (see [262]). We assume that the Cauchy data are axially symmetric and of compact support.

Let us look at the behaviour of the solution at future null infinity $\mathcal{J}^+$. Let ρ, φ be polar coordinates in the plane, and introduce the retarded time coordinate $u = t-\rho$ to explore the fall-off along constant u null hypersurfaces. For large ρ, the function ψ at $u =$ constant admits a power series expansion in ρ^{-1}:

$$\psi(u,\rho) = \frac{f_0(u)}{\sqrt{\rho}} + \frac{1}{\sqrt{\rho}} \sum_{k=1}^{\infty} \frac{f_k(u)}{\rho^k}. \tag{73}$$

The coefficients in this expansion are determined by integrals over the Cauchy data. At $u \gg \rho_0$, ρ_0 being the radius of the disk in the initial Cauchy surface in which the data are non-zero, we obtain

$$f_0(u) = \frac{k_0}{u^{\frac{3}{2}}} + \frac{k_1}{u^{\frac{1}{2}}} + \ldots, \tag{74}$$

where k_0, k_1 are constants which are determined by the data. If the solution happens to be time-symmetric, so that Ψ_1 vanishes, we find $f_0 \sim u^{-\frac{3}{2}}$ for large u. Similarly, we can also study the behaviour of the solution near the timelike infinity i^+ of 3-dimensional Minkowski space by setting $t = U + \kappa\rho$, $\kappa > 1$, and investigating ψ for $\rho \to \infty$ with U and κ fixed. We refer to [262] for details.

In Bondi-type coordinates $(u = t - \rho, \rho, \varphi)$, equation (70) yields

$$d\sigma^2 = e^{2\gamma}(-du^2 - 2du d\rho) + \rho^2 d\varphi^2. \tag{75}$$

The Einstein equations take the form

$$\gamma_{,u} = 2\rho\, \psi_{,u}(\psi_{,\rho} - \psi_{,u}), \qquad \gamma_{,\rho} = \rho\, \psi_{,\rho}^2, \tag{76}$$

and the wave equation on ψ becomes

$$-2\psi_{,u\rho} + \psi_{,\rho\rho} + \rho^{-1}(\psi_{,\rho} - \psi_{,u}) = 0. \tag{77}$$

The asymptotic form of $\psi(t,\rho)$ is given by the expansion (73). Since we can differentiate (73) term by term, the field equations (76) and (77) imply

$$\gamma_{,u} = -2[\dot{f}_0(u)]^2 + \sum_{k=1}^{\infty} \frac{g_k(u)}{\rho^k}, \tag{78}$$

$$\gamma_{,\rho} = \sum_{k=0}^{\infty} \frac{h_k(u)}{\rho^{k+2}}, \tag{79}$$

where the functions f_k, h_k are products of the functions $f_0, f_k, \dot{f}_0, \dot{f}_k$. Integrating (79) and fixing the arbitrary function of u in the result using (78), we obtain

$$\gamma = \gamma_0 - 2\int_{-\infty}^{u} \left[\dot{f}_0(u)\right]^2 du - \sum_{k=1}^{\infty} \frac{h_k(u)}{(k+1)\rho^{k+1}}. \tag{80}$$

Thus, γ also admits an expansion in ρ^{-1}, where the coefficients depend smoothly on u. It is now straightforward to show that the spacetime admits a smooth future null infinity, $\mathcal{J}^+$. Setting $\tilde{\rho} = \rho^{-1}, \tilde{u} = u, \tilde{\varphi} = \varphi$ and rescaling g_{ab} by a conformal factor $\Omega = \tilde{\rho}$, we obtain

$$d\tilde{\sigma}^2 = \Omega^2 d\sigma^2 = e^{2\tilde{\gamma}}(-\tilde{\rho}^2 d\tilde{u}^2 + 2d\tilde{u}d\tilde{\rho}) + d\tilde{\varphi}^2, \tag{81}$$

where $\tilde{\gamma}(\tilde{u}, \tilde{\rho}) = \gamma(u, \tilde{\rho}^{-1})$. Because of (80), $\tilde{\gamma}$ has a smooth extension through $\tilde{\rho} = 0$. Therefore, $\tilde{g}_{ab}$ is smooth across the surface $\tilde{\rho} = 0$. This surface is the future null infinity, $\mathcal{J}^+$. Hence, the (2+1)-dimensional curved spacetime has a smooth (2-dimensional) null infinity. Penrose's picture works for arbitrarily strong initial data Ψ_0, Ψ_1.

Using (81), we find that at $\mathcal{J}^+$ we have:

$$\gamma(u, \infty) = \gamma_0 - 2\int_{-\infty}^{u} \dot{f}_0^2 du. \tag{82}$$

Since one can make sure that $\gamma = 0$ at i^+ [263], one finds the simple result that

$$\gamma_0 = 2\int_{-\infty}^{+\infty} \dot{f}_0^2 du. \tag{83}$$

At spatial infinity (t = constant, $\rho \to \infty$), the metric is given by

$$d\sigma^2 = e^{2\gamma_0}(-dt^2 + d\rho^2) + \rho^2 d\varphi^2. \tag{84}$$

For a non-zero data, constant γ_0 is positive, whence the metric has a "conical singularity" at spatial infinity. This conical singularity, present at spatial infinity, is "radiated out" according to equation (82). The future timelike infinity, i^+, is smooth. In (2+1)-dimensions, modulo some subtleties [262], equation (82) plays the role of the *Bondi mass loss formula* in (3+1)-dimensions, relating the decrease of the total (Bondi) mass-energy at null infinity to the flux of gravitational radiation. We can thus conclude that *cylindrical waves in (2+1)-dimensions give an explicit model of the Bondi–Penrose radiation theory which admits smooth null and timelike infinity for arbitrarily strong initial data.* There is no other such model available. The general results on the existence of $\mathcal{J}$ in 4 dimensions, mentioned at the end of Sect. 1.3, assume weak data.

It is of interest to investigate cylindrical waves also in a (3+1)-dimensional context. The asymptotic behaviour of these waves was discussed by Stachel

[250] many years ago. However, his work deals solely with asymptotic directions, which are perpendicular to the axis of symmetry, i.e. to the $\partial/\partial z$ – Killing vector. Detailed calculations show that, in contrast to the perpendicular directions, where null infinity in the (3+1)-dimensional framework does not exist, it *does* exist in other directions for data of compact support. If the data are not time-symmetric, the fall-off is so slow that (local) null infinity has a *polyhomogeneous* (logarithmic) character [180] – see [263] for details.

We have concentrated on the simplest case of Einstein–Rosen waves. They served as a prototype for developing a general framework to analyze the asymptotic structure of spacetime at null infinity in three spacetime dimensions. This structure has a number of quite surprising features which do not arise in the Bondi–Penrose description in four dimensions [262]. One of the motivations for developing such a framework is to provide a natural point of departure for constructing the stage for asymptotic quantization and the S-matrix theory of an interesting midisuperspace in quantum gravity.

9.2 Cylindrical Waves and Quantum Gravity

As the editors of the Proceedings of the 117th WE Heraeus Seminar on canonical gravity in 1993 [265], Jürgen Ehlers and Helmut Friedrich start their Introduction realistically: "When asking a worker in the field about the progress in quantum general relativity in the last decade, one shouldn't be surprised to hear: 'We understand the problems better'. If it referred to a lesser task, such an answer would sound ironic. But the search for quantum gravity... has been going on now for more than half a century and in spite of a number of ingenious proposals, a satisfactory theory is still lacking..." Although I am following the subject from afar, I believe that one would not be too wrong if one repeated the same words in 1999. However, apart from general theoretical developments, many interesting quantum gravity models have been studied, and exact solutions have played a basic role in them. In particular, in the investigations of (spherical) gravitational collapse and in quantum cosmology based typically on homogeneous cosmological models (cf. Sect. 12.1), one starts from simple classical solutions – see e.g. [266–268] for reviews and [269] for a bibliography up to 1990. A common feature of such models is the reduction of infinitely many degrees of freedom of the gravitational field to a *finite* number. In quantum field theory (such as quantum electrodynamics) a typical object to be quantized is a wave with an infinite number of degrees of freedom. The first radiative solutions of the gravitational field equations which were subject to quantization were the Einstein–Rosen waves. Kuchař [259] applied the methods of canonical quantization of gravity to these waves, using the methods employed earlier in the minisuperspace models, i.e. restricting himself only to geometries (fields) preserving the symmetries.

The Einstein–Rosen cylindrical waves have an *infinite* number ∞^1 of degrees of freedom contained in one polarization, one degree of freedom for each cylindrical surface drawn around the axis of symmetry. Moreover, the slicing of spacetime by spacelike (cylindrically symmetric) hypersurfaces is not fixed completely by the symmetry – an arbitrary cylindrically symmetric deformation of a given slice leads again to an allowed slice. Such a deformation represents an ∞^1 "fingered time". Hence, the resulting space of 3-geometries on cylindrically symmetric slices is infinitely richer than the minisuperspaces of quantum cosmology. The exact Einstein–Rosen waves thus inspired the first construction of what Kuchař [259] called the "*midisuperspace*".

Let us briefly look at the main steps in Kuchař's procedure.[21] The symmetry of the problem implies that the spatial metric has the form

$$g_{11} = e^{\gamma-\Phi}, \qquad g_{22} = R^2 e^{-\Phi}, \qquad g_{33} = e^{\Phi}, \tag{85}$$

where γ, Φ, and R are functions of a single cylindrical coordinate $x^1 = r$ ($x^2 = \varphi$, $x^3 = z$). Similarly the lapse function $N = N(r)$ depends only on r, and the shift vector has the only nonvanishing radial component $N^1 = N^1(r), N^2 = N^3 = 0$. We have adopted here Kuchař's notation. When we put $R = r = \rho$, $\Phi = 2\psi$, $\gamma \to 2\gamma$, $N = e^{\gamma-\Phi}$, and $N^1 = 0$, we recover the standard Einstein–Rosen line element (65); however, in general the radial and time coordinates t and r differ from the canonical Einstein–Rosen radial and time coordinates in which the metric has the standard form (65). The symmetry implies that the canonical momentum π^{ik} is diagonal and expressible by three functions $\pi_\gamma, \pi_R, \pi_\Phi$ of r; for example, $\pi^{11} = \pi_\gamma e^{\Phi-\gamma}$, and similarly for the other components. After the reduction to cylindrical symmetry, the action functional assumes the canonical form

$$S = 2\pi \int_{-\infty}^{\infty} dt \int_0^{\infty} dr (\pi_\gamma \dot{\gamma} + \pi_R \dot{R} + \pi_\Phi \dot{\Phi} - N\mathcal{H} - N^1 \mathcal{H}_1) , \tag{86}$$

in which γ, R, Φ are the canonical coordinates and $\pi_\gamma, \pi_R, \pi_\Phi$ the conjugate momenta (the integration over z has been limited by $z = z_0$ and $z = z_0 + 1$). The superhamiltonian $\mathcal{H}$ and supermomentum $\mathcal{H}_1$ are rather complicated functions of the canonical variables:

$$\mathcal{H} = e^{\frac{1}{2}(\Phi-\gamma)} \left(-\pi_\gamma \pi_R + \tfrac{1}{2} R^{-1} \pi_\Phi^2 + 2R'' - \gamma' R' + \tfrac{1}{2} R \Phi'^2\right) , \tag{87}$$

$$\mathcal{H}_1 = -2\pi_\gamma' + \gamma' \pi_\gamma + R' \pi_R + \Phi' \pi_\Phi . \tag{88}$$

The most important step now is the replacement of the old canonical variables $\gamma, \pi_\gamma, R, \pi_R$ by a new canonical set T, Π_T, R, Π_R through a suitable canonical transformation. We shall write here only one of its components (see [259,270] for the complete transformation):

$$T(r) = T(\infty) + \int_\infty^r [-\pi_\gamma(r)] \, dr. \tag{89}$$

[21] For the basic concepts and ideas of canonical gravity, we refer to e.g. [18,19] and especially to Kuchař's review [270], where the canonical quantization of cylindrical waves is also analyzed.

By integrating the Hamilton equations following from the action (86), rewritten in the new canonical coordinates, one finds that T and R are the Einstein–Rosen privileged time and radial coordinates, i.e. those appearing in the canonical form (65) of the Einstein–Rosen metric (with $T = t, R = \rho$). According to (89), the Einstein–Rosen time can be reconstructed, in a non-local way, from the momentum π_γ, which characterizes the extrinsic curvature of a given hypersurface. In this way, the concept of the *"extrinsic time representation"* entered canonical gravity with cylindrical gravitational waves.

In terms of the new canonical variables, the superhamiltonian and supermomentum become

$$\mathcal{H} = e^{\frac{1}{2}(\Phi-\gamma)} \left(R'\Pi_T + T'\Pi_R + \tfrac{1}{2}\left(R^{-1}\pi_\Phi^2 + R\Phi'^2\right)\right), \tag{90}$$

$$\mathcal{H}_1 = T'\Pi_T + R'\Pi_R + \Phi'\pi_\Phi. \tag{91}$$

Since $\mathcal{H}$ and $\mathcal{H}_1$ are linear in Π_T and Π_R, the classical constraints $\mathcal{H} = 0, \mathcal{H}_1 = 0$ can immediately be resolved with respect to these momenta, conjugate to the "embedding" canonical variables $T(r)$ and $R(r)$:

$$-\Pi_T = \left(R'^2 - T'^2\right)^{-1} \left[\tfrac{1}{2}(R^{-1}\pi_\Phi^2 + R\Phi'^2)R' - \Phi'\pi_\Phi T'\right] = 0, \tag{92}$$

and similarly for Π_R. It is easy to see [259,270] that the constraints have the same form as the constraints for a massless scalar field Φ propagating on a flat background foliated by arbitrary spacelike hypersurfaces $T = T(r), R = R(r)$. The canonical variables Φ, π_Φ represent the true degrees of freedom, and the remaining canonical variables play the role of spacelike embeddings of a Cauchy hypersurface into spacetime.

After turning the canonical momenta Π_T, Π_R, π_Φ, into variational derivatives, e.g. $\Pi_T = -i\delta/\delta T(r)$, one can impose the classical constraints $\mathcal{H} = 0, \mathcal{H}_1 = 0$ as restrictions on the state functional $\Psi(T, R, \Phi)$: $\mathcal{H}\Psi = 0$, $\mathcal{H}_1\Psi = 0$. In particular, the Wheeler-DeWitt equation $\mathcal{H}\Psi = 0$ in the extrinsic time representation assumes the form of a many-fingered time counterpart of an ordinary Schrödinger equation. This reduces to the ordinary Schrödinger equation for a single massless scalar field in Minkowski space if we adopt the standard foliation $T =$ constant (see [259,270] for details).

The described procedure, first realized in the case of the Einstein–Rosen waves, has opened a new route in canonical and quantum gravity. In contrast to the Arnowitt-Deser-Misner approach, in which the gravitational dynamics is described relative to a fixed foliation of spacetime, in this new approach (called "bubble time" dynamics of the gravitational field or the "internal time formalism" [271]) one tries to extract the many-fingered time (i.e. embeddings of Cauchy hypersurfaces) from the gravitational phase space, but does not fix the foliation in the "target manifold" by coordinate conditions. However, the definition of the target manifold by a gauge (coordinate) condition is needed.

This new approach has been so far successfully applied to a few other models (based on exact solutions) with infinite degrees of freedom, for example, plane gravitational waves, bosonic string, and as late as 1994, to spherically symmetric vacuum gravitational fields [272]. The internal time formalism for spacetimes with two Killing vectors was developed in [273] (therein references to previous works can also be found). Recently, canonical transformation techniques have been applied to Hamiltonian spacetime dynamics with a thin spherical null-dust shell [274]. One would like to construct a midisuperspace model of spherical gravitational collapse, or more specifically, a model for Hawking radiation with backreaction. The extensive past work on Hamiltonian approaches to spherically symmetric geometries (see [274] for more than 40 references in this context) have not yet led to convincing insights. The very basic question of existence of the "internal time" formalism in a general situation has been most recently addressed by Hájíček [49]; the existence has been proven, and shown to be related to the choice of gauge.

9.3 Cylindrical Waves: a Miscellany

Chandrasekhar [247] constructed a formalism for cylindrical waves with two polarizations (cf. the metric (66)), similar to that used for the discussion of the collision of plane-fronted waves (Sect. 8.3). He obtained the "cylindrical" Ernst equation and corroborated (following the suggestion of O. Reula) the physical meaning of Thorne's *C-energy* [253] – the expression for energy suggested for cylindrical fields – by defining a Hamiltonian density corresponding to the Lagrangian density from which the Ernst equation can be derived. A brief summary of older work on the mass loss of a cylindrical source radiating out cylindrical waves and its relation to the C-energy is given in [65]. It should be pointed out, however, that although C-energy is a useful quantity, it was constructed by exploiting the local field equations, without direct reference to asymptotics. The physical energy (per unit z length) at both spatial and null infinity, which is the generator of the time translation, is in fact a non-polynomial function of the C-energy. In the weak field limit the two agree, but in strong fields they are quite different [262].

In [256], an exact solution was constructed with which one can study the *dispersion* of waves: a cylindrical wave packet, which though initially impulsive, after reflection at the axis disperses, and develops shock wave fronts when the original wave meets the waves that are still ingoing. Cylindrical waves have been also analyzed in the context of *phase shifts* occurring in gravitational soliton interactions (see [275] and references therein).

An exact explicit solution for cylindrical waves with two degrees of polarization has been obtained [252] from the Kerr solution after transforming the metric into "cylindrical" coordinates and using the substitution $t \to i\tilde{z}, z \to i\tilde{t}, a \to i\tilde{a}$. Both this solution and the well-known Weber-Wheeler-Bonnor pulse [65] have been employed as *test beds in numerical relativity* [276], in particular in the approach which combines a Cauchy code for determining the dynamics of the central source with a characteristic code for determining the behaviour of radiation [251].

In a number of works cylindrical waves have been considered in interaction with *cosmic strings* [254,255]. The strings are usually modelled as infinitely thin conical singularities. Recently Colombeau's theory of generalized functions was used to calculate the distributional curvature at the axis for a time-dependent cosmic string [277].

A somewhat surprising result concerning cosmic strings and radiation theory should also be noted: although an infinite, static cylindrically symmetric string does not, of course, radiate, it generates a nonvanishing (though "non-radiative") contribution to the Bondi news function [278,279]. Recently, the asymptotics at null infinity of cylindrical waves with both polarizations (and, in general, an infinite cosmic string along the axis) has been analyzed in the context of axisymmetric electrovacuum spacetimes with a translational Killing vector at null infinity [280].

Finally, the cylindrically symmetric electrovacuum spacetimes with both polarizations, satisfying certain completeness and asymptotic flatness conditions in spacelike directions have been shown rigorously to imply that strong cosmic censorship holds [258]. This means that for generic (smooth) initial data the maximal globally hyperbolic development of the data is inextendible (no Cauchy horizons as for example, those discussed in Sect. 3.1 for the Reissner–Nordström spacetime arise). This global existence result is nontrivial since with two polarizations and electromagnetic field present, all field equations are nonlinear.

10 On the Robinson–Trautman Solutions

Robinson–Trautman metrics are the general radiative vacuum solutions which admit a geodesic, shearfree and twistfree null congruence of diverging rays. In the standard coordinates the metric has the form [281]

$$ds^2 = 2r^2P^{-2}d\zeta d\bar{\zeta} - 2du\,dr - [\Delta \ln P - 2r(\ln P)_{,u} - 2mr^{-1}]\,du^2, \qquad (93)$$

where ζ is a complex spatial (stereographic) coordinate (essentially θ and φ), r is the affine parameter along the rays, u is a retarded time, m is a function of u (which can be in some cases interpreted as the mass of the system), $\Delta = 2P^2(\partial^2/\partial\zeta\partial\bar{\zeta})$, and $P = P(u,\zeta,\bar{\zeta})$ satisfies the fourth-order Robinson–Trautman equation

$$\Delta\Delta(\ln P) + 12\,m\,(\ln P)_{,u} - 4m_{,u} = 0. \qquad (94)$$

The best candidates for describing radiation from isolated sources are the Robinson–Trautman metrics of type *II* with the 2-surfaces S^2 given by $u, r =$ constant and having spherical topology. The Gaussian curvature of S^2 can be expressed as $K = \Delta \ln P$. If $K =$ constant, we obtain the Schwarzschild solution with mass equal to $K^{-\frac{3}{2}}$.

These spacetimes have attracted increased attention in the last decade – most recently in the work by Chruściel, and Chruściel and Singleton [282]. In these studies the Robinson–Trautman spacetimes have been shown to exist globally for all positive "times", and to converge asymptotically to a Schwarzschild metric. Interestingly, the extension of these spacetimes across the "Schwarzschild-like" event horizon can only be made with a finite degree of smoothness. All these rigorous studies are based on the derivation and analysis of an asymptotic expansion describing the long-time behaviour of the solutions of the nonlinear parabolic equation (94).

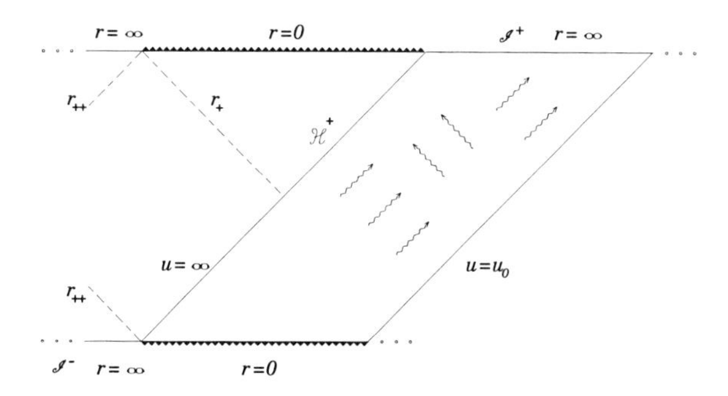

Fig. 11. The evolution of the cosmological Robinson–Trautman solutions with a positive cosmological constant. A black hole with the horizon $\mathcal{H}^+$ is formed; at future infinity $\mathcal{J}^+$ the spacetime approaches a de Sitter spacetime exponentially fast, in accordance with the cosmic no-hair conjecture.

In our recent work [163,283] we studied Robinson–Trautman radiative spacetimes with a positive cosmological constant Λ. The results proving the global existence and convergence of the solutions of the Robinson–Trautman equation (94) can be taken over from the previous studies since Λ does not explicitly enter this equation. We have shown that, starting with arbitrary, smooth initial data at $u = u_0$ (see Fig. 11), these cosmological Robinson–Trautman solutions converge exponentially fast to a Schwarzschild–de Sitter solution at large retarded times ($u \to \infty$). The interior of a Schwarzschild–de Sitter black hole can be joined to an "external" cosmological Robinson–Trautman spacetime across the horizon $\mathcal{H}^+$ with a higher degree of smoothness than in the corresponding case with $\Lambda = 0$. In particular, in the extreme case with $9\Lambda m^2 = 1$, in which the black hole and cosmological horizons coincide, the Robinson–Trautman spacetimes can be extended *smoothly* through $\mathcal{H}^+$ to the extreme Schwarzschild–de Sitter spacetime with the same values of Λ and m. However, such an extension is *not analytic* (and not unique).

We have also demonstrated that the cosmological Robinson–Trautman solutions represent explicit models exhibiting the cosmic no-hair conjecture: a geodesic observer outside of the black hole horizon will see, that inside his past light cone, these spacetimes approach the de Sitter spacetime exponentially fast as he approaches the future (spacelike) infinity $\mathcal{J}^+$. For a freely falling observer the observable universe thus becomes quite bald. This is what the cosmic no-hair conjecture claims. As far as we are aware, these models represent the only exact analytic demonstration of the cosmic no-hair conjecture under the presence of gravitational waves. They also appear to be the only exact examples of black hole formation in nonspherical spacetimes which are not asymptotically flat. Hopefully, these models may serve as tests of various approximation methods, and as test beds in numerical studies of more realistic situations in cosmology.

11 The Boost–Rotation Symmetric Radiative Spacetimes

In this section we would like to describe briefly the only explicit solutions available today which are radiative and represent the fields of finite sources. Needless to say, we cannot hope to find explicit analytic solutions of the Einstein equations without imposing a symmetry. A natural first assumption is axial symmetry, i.e. the existence of a spacelike rotational Killing vector $\partial/\partial\varphi$. However, it appears hopeless to search for a radiative solution with only one symmetry. We are now not interested in colliding plane waves since these do not represent finite sources; we wish our spacetime to be as "asymptotically flat as possible". The unique role of the boost-rotation symmetric spacetimes is exhibited by a theorem, formulated precisely and proved for the vacuum case with hypersurface orthogonal Killing vectors in [284], and generalized to electrovacuum spacetimes with Killing vectors which need not be hypersurface orthogonal in [279] (see also references therein). This theorem roughly states that in *axially* symmetric, locally asymptotically flat spacetimes (in the sense that a null infinity satisfying Penrose's requirements exists, but it need not necessarily exist globally), the only *additional* symmetry that does not exclude radiation is the *boost* symmetry.

In Minkowski spacetime the boost Killing vector has the form

$$\zeta_{boost} = z\frac{\partial}{\partial t} + t\frac{\partial}{\partial z}, \tag{95}$$

so that orbits of symmetry to which the Killing vector is tangent are hyperbolas $z^2 - t^2 = B = \text{constant}$, x, $y = \text{constant}$. Orbits with $B > 0$ are timelike; they can represent worldlines of uniformly accelerated particles in special relativity. Imagine, for example, a charged particle, axially symmetric about the z-axis, moving with a uniform acceleration along this axis. The electromagnetic field produced by such a source will have boost-rotation symmetry.

Figure 12 shows two particles uniformly accelerated in opposite directions along the z-axis. In the space diagram (left), the "string" connecting the particles is also indicated. In the spacetime diagram, the particles' worldlines are shown in bold. Thinner hyperbolas represent the orbits of the boost Killing vector (95) in the regions $t^2 > z^2$ where it is spacelike. In Fig. 13 the corresponding compactified diagram indicates that null infinity cannot be smooth everywhere since it contains four singular points in which particles' worldlines "start" and "end". Notice that in electromagnetism the presence of *two* particles, one moving along $z > 0$, the other along $z < 0$, makes the field symmetric also with respect to inversion $z \to -z$. The electromagnetic field can be shown to be analytic everywhere, except for the places where the particles occur. These two particles move independently of each other, since their worldlines are divided by two null hypersurfaces $z = t, z = -t$. This is analogous to the boost-rotation symmetric spacetimes in general relativity that we are now going to discuss.

Specific examples of solutions representing "uniformly accelerated particles" have been analyzed for 35 years, starting with the first solutions of this type obtained by Bonnor and Swaminarayan [285], and Israel and Khan [286]. In a curved spacetime the "uniform acceleration" is understood with respect to a fictitious Minkowski background, and the "particles" mean singularities or black holes. For a more extensive description of the history of these specific solutions discovered before 1985, see [287]. From a unified point of view, boost-rotation symmetric spacetimes (with hypersurface orthogonal Killing vectors) were defined and treated geometrically in [288]. We refer to this detailed work for rigorous definitions and theorems. Here we shall only sketch some of the general properties and some applications of these spacetimes.

The metric of a general boost-rotation symmetric spacetime in "Cartesian-type" coordinates $\{t, x, y, z\}$ reads:

$$\begin{aligned} ds^2 = & \frac{1}{x^2+y^2}\left[(e^{\lambda}x^2+e^{-\mu}y^2)dx^2+2xy(e^{\lambda}-e^{-\mu})dxdy\right]+ \\ & +\frac{1}{x^2+y^2}(e^{\lambda}y^2+e^{-\mu}x^2)dy^2+\frac{1}{z^2-t^2}(e^{\lambda}z^2-e^{\mu}t^2)dz^2- \\ & -\frac{1}{z^2-t^2}\left[2zt(e^{\lambda}-e^{\mu})dzdt+(e^{\mu}z^2-e^{\lambda}t^2)dt^2\right], \end{aligned} \tag{96}$$

where μ and λ are functions of $\rho^2 = x^2 + y^2$ and $z^2 - t^2$. As a consequence of the vacuum Einstein equations, the function μ must satisfy an equation of the form which is identical to the flat-space wave equation; and function λ is determined in terms of μ by quadrature. Now it can easily be seen that the metric (96) admits axial and boost Killing vectors which have exactly the same form as in Minkowski space, i.e. the axial Killing vector $\partial/\partial\varphi$ and the boost Killing vector (95). In fact, the whole structure of group orbits in boost-rotation symmetric *curved* spacetimes outside the sources (or singularities) is the same as the structure of the orbits generated by the axial and boost Killing vectors in Minkowski space. In particular, the boost Killing vector (95) is timelike in the region $z^2 > t^2$. The invariance of a metric (or of any other field) in a time-direction (determined in a coordinate-free manner by a timelike Killing vector) means stationarity, and of course, we could hardly expect to find radiative properties there. Intuitively, the existence of a timelike Killing vector in the region $z^2 > t^2$ is understandable because there (generalized) uniformly accelerated reference frames can be introduced in which sources are at rest, and the fields are time independent.

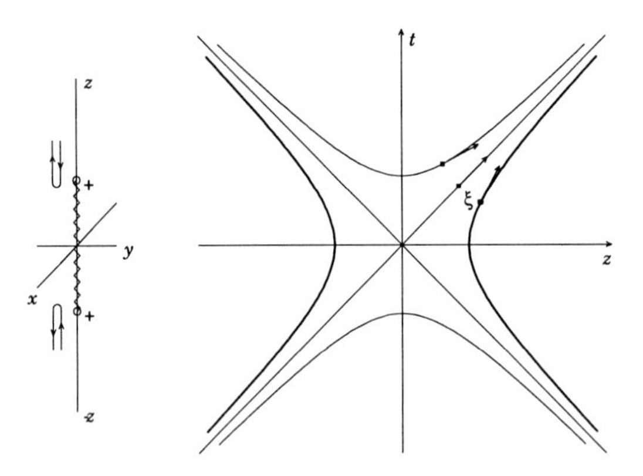

Fig. 12. Two particles uniformly accelerated in opposite directions. The orbits of the boost Killing vector (thinner hyperbolas) are spacelike in the region $t^2 > z^2$.

However, in the other "half" of the spacetime, $t^2 > z^2$, the boost Killing vector (95) is spacelike (see the lines representing orbits of the boost Killing vector in Fig. 12). Hence in this region the metric (96) is nonstationary. Here we expect to discover radiative properties. Indeed, it can be shown that for $t^2 > z^2 + \rho^2$ the metric (96) can *locally* be transformed into the metric of Einstein–Rosen cylindrical waves. Although locally in the whole region $t^2 > z^2$ the metric (96) can be transformed into a radiative metric, the global properties of the boost-rotation symmetric solutions are quite different from those of cylindrical waves. Again, we have to refer to the work [288] for a detailed analysis. Let us only say that the boost-rotation symmetric solutions, if properly defined – with appropriate boundary conditions on functions λ and μ – always admit asymptotically flat null infinity $\mathcal{J}$ at least locally. Starting with arbitrary solutions λ and μ, and adding suitable constants to both λ and μ (Einstein's equations are then still satisfied), we can always guarantee that even *global* $\mathcal{J}$ exists in the sense that it admits smooth spherical sections. For the special type of solutions for λ and μ, *complete* $\mathcal{J}$ satisfies Penrose's requirements, except for four points in which the sources "start" and "end" (cf. Fig. 13). In all cases one finds that the gravitational field in smooth regions of the null infinity is radiative [279,289]. In particular, the leading term of the Riemann curvature tensor, proportional to r^{-1} (where $r^2 = \rho^2 + z^2$), is nonvanishing and has the same algebraic structure as the Riemann tensor of plane waves. This is fully analogous to the asymptotic properties of radiative electromagnetic fields outside finite sources. Recently, general forms of the news functions have been obtained for electrovacuum spacetimes with boost-rotation symmetry and with Killing vectors which need not be hypersurface orthogonal [279].

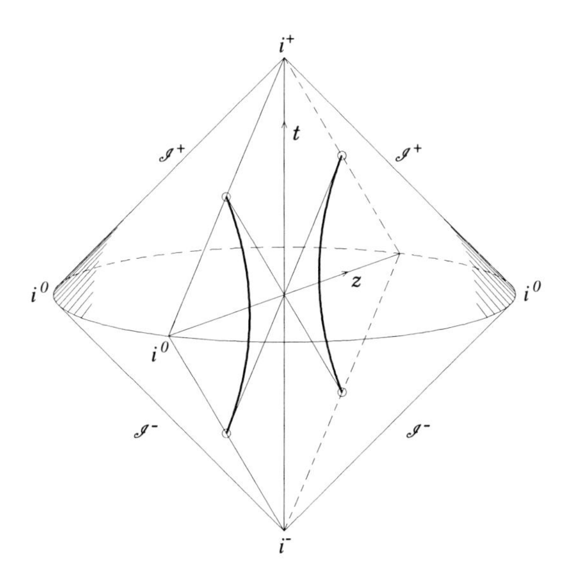

Fig. 13. The Penrose compactified diagram of a boost-rotation symmetric spacetime. Null infinity can admit smooth sections.

It is well known that in general relativity the "causes" of motion are always incorporated in the theory – in contrast to electrodynamics where they need not even be describable by Maxwell's theory. In a general case of the boost-rotation symmetric solutions there exist nodal (conical) singularities of the metric distributed along the z-axis which can be considered as "strings", and cause particles to accelerate. They reveal themselves also at $\mathcal{J}$. However, the distribution of nodes can always be arranged in such a manner that $\mathcal{J}$ admits smooth regular sections as mentioned above.

In exceptional cases, when $\mathcal{J}$ is regular except for four points, either the particles are "self-accelerating" due to their "inner" multipole structure, which has to include a negative mass; or there are more particles distributed along $z > 0$ (and symmetrically along $z < 0$) with the signs and the magnitudes of their masses and accelerations chosen appropriately. (For the concept of a negative mass in general relativity, and the first discussion of a "chasing" pair of a positive and a negative mass particle, see classical papers by Bondi, and Bonnor and Swaminarayan [285].) An infinite number of different analytic solutions representing self-accelerating particles was constructed explicitly [290]. Although a negative mass cannot be bought easily in the shop (as Bondi liked to say), these solutions are the only exact solutions of Einstein's equations available today for which one can find such quantities of physical interest as radiation patterns (angular distribution of gravitational radiation), or total radiation powers [287]. From a mathematical point of view, these solutions represent the only known spacetimes in which *arbitrarily strong* (boost-rotation symmetric) *initial data* can be chosen on a hyperboloidal hypersurface in the region $t^2 > z^2$, which will lead to a complete smooth null infinity and a regular timelike future infinity. With these specific examples one thus does not have to require weak-field initial data as one has to in the work of Friedrich, and Christodoulou and Klainerman, mentioned at the end of Sect. 1.3.

The boost-rotation symmetric radiative spacetimes can be used as test beds for approximation methods or numerical relativity. Bičák, Reilly and Winicour [291] found the explicit boost-rotation symmetric "initial null cone solution", which solves initial hypersurface and evolution equations in "radiative" coordinates employed in the null cone version of numerical relativity. This solution has been used for checking and improving numerical codes for computing gravitational radiation from more realistic sources; a new solution of this type has also been found [292]. Recently, the specific boost-rotation symmetric spacetimes constructed in [290] were used as test beds in the standard version of numerical relativity based on spacelike hypersurfaces [293].

There exist "generalized" boost-rotation symmetric spacetimes which are *not* asymptotically flat, but are of considerable physical interest. They describe accelerated particles in asymptotically "uniform" external fields. One can construct such solutions from asymptotically flat boost-rotation symmetric solutions for the pairs of accelerated particles by a limiting procedure, in which one member of the pair is "removed" to infinity, and its mass parameter is simultaneously increased [294]. Since the resulting spacetimes are not asymptotically flat, their radiative properties are not easy to analyze. Only if the external field is weak will there exist regions in which the spacetimes are approximately flat; and here their radiative properties might be investigated. So far no systematic analysis of these spacetime has been carried out. Nevertheless, they appear to offer the best rigorous examples of the motion of relativistic objects. No nodal singularities or negative masses are necessary to cause an acceleration.

As an eloquent example of such a spacetime consider a charged (Reissner–Nordström) black hole with mass M and charge Q, immersed in an electric field "uniform at infinity", characterized by the field-strength parameter E.

An exact solution of the Einstein–Maxwell equations exists which describes this situation [131]. It goes over into an approximate solution obtained by perturbing the charged black hole spacetime by a weak external electric field which is uniform at infinity [295]. One of the results of the analysis of this solution is very simple: a charged black hole in an electric field starts to accelerate according to Newton's second law, $Ma = QE$, where all the quantities can be determined – and in principle measured – in an approximately flat region of the spacetime from the asymptotic form of the metric. Recall T. S. Eliot again: "There is only the fight to recover what has been lost / And found and lost again and again."

These types of generalized boost-rotation symmetric spacetimes ("generalized C-metrics") have been used by Hawking, Horowitz, Ross, and others [296] in the context of quantum gravity – to describe production of black hole pairs in strong background fields.

Recently, we have studied the spinning C-metric discovered by Plebański and Demiański [297]. Transformations can be found which bring this metric into the canonical form of spacetimes with boost-rotation symmetry [298]. The metric represents two uniformly *accelerated, rotating* black holes, either connected by conical singularity, or with conical singularities extending from each of them to infinity. The spacetime is radiative. No other spacetime of this type, with two Killing vectors which are not hypersurface orthogonal, is available in an explicit form.

12 The Cosmological Models

In light of Karl Popper's belief that "all science is cosmology", it appears unnecessary to justify the choice of solutions for this last section. As in the whole article, these will be primarily vacuum solutions. On the other hand, in light of the light coming from about 10^{11} galaxies, each with about 10^{11} stars, it may seem weird to consider *vacuum* models of the Universe. Indeed, it has become part of the present-day culture that spatially homogeneous and isotropic, expanding Friedmann-Robertson-Walker (FRW) models, filled with uniformly distributed matter, correspond well to basic observational data. In order to achieve a more precise correspondence, it appears sufficient to consider just perturbations of these "standard cosmological models". To explain some "improbable" features of these models such as their isotropy and homogeneity, one finds an escape in inflationary scenarios. These views of a "practical cosmologist" are, for example, embodied in one of the most comprehensive recent treatise on physical cosmology by Peebles [28].

Theoretical (or mathematical) cosmologists, however, point out that more general cosmological models exist which differ significantly from a FRW model at early times, approach the FRW model very closely for a certain epoch, and may diverge from it again in the future. Clearly, the FRW universes represent only a very special class of viable cosmological models, though the simplest and most suitable for interpretations of "fuzzy" cosmological observational data.

Simple exact solutions play a significant role in the evolution of more general models, either as asymptotic or intermediate states. By an "intermediate state" one means the situation when the universe enters and remains in a small neighbourhood of a saddle equilibrium point. A simple example is the Lemaître matter-filled, homogeneous and isotropic model with a nonvanishing cosmological constant (see e.g. [28]), which expands from a dense state (the big bang, or "primeval atom" in Lemaître's 1927 terminology), passes through a quasistatic epoch in which all parameters are close to those of the static Einstein universe (cf. Sect. 1.2), and then the universe expands again. An "asymptotic state" means close either to an initial big bang (or possibly a final big crunch) singularity, or the situation at late times in forever expanding universes. It is easy to see that at late times in indefinitely expanding universes the matter density decreases, and vacuum solutions may become important. However, as we shall discuss below, vacuum models play an important role also close to a singularity, when the matter terms in Einstein's equations are negligible compared to the "velocity terms" (given by the rate of change of scale factors) or to the curvature terms (characterizing the curvature of spacelike hypersurfaces). In particular, the pioneering (and still controversial) work started at the end of the 1950s by Lifshitz and Khalatnikov, and developed later on by Belinsky, Khalatnikov and Lifshitz, has shown that the fact that the presence of matter does not influence the qualitative behaviour of a cosmological model near a singularity has a very general significance (see [299] and [300] for the main original references, and [139] for a brief review).

In gaining an intuition in the analysis of general cosmological singularities, the class of spatially homogeneous anisotropic cosmological models have played a crucial role. These so called *Bianchi models* admit a simply transitive 3-dimensional homogeneity group. Among the Bianchi vacuum models there are special exact explicit solutions, in particular the Kasner and the Bianchi type II solutions, which exhibit some aspects of general cosmological singularities. The Bianchi models have also had an impact on other issues in general relativity and cosmology.

Much work, notably in recent years, has been devoted to the class of both vacuum and matter-filled cosmological solutions which are homogeneous only on 2-dimensional spacelike orbits. Thus they depend on time and on one spatial variable, and can be used to study spatial inhomogeneities as density fluctuations or gravitational waves. The vacuum cosmological models with two spacelike Killing vectors, sometimes called the *Gowdy models*,[22] are

[22] In fact, by Gowdy models, one more often means only the cases with closed group orbits, with two commuting spacelike othogonally-transitive Killing vectors (the surface elements orthogonal to the group orbits are surface-forming).

interpreted as gravitational waves in an expanding (or contracting) universe with compact spatial sections. We shall discuss these two classes separately.

12.1 Spatially Homogeneous Cosmologies

The simplest solutions, the Minkowski, de Sitter, and anti de Sitter spacetimes, which have also been used in cosmological contexts (cf. Sect. 1.3), are 4-*dimensionally homogeneous.* As noted in Sect. 8.1, the vacuum plane waves (see equations (51), (53), (61)) are also homogeneous spacetimes; and since they can be suitably sliced by spacelike hypersurfaces with expanding normal congruence, they can become asymptotic states in homogeneous expanding cosmologies. There exist several important non-vacuum homogeneous spacetimes, for example, the Einstein static universe (cf. Sect. 1.2), and Gödel's stationary, rotating universe (see e.g. [61,301]), famous for the first demonstration that Einstein's equations with a physically permissible matter source are compatible with the existence of closed timelike lines, i.e. with the violation of causality.

Here we shall consider models in which the symmetry group does not make spacetime a homogeneous space, but in which each event in spacetime is contained in a spatial hypersurface that is homogeneous. The standard FRW models represent a special case of such models (they admit, in addition, an isotropy group $SO(3)$ at each point). The general spatially homogeneous solutions comprise of the *Kantowski–Sachs universes* and a much wider class of Bianchi models. By definition, the Bianchi models admit a simply transitive 3-dimensional homogeneity group G_3. There exist special "locally rotationally symmetric" (LRS) Bianchi models which admit a 4-dimensional isometry group G_4 acting on homogeneous spacelike hypersurfaces, but these groups have a simply transitive subgroup G_3. In contrast to this, Kantowski–Sachs spacetimes admit G_4 (acting on homogeneous spacelike hypersurfaces) which does *not* have any simply transitive subgroup G_3; it contains a multiply transitive G_3 acting on 2-dimensional surfaces of constant curvature, $G_4 = I\!R \times SO(3)$. A special case of the vacuum Kantowski–Sachs universe is represented by the Schwarzschild metric inside the horizon (with t and r interchanged). There has been a continuing interest in the Kantowski–Sachs models since their discovery in 1966 [302], to which, as the authors acknowledge, J. Ehlers contributed by his advice. Some of these models had already appeared in the PhD thesis of Kip Thorne in 1965 (see also [303] for magnetic Kantowski–Sachs models). Here, however, we just refer the reader to [304,305] for their classical description, to [306] for a canonical and quantum treatment, and to [307] for the latest discussion of the Kantowski–Sachs quantum cosmologies.

Although the 3-dimensional Lie groups which are simply transitive on homogenous 3-spaces were classified by Bianchi in 1897, the importance of Bianchi's work for constructing vacuum cosmological models was only discovered by Taub in 1951 [197], when the Taub space (cf. Sect. 7) was first given. It is less known that at approximately the same time, if not earlier, the first explicit spatially homogeneous expanding and rotating cosmological models with matter (of the Bianchi type IX) were constructed by Gödel,[23] who first presented his results at the International Congress of Mathematics held at Cambridge (Mass.) from August 30 till September 5, 1950.

An exposition of Bianchi models has been given in a number of places: in the account on relativistic cosmology by Heckmann and Schücking [308] (complementing the chapter on exact solutions by Ehlers and Kundt [53]), in the monographs of Ryan and Shepley [304], and Zel'dovich and Novikov [309], in several comprehensive surveys by MacCallum (see e.g. [305] and [310] for his latest review containing a number of references), most recently, in the book on the dynamical system approach in cosmology (in the Bianchi models in particular) edited by Wainwright and Ellis [311]; and, first but not least, in the classics of Landau and Lifshitz [139]. The Hamiltonian approach initiated by Misner [312] in 1968, and used in, amongst other things, the construction of various minisuperspace models in quantum gravity, has been reviewed by Ryan [266]; for more recent accounts, see several contributions to Misner's Festschrift [313]. An interesting framework which unifies the Hamiltonian approach to the solutions which admit homogeneous hypersurfaces either spacelike (as Bianchi models) or timelike (as static spherical, or stationary cylindrical models) was recently developed by Uggla, Jantzen and Rosquist in [314] (with 115 references on many exact solutions). Herewith we shall only briefly introduce the Bianchi models, note their special role in understanding the character of an initial cosmological singularity, and mention some of the most recent developments not covered by the reviews cited above.

The line element of the Bianchi models can be expressed in the form

$$ds^2 = -dt^2 + g_{ab}(t)\, \omega^a\, \omega^b, \tag{97}$$

where the time-independent 1-forms ω^a $(= E^a_\alpha dx^\alpha)$, $a = 1, 2, 3$, are dual to time-independent[24] spatial frame vectors $\mathbf{E}_a$ (often an arbitrary time-variable $\tilde{t}$ is introduced by $dt = N(\tilde{t})\, d\tilde{t}$, N being the usual lapse function). Both ω^a

[23] Gödel's profound ideas and results in cosmology, and their influence on later developments have been discussed in depth by G. Ellis in his lecture at the Gödel '96 conference in Brno, Czech Republic, where Gödel was born in 1906 (78 years before Gödel, Ernst Mach was born in a place which today belongs to Brno). In the extended written version of Ellis' talk [301] it is indicated that Gödel's work also initiated the investigation of Taub. This may well be true with Gödel's paper on the stationary rotating universe, but Taub's paper on Bianchi models was received by the Annals of Mathematics on May 15, 1959, i.e. before Gödel's lecture on expanding and rotating models at the Congress of Mathematics took place.

[24] The gravitational degrees of freedom are associated with the component (scalar) functions $g_{ab}(t)$ – the so called metric approach. Alternatively, in the orthonormal frame approach, one chooses $g_{ab}(t) = \delta_{ab}$ and describes the evolution by time-dependent forms ω^a. In still another approach one employs the automorphism

and $\mathbf{E}_a$ are group-invariant, commuting with the three Killing fields which generate the homogeneity group. They satisfy the relations

$$d\omega^a = -\frac{1}{2}C^a_{bc}\,\omega^b\wedge\omega^c, \tag{98}$$

$$[\mathbf{E}_a,\mathbf{E}_b] = C^c_{ab}\mathbf{E}^c, \tag{99}$$

where d is the exterior derivative and C^a_{bc} are the structure constants of the Lie algebra of the homogeneity group. The models are classified according to the possible distinct sets of the structure constants. They are first divided into two classes: in class A the trace $C^a_{ba} = 0$, and in class B, $C^a_{ba} \neq 0$. In class A one can choose $C^a_{bc} = n^{(a)}\epsilon_{abc}$ (no summation over a), and classify various symmetry types by parameters $n^{(a)}$ with values $0, \pm 1$. In class B, in addition to $n^{(a)}$, one needs the value of a constant scalar h (related to C^a_{ba}) to characterize types VI$_h$ and VII$_h$ (see e.g. [311]).

The simplest models are the Bianchi I cosmologies in class A with $n^{(a)} = 0$, i.e. $C^a_{bc} = 0$, so that all three Killing vectors (the group generators) commute. They contain the standard Einstein–de Sitter model with flat spatial hypersurfaces (curvature index $k = 0$). In the vacuum case, all Bianchi I models are given by the well-known 1-parameter family of *Kasner metrics* (found in 1921 by E. Kasner and in 1933 by G. Lemaître without considering the Bianchi groups)

$$ds^2 = -dt^2 + t^{2p_1}dx^2 + t^{2p_2}dy^2 + t^{2p_3}dz^2, \tag{100}$$

where

$$p_1 + p_2 + p_3 = 1,\ p_1^2 + p_2^2 + p_3^2 = 1. \tag{101}$$

These metrics were first used to investigate various effects in anisotropic cosmological models. For example, in contrast to standard FRW models with "point-like" initial singularities, the Kasner metrics can permit the so called *"cigar"* and *"pancake" singularities.* To be more specific, consider the congruence of timelike lines with unit tangent vectors n^α orthogonal to constant time hypersurfaces, and define the expansion tensor $\theta_{\alpha\beta}$ by $\theta_{\alpha\beta} = \sigma_{\alpha\beta} + \frac{1}{3}\theta h_{\alpha\beta}$, where $h_{\alpha\beta} = g_{\alpha\beta} + n_\alpha n_\beta$ is a projection tensor, $\sigma_{\alpha\beta} = n_{(\alpha;\beta)} - \frac{1}{3}\theta h_{\alpha\beta}$ is the shear, and $\theta = \theta_\alpha{}^\alpha$. Determining the three spatial eigenvectors of $\theta_{\alpha\beta}$ with the corresponding eigenvalues θ_i $(i = 1, 2, 3)$, one can define the scale factors l_i by the relation $\theta_i = (dl_i/dt)/l_i$, and the Hubble scalar $H = \frac{1}{3}(\theta_1 + \theta_2 + \theta_3)$. In the FRW models, all $l_i \to 0$ at the big bang singularity. In the Kasner models at $t \to 0$ one finds that either two of the l_i go to zero, whereas the third unboundedly increases (a cigar); or one of the l_i tends to zero, while the other two approach a finite value (pancake). Also there is the *"barrel" singularity* in which the two of the l_i go to zero, and the third approaches

of the symmetry group to simplify the spatial metric g_{ab} (see [310,311] for more details).

a finite value. There is an open question as to whether some other possibilities exist [311]. Even in the perfect fluid Kasner model, the approach to the singularity is *"velocity-dominated"* – the "vacuum terms" given by the rates of change of the scale factors dominate the "matter terms" (curvature terms vanish since the Kasner models are spatially flat).

The general vacuum Bianchi type II cosmologies (with one $n^{(a)} = +1$, and the other two vanishing), discovered by Taub in [197], contain two free parameters:

$$ds^2 = -A^2 dt^2 + A^{-2}\, t^{2p_1}(dx + 4p_1 bz\, dy)^2 + A^2(t^{2p_2}\, dy^2 + t^{2p_3}\, dz^2), \quad (102)$$

where

$$A^2 = 1 + b^2 t^{4p_1}, \;\; p_1 + p_2 + p_3 = 1, \;\; p_1^2 + p_2^2 + p_3^2 = 1. \quad (103)$$

If we put the parameter $b = 0$, the metrics (102) become the Kasner solutions (100). Near the big bang the general Bianchi type II solution is asymptotic to a Kasner model. In the future it is asymptotic again to a Kasner model, but with different values of parameters p_i (see e.g. [311]). This fact will be important in the following.

The general Bianchi type V vacuum solutions are also known – these are given by the 1-parameter family of Joseph solutions [311]. The type V models are the simplest metrics in class B (with all $n^{(a)} = 0$ but $C^a_{bc} = 2a_{[b}\delta^a_{c]}$, $a_b =$ constant), and are the simplest Bianchi models which contain the standard FRW open universes ($k = -1$). The Joseph solutions are asymptotic to the specific Kasner solution in the past, and tend to the *"isotropic Milne model"* in the future. This is intuitively understandable since open FRW models, as they expand indefinitely into the future with matter density decreasing, also approach the Milne model. As is well known, the Milne model is just an empty flat (Minkowski) spacetime in coordinates adapted to homogeneous spacelike hypersurfaces (the "mass hyperboloids"), with expanding normals (see e.g. [28]):

$$ds^2 = -d\tau^2 + \tau^2\left[(1+\rho^2)^{-1} d\rho^2 + \rho^2(d\theta^2 + \sin^2\theta\, d\varphi^2)\right], \quad (104)$$

with $\tau = t(1-u^2)^{1/2}$, $\rho = u(1-u^2)$, $u = r/t < 1$, where t, r, θ, φ are standard Minkowski (spherical) coordinates. Because of its significance as an asymptotic solution and its simplicity, the Milne model has been used frequently in pedagogical expositions of relativistic cosmology (see e.g. [28,214]) as well as in cosmological perturbation theory and quantization (see e.g. [315] and references therein). The Milne universe is also an asymptotic state of other Bianchi models such as, for example, the intriguing *Lukash vacuum* type VII$_h$ *solution* [316], which can be interpreted as two monochromatic, circularly polarized waves of time-dependent amplitude travelling in opposite directions on a FRW background, with flat or negative curvature spacelike sections. As was noticed earlier, some indefinitely expanding Bianchi models approach the homogeneous plane wave solutions. Barrow and Sonoda [317] studied the future asymptotic behaviour of the known Bianchi solutions in detail by using nonlinear stability techniques; and in [311] dynamical system methods were used.

From the late 1960s onwards the greatest amount of work was probably devoted to the Bianchi type IX vacuum models, baptized *the Mixmaster universe*[25] by Misner [312]. Type IX models are the most general class A models with all parameters $n^{(a)} = +1$. They are the only Bianchi universes which recollapse. If a perfect fluid is permitted as the matter source, the non-vacuum type IX solutions contain the closed FRW models ($k = +1$) with space sections having spherical topology. As a Bianchi I space admits a group isomorphic with translations in a 3-dimensional Euclidean space, the group of type IX spaces is isomorphic to the group of rotations. None of the pairs of three Killing vectors commute. A general Bianchi IX vacuum solution is not known, but a particular solution is available: the Taub-NUT spacetime, or rather, its spatially homogeneous anisotropic region – the Taub universe (see Sect. 7.2). This fact was, for example, employed in an attempt to understand the limitations of the minisuperspace methods of quantum gravity: by reducing the degrees of freedom to a general Mixmaster universe and then further to the Taub universe one can see what such restrictions imply [318].

The dynamics of general Bianchi cosmologies – and of the Mixmaster models in particular – close to the big bang singularity has been approached with essentially three methods [311]: (i) piecewise approximation methods, (ii) Hamiltonian methods, and (iii) dynamical system methods. In the first method, used primarily by Russian cosmologists (cf. [299,300]), the evolution is considered to be a sequence of periods in which certain terms in the Einstein equations dominate whereas other terms can be neglected. The Hamiltonian methods appeared first in the "Mixmaster paper" by Misner [312], were reviewed by Ryan [266], and more recently by Uggla in [311]. With the Hamiltonian (canonical) approaches, minisuperspace methods entered general relativity (cf. Sect. 9.2 on midisuperspace for cylindrical waves). In this approach, infinitely many degrees of freedom are reduced to a finite number: the state of the universe is described by a "particle" moving inside and reflecting instantaneously from the moving potential walls, which approximate the time-dependent potentials in the Hamiltonian. In the third method one employs the fact that Einstein's equations in the case of homogeneous cosmologies can be put into the form of an autonomous system of first-order (ordinary) differential equations on a finite dimensional space $\mathbb{R}^n$. This is of the form $d\mathbf{x}/dt = \mathbf{f}(\mathbf{x})$, with $\mathbf{x} \in \mathbb{R}^n$ representing a state of the model (for example, the suitably normalized components of the shear σ, the Hubble

[25] The name comes from the fact that, in contrast to a standard FRW model, which has a horizon preventing the equalization of possible initial inhomogeneities over large scales, the horizon in a type IX universe is absent, so that mixing is in principle possible. However, as was shown e.g. in [309], "repeated circumnavigations of the universe by light are impossible in the Mixmaster model".

scalar θ, and parameters related to $n^{(a)}$, can serve as the "components" of $\mathbf{x}$). A study of the orbits $\mathbf{x}(t)$ indicates the behaviour of the model. Dynamical system methods are the focus of the book [311]. They also are the main tools of the monograph [319].

In the case of the Bianchi IX models (either vacuum or with perfect fluid), all three methods imply (though do not supply a rigorous proof) that an approach to the past big bang singularity is composed of an infinite sequence of intervals, in each of which the universe behaves approximately as a specific Kasner model (100). The transition "regimes" between two different subsequent Kasner epochs, in which the contraction proceeds along subsequently different axes, is approximately described by Bianchi type II vacuum solutions (102). This famous and enigmatic *"oscillatory approach to the singularity"* (or "Mixmaster behaviour") has rightly entered the classical literature (cf. e.g. [18,19,139]). It indicates that the big bang singularity (and, similarly, a singularity formed during a gravitational collapse) can be much more complicated than the "point-like" singularity in the standard FRW models. This oscillatory character has been suggested not only by the qualitative methods mentioned above, but also by extensive numerical work (see e.g. [311,320]). So far, however, it has resisted a rigorous proof.

In the "standard" picture of the Mixmaster model it is supposed that the evolution of the Bianchi type IX universe near the singularity can be approximated by a mapping of the so called *Kasner circle* onto itself. This is the unit circle in the $\Sigma_+\Sigma_-$ plane, where $\Sigma_\pm = \sigma_\pm/H$ describes the anisotropy in the Hubble flow (cf. e.g. Fig. 6.2 in [311]). Each point on the circle corresponds to a specific Kasner solution with given fixed values of parameters p_i satisfying the conditions (101). There are three exceptional points on the circle – those at which one of the $p_i = +1$, and the other two vanish. From each non-exceptional point P_1 on the Kasner circle there leads a 1-dimensional unstable orbit given by the vacuum Bianchi II solution (102), which joins P_1 to another point P_2 on the circle, then P_2 is mapped to P_3 , etc. This "Kasner map" in the terminology of [311], called frequently also the BKL (Belinsky-Khalatnikov-Lifshitz) map, describes subsequent changes of Kasner epochs during the oscillatory approach to a singularity. Recent rigorous results of Rendall [321] show that for any *finite* sequence generated by the BKL map, there exists a vacuum Bianchi type IX solution which reproduces the sequence with any required accuracy.[26]

The vacuum Bianchi IX models have been extensively analyzed in the context of deterministic chaos and their stochasticity, attracting the interest of leading experts in these fields [320,322]. Above all, it is the numerical work which strongly suggests that it is impossible to make long-time predictions of

[26] A. Rendall (private communication) reports that the main points of the BKL picture for homogeneous universes have been rigorously confirmed in a recent work of H. Ringström (to be published).

the evolution of the system from the initial data, which is the most significant property of a chaotic system.

Most recently, interest in the Bianchi cosmologies with (homogeneous) magnetic and scalar fields has been revived. Following his previous work with Wainwright and Kerr on magnetic Bianchi VI_0 cosmologies [323], LeBlanc [324] has shown that even in Bianchi I cosmologies one finds an oscillatory approach towards the initial singularity if a magnetic field in a general direction is present. (The points on the Kasner circle are now joined by Rosen magneto-vacuum solutions.) Hence, Mixmaster-like oscillations occur due to the magnetic field degrees of freedom, even in the absence of an anisotropic spatial curvature (present in the vacuum type IX models) – the result anticipated by Jantzen [325] in his detailed work on Hamiltonian methods for Bianchi cosmologies with magnetic and scalar fields. Similar conclusions have also been arrived at in [326] for magnetic Bianchi II cosmologies. (LeBlanc's papers contain some new exact magnetic Bianchi solutions and a number of references to previous work.) Interestingly, in contrast to the magnetic field, scalar fields in general *suppress* the Mixmaster oscillations when approaching the initial singularity [327,328].

The theory of spatially homogeneous, anisotropic models is an elegant, intriguing branch of mathematical physics. It has played an important role in general relativity. The classical monograph of Zel'dovich and Novikov [309], or the new volume of Wainwright and Ellis [311] analyze in detail the possible observational relevance of these models: they point out spacetimes close to FRW cosmologies (at least during an epoch of finite duration) which are compatible with observational data. For the most recent work on Bianchi VII_h cosmologies which are potentially compatible with the highly isotropic microwave background radiation, see [329] (and references therein). Nevertheless, the present status is such that, in contrast to for example the Kerr solution, which is becoming an increasingly strong attractor for practical astrophysicists (cf. Sect. 4.3), the anisotropic models have not really entered (astro)physical cosmology so far. Peebles, for example, briefly comments in [28]: "The homogeneous anisotropic solutions allowed by general relativity are a very useful tool for the study of departures from the Robertson-Walker line element. As a realistic model for our Universe, however, these solutions seem to be of limited interest, for they require very special initial conditions: if the physics of the early universe allowed appreciable shear, why would it not also allow appreciable inhomogeneities?"

An immediate reaction, of course, would be to point out that the FRW models require still more "special initial conditions". However, there appears to be a deeper reason why the oscillatory approach towards a singularity may be of fundamental importance. Belinsky, Khalatnikov and Lifshitz [299,300] employed their piecewise approximation method, and concluded 30 years ago that a singularity in a *general, inhomogeneous* cosmological model is spacelike and locally oscillatory: i.e. in their scenario, the evolution at different spatial point decouples. At each spatial point the universe approaches the singularity as a distinct Mixmaster universe. This view, often criticized by purists, appears now to be gaining an increasing number of converts, even among the most rigorous of relativists. As mentioned above, the homogeneous magnetic Bianchi type VI_0 models, investigated by LeBlanc et al., show Mixmaster behaviour. The Bianchi VI_0 models have, as do all Bianchi models, three Killing vectors, but two of them commute. The models can thus be generalized by relaxing the symmetry connected with the third Killing vector; one can so obtain effectively the inhomogeneous (in one dimension) Gowdy-type spacetimes. Weaver, Isenberg and Berger [330], following this idea of Rendall, analyzed these models numerically, and discovered that the Mixmaster behaviour is reached at different spatial points. The numerical evidence for an oscillatory singularity in a generic vacuum $U(1)$ symmetric cosmologies with the spatial topology of a 3-torus has been found still more recently by Berger and Moncrief [331].

Before turning to the Gowdy models, a last word on the "oscillatory approach towards singularity". I heard E. M. Lifshitz giving a talk on this issue a couple of times, with Ya. B. Zel'dovich in the audience. In discussions after the talk, Zel'dovich, who appreciated much this work (its detailed description is included in [309]), could not resist pointing out that the number of oscillations and Kasner epochs will be very limited (to only about ten) because of quantum effects which arise when some scale of a model is smaller than the Planck length $l_{Pl} \sim 10^{-33}$cm. This, however, seems to make the scenario still more intriguing. If this is confirmed rigorously within classical relativity, how will a future quantum gravity modify this picture?

12.2 Inhomogeneous Cosmologies

Among all of the known vacuum inhomogeneous models, the *Gowdy solutions* [332] have undoubtedly played the most distinct role. They belong to the class of solutions with two commuting spacelike Killing vectors. Within a cosmological context, they form a subclass of a wider class of G_2 *cosmologies* – as are now commonly denoted models which admit an Abelian group G_2 of isometries with orbits being spacelike 2-surfaces. A 2-surface with a 2-parameter isometry group must be a space of constant curvature, and since neither a 2-sphere nor a 2-hyperboloid possess 2-parameter subgroups, it must be intrinsically flat. If the 2-surface is an Euclidean plane or a cylinder, then one speaks about planar or cylindrical universes. Gowdy universes are compact – the group orbits are 2-tori T^2.

The metrics with two spacelike Killing vectors are often called the generalized Einstein–Rosen metrics as, for example, by Carmeli, Charach and Malin [333] in their comprehensive survey of inhomogeneous cosmological models of this type. In dimensionless coordinates (t, z, x^1, x^2), the line element can be written as $(A, B = 1, 2)$

$$ds^2/L^2 = e^F(-dt^2 + dz^2) + \gamma_{AB} dx^A dx^B, \qquad (105)$$

where L is a constant length, F and γ_{AB} depend on t and z only, and thus the spacelike Killing vectors are ${}^{(1)}\xi^\alpha = (0,0,1,0)$, ${}^{(2)}\xi^\alpha = (0,0,0,1)$.

The local behaviour of the solutions of this form is described by the gradient of the "volume element" of the group orbits $W = (|\det(\gamma_{AB})|)^{1/2}$. Classical cylindrical Einstein–Rosen waves (cf. Sect. 9) are obtained if $W_{,\alpha}$ is globally spacelike. In Gowdy models, $W_{,\alpha}$ varies from one region to another.[27]

Considering for simplicity the polarized Gowdy models (when the Killing vectors are hypersurface orthogonal), the metric (105) can be written in diagonal form (cf. equations (62), (63), and (65), (66) in the analogous cases of plane and cylindrical waves)

$$ds^2/L^2 = e^{-2U}\left[e^{2\gamma}(-dt^2 + dz^2) + W^2 dy^2\right] + e^{2U} dx^2, \qquad (106)$$

in which $U(t,z)$ and $\gamma(t,z)$ satisfy wavelike dynamical equations and constraints following from the vacuum Einstein equations; the function $W(t,z)$, which determines the volume element of the group orbit, can be cast into a standard form which depends on the topology of t = constant spacelike hypersurfaces Σ.

As mentioned above, in Gowdy models one assumes these hypersurfaces to be compact. Gowdy [332] has shown that Σ can topologically be (i) a 3-torus $T^3 = S^1 \otimes S^1 \otimes S^1$ and $W = t$ (except for the trivial case when spacetime is identified as a Minkowski space), (ii) a 3-handle (or hypertorus, or "closed wormhole") $S^1 \otimes S^2$ with $W = \sin z \sin t$, or (iii) a 3-sphere S^3, again with $W = \sin z \sin t$. (For some subtle cases not covered by Gowdy, see [334].) As the form of W suggests, in the case of a T^3 topology, the universe starts with a big bang singularity at $t = 0$ and then expands indefinitely, whereas in the other two cases it starts with a big bang at $t = 0$, expands to some maximal volume, and then recollapses to a "big crunch" singularity at $t = \pi$. One can determine exact solutions for metric functions in all three cases in terms of Bessel functions [335]. Hence, for the first time *cosmological models closed by gravitational waves* were constructed. Charach found Gowdy universes with some special electromagnetic fields [336], and other generalized Gowdy models were obtained. We refer to the detailed survey [333] for more information, including the work on canonical and quantum treatments of these models, done at the beginning of the 1970s by Berger and Misner, and for extensive references.

Let us only add a few remarks on some more recent developments in which the Gowdy models have played a role. Gowdy-type models have been used to study the *propagation and collision of gravitational waves with toroidal wavefronts* (as mentioned earlier, 2-tori T^2 are the group orbits in the Gowdy

[27] The same is true in the boost-rotation symmetric spacetimes considered in Sect. 11: the part $t^2 > z^2$ of the spacetimes, where the boost Killing vector is spacelike, can be divided into four different regions, in two of which vector $W_{,\alpha}$ is spacelike, and in the other two timelike – see [288] for details.

cosmologies) in the FRW closed universes with a stiff fluid [337]. In the standard Gowdy spacetimes it is assumed that the "twists" associated with the isometry group on T^2 vanish. In [338] the generalized Gowdy models without this assumption are considered, and their global time existence is proved.

As both interesting and non-trivial models, the Gowdy spacetimes have recently attracted the attention of mathematical and numerical relativists with an increasing intensity, as indicated already at the end of the previous section. Chruściel, Isenberg and Moncrief [339] proved that Gowdy spacetimes developed from a dense subset in the initial data set cannot be extended past their singularities, i.e. in "most" Gowdy models the strong cosmic censorship is satisfied.

On cosmic censorship and spacetime singularities, especially in the context of compact cosmologies, we refer to a review by Moncrief [340], based on his lecture in the GR14 conference in Florence in 1995. The review shows clearly how intuition gained from such solutions as the Gowdy models or the Taub-NUT spaces, when combined with new mathematical ideas and techniques, can produce rigorous results with a generality out of reach until recently. To such results belongs also the very recent work of Kichenassamy and Rendall [341] on the sufficiently general class of solutions (containing the maximum number of arbitrary functions) representing unpolarized Gowdy spacetimes. The new mathematical technique, developed by Kichenassamy [342], the so called Fuchsian algorithm, enables one to construct singular (and nonsingular) solutions of partial differential equations with a large number of arbitrary functions, and thus provide a description of singularities. Applying the Fuchsian algorithm to Einstein's equations for Gowdy spacetimes with topology T^3, Kichenassamy and Rendall have proved that general solutions behave at the (past) singularity in a Kasner-like manner, i.e. they are asymptotically velocity dominated with a diverging Kretschmann (curvature) invariant. One needs an additional magnetic field not aligned with the two Killing vectors of the Gowdy unpolarized spacetimes in order to get a general oscillatory (Mixmaster) approach to a singularity, as shown by the numerical calculations [330] mentioned at the end of the previous section.

Much of the work on exact inhomogeneous vacuum cosmological models has been related to "large perturbations" of Bianchi universes. In [343] the authors confined attention to "plane wave" solutions propagating over Bianchi backgrounds of types I-VII. They found universes which are highly inhomogeneous and "chaotic" at early times, but are transformed into clearly "recognizable" gravitational waves at late times.

Other types of metrics can be considered as exact *"gravitational solitons"* propagating on a cosmological background. These are usually obtained by applying the inverse scattering or "soliton" technique of Belinsky and Zakharov [344] to particular solutions of Einstein's equations as "seeds". For example, Carr and Verdaguer [345] found gravisolitons by applying the technique to the homogeneous Kasner seed. Similarly to previous work [343], their models are very inhomogeneous at early times, but evolve towards homogeneity in a wavelike manner at late times.

More recently, Belinsky [346], by applying a two-soliton inverse scattering technique to a Bianchi type VI_0 solution as a seed, constructed an intriguing solution which he christened as a *"gravitational breather"*, in analogy with the Gordon breather in the soliton theory of the sine-Gordon equation. Gravisolitons and antigravisolitons, characterized by an opposite topological charge, can be heuristically introduced and shown to have an attractive interaction. The breather is a bound state of the gravisoliton and antigravisoliton. Belinsky suggests that a time oscillating breather exists; but a later discussion [347] indicates that the oscillations quickly decay. Alekseev, by employing his generalization of the inverse scattering method to the Einstein–Maxwell theory, obtained exact electrovacuum solutions generalizing Belinsky's breather (see his review [348], containing a general introduction on exact solutions).

Verdaguer [349] prepared a very complete review of solitonic solutions admitting two spacelike Killing vector fields, with the main emphasis on cosmological models. Among various aspects of such solutions, he has noted the role of the Bel-Robinson superenergy tensor in the interpretation of cosmological metrics. This tensor and its higher-order generalizations has also been significantly used in estimates in the proofs of long-time existence theorems [39,340]. Recently, differential conservation laws for large perturbations of gravitational field with respect to a given curved background have been fomulated [350], which found an application in solving equations for cosmological perturbations corresponding to topological defects [351]. They should bring more light also on various solitonic models in cosmology.

13 Concluding Remarks

It is hoped that the preceding pages have helped to elucidate at least one issue: that in such a complicated nonlinear theory as general relativity, it is not possible to ask relevant questions of a general character without finding and thoroughly analyzing specific exact solutions of its field equations. The role of some of the solutions in our understanding of gravity and the universe has been so many-sided that to exhibit this role properly on even more than a hundred pages is not really feasible ...

Although we have concentrated on only (electro)vacuum solutions, there remains a number of such solutions that have also played some role in various contexts, but, owing to the absence of additional space and time, or the presence of the author's ignorance, have not been discussed. Tomimatsu-Sato solutions and their generalizations, static plane and cylindrical metrics, and some algebraically special solutions are examples.

In his review of exact solutions, Ehlers [56] wrote 35 years ago that "it seems desirable to construct material sources for vacuum solutions", and 30 years later Bonnor [64], in his review, expressed a similar view. In the above we have noted only some of the thin disk sources of static and stationary spacetimes in Sect. 6. To find physically reasonable material sources for many of the known vacuum solutions remains a difficult open task. In order to make solutions of Einstein's equations with the right-hand side more tractable, one is often tempted to sacrifice realism and consider materials, again using Bondi's phraseology, which are not easy to buy in the shops. Nevertheless, there are solutions representing spacetimes filled with matter which would certainly belong in a more complete discussion of the role of exact solutions.

For example, one of the simplest, the spherically symmetric Schwarzschild interior solution with an incompressible fluid as matter source, modelling "a star of uniform density", gives surprisingly good estimates of an upper bound on the masses of neutron stars; on a more general level, it supplies an instructive example of relativistic hydrostatics [18]. Many other spherical perfect fluid solutions are listed in [61]. The proof of a very plausible fact that any equilibrium, isolated stellar model which is nonrotating must be spherically symmetric, was finally completed in [352] and [353]. Physically more adequate spherically symmetric static solutions with collisionless matter described by the Boltzmann (Vlasov) equation have been studied [101] (yielding, for example, arbitrarily large central redshifts); and some of their aspects have been recently reviewed from a rigorous, mathematical point of view [354]. Going over to the description of matter in terms of physical fields, we should mention the first spherically symmetric regular solutions of the Einstein–Yang–Mills equations ("non-Abelian solitons" discovered by Bartnik and McKinnon [355] in 1988), and non-Abelian black holes with "hair", which were found soon afterwards. They stimulated a remarkable activity in the search for models in which gravity is coupled with Yang–Mills, Higgs, and Skyrmion fields. Very recently these solutions have been surveyed in detail in the review by Volkov and Gal'tsov [356].

The role of the standard FRW cosmological models on the development of relativity and cosmology can hardly be overemphasized. As for two more recent examples of this influence let us just recall that the existence of cosmological horizons in these models was one of the crucial points which inspired the birth of inflationary cosmology (see e.g. [28]); and the very smooth character of the initial singularity has led Penrose [102] to formulate his Weyl curvature hypothesis, related to a still unclear concept of gravitational entropy. Homogeneous but anisotropic Bianchi models filled with perfect fluid are extensively analyzed in [311]. Very recent studies of Bianchi models with collisionless matter [357] reveal how the matter content can qualitatively alter the character of the model.

A number of Bianchi models approach self-similar solutions. Perfect fluid solutions admitting a homothetic vector, which in this case implies both geometrical and physical self-similarity, have been reviewed most recently by Carr and Coley [358]. In their review various astrophysical and cosmological applications of such solutions are also discussed. Self-similar solutions have played a crucial role in the critical phenomena in gravitational collapse. Since their discovery by Choptuik in 1993, they have attracted much effort, which has revealed quite unexpected facts. In [358] these phenomena are analyzed briefly. For a more comprehensive review, see [359].

Self-similar, spherically symmetric solutions have been very relevant in constructing examples of the formation of naked singularities in gravitational collapse (see [358] for a brief summary and references). In particular, the inhomogeneous, spherically symmetric Lemaître–Bondi–Tolman universes containing dust have been employed in this context. Solutions with null dust should be mentioned as well, especially the spherically symmetric Vaidya solutions: imploding spherical null-dust models have been constructed in which naked singularities arise at their centre (see [32] for summary and references).

The Lemaître–Bondi–Tolman models are the most frequently analyzed inhomogeneous cosmological models which contain the standard FRW dust models as special cases (see e.g. [28,32]). In his recent book Krasiński [360] has compiled and discussed most if not all of these exact inhomogeneous cosmological solutions found so far which can be viewed as "exact perturbations" of the FRW models.

Many solutions known already still wait for their role to be uncovered. The role of many others may forever remain just in their "being". However, even if new solutions of a "Kerr-like significance" will not be obtained in the near future, we believe that one should not cease in embarking upon journeys for finding them, and perhaps even more importantly, for revealing new roles of solutions already known. The roads may not be easy, but with todays equipment like Maple or Mathematica, the speed is increasing. Is there another so explicit way of how to learn more about the rich possibilities embodied in Einstein's field equations?

The most remarkable figure of Czech symbolism, Otokar Březina (1868-1929) has consoling words for those who do not meet the "Kerr-type" metric on the road: "Nothing is lost in the world of the spirit; even a stone thrown away may find its place in the hands of a builder, and a house in flames may save the life of someone who has lost his way...".

Acknowledgements

Interaction with Jürgen Ehlers has been important for me over the years: Thanks to my regular visits to his group, which started seven years before the hardly penetrable barrier between Prague and the West disappeared, I have been in contact with "what is going on" much more than I could have been at home. Many discussions with Jürgen, collaboration and frequent discussions with Bernd Schmidt, and with other members of Munich→Potsdam→Golm relativity group are fondly recalled and appreciated.

For helpful comments on various parts of the manuscript I am grateful to Bobby Beig, Jerry Griffiths, Petr Hájíček, Karel Kuchař, Malcolm MacCallum, Alan Rendall and Bernd Schmidt. For discussions and help with references I thank also Piotr Chruściel, Andy Fabian, Joseph Katz, Jorma Louko, Donald Lynden-Bell, Reinhard Meinel, Vince Moncrief, Gernot Neugebauer, Martin Rees, Carlo Ungarelli, Marsha Weaver, and my Prague colleagues. Peter Williams kindly corrected my worst Czechisms. My many thanks go to Eva Kotalíková for her patience and skill in technical help with the long manuscript. Very special thanks to Tomáš Ledvinka: he prepared all the figures and provided long-standing technical help and admirable speed, without which the manuscript would certainly not have been finished in the required time and form. Support from the Albert Einstein Institute and from the grant No. GAČR 202/99/0261 of the Czech Republic is gratefully acknowledged.

References

1. Feynman, R. (1992) The Character of Physical Law, Penguin books edition, with Introduction by Paul Davies; the original edition published in 1965
2. Hartle, J. B., Hawking, S. W. (1983) Wave function of the Universe, Phys. Rev. **D28**, 2960. For more recent developments, see Page, D. N. (1991) Minisuperspaces with conformally and minimally coupled scalar fields, J. Math. Phys. **32**, 3427, and references therein
3. Kuchař, K. V. (1994) private communication based on unpublished calculations. See also Peleg, Y. (1995) The spectrum of quantum dust black holes, Phys. Lett. **B356**, 462
4. Chandrasekhar, S. (1987) Ellipsoidal Figures of Equilibrium, Dover paperback edition, Dover Publ., Mineola, N. Y.
5. Tassoul, J.-L. (1978) Theory of Rotating Stars, Princeton University Press, Princeton, N. J.
6. Binney, J., Tremaine, S. (1987) Galactic Dynamics, Princeton University Press, Princeton. The idea first appeared in the work of Kuzmin, G. G. (1956) Astr. Zh. **33**, 27
7. Taniguchi, K. (1999) Irrotational and Incompressible Binary Systems in the First post-Newtonian Approximation of General Relativity, Progr. Theor. Phys. **101**, 283. For an extensive review, see Taniguchi, K. (1999) Ellipsoidal Figures of Equilibrium in the First post-Newtonian Approximation of General Relativity, Thesis, Department of Physics, Kyoto University
8. Ablowitz, M. J., Clarkson, P. A. (1991) Solitons, Nonlinear Evolution Equations and Inverse Scattering, London Mathematical Society, Lecture Notes in Mathematics **149**, Cambridge University Press, Cambridge
9. Mason, L. J., Woodhouse, N. M. J. (1996) Integrability, Self-Duality, and Twistor Theory, Clarendon Press, Oxford
10. Atiyah, M. (1998) Roger Penrose – A Personal Appreciation, in The Geometric Universe: Science, Geometry, and the work of Roger Penrose, eds. S. A. Hugget, L. J. Mason, K. P. Tod, S. T. Tsou and N. M. J. Woodhouse, Oxford University Press, Oxford
11. Bičák, J. (1989) Einstein's Prague articles on gravitation, in Proceedings of the 5th M. Grossmann Meeting on General Relativity, eds. D. G. Blair and M. J. Buckingham, World Scientific, Singapore. A more detailed technical account is given in Bičák, J. (1979) Einstein's route to the general theory of relativity (in Czech), Čs. čas. fyz. **A29**, 222

12. Einstein, A. (1912) Relativity and Gravitation. Reply to a Comment by M. Abraham (in German), Ann. der Physik **38**, 1059
13. Einstein, A., Grossmann, M. (1913) Outline of a Generalized Theory of Relativity and of a Theory of Gravitation (in German), Teubner, Leipzig; reprinted in Zeits. f. Math. und Physik **62**, 225
14. Einstein, A., Grossmann, M. (1914) Covariance Properties of the Field Equations of the Theory of Gravitation Based on the Generalized Theory of Relativity (in German), Zeits. f. Math. und Physik **63**, 215
15. Pais, A. (1982) 'Subtle is the Lord...' – The Science and the Life of Albert Einstein, Clarendon Press, Oxford
16. Einstein, A. (1915) The Field Equations of Gravitation (in German), König. Preuss. Akad. Wiss. (Berlin) Sitzungsberichte, 844
17. Corry, L., Renn, J. and Stachel, J. (1997) Belated Decision in the Hilbert-Einstein Priority Dispute, Science **278**, 1270
18. Misner, C., Thorne, K. S. and Wheeler, J. A. (1973) Gravitation, W. H. Freeman and Co., San Francisco
19. Wald, R. M. (1984) General Relativity, The University of Chicago Press, Chicago
20. Einstein, A. (1917) Cosmological Considerations in the General Theory of Relativity (in German), König. Preuss. Akad. Wiss. (Berlin) Sitzungsberichte, 142
21. Prosser, V., Folta, J. eds. (1991) Ernst Mach and the Development of Physics, Charles University – Karolinum, Prague
22. Barbour, J., Pfister, H. eds. (1995) Mach's Principle: From Newton's Bucket to Quantum Gravity, Birkhäuser, Boston-Basel-Berlin
23. Lynden-Bell, D., Katz, J. and Bičák J. (1995) Mach's principle from the relativistic constraint equations, Mon. Not. Roy. Astron. Soc. **272**, 150; Errata: Mon. Not. Astron. Soc. **277**, 1600
24. Hořava, P. (1999) M theory as a holographic field theory, Phys. Rev. **D59**, 046004
25. De Sitter, W. (1917) On Einstein's Theory of Gravitation, and its Astronomical Consequences, Part 3, Mon. Not. Roy. Astron. Soc. **78**, 3; see also references therein
26. Hawking, S. W., Ellis, G. F. R. (1973) The large scale structure of space-time, Cambridge University Press, Cambridge
27. Penrose, R. (1968) Structure of Space-Time, in Batelle Rencontres (1967 Lectures in Mathematics and Physics), eds. C. M. DeWitt and J. A. Wheeler, W. A. Benjamin, New York
28. Peebles, P. J. E. (1993) Principles of Physical Cosmology, Princeton University Press, Princeton
29. Bertotti, B., Balbinot, R., Bergia, S. and Messina, A. eds. (1990) Modern Cosmology in Retrospect, Cambridge University Press, Cambridge. See especially the contributions by J. Barbour, J. D. North, G. F. R. Ellis, and W. C. Seitter and H. W. Duerbeck
30. d'Inverno, R. (1992) Introducing Einstein's Relativity, Clarendon Press, Oxford
31. Geroch, R., Horowitz, G. T. (1979) Global structure of spacetimes, in General Relativity, An Einstein Centenary Survey, eds. S. W. Hawking and W. Israel, Cambridge University Press, Cambridge

32. Joshi, P. S. (1993) Global Aspects in Gravitation and Cosmology, Oxford University Press, Oxford
33. Schmidt, H. J. (1993) On the de Sitter space-time – the geometric foundation of inflationary cosmology, Fortschr. d. Physik **41**, 179
34. Eriksen, E., Grøn, O. (1995) The de Sitter universe models, Int. J. Mod. Phys. **4**, 115
35. Bousso, R. (1998) Proliferation of de Sitter space, Phys. Rev. **D58**, 083511; see also Bousso, R. (1999) Quantum global structure of de Sitter space, Phys. Rev. **D60**, 063503
36. Maldacena, J. (1998) The large N limit of superconformal field theories and supergravity, Adv. Theor. Math. Phys. **2**, 231
37. Balasubramanian, V., Kraus, P. and Lawrence, B. (1999) Bulk versus boundary dynamics in anti-de Sitter spacetime, Phys. Rev. **D59**, 046003
38. Veneziano, G. (1991) Scale factor duality for classical and quantum string, Phys. Lett. **B265**, 287; Gasperini, M., Veneziano, G. (1993) Pre-big bang in string cosmology, Astropart. Phys. **1**, 317. For the most recent review, in which also some answers to the critism of the pre-big-bang scenario and possible observational tests can be found, see Veneziano, G. (1999) Inflating, warming up, and probing the pre-bangian universe, hep-th/9902097
39. Christodoulou, D., Klainerman, S. (1994) The Global Nonlinear Stability of the Minkowski Spacetime, Princeton University Press, Princeton
40. Bičák, J. (1997) Radiative spacetimes: Exact approaches, in Relativistic Gravitation and Gravitational Radiation (Proceedings of the Les Houches School of Physics), eds. J.-A. Marck and J.-P. Lasota, Cambridge University Press, Cambridge
41. Friedrich, H. (1986) On the existence of n-geodesically complete or future complete solutions of Einstein's field equations with smooth asymptotic structure, Commun. Math. Phys. **107**, 587
42. Friedrich, H. (1995) Einstein equations and conformal structure: existence of anti-de Sitter-type space-times, J. Geom. Phys. **17**, 125
43. Friedrich, H. (1998) Einstein's Equation and Geometric Asymptotics, in Gravitation and Relativity: At the turn of the Millenium (Proceedings of the GR-15 conference), eds. N. Dadhich and J. Narlikar, Inter-University Centre for Astronomy and Astrophysics Press, Pune
44. Møller, C. (1972) The theory of Relativity, Second Edition, Clarendon Press, Oxford
45. Synge, J. L. (1960) Relativity: The General Theory, North-Holland, Amsterdam
46. Ehlers, J., Pirani, F. A. E. and Schild, A. (1972) The geometry of free-fall and light propagation, in General Relativity, Papers in Honor of J. L. Synge, ed. L. O. O'Raifeartaigh, Oxford University Press, London
47. Majer, U., Schmidt, H.-J. eds. (1994) Semantical Aspects of Spacetime Theories, BI-Wissenschaftsverlag, Mannheim, Leipzig, Wien
48. Misner, C. (1969) Gravitational Collapse, in Brandeis Summer Institute 1968, Astrophysics and General Relativity, eds. M. S. Chrétien, S. Deser and J. Goldstein, Gordon and Breach, New York
49. Hájíček, P. (1999) Choice of gauge in quantum gravity, in Proc. of the 19th Texas symposium on relativistic astrophysics, Paris 1998, to be published; gr-qc/9903089

50. Ehlers, J. (1981) Christoffel's Work on the Equivalence Problem for Riemannian Spaces and Its Importance for Modern Field Theories of Physics, in E. B. Christoffel: The Influence of His Work on Mathematics and the Physical Sciences, eds. P. L. Butzer, F. Fehér, Birkhäuser Verlag, Basel
51. Karlhede, A. (1980) A review of the geometrical equivalence of metrics in general relativity, Gen. Rel. Grav. **12**, 693
52. Paiva, F. M., Rebouças, M. J. and MacCallum, M. A. H. (1993) On limits of spacetimes – a coordinate-free approach, Class. Quantum Grav. **10**, 1165
53. Ehlers, J., Kundt, K. (1962) Exact Solutions of the Gravitational Field Equations, in Gravitation: an introduction to current research, ed. L. Witten, J. Wiley&Sons, New York
54. Ehlers, J. (1957) Konstruktionen und Charakterisierungen von Lösungen der Einsteinschen Gravitationsfeldgleichungen, Dissertation, Hamburg
55. Ehlers, J. (1962) Transformations of static exterior solutions of Einstein's gravitational field equations into different solutions by means of conformal mappings, in Les Théories Relativistes de la Gravitation, eds. M. A. Lichnerowicz, M. A. Tonnelat, CNRS, Paris
56. Ehlers, J. (1965) Exact solutions, in International Conference on Relativistic Theories of Gravitation, Vol. II, London (mimeographed)
57. Jordan, P., Ehlers, J. and Kundt, W. (1960) Strenge Lösungen der Feldgleichungen der Allgemeinen Relativitätstheorie, Akad. Wiss. Lit. Mainz, Abh. Math. Naturwiss. Kl., Nr. 2
58. Jordan, P., Ehlers, J. and Sachs, R. K. (1961) Beiträge zur Theorie der reinen Gravitationsstrahlung, Akad. Wiss. Lit. Mainz, Abh. Math. Naturwiss. Kl., Nr. 1
59. Chandrasekhar, S. (1986) The Aesthetic Base of the General Theory of Relativity. The Karl Schwarzschild lecture, reprinted in Chandrasekhar, S. (1989) Truth and Beauty, Aesthetics and Motivations in Science, The University of Chicago Press, Chicago
60. Chandrasekhar, S. (1975) Shakespeare, Newton, and Beethoven or Patterns of Creativity. The Nora and Edward Ryerson Lecture, reprinted in Chandrasekhar, S. (1989) Truth and Beauty, Aesthetics and Motivations in Science, The University of Chicago Press, Chicago
61. Kramer, D., Stephani, H., Herlt, E. and MacCallum, M. A. H. (1980) Exact solutions of Einstein's field equations, Cambridge University Press, Cambridge
62. Penrose, R. (1999) private communication; see the paper which will appear in special issue of Class. Quantum Gravity celebrating the anniversary of the Institute of Physics
63. Einstein, A. (1950) Physics and Reality, in Out of My Later Years, Philosophical Library, New York. Originally published in the Journal of the Franklin Institute **221**, No. 3; March, 1936
64. Bonnor, W. B. (1992) Physical Interpretation of Vacuum Solutions of Einstein's Equations. Part I. Time-independent solutions, Gen. Rel. Grav. **24**, 551
65. Bonnor, W. B., Griffiths, J. B. and MacCallum, M. A. H. (1994) Physical Interpretation of Vacuum Solutions of Einstein's Equations. Part II. Time-dependent solutions, Gen. Rel. Grav. **26**, 687
66. Bondi, H., van der Burg, M. G. J. and Metzner, A. W. K. (1962) Gravitational Waves in General Relativity. VII. Waves from Axi-symmetric Isolated Systems, Proc. Roy. Soc. Lond. A **269**, 21

67. Ehlers, J. (1973) Survey of General Relativity Theory, in Relativity, Astrophysics and Cosmology, ed. W. Israel, D. Reidel, Dordrecht
68. Künzle, H. P. (1967) Construction of singularity-free spherically symmetric space-time manifolds, Proc. Roy. Soc. Lond. **A297**, 244
69. Schmidt, B. G. (1967) Isometry groups with surface-orthogonal trajectories, Zeits. f. Naturfor. **22a**, 1351
70. Israel, W. (1987) Dark stars: the evolution of an idea, in 300 years of gravitation, eds. S. W. Hawking and W. Israel, Cambridge University Press, Cambridge
71. Ciufolini, I., Wheeler, J. A. (1995) Gravitation and Inertia, Princeton University Press, Princeton
72. Will, C. M. (1996) The Confrontation between General Relativity and Experiment: A 1995 Update, in General Relativity (Proceedings of the 46th Scottish Universities Summer School in Physics), eds. G. S. Hall and J. R. Pulham, Institute of Physics Publ., Bristol
73. Schneider, P., Ehlers, J. and Falco, E. E. (1992) Gravitational Lenses, Springer-Verlag, Berlin
74. Hawking, S. W. (1973) The Event Horizon, in Black Holes (Les Houches 1972), eds. C. DeWitt and B. S. DeWitt, Gordon and Breach, New York-London-Paris
75. Thorne, K. S., Price, R. H. and MacDonald, D. A. (1986) Black Holes: The Membrane Paradigm, Yale University Press, New Haven
76. Frolov, V., Novikov, I. (1998) Physics of Black Holes, Kluwer, Dordrecht
77. Clarke, C. J. S. (1993) The Analysis of Space-Time Singularieties, Cambridge University Press, Cambridge
78. Boyer, R. H. (1969) Geodesic Killing orbits and bifurcate Killing horizons, Proc. Roy. Soc. (London) **A311**, 245
79. Carter, B. (1972) Black Hole Equilibrium States, in Black Holes (Les Houches 1972), eds. C. De Witt and B. S. De Witt, Gordon and Breach, New York-London-Paris
80. Chruściel, P. T. (1996) Uniqueness of stationary, electro-vacuum black holes revisited, Helv. Phys. Acta **69**, 529
81. Heusler, M. (1996) Black Hole Uniqueness Theorems, Cambridge University Press, Cambridge
82. Wald, R. M. (1994) Quantum Field Theory in Curved Spacetime and Black Hole Thermodynamics, The University of Chicago Press, Chicago
83. Rácz, I., Wald R. M. (1996) Global extensions of spacetimes describing asymptotic final states of black holes, Class. Quantum Grav. **13**, 539
84. Penrose, R. (1980) On Schwarzschild Causality – A Problem for "Lorentz Covariant" General Relativity, in Essays in General Relativity, eds. F. J. Tipler, Academic Press, New York
85. Weinberg, S., Gravitation and Cosmology (1972) J. Wiley, New York (see in particular Ch. 6, part 9)
86. Zel'dovich, Ya. B., Grishchuk, L. P. (1988) The general theory of relativity is correct!, Sov. Phys. Usp. **31**, 666. This very pedagogical paper contains a number of references on the field-theoretical approach to gravity
87. Ehlers, J. (1998) General Relativity as Tool for Astrophysics, in Relativistic Astrophysics, eds. H. Riffert et al., Vieweg, Braunschweig/Wiesbaden
88. Rees, M. (1998) Astrophysical Evidence for Black Holes, in Black Holes and Relativistic Stars, ed. R. M. Wald, The University of Chicago Press, Chicago

89. Menou, K., Quataert, E. and Narayan, R. (1998) Astrophysical Evidence for Black Hole Event Horizons, in Gravitation and Relativity: At the turn of the Millennium (Proceedings of the GR-15 Conference), eds. N. Dadhich and J. Narlikar, Inter-University Centre for Astronomy and Astrophysics Press, Pune; also astro-ph/9712015
90. Carr, B. J. (1996) Black Holes in Cosmology and Astrophysics, in General Relativity (Proceedings of the 46th Scottish Universities Summer School in Physics), eds. G. S. Hall and J. R. Pulham, Institute of Physics Publishing, London
91. Chandrasekhar, S. (1984) The Mathematical Theory of Black Holes, Clarendon Press, Oxford
92. Abramowicz, M. A. (1993) Inertial forces in general relativity, in The Renaissance of General Relativity and Cosmology, eds. G. Ellis, A. Lanza and J. Miller, Cambridge University Press, Cambridge
93. Semerák, O. (1998) Rotospheres in Stationary Axisymmetric Spacetimes, Ann. Phys. (N.Y.) **263**, 133; see also 69 references quoted therein
94. Feynman, R. P., Morinigo, F. B., Wagner W. G. (1995) Feynman lectures on gravitation, Addison-Wesley Publ. Co., Reading, Mass.
95. Shapiro, S. L., Teukolsky, S. A. (1983) Black Holes, White Dwarfs, and Neutron Stars, J. Wiley, New York
96. Frank, J., King, A. and Raine, D. (1992) Accretion Power in Astrophysics, 2nd edition, Cambridge University Press, Cambridge
97. Thorne, K. S. (1998) Probing Black Holes and Relativistic Stars with Gravitational Waves, in Black Holes and Relativistic Stars, ed. R. M. Wald, The University of Chicago Press, Chicago. See also lectures by E. Seidel, J. Pullin, and E. Flanagan, in Gravitation and Relativity: At the turn of the Millennium (Proceedings of the GR-15 Conference), eds. N. Dadhich and J. Narlikar, Inter-University Centre for Astronomy and Astrophysics Press, Pune
98. Pullin, J. (1998) Colliding Black Holes: Analytic Insights, in Gravitation and Relativity: At the turn of the Millennium (Proceedings of the GR-15 Conference), eds. N. Dadhich and J. Narlikar, Inter-University Centre for Astronomy and Astrophysics Press, Pune
99. Graves, J. C., Brill, D. R. (1960) Oscillatory character of Reissner-Nordström metric for an ideal charged wormhole, Phys. Rev. **120**, 1507
100. Boulware, D. G. (1973) Naked Singularities, Thin Shells, and the Reissner-Nordström Metric, Phys. Rev. **D8**, 2363
101. Zel'dovich, Ya. B., Novikov, I. D. (1971) Relativistic Astrophysics, Volume 1: Stars and Relativity, The University of Chicago Press, Chicago
102. Penrose, R. (1979) Singularities and time-asymmetry, in General Relativity, An Einstein Centenary Survey, eds. S. W. Hawking and W. Israel, Cambridge University Press, Cambridge
103. Burko, L., Ori, A. (1997) Introduction to the internal structure of black holes, in Internal Structure of Black Holes and Spacetime Singularities, eds. L. Burko and A. Ori, Inst. Phys. Publ., Bristol, and The Israel Physical Society, Jerusalem
104. Bičák, J., Dvořák, L. (1980) Stationary electromagnetic fields around black holes III. General solutions and the fields of current loops near the Reissner-Nordström black hole, Phys. Rev. **D22**, 2933
105. Moncrief, V. (1975) Gauge-invariant perturbations of Reissner-Nordström black holes, Phys. Rev. **D12**, 1526; see also references therein

106. Bičák, J. (1979) On the theories of the interacting perturbations of the Reissner-Nordström black hole, Czechosl. J. Phys. **B29**, 945
107. Bičák, J. (1972) Gravitational collapse with charge and small asymmetries, I: Scalar perturbations, Gen. Rel. Grav. **3**, 331
108. Price, R. H. (1972) Nonspherical perturbations of relativistic gravitational collapse, I: Scalar and gravitational perturbations, Phys. Rev. **D5**, 2419
109. Price, R. H. (1972) Nonspherical perturbations of relativistic gravitational collapse, II: Integer-spin, zero-rest-mass fields, Phys. Rev. **D5**, 2439
110. Bičák, J. (1980) Gravitational collapse with charge and small asymmetries, II: Interacting electromagnetic and gravitational perturbations, Gen. Rel. Grav. **12**, 195
111. Poisson, E., Israel, W. (1990) Internal structure of black holes, Phys. Rev. **D41**, 1796
112. Bonnor, W. B., Vaidya, P. C. (1970) Spherically Symmetric Radiation of Charge in Einstein-Maxwell Theory, Gen. Rel. Grav. **1**, 127
113. Chambers, C. M. (1997) The Cauchy horizon in black hole-de Sitter spacetimes, in Internal Structure of Black Holes and Spacetime Singularities, eds. L. Burko and A. Ori, Inst. Phys. Publ. Bristol, and The Israel Physical Society, Jerusalem
114. Penrose, R. (1998) The Question of Cosmic Censorship, in Black Holes and Relativistic Stars, ed. R. M. Wald, The University of Chicago Press, Chicago
115. Brady, P. R., Moss, I. G. and Myers, R. C. (1998) Cosmic Censorship: As Strong As Ever, Phys. Rev. Lett. **80**, 3432
116. Hubený, V. E. (1999) Overcharging a Black Hole and Cosmic Censorship, Phys. Rev. **D59**, 064013
117. Bičák, J. (1977) Stationary interacting fields around an extreme Reissner-Nordström black hole, Phys. Lett. **64A**, 279. See also the review Bičák, J. (1982), Perturbations of the Reissner-Nordström black hole, in the Proceedings of the Second Marcel Grossmann Meeting on General Relativity, ed. R. Ruffini, North-Holland, Amsterdam, and references therein
118. Hájíček, P. (1981) Quantum wormholes (I.) Choice of the classical solution, Nucl. Phys. **B185**, 254
119. Aichelburg, P. C., Güven, R. (1983) Remarks on the linearized superhair, Phys. Rev. **D27**, 456; and references therein
120. Schwarz, J. H., Seiberg, N. (1999) String theory, supersymmetry, unification, and all that, Rev. Mod. Phys. **71**, S112
121. Carlip, S. (1995) The (2+1)-dimensional black hole, Class. Quantum Grav. **12**, 2853
122. Myers, R. C., Perry, M. J. (1986) Black holes in higher dimensional spacetimes, Ann. Phys. (N.Y.) **172**, 304
123. Gibbons, G. W., Horowitz, G. T. and Townsend, P. K. (1995) Higher-dimensional resolution of dilatonic black-hole singularities, Class. Quantum Grav. **12**, 297
124. Horowitz, G. T., Teukolsky, S. A. (1999) Black holes, Rev. Mod. Phys. **71**, S180
125. Wald, R. M. (1998) Black Holes and Thermodynamics, in Black Holes and Relativistic Stars, ed. R. M. Wald, The University of Chicago Press, Chicago
126. Horowitz, G. T. (1998) Quantum States of Black Holes, in Black Holes and Relativistic Stars, ed. R. M. Wald, the University of Chicago Press, Chicago

127. Skenderis, K. (1999) Black holes and branes in string theory, hep-th/9901050
128. Ashtekhar, A., Baez, J., Corichi, A. and Krasnov, K. (1998) Quantum Geometry and Black Hole Entropy, Phys. Rev. Lett. **80**, 904
129. Youm, D. (1999) Black holes and solitons in string theory, Physics Reports **316**, Nos. 1-3, 1
130. Chamblin, A., Emparan, R. and Gibbons, G. W. (1998) Superconducting p-branes and extremal black holes, Phys. Rev. **D58**, 084009
131. Ernst, F. J. (1976) Removal of the nodal singularity of the C-metric, J. Math. Phys. **17**, 54; see also Ernst, F. J., Wild, W. J. (1976) Kerr black holes in a magnetic universe, J. Math. Phys. **17**, 182
132. Karas, V., Vokrouhlický, D. (1991) On interpretation of the magnetized Kerr-Newman black hole, J. Math. Phys. **32**, 714
133. Kerr, R. P. (1963) Gravitational field of a spinning mass as an example of algebraically special metrics, Phys. Rev. Lett. **11**, 237
134. Stewart, J., Walker, M. (1973) Black holes: the outside story, in Springer tracts in modern physics, Vol. **69**, Springer-Verlag, Berlin
135. Thorne, K. S. (1980) Multipole expansions of gravitational radiation, Rev. Mod. Phys. **52**, 299
136. Hansen, R. O. (1974) Multipole moments of stationary space-times, J. Math. Phys. **15**, 46
137. Beig, R., Simon, W. (1981) On the multipole expansion for stationary space-times, Proc. Roy. Soc. Lond. **A376**, 333
138. de Felice, F., Clarke, C. J. S. (1990) Relativity on curved manifolds, Cambridge University Press, Cambridge
139. Landau, L. D., Lifshitz, E. M. (1962) The Classical Theory of Fields, Pergamon Press, Oxford
140. O'Neill, B. (1994) The Geometry of Kerr Black Holes, A. K. Peters, Wellesley
141. Katz, J., Lynden-Bell, D. and Bičák, J. (1998) Instantaneous inertial frames but retarded electromagnetism in rotating relativistic collapse, Class. Quantum Grav. **15**, 3177
142. Semerák, O. (1996) Photon escape cones in the Kerr field, Helv. Phys. Acta **69**, 69
143. Bičák, J., Stuchlík, Z. (1976) The fall of the shell of dust onto a rotating black hole, Mon. Not. Roy. Astron. Soc. **175**, 381
144. Bičák, J., Semerák, O. and Hadrava, P. (1993) Collimation effects of the Kerr field, Mon. Not. Roy. Astron. Soc. **263**, 545
145. Newman, E. T., Couch, E., Chinnapared, K., Exton, A., Prakash, A. and Torrence, R. (1965) Metric of a rotating charged mass, J. Math. Phys. **6**, 918
146. Garfinkle, D., Traschen, J. (1990) Gyromagnetic ratio of a black hole, Phys. Rev. **D42**, 419
147. Bardeen, J. M. (1973) Timelike and Null Geodesics in the Kerr Metric, in Black Holes, eds. C. DeWitt and B. S. DeWitt, Gordon and Breach, New York
148. Rindler, W. (1997) The case against space dragging, Phys. Lett. **A233**, 25
149. Jantzen, R. T., Carini, P. and Bini, D. (1992) The Many Faces of Gravitoelectromagnetism, Ann. Phys. (N.Y.) **215**, 1; see also the review (1999) The Inertial Forces / Test Particle Motion Game, in the Proceedings of the 8th M. Grossmann Meeting on General Relativity, ed. T. Piran, World Scientific, Singapore

150. Karas, V., Vokrouhlický, D. (1994) Relativistic precession of the orbit of a star near a supermassive rotating black hole, Astrophys. J. **422**, 208
151. Blandford, R. D., Znajek, R. L. (1977) Electromagnetic extraction of energy from Kerr black holes, Mon. Not. Roy. Astron. Soc. **179**, 433. See also Blandford, R. (1987) Astrophysical black holes, in 300 years of gravitation, eds. S. W. Hawking and W. Israel, Cambridge University Press, Cambridge
152. Bičák, J., Janiš, V. (1985) Magnetic fluxes across black holes, Mon. Not. Roy. Astron. Soc. **212**, 899
153. Punsly, B., Coroniti, F. V. (1990) Relativistic winds from pulsar and black hole magnetospheres, Astrophys. J. **350**, 518. See also Punsly, B. (1998) High-energy gamma-ray emission from galactic Kerr-Newman black holes. The central engine, Astrophys. J. **498**, 640, and references therein
154. Abramowicz, M. (1998) private communication
155. Mirabel, I. F., Rodríguez, L. F. (1998) Microquasars in our Galaxy, Nature **392**, 673
156. Futterman, J. A. H., Handler, F. A. and Matzner, R. A. (1988) Scattering from black holes, Cambridge University Press, Cambridge
157. Bičák, J., Dvořák, L. (1976) Stationary electromagnetic fields around black holes II. General solutions and the fields of some special sources near a Kerr black hole, Gen. Rel. Grav. **7**, 959
158. Sasaki, M., Nakamura, T. (1990) Gravitational Radiation from an Extreme Kerr Black Hole, Gen. Rel. Grav. **22**, 1551; and references therein
159. Krivan, W., Price, R. H. (1999) Formation of a rotating Black Hole from a Close-Limit Head-On Collision, Phys. Rev. Lett. **82**, 1358
160. Campanelli, M., Lousto, C. O. (1999) Second order gauge invariant gravitational perturbations of a Kerr black hole, Phys. Rev. **D59**, 124022
161. Fabian, A. C. (1999) Emission lines: signatures of relativistic rotation, in Theory of Accretion Disks, eds. M. Abramowicz, G. Björnson, J. Pringle, Cambridge University Press, Cambridge
162. Ipser, J. R. (1998) Low-Frequency Oscillations of Relativistic Accretion Disks, in Relativistic Astrophysics, eds. H. Riffert et al., Vieweg, Braunschweig, Wiesbaden
163. Bičák, J., Podolský, J. (1997) The global structure of Robinson-Trautman radiative space-times with cosmological constant, Phys. Rev. **D55**, 1985
164. Hartle, J. B., Hawking, S. W. (1972) Solutions of the Einstein-Maxwell equations with many black holes, Commun. Math. Phys. **26**, 87
165. Heusler, M. (1997) On the Uniqueness of the Papapetrou-Majumdar metric, Class. Quantum Grav. **14**, L129
166. Chruściel, P. T. (1999) Towards the classification of static electro-vacuum space-times containing an asymptotically flat spacelike hypersurface with compact interior, Class. Quantum Grav. **16**, 689. See also Chruściel's very general result for the vacuum case in the preceding paper: The classification of static vacuum space-times containing an asymptotically flat spacelike hypersurface with compact interior, Class. Quantum Grav. **16**, 661
167. Kramer, D., Neugebauer, G. (1984) Bäcklund Transformations in General Relativity, in Solutions of Einstein's Equations: Techniques and Results, eds. C. Hoenselaers and W. Dietz, Lecture Notes in Physics 205, Springer-Verlag, Berlin
168. Bičák, J., Hoenselaers, C. (1985) Two equal Kerr-Newman sources in stationary equilibrium, Phys. Rev. **D31**, 2476

169. Weinstein, G. (1996) N-black hole stationary and axially symmetric solutions of the Einstein/Maxwell equations, Comm. Part. Diff. Eqs. **21**, 1389
170. Dietz, W., Hoenselaers, C. (1982) Stationary System of Two Masses Kept Apart by Their Gravitational Spin-Spin Interaction, Phys. Rev. Lett. **48**, 778; see also Dietz, W. (1984) HKX-Transformations: Some Results, in Solutions of Einstein's Equations: Techniques and Results, eds. C. Hoenselaers and W. Dietz, Lecture Notes in Physics 205, Springer-Verlag, Berlin
171. Kastor, D., Traschen, J. (1993) Cosmological multi-black-hole solutions, Phys. Rev. **D47**, 5370
172. Brill, D. R., Horowitz, G. T., Kastor, D. and Traschen, J. (1994) Testing cosmic censorship with black hole collisions, Phys. Rev. **D49**, 840
173. Welch, D. L. (1995) Smoothness of the horizons of multi-black-hole solutions, Phys. Rev. **D52**, 985
174. Brill, D. R., Hayward, S. A. (1994) Global structure of a black hole cosmos and its extremes, Class. Quantum Grav. **11**, 359
175. Ida, D., Nakao, K., Siino, M. and Hayward, S. A. (1998) Hoop conjecture for colliding black holes, Phys. Rev. **D58**, 121501
176. Scott, S. M., Szekeres, P. (1986) The Curzon singularity I: spatial section, Gen. Rel. Grav. **18**, 557; The Curzon singularity II: global picture, Gen. Rel. Grav. **18**, 571
177. Bičák, J., Lynden-Bell, D. and Katz, J. (1993) Relativistic disks as sources of static vacuum spacetimes, Phys. Rev. **D47**, 4334
178. Bičák, J., Lynden-Bell, D. and Pichon, C. (1993) Relativistic discs and flat galaxy models, Mon. Not. Roy. Astron. Soc. **265**, 26
179. Evans, N. W., de Zeeuw, P. T. (1992) Potential-density pairs for flat galaxies, Mon. Not. Roy. Astron. Soc. **257**, 152
180. Chruściel, P., MacCallum, M. A. H. and Singleton, P. B. (1995) Gravitational waves in general relativity XIV. Bondi expansions and the 'polyhomogeneity' of $\mathcal{J}$, Phil. Trans. Roy. Soc. Lond. **A350**, 113
181. Semerák, O., Zellerin, T. and Žáček, M. (1999) The structure of superposed Weyl fields, Mon. Not. Roy. Astron. Soc., **308**, 691 and 705
182. Lemos, J. P. S., Letelier, P. S. (1994) Exact general relativistic thin disks around black holes, Phys. Rev. **D49**, 5135
183. González, G. A., Letelier, P. S. (1999) Relativistic Static Thin Disks with Radial Stress Support, Class. Quantum Grav. **16**, 479
184. Letelier, P. S. (1999) Exact General Relativistic Disks with Magnetic Fields, gr-qc/9907050
185. Krasiński, A. (1978) Sources of the Kerr metric, Ann. Phys. (N.Y.) **112**, 22
186. McManus, D. (1991) A toroidal source for the Kerr black hole geometry, Class. Quantum Grav. **8**, 863
187. Bardeen, J. M., Wagoner, R. V. (1971) Relativistic disks. I. Uniform rotation, Astrophys. J. **167**, 359
188. Bičák, J., Ledvinka, T. (1993) Relativistic Disks as Sources of the Kerr Metric, Phys. Rev. Lett. **71**, 1669. See also (1993) Sources for stationary axisymmetric gravitational fields, Max-Planck-Institute for Astrophysics, Green report MPA 726, Munich
189. Pichon, C., Lynden-Bell, D. (1996) New sources for Kerr and other metrics: rotating relativistic discs with pressure support, Mon. Not. Roy. Astron. Soc. **280**, 1007

190. Barrabès, C., Israel, W. (1991) Thin shells in general relativity and cosmology: the lightlike limit, Phys. Rev. **D43**, 1129
191. Ledvinka, T., Bičák, J. and Žofka, M. (1999) Relativistic disks as sources of Kerr-Newman fields, in Proc. 8th M. Grossmann Meeting on General Relativity, ed. T. Piran, World Sci., Singapore
192. Neugebauer, G., Meinel, R. (1995) General Relativistic Gravitational Fields of a Rigidly Rotating Disk of Dust: Solution in Terms of Ultraelliptic Functions, Phys. Rev. Lett. **75**, 3046
193. Neugebauer, G., Kleinwächter, A. and Meinel, R. (1996) Relativistically rotating dust, Helv. Phys. Acta **69**, 472
194. Meinel, R. (1998) The rigidly rotating disk of dust and its black hole limit, in Proc. of the Second Mexican School on Gravitation and Mathematical Physics, eds. A. Garcia et al., Science Network Publishing, Konstanz, gr-qc/9703077
195. Breitenlohner, P., Forgács, P. and Maison, D. (1995) Gravitating Monopole Solutions II, Nucl. Phys. **442B**, 126
196. Misner, Ch. (1967) Taub-NUT Space as a Counterexample to Almost Anything, in Relativity Theory and Astrophysics 1, Lectures in Applied Mathematics, Vol. 8, ed. J. Ehlers, American Math. Society, Providence, R. I.
197. Taub, A. H. (1951) Empty space-times admitting a three parameter group of motions, Ann. Math. **53**, 472
198. Newman, E., Tamburino, L. and Unti, T. (1963) Empty-space generalization of the Schwarzschild metric, J. Math. Phys. **4**, 915
199. Lynden-Bell, D., Nouri-Zonoz, M. (1998) Classical monopoles: Newton, NUT space, gravomagnetic lensing, and atomic spectra, Rev. Mod. Phys. **70**, 427
200. Geroch, R. (1971) A method for generating solutions of Einstein's equations, J. Math. Phys. **12**, 918 and J. Math. Phys. **13**, 394
201. Ehlers, J. (1997) Examples of Newtonian limits of relativistic spacetimes, Class. Quantum Grav. **14**, A119
202. Wheeler, J. A. (1980) The Beam and Stay of the Taub Universe, in Essays in General Relativity, eds. F. J. Tipler, Academic Press, New York
203. Hájíček, P. (1971) Extension of the Taub and NUT spaces and extensions of their tangent bundles, Commun. Math. Phys. **17**, 109; Bifurcate spacetimes, J. Math. Phys. **12**, 157; Causality in non-Hausdorff spacetimes, Commun. Math. Phys. **21**, 75
204. Thorne, K. S. (1993) Misner Space as a Prototype for Almost Any Pathology, in Directions in General Relativity, Vol. 1, eds. B. L. Hu, M. P. Ryan and C. V. Vishveshwara, Cambridge University Press, Cambridge
205. Gibbons, G. W., Manton, N. S. (1986) Classical and Quantum Dynamics of BPS monopoles, Nuclear Physics **B274**, 183
206. Kraan T. C., Baal P. (1998) Exact T-duality between calorons and Taub – NUT spaces, INLO-PUB-4/98, hep-th/9802049
207. Bičák, J., Podolský, J. (1999) Gravitational waves in vacuum spacetimes with cosmological constant. I. Classification and geometrical properties of non-twisting type N solutions. II. Deviation of geodesics and interpretation of non-twisting type N solutions, J. Math. Phys. **44**, 4495 and 4506
208. Aichelburg, P. C., Balasin, H. (1996) Symmetries of pp-waves with distributional profile, Class. Quantum Grav. **13**, 723
209. Aichelburg, P. C., Balasin, H. (1997) Generalized symmetries of impulsive gravitational waves, Class. Quantum Grav. **14**, A31

210. Aichelburg, P. C., Sexl, R. U. (1971) On the gravitational field of a massless particle, Gen. Rel. Grav. **2**, 303
211. Penrose, R. (1972) The geometry of impulsive gravitational waves, in General Relativity, Papers in Honour of J. L. Synge, ed. L. O'Raifeartaigh, Clarendon Press, Oxford
212. Griffiths, J. B. (1991) Colliding Plane Waves in General Relativity, Clarendon Press, Oxford
213. Bondi, H., Pirani, F. A. E. and Robinson, I. (1959) Gravitational waves in general relativity. III. Exact plane waves, Proc. Roy. Soc. Lond. A **251**, 519
214. Rindler, W. (1977) Essential Relativity (2nd edition), Springer, New York-Berlin
215. Penrose, R. (1965) A remarkable property of plane waves in general relativity, Rev. Mod. Phys. **37**, 215
216. Lousto, C. O., Sánchez, N. (1989) The ultrarelativistic limit of the Kerr-Newman geometry and particle scattering at the Planck scale, Phys. Lett. **B232**, 462
217. Ferrari, V., Pendenza, P. (1990) Boosting the Kerr Metric, Gen. Rel. Grav. **22**, 1105
218. Balasin, H., Nachbagauer, H. (1995) The ultrarelativistic Kerr-geometry and its energy-momentum tensor, Class. Quantum Grav. **12**, 707
219. Podolský, J., Griffiths, J. B. (1998) Boosted static multipole particles as sources of impulsive gravitational waves, Phys. Rev. **D58**, 124024
220. Hotta, M., Tanaka, M. (1993) Shock-wave geometry with non-vanishing cosmological constant, Class. Quantum Grav. **10**, 307
221. Podolský, J., Griffiths, J. B. (1997) Impulsive gravitational waves generated by null particles in de Sitter and anti-de Sitter backgrounds, Phys. Rev. **D56**, 4756
222. D'Eath, P. D. (1996) Black Holes: Gravitational Interactions, Clarendon Press, Oxford
223. 't Hooft, G. (1987) Graviton dominance in ultra-high-energy scattering, Phys. Lett. **B198**, 61
224. Fabbrichesi, M., Pettorino, R., Veneziano, G. and Vilkovisky, G. A. (1994) Planckian energy scattering and surface terms in the gravitational action, Nucl. Phys. **B419**, 147
225. Kunzinger, M., Steinbauer, R. (1999) A note on the Penrose junction conditions, Class. Quantum Grav. **16**, 1255
226. Kunzinger, M., Steinbauer, R. (1999) A rigorous solution concept for geodesic and geodesic deviation equations in impulsive gravitational waves, J. Math. Phys. **40**, 1479
227. Podolský, J., Veselý, K. (1998) Chaotic motion in pp-wave spacetimes, Class. Quantum Grav. **15**, 3505
228. Levin, O., Peres, A. (1994) Quantum field theory with null-fronted metrics, Phys. Rev. **D50**, 7421
229. Klimčík, C. (1991) Gravitational waves as string vacua I, II, Czechosl. J. Phys. **41**, 697 (see also references therein)
230. Gibbons, G. W. (1999) Two loop and all loop finite 4-metrics, Class. Quantum Grav. **16**, L 71
231. Bičák, J., Pravda, V. (1998) Curvature invariants in type N spacetimes, Class. Quantum Grav. **15**, 1539

232. Pravda, V. (1999) Curvature invariants in type-*III* spacetimes, Class. Quantum Grav. **16**, 3321
233. Khan, K. A., Penrose, R. (1971) Scattering of two impulsive gravitational plane waves, Nature **229**, 185
234. Szekeres, P. (1970) Colliding gravitational waves, Nature **228**, 1183
235. Szekeres, P. (1972) Colliding plane gravitational waves, J. Math. Phys. **13**, 286
236. Yurtsever, U. (1988) Structure of the singularities produced by colliding plane waves, Phys. Rev. **D38**, 1706
237. Hauser, I., Ernst, F. J. (1989) Initial value problem for colliding gravitational waves – I/II, J. Math. Phys. **30**, 872 and 2322; (1990) and (1991) Initial value problem for colliding gravitational waves – III/IV, J. Math. Phys. **31**, 871 and **32**, 198;
238. Hauser, I., Ernst, F. J. (1999) Group structure of the solution manifold of the hyperbolic Ernst equation – general study of the subject and detailed elaboration of mathematical proofs, 216 pages, gr-qc/9903104
239. Nutku, Y., Halil, M. (1977) Colliding impulsive gravitational waves, Phys. Rev. Lett. **39**, 1379
240. Matzner, R., Tipler, F. J. (1984) Methaphysics of colliding self-gravitating plane waves, Phys. Rev. **D29**, 1575
241. Chandrasekhar, S., Ferrari, V. (1984) On the Nutku-Halil solution for colliding impulsive gravitational waves, Proc. Roy. Soc. Lond. **A396**, 55
242. Chandrasekhar, S., Xanthopoulos, B. C. (1986) A new type of singularity created by colliding gravitational waves, Proc. Roy. Soc. Lond. **A408**, 175
243. Chandrasekhar, S., Xanthopoulos, B. C. (1985) On colliding waves in the Einstein-Maxwell theory, Proc. Roy. Soc. Lond. **A398**, 223
244. Bičák, J. (1989) Exact radiative space-times, in Proceedings of the fifth Marcel Grossmann Meeting on General Relativity, eds. D. Blair and M. J. Buckingham, World Scientific, Singapore
245. Yurtsever, U. (1987) Instability of Killing-Cauchy horizons in plane-symmetric spacetimes, Phys. Rev. **D36**, 1662
246. Yurtsever, U. (1988) Singularities in the collisions of almost-plane gravitational waves, Phys. Rev. **D38**, 1731
247. Chandrasekhar, S. (1986) Cylindrical waves in general relativity, Proc. Roy. Soc. Lond. **A408**, 209
248. Einstein, A., Rosen, N. (1937) On Gravitational Waves, J. Franklin Inst. **223**, 43
249. Beck, G. (1925) Zur Theorie binärer Gravitationsfelder, Z. Phys. **33**, 713
250. Stachel, J. (1966) Cylindrical Gravitational News, J. Math. Phys. **7**, 1321
251. d'Inverno, R. (1997) Combining Cauchy and characteristic codes in numerical relativity, in Relativistic Gravitation and Gravitational Radiation (Proceedings of the Les Houches School of Physics), eds. J.-A. Marck and J.-P. Lasota, Cambridge University Press, Cambridge
252. Piran, T., Safier, P. N. and Katz, J. (1986) Cylindrical gravitational waves with two degrees of freedom: An exact solution, Phys. Rev. **D34**, 331
253. Thorne, K. S. (1965) C-energy, Phys. Rev. **B138**, 251
254. Garriga, J., Verdaguer, E. (1987) Cosmic strings and Einstein-Rosen waves, Phys. Rev. **D36**, 2250
255. Xanthopoulos, B. C. (1987) Cosmic strings coupled with gravitational and electromagnetic waves, Phys. Rev. **D35**, 3713

256. Chandrasekhar, S., Ferrari, V. (1987) On the dispersion of cylindrical impulsive gravitational waves, Proc. Roy. Soc. Lond. **A412**, 75
257. Tod, K. P. (1990) Penrose's quasi-local mass and cylindrically symmetric spacetimes, Class. Quantum Grav. **7**, 2237
258. Berger, B. K., Chruściel, P. T. and Moncrief, V. (1995) On "Asymptotically Flat" Space-Times with G_2-Invariant Cauchy Surfaces, Ann. Phys. (N.Y.) **237**, 322
259. Kuchař, K. V. (1971) Canonical quantization of cylindrical gravitational waves, Phys. Rev. **D4**, 955
260. Ashtekar, A., Pierri, M. (1996) Probing quantum gravity through exactly soluble midisuperspaces 1, J. Math. Phys. **37**, 6250
261. Korotkin, D., Samtleben, H. (1998) Canonical Quantization of Cylindrical Gravitational Waves with Two Polarizations, Phys. Rev. Lett. **80**, 14
262. Ashtekar, A., Bičák, J. and Schmidt, B. G. (1997) Asymptotic structure of symmetry-reduced general relativity, Phys. Rev. **D55**, 669
263. Ashtekar, A., Bičák, J. and Schmidt, B. G. (1997) Behaviour of Einstein-Rosen waves at null infinity, Phys. Rev. **D55**, 687
264. Penrose, R. (1963) Asymptotic properties of fields and space-times, Phys. Rev. Lett. **10**, 66; (1965) Zero rest-mass fields including gravitation: asymptotic behaviour, Proc. Roy. Soc. Lond. **A284**, 159
265. Ehlers, J., Friedrich, H. eds. (1994) in Canonical Gravity: From Classical to Quantum, Springer-Verlag, Berlin-Heidelberg
266. Ryan, M. (1972) Hamiltonian Cosmology, Springer-Verlag, Berlin
267. MacCallum, M. A. H. (1975) Quantum Cosmological Models, in Quantum Gravity, eds. C. J. Isham, R. Penrose and D. W. Sciama, Clarendon Press, Oxford
268. Halliwell, J. J. (1991) Introductory Lectures on Quantum Cosmology, in Quantum Cosmology and Baby Universes, eds. S. Coleman, J. Hartle, T. Piran and S. Weinberg, World Scientific, Singapore
269. Halliwell, J. J. (1990) A Bibliography of Papers on Quantum Cosmology, Int. J. Mod. Phys. **A5**, 2473
270. Kuchař, K. V. (1973) Canonical Quantization of Gravity, in Relativity, Astrophysics and Cosmology, ed. W. Israel, Reidel, Dordrecht
271. Kuchař, K. V. (1992) Time and Interpretations of Quantum Gravity, in Proceedings of the 4th Canadian Conference on General Relativity and Relativistic Astrophysics, eds. G. Kunstatter, D. Vincent and J. Williams, World Scientific, Singapore
272. Kuchař, K. V. (1994) Geometrodynamics of Schwarzschild black holes, Phys. Rev. **D50**, 3961
273. Romano, J. D., Torre, C. G. (1996) Internal Time Formalism for Spacetimes with Two Killing Vectors, Phys. Rev. **D53**, 5634. See also Torre, C. G. (1998) Midi-superspace Models of Canonical Quantum Gravity, gr-qc/9806122
274. Louko, J., Whiting, B. F. and Friedman, J. L. (1998) Hamiltonian spacetime dynamics with a spherical null-dust shell, Phys. Rev. **D57**, 2279
275. Griffiths, J. B., Miccicho, S. (1997) The Weber-Wheeler-Bonnor pulse and phase shifts in gravitational soliton interactions, Phys. Lett. **A233**, 37
276. Piran, T., Safier, P. N. and Stark, R. F. (1985) General numerical solution of cylindrical gravitational waves, Phys. Rev. **D32**, 3101
277. Wilson, J. P. (1997) Distributional curvature of time dependent cosmic strings, Class. Quantum Grav. **14**, 3337

278. Bičák, J., Schmidt, B. G. (1989) On the asymptotic structure of axisymmetric radiative spacetimes, Class. Quantum Grav. **6**, 1547
279. Bičák, J., Pravdová, A. (1998) Symmetries of asymptotically flat electrovacuum spacetimes and radiation, J. Math. Phys. **39**, 6011
280. Bičák, J., Pravdová, A. (1999) Axisymmetric electrovacuum spacetimes with a translational Killing vector at null infinity, Class. Quantum Grav. **16**, 2023
281. Robinson, I., Trautman, A. (1962) Some spherical gravitational waves in general relativity, Proc. Roy. Soc. Lond. **A265**, 463 ; see also [61]
282. Chruściel, P. T. (1992) On the global structure of Robinson-Trautman spacetimes, Proc. Roy. Soc. Lond. A **436**, 299; Chruściel, P. T., Singleton, D. B. (1992) Non-Smoothness of Event Horizons of Robinson-Trautman Black Holes, Commun. Math. Phys. **147**, 137, and references therein
283. Bičák, J., Podolský, J. (1995) Cosmic no-hair conjecture and black-hole formation: An exact model with gravitational radiation, Phys. Rev. **D52**, 887
284. Bičák, J., Schmidt, B. G. (1984) Isometries compatible with gravitational radiation, J. Math. Phys. **25**, 600
285. Bonnor, W. B., Swaminarayan, N. S. (1964) An exact solution for uniformly accelerated particles in general relativity, Zeit. f. Phys. **177**, 240. See also the original paper on negative mass in general relativity by Bondi, H. (1957) Rev. Mod. Phys. **29**, 423
286. Israel, W., Khan, K. A. (1964) Collinear particles and Bondi dipoles in general relativity, Nuov. Cim. **33**, 331
287. Bičák J. (1985) On exact radiative solutions representing finite sources, in Galaxies, axisymmetric systems and relativity (Essays presented to W. B. Bonnor on his 65th birthday), ed. M. A. H. MacCallum, Cambridge University Press, Cambridge
288. Bičák, J., Schmidt, B. G. (1989) Asymptotically flat radiative space-times with boost-rotation symmetry: the general structure, Phys. Rev. **D40**, 1827
289. Bičák J. (1987) Radiative properties of spacetimes with the axial and boost symmetries, in Gravitation and Geometry (A volume in honour of Ivor Robinson), eds. W. Rindler and A. Trautman, Bibliopolis, Naples
290. Bičák, J., Hoenselaers, C. and Schmidt, B. G. (1983) The solutions of the Einstein equations for uniformly accelerated particles without nodal singularities II. Self-accelerating particles, Proc. Roy. Soc. Lond. **A390**, 411
291. Bičák, J., Reilly, P. and Winicour, J. (1988) Boost rotation symmetric gravitational null cone data, Gen. Rel. Grav. **20**, 171
292. Gómez R., Papadopoulos P. and Winicour J. (1994) J. Math. Phys. **35**, 4184
293. Alcubierre, M., Gundlach, C. and Siebel, F. (1997) Integration of geodesics as a test bed for comparing exact and numerically generated spcetimes, in Abstracts of Plenary Lectures and Contributed Papers (GR15), Inter-University Centre for Astronomy and Astrophysics Press, Pune
294. Bičák, J., Hoenselaers, C. and Schmidt B.G., (1983) The solutions of the Einstein equations for uniformly accelerated particles without nodal singularities I. Freely falling particles in external fields, Proc. Roy. Soc. Lond. **A390**, 397
295. Bičák, J. (1980) The motion of a charged black hole in an electromagnetic field, Proc. Roy. Soc. Lond. **A371**, 429
296. Hawking, S. W., Horowitz, G. T. and Ross, S. F. (1995) Entropy, area, and black hole pairs, Phys. Rev. **D51**, 4302; Mann, R. B., Ross, S. F. (1995) Cosmological production of charged black hole pairs, Phys. Rev. **D52**, 2254;

Hawking, S. W., Ross, S. F. (1995) Pair production of black holes on cosmic strings, Phys. Rev. Lett. **75**, 3382
297. Plebański, J., Demiański, M. (1976) Rotating, charged and uniformly accelerating mass in general relativity, Ann. Phys. (N.Y.) **98**, 98
298. Bičák, J., Pravda, V. (1999) Spinning C-metric: radiative spacetime with accelerating, rotating black holes, Phys. Rev. **D60**, 044004
299. Belinsky, V. A., Khalatnikov, I. M. and Lifshitz, E. M. (1970) Oscillatory approach to a singular point in the relativistic cosmology, Adv. in Phys. **19**, 525
300. Belinsky, V. A., Khalatnikov, I. M. and Lifshitz, E. M. (1982) A general solution of the Einstein equations with a time singularity, Adv. in Phys. **31**, 639
301. Ellis, G. F. R. (1996) Contributions of K. Gödel to Relativity and Cosmology, in Gödel '96: Logical Foundations of Mathematics, Computer Science and Physics – Kurt Gödel's Legacy, ed. P. Hájek, Springer-Verlag, Berlin-Heidelberg; see also preprint 1996/7 of the Dept. of Math. and Appl. Math., University of Cape Town
302. Kantowski, R., Sachs, R. K. (1966) Some Spatially Homogenous Anisotropic Relativistic Cosmological Models, J. Math. Phys. **7**, 443
303. Thorne, K. S. (1967) Primordial element formation, primordial magnetic fields, and the isotropy of the universe, Astrophys. J. **148**, 51
304. Ryan, M. P., Shepley, L. C. (1975) Homogeneous Relativistic Cosmologies, Princeton University Press, Princeton
305. MacCallum, M. A. H. (1979) Anisotropic and inhomogeneous relativistic cosmologies, in General Relativity (An Einstein Centenary Survey), eds. S. W. Hawking and W. Israel, Cambridge University Press, Cambridge
306. Obregón, O., Ryan, M. P. (1998) Quantum Planck size black hole states without a horizon, Modern Phys. Lett. A **13**, 3251; see also references therein
307. Nojiri, S., Obregón, O., Odintsov, S. D. and Osetrin, K. E. (1999) (Non)singular Kantowski-Sachs universe from quantum spherically reduced matter, Phys. Rev. **D60**, 024008
308. Heckmann, O., Schücking, E. (1962) Relativistic Cosmology, in Gravitation: an introduction to current research, ed. L. Witten, J. Wiley and Sons, New York
309. Zel'dovich, Ya. B., Novikov, I. D. (1983) Relativistic Astrophysics, Volume 2: The Structure and Evolution of the Universe, The University of Chicago Press, Chicago
310. MacCallum, M. A. H. (1994) Relativistic cosmologies, in Deterministic Chaos in General Relativity, eds. D. Hobill, A. Burd and A. Coley, Plenum Press, New York
311. Wainwright, J., Ellis, G. F. R. eds. (1997) Dynamical Systems in Cosmology, Cambridge University Press, Cambridge
312. Misner, C. W. (1969) Mixmaster universe, Phys. Rev. Lett. **22**, 1071
313. Hu, B. L., Ryan, M. P. and Vishveshwara, C. V. eds. (1993) Directions in General Relativity, Vol. 1 (Papers in honor of Charles Misner), Cambridge University Press, Cambridge
314. Uggla, C., Jantzen, R. T. and Rosquist, K. (1995) Exact hypersurface-homogeneous solutions in cosmology and astrophysics, Phys. Rev. **D51**, 5522
315. Tanaka, T., Sasaki, M. (1997) Quantized gravitational waves in the Milne universe, Phys. Rev. **D55**, 6061

316. Lukash, V. N. (1975) Gravitational waves that conserve the homogeneity of space, Sov. Phys. JETP **40**, 792
317. Barrow, J. D., Sonoda, D. H. (1986) Asymptotic stability of Bianchi type universes, Physics Reports **139**, 1
318. Kuchař, K. V., Ryan, M. P. (1989) Is minisuperspace quantization valid?: Taub in Mixmaster, Phys. Rev. **D40**, 3982. The approach was first used in Kuchař, K. V., Ryan, M. P. (1986) Can Minisuperspace Quantization be Justified?, in Gravitational Collapse and Relativity, eds. H. Sato and T. Nakamura, World Scientific, Singapore
319. Bogoyavlenski, O. I. (1985) Methods in the Qualitative Theory of Dynamical Systems in Astrophysics and Gas Dynamics, Springer-Verlag, Berlin
320. Hobill, D., Burd, A. and Coley, A. eds. (1994) Deterministic Chaos in General Relativity, Plenum Press, New York
321. Rendall, A. (1997) Global dynamics of the Mixmaster model, Class. Quantum Grav. **14**, 2341
322. Khalatnikov, I. M., Lifshitz, E. M., Khamin, K. M., Shehur, L. N. and Sinai, Ya. G. (1985) On the Stochasticity in Relativistic Cosmology, J. of Statistical Phys. **38**, 97
323. LeBlanc, V. G., Kerr, D. and Wainwright, J. (1995) Asymptotic states of magnetic Bianchi VI_0 cosmologies, Class. Quantum Grav. **12**, 513
324. LeBlanc, V. G. (1977) Asymptotic states of magnetic Bianchi I cosmologies, Class. Quantum Grav. **14**, 2281
325. Jantzen, R. T. (1986) Finite-dimensional Einstein-Maxwell-scalar field system, Phys. Rev. **D33**, 2121
326. LeBlanc, V. G. (1998) Bianchi II magnetic cosmologies, Class. Quantum Grav. **15**, 1607
327. Belinsky, V. A., Khalatnikov, I. M. (1973) Effect of scalar and vector fields on the nature of the cosmological singularity, Soviet Physics JETP **36**, 591
328. Berger, B. K. (1999) Influence of scalar fields on the approach to a cosmological singularity, gr-qc/9907083
329. Wainwright, J., Coley, A. A., Ellis, G. F. R. and Hancock, M. (1998) On the isotropy of the Universe: do Bianchi VII_h cosmologies isotropize? Class. Quantum Grav. **15**, 331
330. Weaver, M., Isenberg, J. and Berger, B. K. (1998) Mixmaster Behavior in Inomogeneous Cosmological Spacetimes, Phys. Rev. Lett. **80**, 2984
331. Berger, B. K., Moncrief, V. (1998) Evidence for an oscillatory singularity in generic $U(1)$ cosmologies on $T^3 \times R$, Phys. Rev. **D58**, 064023
332. Gowdy, R. H. (1971) Gravitational Waves in Closed Universes, Phys. Rev. Lett. **27**, 826; Gowdy, R. H. (1974) Vacuum Spacetimes with Two-Parameter Spacelike Isometry Groups and Compact Invariant Hypersurfaces: Topologies and Boundary Conditions, Ann. Phys. (N.Y.) **83**, 203
333. Carmeli, M., Charach, Ch. and Malin, S. (1981) Survey of cosmological models with gravitational scalar and electromagnetic waves, Physics Reports **76**, 79
334. Chruściel, P. T. (1990) On Space-Times with $U(1)\times U(1)$ Symmetric Compact Cauchy Surfaces, Ann. Phys. (N. Y.) **202**, 100
335. Gowdy, R. H. (1975) Closed gravitational-wave universes: Analytic solutions with two-parameter symmetry, J. Math. Phys. **16**, 224
336. Charach, Ch. (1979) Electromagnetic Gowdy universe, Phys. Rev. **D19**, 3516
337. Bičák, J., Griffiths, J. B. (1996) Gravitational Waves Propagating into Friedmann-Robertson-Walker Universes, Ann. Phys. (N.Y) **252**, 180

338. Berger, B. K., Chruściel, P. T., Isenberg, J. and Moncrief, V. (1997) Global Foliations of Vacuum Spacetimes with T^2 Isometry, Ann. Phys. (N.Y.) **260**, 117
339. Chruściel, P. T., Isenberg, J. and Moncrief, V. (1990) Strong cosmic censorship in polarized Gowdy spacetimes, Class. Quantum Grav. **7**, 1671
340. Moncrief, V. (1997) Spacetime Singularities and Cosmic Censorship, in Proc. of the 14th International Conference on General Relativity and Grativation, eds. M. Francaviglia, G. Longhi, L. Lusanna and E. Sorace, World Scientific, Singapore
341. Kichenassamy, S., Rendall, A. D. (1998) Analytic description of singularities in Gowdy spacetimes, Class. Quantum Grav. **15**, 1339
342. Kichenassamy, S. (1996) Nonlinear Wave Equations, Marcel Dekker Publ. New York
343. Adams, P. J., Hellings, R. W., Zimmermann, R. L., Farhoosh, H., Levine, D. I. and Zeldich, S. (1982) Inhomogeneous cosmology: gravitational radiation in Bianchi backgrounds, Astrophys. J. **253**, 1
344. Belinsky, V., Zakharov, V. (1978) Integration of the Einstein equations by means of the inverse scattering problem technique and construction of exact soliton solutions, Sov. Phys. JETP **48**, 985
345. Carr, B. J., Verdaguer, E. (1983) Soliton solutions and cosmological gravitational waves, Phys. Rev. **D28**, 2995
346. Belinsky, V. (1991) Gravitational breather and topological properties of gravisolitons, Phys. Rev. **D44**, 3109
347. Kordas, P. (1993) Properties of the gravibreather, Phys. Rev. **D48**, 5013
348. Alekseev, G. A. (1988) Exact solutions in the general theory of relativity, Proceedings of the Steklov Institute of Mathematics, Issue 3, p. 215
349. Verdaguer, E. (1993) Soliton solutions in spacetimes with spacelike Killing fields, Physics Reports **229**, 1
350. Katz, J., Bičák, J. and Lynden-Bell, D. (1997) Relativistic conservation laws and integral constraints for large cosmological perturbations, Phys. Rev. **D55**, 5957
351. Uzan, J. P., Deruelle, M. and Turok, N. (1998) Conservation laws and cosmological perturbations in curved universes, Phys. Rev. **D57**, 7192
352. Beig, R., Simon, W. (1992) On the Uniqueness of Static Perfect-Fluid Solutions in General Relativity, Commun. Math. Phys. **144**, 373
353. Lindblom, L., Masood-ul-Alam (1994) On the Spherical Symmetry of Static Stellar Models, Commun. Math. Phys. **162**, 123
354. Rendall, A. (1997) Solutions of the Einstein equations with matter, in Proc. of the 14th International Conference on General Relativity and Gravitation, eds. M. Francaviglia, G. Longhi, L. Lusanna and E. Sorace, World Scientific, Singapore
355. Bartnik, R., McKinnon, J. (1988) Particlelike Solutions of the Einstein-Yang-Mills Equations, Phys. Rev. Lett. **61**, 141
356. Volkov, M. S., Gal'tsov, D. V. (1999) Gravitating Non-Abelian Solitons and Black Holes with Yang-Mills Fields, Physics Reports **319**, 1
357. Rendall, A. D., Tod, K. P. (1999) Dynamics of spatially homogeneous solutions of the Einstein-Vlasov equations which are locally rotationally symmetric, Class. Quantum Grav. **16**, 1705
358. Carr, B. J., Coley, A. A. (1999) Self-similarity in general relativity, Class. Quantum Grav. **16**, R 31

359. Gundlach, C. (1998) Critical Phenomena in Gravitational Collapse, Adv. Theor. Math. Phys. **2**, 1
360. Krasiński, A. (1997) Inhomogeneous Cosmological Models, Cambridge University Press, Cambridge

The Cauchy Problem for the Einstein Equations

Helmut Friedrich and Alan Rendall

Max-Planck-Institut für Gravitationsphysik, Am Mühlenberg 1,
14476 Golm, Germany

Abstract. Various aspects of the Cauchy problem for the Einstein equations are surveyed, with the emphasis on local solutions of the evolution equations. Particular attention is payed to giving a clear explanation of conceptual issues which arise in this context. The question of producing reduced systems of equations which are hyperbolic is examined in detail and some new results on that subject are presented. Relevant background from the theory of partial differential equations is also explained at some length.

1 Introduction

One of the most striking differences between the Newtonian theory of gravity and its successor, general relativity, is that in the latter the gravitational field acquires its own dynamical properties. Its time evolution is complicated even in the absence of matter. This contrasts with the fact that in the Newtonian theory the field vanishes when no matter is present. The field equation, namely the Poisson equation, together with the boundary condition that the field vanishes at infinity, which is an essential part of the theory, combine to give this result. The Einstein equations, the field equations of general relativity, allow idealized situations which represent gravitational waves in an otherwise empty universe, without any material sources. This reflects the different mathematical nature of the equations involved in these two cases. The Poisson equation is elliptic while the Einstein equations are essentially hyperbolic in nature. The meaning of the term 'essential' in this context is not simple and explaining it is a major theme in the following.

In order to understand the special theoretical difficulties connected with the Einstein equations and what mathematical approaches may be appropriate to overcome them, it is useful to compare gravitation with electromagnetism. The motion of charged matter can be described within the full Maxwell theory. However, there is also another possibility, which is used when relativistic effects are small, for instance in many situations in plasma physics. Here dynamical matter is coupled to the electrostatic field generated by this matter at any given time. In this quasi-static model the electric field follows the sources in a passive way while in the full theory there are propagating degrees of freedom. As in the case of gravitation, the elliptic equation in the non-relativistic theory (the Poisson equation again) is replaced by a system of hyperbolic equations, the Maxwell equations.

The fact that the Maxwell equations are so tractable is due to their linearity, a convenient feature not shared by the Einstein equations. The theory of linear partial differential equations in general, and of linear hyperbolic equations in particular, are much better developed than the corresponding nonlinear theories. As a side remark it may be noted that the combined equations describing electromagnetic fields together with their sources are nonlinear and that in that context serious theoretical problems, such as that of describing radiation damping, do appear.

Solutions of hyperbolic equations can be uniquely determined by their values on a suitable initial hypersurface. The Cauchy problem is the task of establishing a one to one correspondence between solutions and initial data, and studying further properties of this correspondence. The solution determined by a particular initial datum may be *global*, i.e. defined on the whole space where the equations are defined, or *local*, i.e. only defined on a neighbourhood of the initial hypersurface. 'Local' and 'global' could be called local and global *in time* since in the case where a preferred time coordinate is present that is exactly what they mean.

From what has been said so far we see that in studying the Einstein equations we are faced with a system of nonlinear hyperbolic equations. Among nonlinear hyperbolic equations in physics, those which have been studied most extensively are the Euler equations, and so we may hope to get some insights from that direction. At the same time, it is wise to be careful not to treat the analogy too uncritically, since the status of the Euler and Einstein equations is very different. The Euler equations are phenomenological in nature and much is understood about how they arise from models on a more fundamental level. The Einstein equations have been thought of as representing fundamental physics for most of their history and the recent idea that they arise as a formal limiting case in string theory will require, at the very least, a lot more work before it can offer a solid alternative to this. To return to the Euler equations, one of their well-known features is the formation of shocks. While there is no indication of a directly analogous phenomenon for the Einstein equations, it does draw attention to a fundamental fact. For linear hyperbolic equations it is in general possible to solve the Cauchy problem globally, i.e. to show the existence of a global solution corresponding to each initial datum. For nonlinear hyperbolic equations this is much more difficult and whether it can be done or not must be decided on a case by case basis. The formation of shocks in solutions of the Euler equations is an example of the difficulties which can occur. A general theory for general equations can only be hoped for in the case of the local (in time) Cauchy problem.

For the Einstein equations we must expect to encounter the problem that solutions of the Cauchy problem for nonlinear hyperbolic equations do not exist globally. In the case of the Einstein equations this issue is clouded by the fact that the distinction between local and global solutions made above does not apply. To define the notions of local and global we used the concept of the space where the equations are defined. In other words we used a background space. As we will see in more detail later, in the case of the Einstein equations there is no background space; the space-time manifold is part of the solution. It is better in this case to talk only about local and global *properties* of solutions and not about local and global *solutions.* We may then loosely use the words 'local' and 'global' to refer to all aspects of the Cauchy problem which refer to local and global properties of solutions, respectively. Solutions of the Einstein equations present global features such as the formation of black holes which are peculiar to this system and which are made possible by the lack of a background space.

In this article we are not concerned with general systems of nonlinear hyperbolic equations, but with a particular one, which is given to us by general relativity. Actually, when the coupling to matter fields is taken into account, we do get a variety of hyperbolic systems. Nevertheless, we might hope that in at least some situations of interest, such as gravitational collapse, the dynamics of the gravitational field would dominate the qualitative behaviour and let the effects of the particular matter model fade into the background. In any case, it is useful to retain the distinction between the local and global Cauchy problems. The global problem is what we want to solve, but the local problem is a natural first step. Our original plan was to cover both topics, but along the way we discovered that the first step is already so rich that on grounds of time and space we have relegated global questions to passing remarks. Along the way we stumbled over a variety of 'well-known' things which turned out not to be known at all, or even to be false.

The theory of the Cauchy problem allows us to formulate and establish relativistic causality within general relativity. Another basic function of the solution of the Cauchy problem is to parametrize solutions of the field equations in a useful way. To single out the class of solutions relevant to the description of a given physical situation, we can single out an appropriate subclass of initial data, which is often simpler to do. This does not mean that by identifying this class of solutions we have solved the problem. Rather it means that we have found a class of problems which may be of physical relevance and which it is therefore desirable to investigate mathematically. In the end the central mathematical problem is to discover some of the global properties of the solutions being studied. As has already been said, this article is almost entirely restricted to local questions, so that we will say little more about this point. However we will return to it in the last section.

The point of view of the Cauchy problem can also be used to throw light on various other issues. For instance, it provides a framework in which it can be shown that certain approximation methods used in general relativity really provide approximations to solutions of the Einstein equations. It can help to provide an analytical basis for the development of efficient and reliable numerical schemes to solve the Einstein equations. The Cauchy problem can furnish examples which throw light on general conjectures about solutions of the Einstein equations. It can be used to investigate whether certain properties of explicit solutions of the Einstein equations of importance in general relativity are stable to perturbations.

The structure of the article will now be outlined. In the second section a number of fundamental concepts are introduced and critically reviewed. In particular, the splitting of the field equations into evolution equations and constraints is described. The rest of the article concentrates on the evolution equations. The question of how to solve the constraints on the initial hypersurface is not considered further. The second section presents the basic elements which go into solving the local Cauchy problem. It discusses in particular the question of gauge freedom and how to show that the constraints are satisfied everywhere, given that they are satisfied on the initial hypersurface (propagation of the constraints). These are aspects of the question of *hyperbolic reduction*, i.e. how questions about the Cauchy problem for the Einstein equations can be reduced to questions about the Cauchy problem for hyperbolic equations.

The solution of the Cauchy problem relies on the use of techniques from the theory of partial differential equations (PDE). The third section presents some of the relevant techniques and attempts to explain some of the important concepts which play a role in the theory of the Cauchy problem for hyperbolic equations. We have chosen to concentrate on symmetric hyperbolic systems rather than on other kinds of hyperbolic equations such as nonlinear wave equations, which could also be used as a basis for studying the Cauchy problem for the Einstein equations. It will be seen that symmetric hyperbolic systems provide a very flexible tool. A comparative discussion of various notions of hyperbolicity is also given.

In the fourth section we return to the question of hyperbolic reduction. Different ways of reducing the Einstein equations in 3+1 form are presented and compared. The ADM equations, which were already introduced in Sect. 2, are discussed further. A form of these equations which has proved successful in numerical calculations is analysed. A number of other illustrative examples are treated in detail. The first example is that of the Einstein–Euler system for a self-gravitating perfect fluid. The case of dust, which is significantly different from that of a fluid with non-vanishing pressure, and which leads to serious problems in some approaches, is successfully handled. The second example is a variant of the pure Cauchy problem, namely the initial boundary value problem. The third example is the Einstein–Dirac system, which gives rise to particular problems of its own.

The fifth section starts with a discussion of the proofs of the basic statements of local existence, uniqueness and stability for the Einstein equations, based on the PDE theory reviewed in the third section and a particular hyperbolic reduction already introduced in Sect. 2. Then the extension of these results to the case where different kinds of matter are present is sketched. Existence, uniqueness and stability constitute together the statement that the Cauchy problem is well-posed. The significance of this is underlined by an example of a situation where the Cauchy problem is ill-posed. Finally it is shown how existence and uniqueness imply the inheritance of symmetries of the initial data by the corresponding solutions.

The choice of topics covered in this article is heavily dependent on the interests of the authors. In the last section we list a number of the important topics which were not discussed. While this list is also influenced by personal taste, we hope that it will provide the interested reader with the opportunity to form a balanced view of the subject.

2 Basic Observations and Concepts

In this section we give an introductory survey of various aspects of the field equations. Most of them will be discussed again, in greater detail, in later sections.

Gravitational fields represented by isometric space-times must be considered as physically equivalent and therefore the field equations for the metric must have the property that they determine isometry classes of solutions, not specific coordinate representations. This is achieved by the covariant Einstein equations

$$R_{\mu\nu} - \frac{1}{2}\,R\,g_{\mu\nu} + \lambda\,g_{\mu\nu} = \kappa\,T_{\mu\nu}.$$

It is often said that many of the specific features of the Einstein equations are related to this covariance. One may wonder, however, what is so special about it, since the wave equation, the Yang–Mills equations, the Euler equations, just to name a few examples, are also covariant. The difference lies in the fundamental nature of the metric field. While the examples just quoted are defined with respect to some background structure, namely a given Lorentz space (in the case of the Yang–Mills equations in four dimensions a conformal structure suffices), the Einstein equations are designed to determine Lorentz manifolds without introducing any extraneous structures. Consequently, the solutions of the equations themselves provide the background on which the equations are to be solved.

This makes it evident that Einstein's equations can at best be quasi-linear. That they are in fact quasi-linear can be seen from their explicit expression. Writing the equation in the form

$$R_{\mu\nu} - \lambda\,g_{\mu\nu} = \kappa\,(T_{\mu\nu} - \frac{1}{2}\,T\,g_{\mu\nu}), \tag{2.1}$$

the principal part of the differential operator of second order which acts on the metric coefficients is given by the left hand side. In arbitrary coordinates we have

$$R_{\mu\nu} = -\frac{1}{2} g^{\lambda\rho} \left\{ \frac{\partial^2 g_{\mu\nu}}{\partial x^\lambda \, \partial x^\rho} + \frac{\partial^2 g_{\lambda\rho}}{\partial x^\mu \, \partial x^\nu} - \frac{\partial^2 g_{\mu\rho}}{\partial x^\lambda \, \partial x^\nu} - \frac{\partial^2 g_{\rho\nu}}{\partial x^\mu \, \partial x^\lambda} \right\} + Q_{\mu\nu}(g, \partial g), \tag{2.2}$$

where Q denotes a rational function of the metric coefficients and their first order derivatives. We see that the equations are quasi-linear but not better, i.e. they are linear in the derivatives of highest order but these derivatives come with coefficients which are given by the unknown itself. The Euler equations are also quasi-linear in this sense. However, they are studied in general on an independent background space-time in terms of which their solutions may be analysed.

2.1 The Principal Symbol

Consider a system of k partial differential equations of order m for an $\mathbb{R}^k$-valued unknown u defined on an open subset U of $\mathbb{R}^n$. Suppose that in coordinates x^μ, $\mu = 1, \ldots, n$ on U it takes the form

$$P\,u \equiv \sum_{|\alpha| \leq m} A^\alpha \, D^\alpha \, u = f. \tag{2.3}$$

Here the A^α denote smooth real $k \times k$-matrix-valued functions, $f(x)$ is a smooth $\mathbb{R}^k$-valued function, the $\alpha = (\alpha_1, \ldots, \alpha_n)$, with non-negative integers α_j, are multi-indices, $|\alpha| = \alpha_1 + \ldots + \alpha_n$, and $D^\alpha = \partial^{\alpha_1}_{x^1} \ldots \partial^{\alpha_n}_{x^n}$. Assume, first, that the equations are linear, so that $A^\alpha = A^\alpha(x)$.

The question of the formal solvability of the equation leads to the important notion of a characteristic. Suppose that $H = \{\Phi = \text{const.}, d\Phi \neq 0\}$ is a hypersurface of U, defined in terms of a smooth function Φ. A *Cauchy data set on* H for equation (2.3) consists of a set of functions $u_0, u_1, \ldots, u_{(m-1)}$ on H. Let the coordinates x^μ be chosen such that $\Phi = x^1$ near H. Interpreting the functions u_i as the derivatives $\partial^i_{x^1} u$ of a solution u to our equation, defined in a neighbourhood of H, we ask whether equation (2.3) allows the function $\partial^m_{x^1} u$ to be determined uniquely on H from the Cauchy data. Since all functions $D^\alpha \, u$ with $\alpha_1 \leq m - 1$, can obviously be derived from the data, it follows that we can solve the equation on H for $\partial^m_{x^1} \, u$ if and only if the matrix $A^{(m,0,\ldots,0)}$ is invertible at all points of H. Using for any covector ξ_μ and multi-index α the notation $\xi^\alpha = \xi_1^{\alpha_1} \ldots \xi_n^{\alpha_n}$, we observe that with our assumptions $A^{(m,0,\ldots,0)} = \sum_{|\alpha|=m} A^\alpha \, D^\alpha \, \Phi$. The invertibility of the matrix on the right hand side is independent of the coordinates chosen to represent equation (2.3) and the function Φ used to represent H. This follows from the transformation law of the coefficients of the differential equation and of covectors under coordinate transformations.

We are thus led to introduce the following covariant notions. For a covector ξ at a given point $x \in M$, we define the *principal symbol* of (2.3) at (x, ξ) as the matrix $\sigma(x,\xi) = \sum_{|\alpha|=m} A^{\alpha}(x)\,\xi^{\alpha}$ (or the associated linear map). The reader should be warned that a definition with ξ^i replaced by $i\xi^i$, where i is the imaginary unit, is often used, since it fits well with the Fourier transform. The latter convention will not be used here. A hypersurface H, represented by a function Φ as above, is called *nowhere characteristic* for (2.3), if $\det(\sigma(x, d\,\Phi)) \neq 0$ for $x \in H$ and we say that H is a *characteristic hypersurface*, or simply a *characteristic* for (2.3), if $\det(\sigma(x, d\,\Phi(x))) = 0$ for $x \in H$.

Given Cauchy data on a hypersurface H which is nowhere characteristic, we can determine a formal expansion of a possible solution u in terms of x^1 on H by taking formal derivatives of (2.3) and solving for $\partial^i_{x^1} u$ on H, $i = m, m+1, \ldots$. Conversely, if H is characteristic, Cauchy data cannot be prescribed freely on H, because the equation implies relations among the Cauchy data on H. We refer to these relations as the *inner equations* induced by the equations on the characteristic.

If the rank of the matrix $\sigma(x, d\,\Phi(x))$ is $k - j$ with some positive integer j, there are j such relations. The principal symbol is not sufficient to describe the precise nature of these relations. The complete information of (2.3) is needed for this.

If equation (2.3) is quasi-linear, so that the coefficients of the equation do not only depend on the points of M but also on the unknown and its derivatives of order less than m, we have to proceed slightly differently. Suppose u is a solution of the quasi-linear equation (2.3). For given covector ξ at the point $x \in M$ we define the *principal symbol* of (2.3) *with respect to* u as the matrix $\sigma(x,\xi) = \sum_{|\alpha|=m} A^{\alpha}(x, u(x), \ldots, D^{\beta}\,u(x)|_{|\beta|=m-1})\,\xi^{\alpha}$ and use it to define, by the condition above, the *characteristics of* (2.3) *with respect to* u. Thus for quasi-linear equations the characteristics depend on the solution.

We return to the Einstein equations. Assuming that (M, g) is a solution of (2.1) of dimension n, we find that for given covector ξ at $x \in M$ the principal symbol of the operator in (2.2) defines the linear map

$$k_{\mu\nu} \to (\sigma \cdot k)_{\mu\nu} = -\frac{1}{2}\,g^{\lambda\rho}(x)\,\{k_{\mu\nu}\,\xi_{\lambda}\,\xi_{\rho} + k_{\lambda\rho}\,\xi_{\mu}\,\xi_{\nu} - k_{\mu\rho}\,\xi_{\lambda}\,\xi_{\nu} - k_{\lambda\nu}\,\xi_{\mu}\,\xi_{\rho}\}\,, \tag{2.4}$$

of the set of symmetric covariant tensors at x into itself. If $k_{\mu\nu}$ is in its kernel, we have

$$0 = k_{\mu\nu}\,\xi_{\rho}\,\xi^{\rho} + g^{\rho\lambda}\,k_{\rho\lambda}\,\xi_{\mu}\,\xi_{\nu} - k_{\mu\rho}\,\xi^{\rho}\,\xi_{\nu} - k_{\nu\rho}\,\xi^{\rho}\,\xi_{\mu}, \tag{2.5}$$

from which we see that tensors of the form $k_{\mu\nu} = \xi_{\mu}\,\eta_{\nu} + \xi_{\nu}\,\eta_{\mu}$, with arbitrary covector η, generate an n-dimensional subspace of the kernel. This subspace coincides with the kernel if $\xi_{\rho}\,\xi^{\rho} \neq 0$. If $\xi_{\rho}\,\xi^{\rho} = 0$, $\xi_{\rho} \neq 0$, equation (2.5) takes the form $\xi_{\mu}\,\eta_{\nu} + \xi_{\nu}\,\eta_{\mu} = 0$, with some covector η_{μ}. Since $\xi_{\mu} \neq 0$ we must have $\eta_{\mu} = 0$ or, equivalently, $\frac{1}{2}\,g^{\rho\lambda}\,k_{\rho\lambda}\,\xi_{\nu} = k_{\rho\nu}\,\xi^{\rho}$. The solutions, given by $k_{\mu\nu} = h_{\mu\nu} + a\,(\xi_{\mu}\,\eta^*_{\nu} + \xi_{\nu}\,\eta^*_{\mu})$ with $a \in \mathbb{R}$, η^*_{μ} a fixed covector with $\xi^{\mu}\,\eta^*_{\mu} \neq 0$, and $h_{\mu\nu}$ satisfying $\xi^{\mu}\,h_{\mu\nu} = 0$, $g^{\mu\nu}\,h_{\mu\nu} = 0$, span a space of dimension $\frac{n\,(n-1)}{2}$, which is strictly larger than n if $n \geq 4$.

It follows that any hypersurface is characteristic for the Einstein equations. In the case of spacelike or timelike hypersurfaces n relations are implied on Cauchy data sets, in the case of null hypersurfaces there are $\frac{n\,(n-3)}{2}$ additional relations. The fact that the rank of the principal symbol drops further for covectors ξ which are null has the immediate consequence that the inner equations induced on null hypersurfaces are completely different from the inner equations induced on nowhere characteristic hypersurfaces. This is related to the fact that null hypersurfaces (and appropriate generalizations admitting caustics) may represent wave fronts, swept out by high frequency perturbations of the field. The propagation of the latter is governed by the inner equations induced on these hypersurfaces (cf. [27] and, for a modern discussion, also [54], [88] for the mathematical aspects of this statement).

The fact that on any hypersurface the tensors of the form $k_{\mu\nu} = \xi_\mu\,\eta_\nu + \xi_\nu\,\eta_\mu$ are in the kernel of the principal symbol map at (x, ξ) can be deduced by an abstract argument which uses only the covariance and the quasi-linearity of the equations (cf. [12], [64]). This emphasizes again the special role of null hypersurfaces. Assume that ϕ_τ is the flow of a vector field X. If we denote by ϕ_τ^* the associated pull-back operation, we have

$$\phi_\tau^*(\mathrm{Ric}[g]) = \mathrm{Ric}[\phi_\tau^*(g)].$$

Taking derivatives with respect to τ we obtain

$$\mathcal{L}_X\,\mathrm{Ric}[g] = \mathrm{Ric}'_g[\mathcal{L}_X\,g],$$

where the right hand side denotes the (at g) linearized Ricci operator applied to the Lie derivative $\mathcal{L}_X\,g_{\mu\nu} = \nabla_\mu\,X_\nu + \nabla_\nu\,X_\mu$ of the metric. Considering this equation as a differential equation for X, the principal part is given by the terms of third order appearing on the right hand side. Since the equation is an identity, holding for all vector fields, these terms must cancel. It follows that the expressions $\xi_\mu\,X_\nu + \xi_\nu\,X_\mu$ are in the kernel of the principal symbol of the linearized and thus also of the non-linear Ricci operator.

For the further discussion it will be convenient to use coordinates related to a given hypersurface. For simplicity and since it is the case of most interest to us, we will assume the latter to be spacelike and denote it by S. Let T be a non-vanishing *time flow vector field* transverse to S (but not necesssarily timelike) and $t = x^0$ a function with $\{t = 0\} = S$ and $< dt, T > = 1$. We choose local coordinates x^a, $a = 1, 2, 3$, on S and extend them to a neighbourhood of S such that $< dx^a, T > = 0$. In these coordinates $T = \partial_t$ and the metric has the *ADM representation*

$$g = -(\alpha\,dt)^2 + h_{ab}\,(\beta^a\,dt + dx^a)\,(\beta^b\,dt + dx^b), \tag{2.6}$$

where $h = h_{ab}\,dx^a\,dx^b$ represents the induced Riemannian metric on the slice S. To ensure the non-degeneracy of the metric we assume the *lapse function* α to be positive. We write $\beta = \beta^\mu\,\partial_\mu = \beta^a\,\partial_a$ for the *shift vector field*, which is tangent to the time slices. We assume further that the metric h induced on the time slices is Riemannian and we denote its contravariant form by h^{ab}, so that $h^{ab}\, h_{bc} = \delta^a{}_c$.

The (future-directed) unit normal to the time slices is given by $n^\mu = \frac{1}{\alpha}\,(\delta^\mu{}_0 - \beta^\mu)$, whence $n_\mu = -\alpha\,\delta^0{}_\mu$. We write $h^\mu{}_\nu = g^\mu{}_\nu + n^\mu\, n_\nu$ for the orthogonal projector onto the time slices, $h_{\mu\nu} = g_{\mu\nu} + n_\mu\, n_\nu$ for the 4-dimensional representation of the interior metric on the time slices, and use g to perform index shifts. The second fundamental form (extrinsic curvature) of the time slices is given by

$$\chi_{\mu\nu} \equiv h_\mu{}^\lambda\, h_\nu{}^\rho\, \nabla_\lambda\, n_\rho = \frac{1}{2}\, h_\mu{}^\lambda\, h_\nu{}^\rho\, \mathcal{L}_n\, g_{\lambda\rho} = \frac{1}{2}\, \mathcal{L}_n\, h_{\mu\nu}.$$

For tensor fields $t_\mu \ldots{}^\nu$ intrinsic to the time slices, i.e. having vanishing contractions with n, the h-covariant derivative D on a fixed time slice is given by

$$D_\rho\, t_\mu \ldots{}^\nu = h^\lambda{}_\rho\, h^\phi{}_\mu \ldots h_\psi{}^\nu\, \nabla_\lambda\, t_\phi \ldots{}^\psi.$$

Finally, we write $n(f) = n^\mu\, \partial_\mu\, f$, note that $a_\mu = n^\nu\, \nabla_\nu\, n_\mu = D_\mu \log(\alpha)$, and use the coefficients $\gamma_\mu{}^\nu{}_\rho = \Gamma_\lambda{}^\eta{}_\pi\, h^\lambda{}_\mu\, h^\nu{}_\eta\, h^\pi{}_\rho$ which represent the covariant derivative D in the sense that the components $\gamma_a{}^b{}_c$ are the Christoffel symbols of the metric h_{ab}. The connection coefficients of $g_{\mu\nu}$ can then be written

$$\begin{aligned}\Gamma_\mu{}^\nu{}_\rho &= n_\mu\, n^\nu\, n_\rho\, n(\log\alpha) + n_\mu\, a^\nu\, n_\rho - a_\mu\, n^\nu\, n_\rho - n_\mu\, n^\nu\, a_\rho \\ &+ n_\mu\, n_\rho\, \frac{1}{\alpha}\, \beta^\nu{}_{,\pi}\, n^\pi - n_\mu\, \frac{1}{\alpha}\, \beta^\nu{}_{,\pi}\, h^\pi{}_\rho - n_\rho\, \frac{1}{\alpha}\, \beta^\nu{}_{,\pi}\, h^\pi{}_\mu \\ &+ n^\nu\, \chi_{\mu\rho} - n_\mu\, \chi^\nu{}_\rho - \chi_\mu{}^\nu\, n_\rho + \gamma_\mu{}^\nu{}_\rho.\end{aligned} \tag{2.7}$$

2.2 The Constraints

To isolate the geometrically relevant information contained in the Cauchy data, we reduce the coordinate freedom tranverse to S by assuming T to be a geodesic unit vector field normal to S. The metric then takes the form

$$g = -d\,t^2 + h_{ab}\, d\,x^a\, d\,x^b, \tag{2.8}$$

and the pull-back of the second fundamental to the time slices is given by

$$\chi_{ab} = \frac{1}{2}\, \partial_t\, h_{ab}. \tag{2.9}$$

Expressions (2.8) and (2.9) suggest that the *essential Cauchy data* for the metric field are given by the induced metric h_{ab} and the second fundamental form χ_{ab} on the spacelike hypersurface S. Since coordinate transformations are unrestricted, the n inner relations induced on general Cauchy data on a spacelike hypersurface S must be conditions on the essential Cauchy data.

If we contract the covector ξ, which corresponds in our context to the normal n of S, with the principal symbol of the Einstein tensor $G_{\mu\nu} = R_{\mu\nu} - \frac{1}{2} R\, g_{\mu\nu}$ and evaluated on k, we get

$$\xi^{\mu} \left((\sigma \cdot k)_{\mu\nu} - \frac{1}{2}\, g_{\mu\nu}\, g^{\lambda\rho}\, (\sigma \cdot k)_{\lambda\rho} \right) = 0.$$

This identity indicates the combinations of the equations which contain only derivatives of first order in directions transverse to S. Indeed, if we express the equations

$$0 = Z^{H} \equiv n^{\mu}\, n^{\nu}\, (G_{\mu\nu} + \lambda\, g_{\mu\nu} - \kappa\, T_{\mu\nu}),$$
$$0 = Z^{M}_{\nu} \equiv n^{\mu}\, h_{\nu}{}^{\rho}\, (G_{\mu\rho} + \lambda\, g_{\mu\rho} - \kappa\, T_{\mu\rho}),$$

on S in terms of h and χ, write $\rho = n^{\mu}\, n^{\nu}\, T_{\mu\nu}$, $j_{\nu} = -n^{\mu}\, h^{\rho}{}_{\nu}\, T_{\mu\rho}$, and pull-back to S, we get the *constraint equations on spacelike hypersurfaces*, i.e. the *Hamiltonian constraint*

$$0 = 2\, Z^{H} = r - \chi_{ab}\, \chi^{ab} + (\chi_{a}{}^{a})^{2} - 2\,\lambda - 2\,\kappa\,\rho, \tag{2.10}$$

and the *momentum constraint*

$$0 = Z^{M}_{b} = D_{a}\, \chi_{b}{}^{a} - D_{b}\, \chi_{a}{}^{a} + \kappa\, j_{b}, \tag{2.11}$$

where r denotes the Ricci scalar and D the connection of h.

These equations have the following geometric meaning. On a spacelike hypersurface S the curvature tensors $R^{\mu}{}_{\nu\rho\eta}$ and $r^{\mu}{}_{\nu\rho\eta}$ of g and h respectively and the second fundamental form $\chi_{\mu\nu}$ are related, irrespective of any field equation, by the *Gauss equation*

$$r^{\mu}{}_{\nu\rho\eta} = h^{\mu}{}_{\lambda}\, R^{\lambda}{}_{\pi\phi\psi}\, h^{\pi}{}_{\nu}\, h^{\phi}{}_{\rho}\, h^{\psi}{}_{\eta} - \chi^{\mu}{}_{\rho}\, \chi_{\nu\eta} + \chi^{\mu}{}_{\eta}\, \chi_{\nu\rho} \tag{2.12}$$

and the *Codazzi equation*

$$D_{\mu}\, \chi_{\nu\eta} - D_{\nu}\, \chi_{\mu\eta} = -n_{\lambda}\, R^{\lambda}{}_{\pi\phi\psi}\, h^{\pi}{}_{\eta}\, h^{\phi}{}_{\mu}\, h^{\psi}{}_{\nu}. \tag{2.13}$$

The constraint equations follow from (2.12) and (2.13) by contractions, the use of the field equations, and pull-back to S. Thus the constraints represent the covariant condition for the isometric embeddibility of an *initial data set* $(S, h_{ab}, \chi_{ab}, \rho, j_{a})$ into a solution of the Einstein equations.

We note here that the constraints (2.10) and (2.11), which are analogues of the constraints of Maxwell's equations, have important physical consequences. One of the most important of these is the positivity of the mass which can be associated with an asymptotically flat initial data set (subject to reasonable conditions) [82], [94].

2.3 The Bianchi Identities

Before analysing the structure of the field equations further, we note some important identities. The Riemann tensor Riem$[g]$ of the metric g, given by

$$R^{\mu}{}_{\nu\lambda\rho} = \partial_{\lambda} \Gamma_{\rho}{}^{\mu}{}_{\nu} - \partial_{\rho} \Gamma_{\lambda}{}^{\mu}{}_{\nu} + \Gamma_{\lambda}{}^{\mu}{}_{\delta} \Gamma_{\rho}{}^{\delta}{}_{\nu} - \Gamma_{\rho}{}^{\mu}{}_{\delta} \Gamma_{\lambda}{}^{\delta}{}_{\nu}, \tag{2.14}$$

where $\Gamma_{\rho}{}^{\mu}{}_{\nu}$ denotes the Christoffel symbols of $g_{\mu\nu}$, has the covariance property

$$\mathrm{Riem}[\phi^*(g)] = \phi^*(\mathrm{Riem}[g]),$$

where ϕ denotes a diffeomorphism of M into itself. Two important identities are a direct (cf. [64]) consequence of this, the *first Bianchi identity*

$$R^{\mu}{}_{\lambda\nu\rho} + R^{\mu}{}_{\rho\lambda\nu} + R^{\mu}{}_{\nu\rho\lambda} = 0, \tag{2.15}$$

and the *second Bianchi identity*

$$\nabla_{\mu} R^{\gamma}{}_{\lambda\nu\rho} + \nabla_{\rho} R^{\gamma}{}_{\lambda\mu\nu} + \nabla_{\nu} R^{\gamma}{}_{\lambda\rho\mu} = 0. \tag{2.16}$$

The latter implies the further identities

$$\nabla_{\mu} R^{\mu}{}_{\nu\lambda\rho} = \nabla_{\lambda} R_{\nu\rho} - \nabla_{\rho} R_{\nu\lambda}, \tag{2.17}$$

$$\nabla^{\mu} R_{\mu\nu} - \frac{1}{2} \nabla_{\nu} R = 0. \tag{2.18}$$

The second Bianchi identity will serve us two quite different purposes. Firstly, it will allow us to resolve certain problems arising from the degeneracy of the principal symbol considered above (it is the integrability condition which allows us to show the propagation of suitably chosen gauge conditions and the preservation of the constraints). Secondly, it will provide us with alternative representations of the field equations.

2.4 The Evolution Equations

In this section we shall discuss a few basic ideas about the evolution problem. Our observations about the constraints and the decomposition of

$$Z_{\mu\nu} \equiv G_{\mu\nu} + \lambda\, g_{\mu\nu} - \kappa\, T_{\mu\nu},$$

given by

$$Z_{\mu\nu} = Z^S_{\mu\nu} - n_{\nu}\, Z^M_{\mu} - n_{\mu}\, Z^M_{\nu} + n_{\mu}\, n_{\nu}\, Z^H, \tag{2.19}$$

with $Z^S_{\mu\nu} = h_{\mu}{}^{\lambda}\, h^{\rho}_{\nu}\, Z_{\lambda\rho}$, suggest that the basic information on the evolution equations should be contained in

$$Z^S_{\mu\nu} = 0,$$

or any combination of it with the constraints. To obtain simple expressions in terms of the field $h_{\mu\nu}$ and the second fundamental form $\chi_{\mu\nu}$, defined by the generalization

$$\mathcal{L}_T h_{\mu\nu} = 2\,\alpha\,\chi_{\mu\nu} + \mathcal{L}_\beta\, h_{\mu\nu}, \tag{2.20}$$

of (2.9), we consider the equation

$$0 = Z^S_{\mu\nu} - \frac{1}{2}\,h_{\mu\nu}\,(h^{\lambda\rho}\,Z^S_{\lambda\rho} - Z^H) \tag{2.21}$$

$$= \frac{1}{\alpha}\,(\mathcal{L}_T\,\chi_{\mu\nu} - \mathcal{L}_\beta\,\chi_{\mu\nu} - D_\mu\,D_\nu\,\alpha) + r_{\mu\nu} + \chi_\rho{}^\rho\,\chi_{\mu\nu} - 2\,\chi_{\rho\mu}\,\chi_\nu{}^\rho$$

$$-\lambda\,h_{\mu\nu} - \kappa\,h_\mu{}^\rho\,h_\nu{}^\lambda\,(T_{\rho\lambda} - \frac{1}{2}\,T\,g_{\rho\lambda}).$$

Together with (2.20) it should be regarded as an evolution equation for the fields $h_{\mu\nu}$, $\chi_{\mu\nu}$.

It then appears natural to analyse the general solution of the Einstein equations by the following procedure: Find initial data, i e. a solution h_{ab}, χ_{ab} of the constraints, on the slice $S = \{t = 0\}$. Then find the solution h_{ab}, χ_{ab} of the equations

$$\partial_t\,h_{ab} = 2\,\alpha\,\chi_{ab} + \mathcal{L}_\beta\,h_{ab}, \tag{2.22}$$

$$\partial_t\,\chi_{ab} = -\alpha\,(r_{ab} + \chi_c{}^c\,\chi_{ab} - 2\,\chi_{ac}\,\chi_b{}^c) \tag{2.23}$$

$$+D_a\,D_b\,\alpha + \mathcal{L}_\beta\,\chi_{ab} + \alpha\,(\lambda\,h_{ab} + \kappa\,(T_{ab} - \frac{1}{2}\,T\,h_{ab})),$$

equivalent to (2.20) and (2.21), which induces these data on S. The first step will not be considered further in this article; we shall give some relevant references on the problem of solving the constraint equations in Sect. 6. Here we want to comment on the second step, which raises several questions.

i) What determines the functions α, β^a? Is it possible to prescribe them, at least locally near S, as arbitrary functions $\alpha = \alpha(t, x^c)$, $\beta^a = \beta^a(t, x^c)$ of the coordinates, possibly with the restriction $\alpha^2 - \beta_c\,\beta^c > 0$, which would make ∂_t timelike? We could give a positive answer to this question, if, starting from a representation of the metric (2.6) in terms of some coordinate system $x^{\mu'}$ with $t' = x^{0'} = 0$ on S, we could always find a coordinate transformation $t = t(x^{\mu'})$, $x^a = x^a(x^{\mu'})$ with $t(0, x^{a'}) = 0$, which casts the metric into the desired form, i.e. achieves

$$-\frac{1}{\alpha^2(t, x^c)} = g^{00} = g^{\mu'\nu'}(x^{\lambda'})\,t_{,\mu'}\,t_{,\nu'}, \tag{2.24}$$

$$\frac{1}{\alpha^2(t, x^c)}\,\beta^a(t, x^c) = g^{0a} = g^{\mu'\nu'}(x^{\lambda'})\,t_{,\mu'}\,x^a{}_{,\nu'}. \tag{2.25}$$

If the left hand side of the first equation only depended on t and $x^{\mu'}$, the standard theory of first order PDE's for a single unknown could be applied to this equation and second equation would essentially reduce to an ODE. However, in general this theory does not apply, because the dependence of the function α on x^a introduces a coupling to the second equation.

ii) Suppose we could prescribe lapse and shift arbitrarily. Could we then show the existence of a (unique) solution h_{ab}, χ_{ab} of the initial value problem for the equations (2.22) and (2.23) (possibly coupled to some matter equations)?

iii) Suppose we could answer the last question positively, would the resulting solution to (2.22) and (2.23) then satisfy the constraints (2.10) and (2.11) on the slices $\{t = \text{const.}\}$? Only then would we know that the metric $g_{\mu\nu}$, obtained from our fields h_{ab}, χ_{ab}, α β^c by (2.6), is a solution of the Einstein equations.

We can answer question (iii) as follows. Suppose equation (2.21) is satisfied on a set $M =]-a, c[\times S$, with a, $c > 0$, and the solution induces the given data on $\{0\} \times S$, which we assume to be identified with S in the obvious way. Since (2.21) is equivalent to $Z^S_{\mu\nu} - h_{\mu\nu}\, Z^H = 0$, we can write on M by (2.19)

$$G_{\mu\nu} + \lambda\, g_{\mu\nu} - \kappa\, T_{\mu\nu} = -n_\nu\, Z^M_\mu - n_\mu\, Z^M_\nu + \{2\, n_\mu\, n_\nu + g_{\mu\nu}\}\, Z^H.$$

Taking the divergence, using the contracted Bianchi identity (2.18), assuming that the matter field equations have been given such as to ensure $\nabla^\mu\, T_{\mu\nu} = 0$, and splitting into normal and tangential parts, we get the equations

$$n^\mu\, \nabla_\mu\, Z^H - h^{\mu\nu}\, D_\mu\, Z^M_\nu = 2\, Z^M_\nu\, n^\mu\, \nabla_\mu\, n^\nu - 2\, Z^H\, \nabla_\mu\, n^\mu,$$

$$n^\mu\, \nabla_\mu\, Z^M_\nu - D_\nu\, Z^H = -Z^M_\mu\, \nabla^\mu\, n_\nu - Z^M_\nu\, \nabla_\mu\, n^\mu$$
$$+Z^M_\rho\, n_\nu\, n^\mu\, \nabla_\mu\, n^\rho + 2\, Z^H\, n^\mu\, \nabla_\mu\, n_\nu,$$

which imply *subsidiary equations*, satisfied by Z^H and Z^M_a,

$$\left\{ \begin{pmatrix} \frac{1}{\alpha} & 0 \\ 0 & \frac{1}{\alpha}\, h^{ac} \end{pmatrix} \partial_0 + \begin{pmatrix} -\frac{1}{\alpha}\, \beta^d & -h^{cd} \\ -h^{ad} & -\frac{1}{\alpha}\, h^{ac}\, \beta^d \end{pmatrix} \partial_d \right\} \begin{pmatrix} Z^H \\ Z^M_c \end{pmatrix} = \begin{pmatrix} h \\ h^a \end{pmatrix}, \tag{2.26}$$

where h, h^a denote linear functions of Z^H, Z^M_a. Since it is a system for $v = {}^t(Z^H, Z^M_c)$ of the form

$$A^\mu\, \partial_\mu\, v + B\, v = 0,$$

with symmetric matrices A^μ and a positive definite matrix A^0, it is *symmetric hyperbolic* (cf. Sect. 3.1). Moreover, it has *characteristic polynomial*

$$\det(A^\mu\, \xi_\mu) = -\det(h^{ab})\, (n^\rho\, \xi_\rho)^2\, g^{\mu\nu} \xi_\mu\, \xi_\nu, \tag{2.27}$$

which implies that its characteristics are hypersurfaces which are timelike or null with respect to the metric $g_{\mu\nu}$.

Consequently, if $S' = \{\phi = 0, \phi_{,\mu} \neq 0\} \subset M$, with $\phi \in C^\infty(M)$, is a spacelike hypersurface, the matrix $A^\mu\, \phi_{,\mu}$ is positive definite on S'. Suppose S', S'' are two spacelike hypersurfaces which intersect at their common

2–dimensional boundary Z and bound a compact 'lens-shaped region' in M. Then it follows from the discussion of symmetric hyperbolic systems in Sect. 3.1 that the fields Z^H, Z^M_c must vanish on S'' if they vanish on S'.

To make a precise statement about the consequences of this property, we need the important notion of the domain of dependence. Let us assume that there is given a time orientation on (M, g). If U is a closed subset of M we define the *future (past) domain of dependence of* U *in* M as the set of points $x \in M$ such that any g–non-spacelike curve in M through x which is inextendible in the past (future) intersects U. We denote this set by $D^+(U)$ (resp. $D^-(U)$).

It can be shown that the result about lens-shaped regions referred to above and the fact that the fields Z^H, Z^M_c vanish on S imply that Z^H and Z^M_c, whence also Z^M_μ, vanish on the *domain of dependence* $D(S) = D^+(S) \cup D^-(S)$ of S in M. This shows the *preservation of the constraints* under the evolution defined by equations (2.22) and (2.23) and the prescribed lapse and shift.

Questions (i), (ii) are more delicate. Let us assume that the coefficients $g^{\mu'\nu'}(x^{\lambda'})$ are real analytic functions for $x^{\lambda'}$ in an open subset V of $\mathbb{R}^4$ with $V \cap \{x^{0'} = 0\} \neq \emptyset$, and that $\alpha > 0$ and β^c are real analytic functions of t and x^a. Then equations (2.24) and (2.25) can be written in the form

$$t_{,0'} = F_1(t, x^a, t_{,c'}, x^a{}_{,c'}, x^{\mu'}), \quad x^a{}_{,0'} = F_2(t, x^a, t_{,c'}, x^a{}_{,c'}, x^{\mu'}),$$

with functions F_1, F_2 which are real analytic for $(t, x^a, t_{,c'}, x^a{}_{,c'}, x^{\mu'}) \in \mathbb{R}^{16} \times V$. Thus, by the theorem of Cauchy–Kowalewskaya (cf. [30]), the differential problem considered in question (i) can be solved in a neighbourhood of the set $\{x^{0'} = 0\} \subset V$. Using the covariance of equations (2.24), (2.25), it follows that given a real analytic Lorentz space (M, g), an analytic spacelike hypersurface S in M with coordinates x^a on S, and analytic functions $\alpha = \alpha(t, x^c) > 0$, $\beta^a = \beta^a(t, x^c)$ there exist unique real analytic coordinates t, x^a on some neighbourhood of S in M such that $t = 0$ on S and the lapse and shift in the expression of g in these coordinates are given by α and β^a.

Assume now that the 3–manifold S, the initial data h_{ab}, χ_{ab} solving the constraints on S, as well as the functions $\alpha(t, x^c) > 0$, $\beta^a(t, x^c)$ are real analytic and assume for simplicity that $T_{\mu\nu} = 0$. Then we can derive from (2.22) and (2.23) a differential system for $u = (h_{ab}, k_{abc} \equiv h_{ab,c}, \chi_{ab})$ of the form $\partial_t u = H(u, t, x^a)$ with a function H which is real analytic where $\det(h_{ab}) \neq 0$. Again the theorem of Cauchy–Kowalewskaya tells us that this system, whence also (2.22) and (2.23), has a unique real analytic solution on $\mathbb{R} \times S$ near $\{0\} \times S$ for the data which are given on $\{0\} \times S$ after identifying the latter in the obvious way with S. By our discussion of (iii) we know that we thus obtain a unique analytic solution to the full Einstein equations.

It should be noted that the solution obtained in this way depends a priori not only on the data but also on the chosen lapse and shift. That the latter do in fact only affect the coordinate representation of the solution can be seen as follows. Given another set of analytic functions $\alpha'(t', x'^c) > 0$, $\beta'^a(t', x'^c)$, we can either find, as remarked above, a coordinate transformation $t \to t'$, $x^a \to x'^a$ such that the given metric has the new values of lapse and shift in the new coordinates , or we can deal with the initial value problem for (2.22) and (2.23) with the new lapse and shift. However, due to the uniqueness of this solution it must coincide with the first solution in its new coordinate representation.

It is a remarkable fact that in the course of solving the Einstein equations we can prescribe rather arbitrarily four functions α, β^a (in the analytic case) which are considered first as functions on some abstract $\mathbb{R}^4$ but which, once the solution has been constructed, acquire the meaning of lapse and shift for the coordinate expression of the metric. Since the coordinates in which they have this meaning are defined by α, β^a implicitly (via the field equations), we refer to these functions as the *gauge source functions* of our procedure (we shall see below that, depending on the chosen equations, quite different objects can play the role of gauge source functions) and to the act of prescribing these functions as imposing a *gauge condition.* The considerations above show also that *the manifold on which the solution is constructed must be regarded as part of the solution.* The transition functions relating the different coordinates we have considered, as well that the domains of definition of these coordinate systems themselves, are determined by the gauge source functions, the field equations, and the initial data.

We have seen that in the case where the data and the given lapse and shift are real analytic, we can answer our questions in a satisfactory way. However, the assumption is not satisfactory. This is not meant to say anything against analytic solutions. In fact, most of the 'exact solutions' which are the source of our intuition for general relativistic phenomena are (piecewise) analytic. However, we should not restrict to analyticity in principle. One reason is that it would be in conflict with one of the basic tenets of general relativity. Given two non-empty open subsets U, V of a connected space-time (M, g) such that no point of U can be connected by a causal curve with a point of V, any process in U should be independent of what happens in V. However, if the space-time is analytic, the field in V is essentially fixed by the behaviour of g in U. For instance, we would not be able to study the evolution of data h_{ab}, χ_{ab} on a 3−manifold S where h_{ab} is conformally flat in some open subset of S but not in another one.

Therefore, Einstein's equations should allow us to discuss the existence and uniqueness of solutions, and also the continuous dependence of the latter on the data (stability), in classes of functions which are C^∞ or of even lower smoothness. In other words, Einstein's equations should imply evolution equations for which the Cauchy problem is *well posed* (cf. [45], [49], [91]). Whether an initial value problem is well posed cannot be decided on the level of analytic solutions and with the methods used to prove the Cauchy–Kowalewskaya theorem. On this level there is no basic distinction between initial value problems based on spacelike and those based on timelike hypersurfaces, though in the latter case the stability property is known not to be satisfied. Thus we are led to search for evolution equations which satisfy some 'hyperbolicity condition', i.e. a condition (essentially) on the algebraic structure of the equations which entails the well-posedness of the Cauchy problem.

A number of different hyperbolicity conditions are known, all of them having in common that they require the equations to admit at each point a maximal number of 'real' characteristics: if the equations have a local expression of the form (2.3), then the operator P is hyperbolic at $x \in U$ only if there is a covector $\zeta \in T^*_x M \setminus \{0\}$ such that every 2−dimensional plane in the cotangent space $T^*_x M$ containing ζ intersects the *conormal cone* $\{det(\sum_{|\alpha|=m} A^\alpha \xi^\alpha) = 0\}$ in $k \times m$ real lines (counting multiplicities) (cf. [27]). But notice that this condition alone does not ensure the well-posedness of the Cauchy problem. Some further remarks about different notions of hyperbolicity can be found in Sect. 3.3.

To analyse the situation in the case of (2.22) and (2.23), we solve (2.22) for χ_{ab} and insert into (2.23) to obtain a system of second order for h_{ab} which takes the form

$$\begin{aligned} &\frac{1}{\alpha^2}\{\partial_t^2 h_{ab} - \beta^c \partial_c \partial_t h_{ab} - \beta^c \partial_t \partial_c h_{ab} + \beta^c \beta^d \partial_c \partial_d h_{ab}\} \\ &-h^{cd}(\partial_c \partial_d h_{ab} + \partial_a \partial_b h_{cd} - \partial_a \partial_c h_{bd} - \partial_b \partial_c h_{ad}) \\ &-\frac{2}{\alpha}\partial_a \partial_b \alpha - \frac{2}{\alpha}\left(h_{c(a}\partial_{b)}\partial_t \beta^c - \beta^d h_{c(a}\partial_{b)}\partial_d \beta^c\right) \\ &= \quad \text{terms of lower order in} \quad h_{ab},\ \alpha,\ \beta^c. \end{aligned} \tag{2.28}$$

To analyse the characteristic polynomial, we have to know how α and β^c are related to the solution h_{ab}. Suppose, for simplicity, that α and β^c are given functions.. Then we have to calculate for given covector $\xi_\mu \neq 0$ the determinant of the linear map

$$k_{ab} \mapsto \bar{k}_{ab} = -g^{\mu\nu}\xi_\mu \xi_\nu k_{ab} - h^{cd} k_{cd} \xi_a \xi_b + 2\xi^c k_{c(a}\xi_{b)},$$

of symmetric tensors, where we set $\xi^a = h^{ab}\xi_b$. Denoting by $A'(\xi)$ the linear transformation which maps the independent components k_{ab}, $a \leq b$, onto the $\bar{k}_{ab}$, $a \leq b$, we find

$$\det A'(\xi) = -(n^\mu \xi_\mu)^6 (g^{\mu\nu}\xi_\mu \xi_\nu)^3.$$

Thus, if we consider lapse and shift as given, the conormal cone of system (2.28) satisfies the condition required by hyperbolicity. Moreoever, the characteristics are timelike or null as one would expect for the evolution equations of a theory which is founded on the idea that physical processes propagate on or inside the light cone of the metric field.

However, in spite of the fact that these equations have been used for a long time and in various contexts, and in spite of the naturality of equation

(2.23) which appears to be indicated by the hyperbolicity of the subsidiary equations, it is not known whether the Cauchy problem for equations (2.28) is well posed. It appears that we need to make use of the constraint equations to obtain suitable evolution equations.

Therefore we proceed along a different route. While the form of the constraint equations induced on a given hypersurface is unique, there is a huge freedom to modify our evolution equations. We can try to bring the principal part of (2.28) into a suitable form by using the constraints, by assuming lapse and shift to be functionals of the metric, or by subjecting them to some equations. Then gauge source functions of quite a different nature may appear. Our choice in this section is motivated by the following two observations ([65], [16], cf. also [36]):

(i) Suppose S is some spacelike hypersurface of some Lorentz space (M, g) and x^a are coordinates on some open subset U of S. Let $F^{\mu'} = F^{\mu'}(x^{\nu'})$ be four smooth real functions defined on $\mathbb{R}^4$. Then there exist coordinates $x^{\nu'}$ on some neighbourhood of U in M with $x^{0'} = 0$, $x^{a'} = x^a$ on U and such that the Christoffel coefficients of g in these coordinates satisfy the relations $\Gamma^{\mu'}(x^{\nu'}) = F^{\mu'}(x^{\nu'})$, where $\Gamma^{\mu'} = g^{\lambda' \rho'} \Gamma_{\lambda'}{}^{\mu'}{}_{\rho'}$.

Obviously, one can construct coordinates $x^{\nu'}$ on some neighbourhood of U by solving Cauchy problems for the semi-linear wave equations $\nabla_\mu \nabla^\mu x^{\nu'} = -F^{\mu'}(x^{\rho'})$ with Cauchy data on U which are consistent with our requirements. When the wave equations are expressed in these coordinates the relations above result.

(ii) The 4–dimensional Ricci tensor can be written in the form

$$R_{\mu\nu} = -\frac{1}{2} g^{\lambda\rho} g_{\mu\nu,\lambda\rho} + \nabla_{(\mu}\Gamma_{\nu)} \tag{2.29}$$

$$+\Gamma_\lambda{}^\eta{}_\mu g_{\eta\delta} g^{\lambda\rho} \Gamma_\rho{}^\delta{}_\nu + 2\Gamma_\delta{}^\lambda{}_\eta g^{\delta\rho} g_{\lambda(\mu} \Gamma_{\nu)}{}^\eta{}_\rho,$$

where we set

$$\Gamma_\nu = g_{\nu\mu} \Gamma^\mu, \quad \nabla_\mu \Gamma_\nu = \partial_\mu \Gamma_\nu - \Gamma_\mu{}^\lambda{}_\nu \Gamma_\lambda.$$

Thus, if we consider the Γ_ν as given functions, the Einstein equations take the form of a system of wave equations for the metric coefficients.

Before we show that these observations lead to a short and elegant argument for the well-posedness of the initial value problem for Einstein's equations (assuming well behaved matter equations), we indicate how the form of equations (2.29) relates to our previous considerations.

From the expressions (2.7) we get the relations

$$\partial_t \alpha - \alpha_{,a} \beta^a = \alpha^2 (\chi - n_\nu \Gamma^\nu), \tag{2.30}$$

$$\partial_t \beta^a - \beta^a{}_{,b} \beta^b = \alpha^2 (\gamma^a - D^a \log \alpha - h^a{}_\nu \Gamma^\nu), \tag{2.31}$$

which indicate that prescribing the functions Γ^ν may fix the evolution of lapse and shift. Writing the 3–Ricci tensor similarly to (2.29)

$$r_{ab} = -\frac{1}{2} h^{cd} h_{ab,cd} + D_{(a}\gamma_{b)} + \gamma_c{}^d{}_a h_{fd} h^{ce} \gamma_e{}^f{}_b + 2\gamma_c{}^d{}_e h^{cf} h_{d(a}\gamma_{b)}{}^e{}_f,$$

with

$$\gamma_a = h_{ab}\gamma^b = h_{ab} h^{cd} \gamma_c{}^b{}_d = h^{cd}(h_{ac,d} - \frac{1}{2} h_{cd,a}), \tag{2.32}$$

we obtain (2.28) in the form

$$\frac{1}{\alpha^2}(\partial_t^2 h_{ab} - 2 h_{ab,tc}\beta^c + h_{ab,cd}\beta^c\beta^d) - h^{cd} h_{ab,cd}$$

$$-\frac{2}{\alpha^2} D_{(a}\left[h_{b)c}(\partial_t\beta^c - \beta^d\partial_d\beta^c)\right] + 2 D_{(a}\gamma_{b)} - 2 D_a D_b \log\alpha$$

$$= \quad \text{terms of lower order in} \quad h_{ab},\ \alpha,\ \beta^c.$$

Using (2.31) to replace the terms of second order in the second line, we get

$$-g^{\mu\nu} h_{ab,\mu\nu} = -2 D_{(a}\{h_{b)c} h^c{}_\mu \Gamma^\mu\} \tag{2.33}$$

$$+\frac{2}{\alpha^2}\{D_a\alpha D_b\alpha + \chi_{ab}(\partial_t\alpha - \mathcal{L}_\beta\alpha) + 2 D_{(a}\alpha h_{b)c}(\partial_t\beta^c - \beta^c{}_{,d}\beta^d)$$

$$+2\beta^c{}_{,(a} h_{b)c,t} + h_{c(a}\beta^d{}_{,b)}\beta^c{}_{,d} - \beta^d{}_{,(a}\mathcal{L}_\beta h_{b)d}\}$$

$$+2\,\{2\chi_{ac}\chi_b{}^c - \chi_c{}^c\chi_{ab} - \gamma_c{}^d{}_a h_{fd} h^{ce}\gamma_e{}^f{}_b$$

$$-2\gamma_c{}^d{}_e h^{cf} h_{d(a}\gamma_{b)}{}^e{}_f + \lambda h_{ab} + \kappa(T_{ab} - \frac{1}{2} T h_{ab})\Big\},$$

which can be read as a wave equation for h_{ab}. From (2.30) and (2.22) we get

$$\frac{1}{\alpha^2}(\partial_t - \beta^c\partial_c)^2\alpha = (\partial_t - \beta^c\partial_c)\chi + 2\alpha\chi^2 - 5\alpha^2\chi n_\nu\Gamma^\nu \tag{2.34}$$

$$+3\alpha^3(n_\nu\Gamma^\nu)^2 - \alpha(\partial_t - \beta^c\partial_c)(n_\nu\Gamma^\nu).$$

Taking the trace of (2.23) and using (2.22) and the Hamiltonian constraint (2.10), we get

$$(\partial_t - \beta^c\partial_c)\chi = D_a D^a\alpha - \alpha\left\{\chi_{ab}\chi^{ab} - \lambda + \kappa\frac{1}{2}(\rho + h^{ab} T_{ab})\right\}, \tag{2.35}$$

whence, using (2.34), the wave equation

$$\frac{1}{\alpha^2}(\partial_t - \beta^c\partial_c)^2\alpha - D_a D^a\alpha = -\alpha\,\{\chi_{ab}\chi^{ab} - \lambda \tag{2.36}$$

$$+\kappa\frac{1}{2}(\rho + h^{ab} T_{ab}) + 5\alpha\chi n_\nu\Gamma^\nu - 3\alpha^2(n_\nu\Gamma^\nu)^2 + (\partial_t - \beta^c\partial_c)(n_\nu\Gamma^\nu)\Big\}.$$

From (2.32), (2.22), and the momentum constraint (2.11) follows

$$(\partial_t - \beta^c \partial_c) \gamma^a = h^{cd} \beta^a{}_{,cd} + \alpha D^a \chi + 2 \alpha \kappa j^a \tag{2.37}$$
$$+2 \chi^{ac} D_c \alpha - \chi D^a \alpha - 2 \alpha \chi^{cd} \gamma_c{}^a{}_d - \beta^a{}_{,c} \gamma^c,$$

which implies together with (2.30), (2.31), and (2.22) the wave equation

$$\frac{1}{\alpha^2} (\partial_t - \beta^c \partial_c)^2 \beta^a - h^{cd} \beta^a{}_{,cd} = 2 \alpha \kappa j^a \tag{2.38}$$
$$+4 (\chi^{ac} - \chi h^{ac}) D_c \alpha - 2 \alpha (\chi^{bc} - \chi h^{bc}) \gamma_b{}^a{}_c - \beta^a{}_{,c} \gamma^c + \beta^a{}_{,c} D^c \log \alpha$$
$$-2 \alpha n_\nu \Gamma^\nu (\gamma^a - D^a \log \alpha - h^a{}_\nu \Gamma^\nu) - 2 \alpha \chi h^a{}_\nu \Gamma^\nu$$
$$+D^a(\alpha n_\nu \Gamma^\nu) - (\partial_t - \beta^c \partial_c)(h^a{}_\nu \Gamma^\nu).$$

Equations (2.33), (2.36), and (2.38) form a hyperbolic system for the fields h_{ab}, α, β^a if we consider the functions Γ^μ as given. We note that besides (2.23) we used the constraints as well as (2.7) to derive this system.

We return to the evolution problem. Suppose we are given smooth data h_{ab}, χ_{ab}, ρ, j_a, i. e. a solution of the constraints, on some 3–dimensional manifold S, which, for simplicity, we assume to be diffeomorphic to $\mathbb{R}^3$ and endowed with global coordinates x^a. Following our previous considerations, we set $M = \mathbb{R} \times S$, denote by $t = x^0$ the natural coordinate on $\mathbb{R}$ and extend the x^a in the obvious way to M. We embed our initial data set into M by identifying S diffeomorphically with $\{0\} \times S$ (the need for this embedding shows that it is in general not useful to restrict the choice of the coordinates x^a).

We now choose four smooth real functions $F_\mu(x^\lambda)$ on M which will be assigned the role of gauge source functions in the following procedure. As suggested by (2.29) and the preceeding discussion, we study the Cauchy problem for the *reduced equations*

$$-\frac{1}{2} g^{\lambda\rho} g_{\mu\nu,\lambda\rho} + \nabla_{(\mu} F_{\nu)} + \Gamma_\lambda{}^\eta{}_\mu g_{\eta\delta} g^{\lambda\rho} \Gamma_\rho{}^\delta{}_\nu \tag{2.39}$$
$$+2 \Gamma_\delta{}^\lambda{}_\eta g^{\delta\rho} g_{\lambda(\mu} \Gamma_{\nu)}{}^\eta{}_\rho = -\lambda g_{\mu\nu} + \kappa (T_{\mu\nu} - \frac{1}{2} T g_{\mu\nu}).$$

Since we do not want to get involved in this section with details of the matter equations, we assume the Cauchy problem for them to be well posed and the energy-momentum tensor to be divergence free as a consequence of these equations (one of the remarks following formula (4.70) shows that the situation can be occasionally somewhat more subtle).

To prepare Cauchy data for (2.39), we choose on S a smooth positive lapse function α and a smooth shift vector field β^a, which determine with the datum h_{ab} the ADM representation (2.6) for $g_{\mu\nu}$ on S. Using now h_{ab}, χ_{ab}, α, β^a in equations (2.30), (2.31) (with Γ^ν replaced by $g^{\nu\mu} F_\mu$), and (2.22), we obtain the corresponding datum $\partial_t g_{\mu\nu}$ on S.

It will be shown in Sect. 5 that the Cauchy problem for (2.39) and the initial data above is well posed. Thus there exists a smooth solution $g_{\mu\nu}$ to it on some neighbourhood M' of S. To assure uniqueness, we assume that M' coincides with the domain of dependence of S in M' with respect to $g_{\mu\nu}$.

With the functions Γ_μ calculated from $g_{\mu\nu}$, the reduced equation (2.39) takes the form

$$G_{\mu\nu} + \lambda\, g_{\mu\nu} - \kappa\, T_{\mu\nu} = \nabla_{(\mu}\, D_{\nu)} - \frac{1}{2}\, g_{\mu\nu}\, \nabla_\rho\, D^\rho$$

with $D_\mu = \Gamma_\mu - F_\mu$. To see that our *gauge conditions are preserved* under the evolution defined by (2.39) and our gauge source functions, we show that $D_\mu = 0$ on M'.

Using the Bianchi identity (2.18) and our assumptions on the matter equations, we get, by taking the divergence of the equation above, the subsidiary equation

$$\nabla_\mu\, \nabla^\mu\, D_\nu + R^\mu{}_\nu\, D_\mu = 0.$$

Since we used (2.30) and (2.31) to determine the initial data, we know that $D_\mu = 0$ on S. Moreover, equations (2.36)and (2.38) may be used, on the one hand, to calculate $\partial_t\, \Gamma_\mu$ from $g_{\mu\nu}$, but they are, on the other hand, contained in (2.39), with Γ_μ replaced by F_μ. This implies that $\partial_t\, D_\mu = 0$ on S. Using the uniqueness property for systems of wave equations discussed in Sect. 5.1, we conclude that D_μ vanishes on M' and $g_{\mu\nu}$ solves indeed Einstein's equations (2.1) on M'.

We refer to the process of reducing the initial value problem for Einstein's equations to an initial value problem of a hyperbolic system as a *hyperbolic reduction*. The argument given here was developed for the first time in [16] with the 'harmonicity assumption' $\Gamma_\mu = 0$. We prefer to keep the complete freedom to specify the gauge source functions, because some important and complicated problems are related to this. Our derivation of the system (2.33), (2.36), (2.38) illustrates the intricate relations between equation (2.23), the constraints, and the conservation of the gauge condition in our case. Though other such arguments may differ in details, the overall structure of all hyperbolic reduction procedures are similar.

2.5 Assumptions and Consequences

If one wants to investigate single solutions or general classes of solutions to the Einstein equations by an abstract analysis of the hyperbolic reduced equations, there are certain properties which need to be 'put in by hand' and others which are determined by the field equations.

The properties in the first class depend largely on the type of physical systems which are to be modelled and on the structure of the hypersurface on which data are to be prescribed. The latter could be spacelike everywhere as in the Cauchy problem considered above. One may wish in this case to consider cosmological models with compact time slices, one of which is the initial hypersurface, one may wish to model the field of an isolated system, in which case the initial hypersurface should include a domain extending to infinity with the field satisfying certain fall-off conditions, or one may wish to consider initial hypersurfaces with 'inner boundaries' whose nature may depend on various geometrical and physical considerations. Other cases of interest are the 'initial-boundary value problem'(cf. Sect. 4.3), where data are given on a spacelike and a timelike hypersurface which intersect in some spacelike 2-surface, the 'characteristic initial value problem' where data are given on two null hypersurfaces which intersect in some spacelike two-surface, or various other combinations of hypersurfaces. Finally, we have to make a choice of matter model which may introduce specific problems which are independent of the features of Einstein's equations considered above. All these considerations will affect the problem of finding appropriate solutions to the corresponding constraint equations and the nature of this problem depends to a large extent on the nature of the bounding hypersurface. We note that the fall-off behaviour of the data on spacelike hypersurfaces can only partly be specified freely, other parts being determined by the constraint equations.

There are conditions which we just assume because we cannot do better. For instance, one may wish to analyse solutions which violate some basic causality conditions (cf. [52]) or which admit various identifications in the future and past of the initial hypersurface. Such violations or identifications introduce compatibility conditions on the data which cannot be controlled locally. At present there are no techniques available to analyse such situations in any generality. In the context of the Cauchy problem we shall only consider solutions (M, g) of the Einstein equations which are *globally hyperbolic* and such that the initial hypersurface S is a *Cauchy hypersurface* of the solution, i.e. every inextendible non-spacelike curve in M intersects S exactly once [52].

The class of properties which need to be inferred from the structure of the data and the field equations includes in general everything which has to do with the long time evolution of the field: the development of singularities and horizons or of asymptotic regimes where the field becomes in any sense weak and possibly approximates the Minkowski field. This does not, of course, preclude the possibility of making assumptions on the singular or asymptotic behaviour of the field and analysing the consistency of these assumptions with the field equations. But in the end we would like to characterize the long time behaviour of the field in terms of the initial data.

3 PDE Techniques

3.1 Symmetric Hyperbolic Systems

In order to obtain results on the existence, uniqueness and stability of solutions of the Einstein equations it is necessary to make contact with the theory of partial differential equations. A good recent textbook on this theory is [32]. One part of the theory which can usefully be applied to the Einstein equations is the theory of symmetric hyperbolic systems. This will be discussed in some detail in the following. The aim is not to give complete proofs of the results of interest in general relativity, but to present some arguments which illustrate the essential techniques of the subject. It is important to note that the use of symmetric hyperbolic systems is not the only way of proving local existence and uniqueness theorems for the Einstein equations. The original proof [16] used second order hyperbolic equations. Other approaches use other types of equations such as mixed hyperbolic-elliptic systems (see, e.g. [25], Chap. 10). The basic tools used to prove existence are the same in all cases. First a family of approximating problems is set up and solved. Of course, in order that this be helpful, the approximating problems should be easier to solve than the original one. Then energy estimates, whose definition is discussed below, are used to show that the solutions of the approximating problems converge to a solution of the original problem in a certain limit.

Now some aspects of the theory of symmetric hyperbolic systems will be discussed. We consider a system of equations for k real variables which are collected into a vector-valued function u. The solution will be defined on an appropriate subset of $\mathbb{R} \times \mathbb{R}^n$. (The case needed for the Einstein equations is $n = 3$.) A point of $\mathbb{R} \times \mathbb{R}^n$ will be denoted by (t, x). The equations are of the form:

$$A^0(t,x,u)\partial_t u + A^i(t,x,u)\partial_i u + B(t,x,u) = 0$$

This system of equations is called symmetric hyperbolic if the matrices A^0 and A^i are symmetric and A^0 is positive definite. This system is quasi-linear, which means that it is linear in its dependence on the first derivatives. The notion of symmetric hyperbolicity can be defined more generally, but here we restrict to the quasi-linear case without further comment. It is relatively easy to prove a uniqueness theorem for solutions of a symmetric hyperbolic system and so we will do this before coming to the more complicated existence proofs. A hypersurface S is called spacelike with respect to a solution u of the equation if for any 1-form n_α conormal to S (i.e. vanishing on vectors tangent to S) the expression $A^\alpha(t,x,u)n_\alpha$ is positive definite. This definition has a priori nothing to do with the usual sense of the word spacelike in general relativity. However, as will be seen below, the two concepts are closely related in some cases. Define a *lens-shaped region* to be an open subset G of $\mathbb{R} \times \mathbb{R}^n$ with compact closure whose boundary is the union of a subset S_0 of the hypersurface $t = 0$ with smooth boundary ∂S_0 and a spacelike hypersurface S_1 with a boundary which coincides with ∂S_0. It will be shown that if G is a lens-shaped region with respect to solutions u_1 and u_2 of a symmetric hyperbolic system and if the two solutions agree on S_0 then they agree on all of G. This statement is subject to differentiability requirements.

Consider a symmetric hyperbolic system where A^α and B are C^1 functions of their arguments. Let u_1 and u_2 be two solutions of class C^1. Let G be a lens-shaped region with respect to u_1 and u_2 whose boundary is the union of hypersurfaces S_0 and S_1 as before. There exist continuous functions M^α and N such that:

$$\begin{aligned} A^\alpha(t,x,u_1) - A^\alpha(t,x,u_2) &= M^\alpha(t,x,u_1,u_2)(u_1-u_2) \\ B(t,x,u_1) - B(t,x,u_2) &= N(t,x,u_1,u_2)(u_1-u_2) \end{aligned}$$

This sharp form of the mean value theorem is proved for instance in [50]. It follows that:

$$A^\alpha(u_1)\partial_\alpha(u_1-u_2) + [M^\alpha(u_1,u_2)(\partial_\alpha u_2) + N(u_1,u_2)](u_1-u_2) = 0$$

Here the dependence of functions on t and x has not been written out explicitly. This equation for $u_1 - u_2$ can be abbreviated to

$$A^\alpha(u_1)\partial_\alpha(u_1-u_2) = Q(u_1-u_2)$$

where Q is a continuous function of t and x. It follows that

$$\begin{aligned} &\partial_\alpha(\langle e^{-kt}A^\alpha(u_1)(u_1-u_2), u_1-u_2\rangle) \\ &= e^{-kt}\langle[-kA^0(u_1) + (\partial_\alpha A^\alpha)(u_1) + 2Q](u_1-u_2), u_1-u_2\rangle \end{aligned}$$

for any constant k. Now apply Stokes' theorem to this on the region G. This gives an equation of the form $I_1 = I_0 + I_G$ where I_0, I_1 and I_G are integrals over S_0, S_1 and G respectively. Because S_1 is spacelike, I_1 is non-negative. If k is chosen large enough the matrix $P = kA^0(u_1) - (\partial_\alpha A^\alpha)(u_1) - 2Q$ is uniformly positive definite on G. This means that there exists a constant $C > 0$ such that $\langle v, P(t,x,u_1)v\rangle \geq C\langle v,v\rangle$ for all $v \in \mathbb{R}^k$ and (t,x) in G. Thus if $u_1 - u_2$ is not identically zero on G, then the volume integral I_G can be made negative by an appropriate choice of k. If u_1 and u_2 have the same initial data then $I_0 = 0$. In this case $I_0 + I_G$ is negative and we get a contradiction. Thus in fact $u_1 - u_2 = 0$.

This argument shows that locally in a neighbourhood of the initial hypersurface S the solution of a symmetric hyperbolic system at a given point is determined by initial data on a compact subset of S. For any point near enough to S is contained in a lens-shaped region. In this context we would like to introduce the concept of domain of dependence for a symmetric hyperbolic system. There are problems with conflicting terminology here, which we will attempt to explain now. If a given subset E of S is chosen, then the *domain of dependence* $\tilde{D}(E)$ of E is the set of all points of $\mathbb{R}\times\mathbb{R}^n$ such that the value of a solution of the symmetric hyperbolic system at that point is determined by the restriction of the initial data to E. Comparing with the definition of the domain of dependence for the Einstein equations given in Sect. 2.4 we find a situation similar to that seen already for the term 'spacelike'. The applications of the term to the Einstein equations and to symmetric hyperbolic systems might seem at first to be completely unrelated, but in fact there is a close relation. This fact was already alluded to in Sect. 2.4. Even for hyperbolic equations there is another ambiguity in terminology. The definition for symmetric hyperbolic systems conflicts with terminology used in some places in the literature [27]. Often the set which is called domain of dependence here is referred to as the domain of determinacy of E while E is said to be a domain of dependence for a point p if p lies in what we here call the domain of dependence. In the following we will never use the terminology of [27]. However, we felt it better to warn the reader of the problems than to pass over them in silence.

The relationship of the general definition of the domain of dependence for symmetric hyperbolic systems adopted in this article with the domain of dependence in general relativity will now be discussed in some more detail. As will be explained later, the harmonically reduced Einstein equations can be transformed to a symmetric hyperbolic system by introducing the first derivatives of the metric as new variables. The comparison we wish to make is between the domain of dependence $\tilde{D}(S)$ defined by this particular symmetric hyperbolic system and the domain of dependence $D(S)$ as defined in Sect. 2.4. The first important point is that a hypersurface is spacelike with respect to the symmetric hyperbolic system if and only if it is spacelike in the usual sense of general relativity, i.e. if the metric induced on it by the space-time metric is positive definite. Given this fact, we see that the notion of lens-shaped region used in this section coincides in the case of this particular symmetric hyperbolic system derived from the Einstein equations with the concept introduced in Sect. 2.4.

The above uniqueness statement will now be compared for illustrative purposes with the well-known uniquess properties of solutions of the wave equation in Minkowski space. Here we see a simplified form of the situation which has just been presented in the case of the Einstein equations. The wave equation, being second order, is not symmetric hyperbolic. However, it can be reduced to a symmetric hyperbolic system by introducing first derivatives of the unknown as new variables in a suitable way. It then turns out that once again the concept of a spacelike hypersurface, as introduced for symmetric hyperbolic systems, coincides with the usual (metric) notion for hypersurfaces in Minkowski space. It is well known that if initial data for the wave equation is given on the hypersurface $t = 0$ in Minkowski space, and if p is a point in the region $t > 0$, then the solution at p is determined uniquely by the data within the intersection of the past light cone of p with the hypersurface $t = 0$. Once the wave equation has been reduced to symmetric hyperbolic form this statement can be deduced from the above uniqueness theorem for symmetric hyperbolic systems. It suffices to note the simple geometric fact that any point in the region $t > 0$ strictly inside the past light cone of p is contained in a lens-shaped region with the same property. Thus the solution is uniquely determined inside the past light cone of p and hence, by continuity, at p.

The existence of the domain of dependence has the following consequence. If two sets u_1^0 and u_2^0 of initial data on the same initial hypersurface S coincide on an open subset U of S, then the corresponding domains of dependence $\tilde{D}_1(U)$ and $\tilde{D}_2(U)$, as well as the corresponding solutions u_1 and u_2 on them, coincide. In other words, on $\tilde{D}_1(U)$ the solution u_1 is independent of the extension of u_1^0 outside U. It follows in particular that there is no need to impose boundary conditions or fall-off conditions on u_1 'on the edge of S' or some periodicity condition to determine the solution locally near a point of S. This is referred to as the localization property of symmetric hyperbolic systems since it means that the theory is not dependent on knowing how the initial data behave globally in space. In the following data on $\mathbb{R}^n$ will be considered which are periodic in the spatial coordinates. This is equivalent to replacing $\mathbb{R}^n$ as domain of definition of the unknown by the torus T^n obtained by identifying the Cartesian coordinates in $\mathbb{R}^n$ modulo 2π. In order to prove statements about solutions of a symmetric hyperbolic system, something must be assumed about the regularity of the coefficients. This will in particular imply that A^0 is continuous. The continuity of A^0 and the compactness of the torus then show together that $A^0(t, x, u)$ is uniformly positive definite on any finite closed time interval for any given continuous function u defined on this interval.

The general strategy of the existence proof we will present here will now be discussed in more detail. This proof is essentially that given in [90]. The first detailed proof of an existence theorem for general quasi-linear symmetric hyperbolic systems was given by Kato [60] using the theory of semigroups. The proof here uses less sophisticated functional analysis, but the basic pattern of approximations controlled by energy estimates is the same in both cases. For simplicity we will restrict to the special case where A^0 is the identity matrix. Once the proof has been carried out in that case the differences which arise in the general case will be pointed out. Assuming now that A^0 is the identity, the equation can be written in the form:

$$\partial_t u = -A^i(t, x, u)\partial_i u - B(t, x, u)$$

This looks superficially like an ordinary differential equation in an infinite dimensional space of functions of x. Unfortunately, this point of view is not directly helpful in proving local existence. The essential point is that if $A^i(t, x, u)\partial_i$ is thought of as an operator on a space of functions of finite differentiability, this operator is unbounded. The strategy is now to replace the unbounded operator by a bounded one. In fact we even go further and replace the infinite dimensional space by a finite dimensional one. This is done by the introduction of a mollifier (smoothing operator) at appropriate places in the equation. The mollifier contains a parameter ϵ. The resulting family of equations depending on ϵ defines the family of approximating problems referred to above in this particular case. The approximating problems can be solved using the standard theory of ordinary differential equations. It then remains to show that the solutions to the approximating problems converge to a solution of the original problem as ϵ tends to zero.

A convenient mollifier on the torus can be constructed using the Fourier transform. We now recall some facts concerning the Fourier transform on the torus. If u is a continuous complex-valued function on T^n then its Fourier coefficients are defined by:

$$\hat{u}(\xi) = \int_{T^n} u(x) e^{-i\langle x,\xi\rangle} dx$$

Here ξ is an element of Z^n, i.e. a sequence of n integers. In fact this formula makes sense for any square integrable function on the torus. It defines a linear mapping from the space $L^2(T^n)$ of complex square integrable functions on the torus to the space of all complex-valued functions on Z^n. Its image is the space of square summable functions $L^2(Z^n)$ and the L^2 norm of u is equal to that of $\hat{u}$ up to a constant factor. In particular the Fourier transform defines an isomorphism of $L^2(T^n)$ onto $L^2(Z^n)$. Thus to define an operator on $L^2(T^n)$ it suffices to define the operator on $L^2(Z^n)$ which corresponds to it under the Fourier transform. These statements follow from the elementary theory of Hilbert spaces once it is known that trigonometric polynomials (i.e. finite linear combinations of functions of the form $e^{i\langle\xi,x\rangle}$) can be used to approximate all continuous functions on the torus in the sense of uniform convergence. This is worked out in detail for the case $n = 1$ in [80], Chap. 4. The case of general n is not much different, once the approximation property of trigonometric polynomials is known in that context. This follows, for instance, from the Stone–Weierstrass theorem ([81], p. 122).

Let ϕ be a smooth real-valued function with compact support on $\mathbb{R}^n$ which is identically one in a neighbourhood of the origin and satisfies $0 \leq \phi(\xi) \leq 1$ for all ξ and $\phi(-\xi) = \phi(\xi)$. For a positive real number ϵ let $\phi_\epsilon(\xi) = \epsilon^{-n}\phi(\xi/\epsilon)$. If we identify Z^n with the set of points in $\mathbb{R}^n$ with integer coordinates it is possible to define functions ϕ_ϵ on Z^n by restriction. The mollifier J_ϵ is defined by

$$\widehat{J_\epsilon u} = \phi_\epsilon \hat{u}$$

for any square integrable complex-valued function u on the torus. It is a bounded self-adjoint linear operator on $L^2(T^n)$. In fact $\|J_\epsilon\| = 1$ for all ϵ. The symmetry condition imposed on ϕ ensures that if u is real, $J_\epsilon u$ is also real. Note that, for given ϵ, $\widehat{J_\epsilon u}$ is only non-zero at a finite number of points, the number of which depends on ϵ. It follows that $J_\epsilon u$ is a trigonometric polynomial and that the space of trigonometric polynomials which occurs for fixed ϵ and all possible functions u is finite dimensional. As a consequence the image of $L^2_{\mathbb{R}}(T^n)$, the subspace of $L^2(T^n)$ consisting of real-valued functions, under the operator J_ϵ is a finite dimensional space V_ϵ. As ϵ tends to zero the dimension of V_ϵ tends to infinity and this is why these spaces can in a certain sense approximate the infinite-dimensional space $L^2_{\mathbb{R}}(T^n)$.

The original problem will now be approximated by problems involving ordinary differential equations on the finite-dimensional spaces V_ϵ. If the initial datum for the original problem is u^0, the initial datum for the approximating problem will be $u_\epsilon^0 = J_\epsilon u^0$. The equation to be solved in the approximating problem is:

$$\partial_t u_\epsilon = -J_\epsilon[A^i(t,x,u_\epsilon)J_\epsilon \partial_i u_\epsilon + B(t,x,u_\epsilon)] = F(u_\epsilon) \tag{3.1}$$

Here the unknown u_ϵ takes values in the vector space V_ϵ. In order to apply the standard existence and uniqueness theory for ordinary differential equations to this it suffices to check that the right hand side is a smooth function of u_ϵ provided A^i and B are smooth functions of their arguments. Let us first check that F is continuous. If u_n is a sequence of elements of V_ϵ which converges to u then $J_\epsilon \partial_i u_n$ converges to $J_\epsilon \partial_i u$. The sequence of functions $A^i(t,x,u_n)J_\epsilon \partial_i u_n + B(t,x,u_n)$ converges uniformly to $A^i(t,x,u)J_\epsilon \partial_i u + B(t,x,u)$. To see that F is continous it remains to show that if $v_n \to v$ uniformly $J_\epsilon v_n \to J_\epsilon v$. If $v_n \to v$ uniformly then the convergence also holds in the L^2 norm. This implies that $\hat{v}_n \to \hat{v}$ in $L^2(Z^n)$. Hence $\phi_\epsilon \hat{v}_n \to \phi_\epsilon \hat{v}$ pointwise. From this it can be concluded that $J_\epsilon v_n \to J_\epsilon v$. This completes the proof of the continuity of F. Differentiability can be shown in a similar way. The main step is to show that if u and v are elements of V_ϵ then $\lim_{s\to 0} s^{-1}[A^i(t,x,u+sv) - A^i(t,x,u)]$ exists (in the sense of uniform convergence). By smoothness of A^i it does exist and is equal to $D_3 A(t,x,u)v$, where D_3 denotes the derivative with respect to the third argument. In this way the existence of the first derivative of F can be demonstrated and the explicit expression for the derivative shows that it is continuous. Higher derivatives can be handled similarly. Hence F is a smooth function from V_ϵ to V_ϵ. The standard theory of ordinary differential equations now gives the existence of a unique solution of the approximating problem for each positive value of ϵ. Existence is only guaranteed for a short time T_ϵ and at this point in the argument it is not excluded that T_ϵ could tend to zero as $\epsilon \to 0$. It will now be seen that this can be ruled out by the use of energy estimates.

The basic energy estimate involves computing the time derivative of the energy functional defined by

$$\mathcal{E} = \int_{T^n} |u_\epsilon|^2 dx \tag{3.2}$$

Since the functions u_ϵ are smooth and defined on a compact manifold differentiation under the integral is justified.

$$\begin{aligned} d\mathcal{E}/dt &= 2\int \langle u_\epsilon, \partial_t u_\epsilon \rangle \\ &= -2\int \langle u_\epsilon, J_\epsilon[A^i J_\epsilon \partial_i u_\epsilon + B] \rangle \end{aligned}$$

Using the fact that J_ϵ commutes with the operators ∂_i and is self-adjoint, gives

$$\int \langle u_\epsilon, J_\epsilon[A^i J_\epsilon \partial_i u_\epsilon] \rangle = \int \langle J_\epsilon u_\epsilon, A^i \partial_i J_\epsilon u_\epsilon \rangle$$

Using Stokes theorem and the symmetry of A^i gives

$$\int \langle J_\epsilon u_\epsilon, A^i \partial_i J_\epsilon u_\epsilon\rangle = -\int \langle \partial_i A^i J_\epsilon u_\epsilon, J_\epsilon u_\epsilon\rangle - \int \langle A^i J_\epsilon \partial_i u_\epsilon, J_\epsilon u_\epsilon\rangle$$

It follows that

$$\int \langle u_\epsilon, J_\epsilon[A^i J_\epsilon \partial_i u_\epsilon]\rangle = -\frac{1}{2}\int \langle \partial_i A^i J_\epsilon u_\epsilon, J_\epsilon u_\epsilon\rangle$$

Substituting this in equation (3.2) gives:

$$\partial_t \mathcal{E} = \int \langle (\partial_i A^i) J_\epsilon u_\epsilon - 2B, J_\epsilon u_\epsilon\rangle$$

Now

$$\|(\partial_i A^i) J_\epsilon u_\epsilon - 2B\|_{L^2} \leq \|\partial_i A^i\|_{L^\infty} \|u_\epsilon\|_{L^2} + 2\|B\|_{L^2}$$

Hence it follows by the Cauchy-Schwarz inequality that

$$\partial_t \mathcal{E} \leq \|\partial_i A^i\|_{L^\infty} \mathcal{E} + 2\|B\|_{L^2} \mathcal{E}^{1/2}$$

This is the fundamental energy estimate. Note that this computation is closely related to that used to prove uniqueness for solutions of symmetric hyperbolic systems above.

The existence proof also requires higher order energy estimates. To obtain these, first differentiate the equation one or more times with respect to the spatial coordinates. Higher order energy functionals are defined by

$$\mathcal{E}_s = \sum_{|\alpha|\leq s} \int_{T^n} (D^\alpha u_\epsilon)^2 dx$$

The square root of the energy functional $\mathcal{E}_s$ is a norm which defines the Sobolev space H^s. It is because of the energy estimates that Sobolev spaces play such an important role in the theory of hyperbolic equations. Differentiating the equation for (3.1 gives

$$\begin{aligned}\partial_t(D^\alpha u_\epsilon) = &-J_\epsilon[A^i(t,x,u_\epsilon) J_\epsilon \partial_i D^\alpha u_\epsilon] - J_\epsilon[D^\alpha(A^i(t,x,u_\epsilon) J_\epsilon \partial_i u) \\ &-A^i(t,x,u_\epsilon) J_\epsilon \partial_i D^\alpha u_\epsilon] - J_\epsilon D^\alpha(B(t,x,u_\epsilon))\end{aligned} \tag{3.3}$$

Differentiating the expression for $\mathcal{E}_s$ with respect to time causes the quantity $\partial_t(D^\alpha u_\epsilon)$ to appear. This can be substituted for using the equation (3.3). Stokes theorem can be used to eliminate the highest order derivatives, as in the basic energy estimate. To get an inequality for $\mathcal{E}_s$ similar to that derived above for $\mathcal{E}$ it remains to obtain L^2 estimates for the quantities

$$D^\alpha(A^i(t,x,u_\epsilon) J_\epsilon \partial_i u_\epsilon) - A^i(t,x,u_\epsilon) J_\epsilon \partial_i D^\alpha u_\epsilon$$

and

$$D^\alpha B(t,x,u_\epsilon)$$

This can be done by means of the Moser inequalities, which will be stated without proof. (For the proofs see [90].)

The first Moser estimate says that there exists a constant $C > 0$ such that for all bounded functions f, g on T^3 belonging to the Sobolev space H^s the following inequality holds:

$$\|D^\alpha(fg)\|_{L^2} \leq C(\|f\|_{L^\infty}\|D^s g\|_{L^2} + \|D^s f\|_{L^2}\|g\|_{L^\infty})$$

Here $s = |\alpha|$ and for a given norm $\|D^s g\|$ is shorthand for the maximum of $\|D^\alpha g\|$ over all multiindices α with $|\alpha| = s$. The second estimate says that there exists a constant $C > 0$ such that for all bounded functions f, g on T^3 such that the first derivatives of g are bounded, f is in H^s and g is in H^{s-1} the following inequality holds:

$$\|D^\alpha(fg) - fD^\alpha g\|_{L^2} \leq C(\|D^s f\|_{L^2}\|g\|_{L^\infty} + \|Df\|_{L^\infty}\|D^{s-1}g\|_{L^2})$$

The third estimate concerns composition with nonlinear functions. Let F be a smooth function defined on an open subset of $\mathbb{R}^k$ and taking values in $\mathbb{R}^k$. There exists a constant $C > 0$ such that for all functions f on T^3 taking values in a fixed compact subset K of U and belonging to the Sobolev space H^s and any multiindex α with $s = |\alpha| \geq 1$ the following inequality holds:

$$\|D^\alpha(F(f))\|_{L^2} \leq C\|Df\|_{L^\infty}^{s-1}\|D^s f\|_{L^2}$$

The result of using the Moser inequalities in the expression for $d\mathcal{E}_s/dt$ is an estimate of the form:

$$\partial_t \mathcal{E}_s \leq C\mathcal{E}_s$$

where the constant C depends on a compact set K in which u_ϵ takes values, as above, and the L^∞ norm of the first derivatives of u_ϵ. This can be estimated by a C^1 function of the C^1 norm of u_ϵ. By the Sobolev embedding theorem (see e.g. [63]), there is a constant C such that $\|u_\epsilon\|_{C^1} \leq C\|u_\epsilon\|_{H^s}$ for any $s > n/2 + 1$. Thus we obtain a differential inequality of the form

$$\partial_t \mathcal{E}_s \leq f(\mathcal{E}_s)$$

for a C^1 function f. It follows that the quantity $\mathcal{E}_s(t)$ satisfies the inequality

$$\mathcal{E}_s(t) \leq z(t)$$

where $z(t)$ is the solution of the differential equation $dz/dt = f(z)$ with initial value $\mathcal{E}(0)$ at $t = 0$. (A discussion of comparison arguments of this type can be found in [51], Chap. 3.) Since the function z remains finite on some time interval $[0, T]$, the same must be true for $\mathcal{E}_s$ for any $s > n/2 + 1$ and the solutions u_ϵ exist on the common interval $[0, T]$ for all ϵ. Morover $\|u_\epsilon(t)\|_{H^s}$ is bounded independently of ϵ for all $t \in [0, T]$. Putting this in the equation shows that $\|\partial_t u_\epsilon(t)\|_{H^{s-1}}$ is also bounded.

The proof of existence of a solution with the given initial datum now follows by some functional analysis. The main point is that the boundedness of a sequence of functions in some norm often implies the existence of a sequence which is convergent in another topology. The reader who does not possess and does not wish to acquire a knowledge of functional analysis may wish to skip the rest of this paragraph. The space defined by the L^∞ norm of $\|u(t)\|_{H^s}$ is a Banach space and is also the dual of a separable Banach space. The family u_ϵ is bounded in this space and hence, by the Banach–Alaoglu theorem (see [81], pp. 68-70) there is a sequence ϵ_k tending to zero such that $u_{\epsilon_k}(t)$ converges uniformly in the weak topology of $H^s(T^n)$ to some function u which is continuous in t with values in $H^s(T^n)$ with repect to the weak topology. The same argument can be applied to the family $\partial_t u_\epsilon$. As a result, after possibly passing to a subsequence, $\partial_t u_{\epsilon_k}(t)$ converges uniformly in the weak topology of $H^{s-1}(T^n)$. On the other hand, since T^n is compact, the inclusion of $H^s(T^n)$ in $H^{s-1}(T^n)$ is compact. (This is Rellich's Lemma. See e.g. [47], p. 88) Thus, by the vector-valued Ascoli theorem [29] it can be assumed, once again passing to a subsequence if necessary, that $u_{\epsilon_k}(t)$ converges uniformly in the norm topology of $H^{s-1}(T^n)$. If $s > n/2 + 2$ then the Sobolev embedding theorem can be used to deduce that it converges uniformly in the norm of $C^1(T^n)$. Once this is known it follows directly that the expression on the right hand side of the approximate equation converges uniformly to that on the right hand side of the exact equation. On the other hand $\partial_t u_k$ converges to $\partial_t u$ in the sense of distributions and so u satifies the original equation in the sense of distributions. The equation then implies that the solution is C^1 and is a classical solution. It also has the desired initial datum.

Remarks

1. The existence and uniqueness theorem whose proof has just been sketched says that given an initial datum u_0 in $H^s(T^n)$ with $s > n/2 + 2$ there is a solution for this datum which is a bounded measurable function $u(t)$ with values in $H^s(T^n)$ and is such that $\partial_t u$ is a bounded measurable function with values in $H^{s-1}(T^n)$. With some further work it can be shown that $n/2 + 2$ can be replaced by $n/2 + 1$ and that the bounded measurable functions are in fact continuous with values in the given space [90].
2. The time of existence of a solution which is continuous with values in $H^s(T^n)$ depends only on a bound for the norm of the initial data in $H^s(T^n)$.
3. The analogues of the higher energy estimates whose derivation was sketched here for the solutions of the approximating problems hold for the solution itself [90]. This implies that as long as the C^1 norm of $u(t)$ remains bounded the H^s norm also remains bounded. This leads to the following continuation criterion. If $u(t)$ is a solution on an interval $[0, T)$ and if $\|u(t)\|_{C^1}$ is bounded independently of t by a constant $C > 0$ then the solution exists on a longer time interval. For there is some T_1 such that a solution corresponding to any data u_0 with $\|u_0\|_{H^s} \leq C$ exists on $[0, T_1)$. Considering setting data at times

just preceding T, and using uniqueness to show that the solutions obtained fit together to give a single solution shows the existence of a solution of the original problem on the interval $[0, T + T_1)$.
4. The previous remark implies the existence of C^∞ solutions corresponding to C^∞ data. For suppose that a datum u_0 of class C^∞ is given. Then u_0 belongs to $H^s(T^n)$ for each s. Thus there is a corresponding solution on an interval $[0, T_s)$ which is continuous with values in H^s. Assume that T_s has been chosen as large as possible. It is not possible that $T_s \to 0$ as $s \to \infty$. For the boundedness of the solution in H^s, $s > n/2 + 1$ implies its boundedness in C^1 and this in turn implies that the solution can be extended to a longer time interval. Thus in fact T_s does not depend on s. Using the equation then shows that the solution is C^∞.
5. Analogous statements to all of the above can be obtained for the case where A^0 is not the identity with similar techniques. The essential point is to modify the expressions for the energy functionals using A^0. For instance the basic energy for the approximating problems is given by $\mathcal{E} = \int \langle A^0 u_\epsilon, u_\epsilon \rangle$.
6. Our treatment here differs slightly from that of Taylor [90] by the use of a reduction to finite-dimensional Banach spaces. Which variant is used is a matter of taste.

The statement made in the second remark is a part of what is referred to as Cauchy stability. This says that when the initial data is varied the corresponding solution changes in a way which depends continuously on the data. This assertion applies to solutions defined on a fixed common time interval $[0, T]$. The above remark shows that a common time interval of this kind can be found for data which are close to a given datum in $H^s(T^n)$. A related statement is that if the coefficients of the equation and the data depend smoothly on a parameter, then for a compact parameter interval the corresponding solutions for different parameter values exist on a common time interval $[0, T]$ and their restrictions to this time interval depend smoothly on the parameter.

3.2 Symmetric Hyperbolic Systems on Manifolds

In the last section it was shown how existence and uniqueness statements can be obtained for solutions of symmetric hyperbolic systems with data on a torus. It was also indicated that these imply results for more general settings by means of the domain of dependence. In this section some more details concerning these points will be provided. First it is necessary to define what is meant by a symmetric hyperbolic system on a manifold M. (In the context of the last section $M = T^3 \times I$.) In general this will be an equation for sections of a fibre bundle E over M which, for simplicity, we will take to be a vector bundle. The easiest definition is that a symmetric hyperbolic system is an equation of the form $P(u) = 0$, where P is a nonlinear differential operator on M which in any local coordinate system satisfies the definition given previously. For the general invariant definition of a differential operator on sections of a bundle the reader is referred to [71]. The object defined in local coordinates by $\sigma(\xi) = A^\alpha \xi_\alpha$ is called the principal symbol of the differential operator (cf. Sect. 2.1). Here A^α is supposed to be evaluated on a given section u. The principal symbol is defined invariantly by the differential operator. Let $L(V)$ be the bundle of linear mappings from V to itself. Then the invariantly defined principal symbol is a section of the pullback of the bundle $L(V)$ to the cotangent bundle T^*M. Given a local trivialization of V it can be identified locally with a $k \times k$ matrix-valued function on the cotangent bundle. Here k is the dimension of the fibre of V, i.e. the number of unknowns in the local coordinate representation.

In order to define the condition of symmetry in the definition of symmetric hyperbolic systems in an abstract context, it is necessary to introduce a Riemannian metric on the bundle V. This metric is not visible in the local coordinate representation, since in that context this role is played by the flat metric whose components in the given coordinates are given by the Kronecker delta. The symmetry condition is expressed in terms of the principal symbol and the chosen inner product as:

$$\langle \sigma(\xi) v, w \rangle = \langle v, \sigma(\xi) w \rangle$$

for all sections v, w of V and all covectors ξ. The positivity of A^0 is replaced by the condition that for some covector ξ the quadratic form associated to $\sigma(\xi)$ via the metric on V (by lowering the index) is positive definite. A hypersurface such that all its non-vanishing conormal vectors have this property is called spacelike.

Suppose now we have a symmetric hyperbolic system on a manifold M, a submanifold S of M and initial data such that S is spacelike. The aim is to show the existence of a solution on an open neighbourhood U of S in M with the prescribed data and that this is the unique solution on U with this property. Choose a covering of S by open sets U_α with the property that for each α the closure of this set is contained in a chart domain over which the bundle V can be trivialized globally. The problem of solving the equations on some open region in M with data the restriction of the original initial data to U_α can be solved by means of the results already discussed. For the local coordinates can be used to identify U_α with an open subset U'_α of a torus T^n and the initial data can be extended smoothly to the whole torus. The equation itself can also be transferred and extended. Corresponding to the extended data there exists a solution of the equations on $T^n \times I$ for some interval I. The coordinates can be used to transfer (a restriction of) the solution to an open subset W_α of M which is an open neighbourhood of U_α. The domain of dependence can then be used to show that there is an open neighbourhood where the solutions on the different W_α agree on the intersections. More specifically, on an intersection of this kind the two in principle different solutions can be expressed in terms of the same local coordinates and trivialization. By construction the coordinate representation of the data is the same in both cases. Thus the local uniqueness theorem for symmetric hyperbolic systems implies that both solutions are equal in a neighbourhood of the initial hypersurface. Thus the solutions on different local patches fit together to define the desired solution. Moreover the domain of dependence argument shows that it is unique on its domain of definition.

3.3 Other Notions of Hyperbolicity

There are two rather different aspects of the concept of hyperbolicity. The one is an intuitive idea of what good properties a system of equations should have in order to qualify for the name hyperbolic. The archetypal hyperbolic equation is the wave equation and so the desired properties are generalizations of properties of solutions of the wave equation. The first property is that the system should have a well-posed initial value problem. This means that there should exist solutions corresponding to appropriate initial data, that these should be uniquely determined by the data, and that they should depend continuously on the data in a suitable sense. The second property is the existence of a finite domain of dependence. This means that the value of solutions at a given point close to the initial hypersurface should depend only on data on a compact subset of the initial hypersurface.

The first aspect of hyperbolicity which has just been presented is a list of wishes. The second aspect is concerned with the question, how these wishes can be fulfilled. The idea is to develop criteria for hyperbolicity. For equations satisfying one of these criteria, the desired properties can be proved once and for all. Then all the user who wants to check the hyperbolicity of a given system has to do is to check the criteria, which are generally more or less algebraic in nature. An example is symmetric hyperbolicity, which we have already discussed in detail.

Symmetric hyperbolicity can be applied to very many problems in general relativity, but there are also other notions of hyperbolicity which have been applied and cases where no proof using symmetric hyperbolic systems is available up to now. We will give an overview of the situation, without going into details. The concepts which will be discussed are strict hyperbolicity, strong hyperbolicity and Leray hyperbolicity. This multiplicity of definitions is a consequence of the fact that there is no one ideal criterion which covers all cases.

There is unfortunately to our knowledge no fully general treatment of these matters in the mathematical literature. A general theory of this kind cannot be given in the context of this article, and we will refrain from stating any general theorems in this section. We will simply indicate some of the relations between the different types of hyperbolicity and give some references where more details can be found.

The discussion will use some ideas from the theory of pseudodifferential operators. For introductions to this theory see [88], [89], [3]. One important tool is an operator Λ which defines an isomorphism from functions in the Sobolev space H^s to functions in the Sobolev space H^{s-1}. Powers of this operator can be used to associate functions with a given degree of differentiability (in the sense of Sobolev spaces) with functions with a different degree of differentiability. The operator Λ is non-local. On the torus it could be chosen to be the operator $(1 - \Delta)^{1/2}$, defined in an appropriate way.

The concept of strict hyperbolicity is easily defined. Suppose that we have a system of N equations of order m. Suppose there is a covector τ such that for any covector ξ not proportional to τ the equation $\det \sigma(x, \tau + \lambda\xi)$ has Nm distinct real solutions λ. In this case we say that all characteristics are real and distinct and that the system is strictly hyperbolic. The weakness of this criterion is due to the fact that it is so often not satisfied for systems of physical interest. For example, the system for two functions u and v given by the wave equation for each of them is not strictly hyperbolic. The wave equation itself is strictly hyperbolic but simply writing it twice side by side leads to characteristics which are no longer distinct, so that strict hyperbolicity is lost.

Given a strictly hyperbolic system of order m with unknown u, let $u_i = \Lambda^{-i}\partial_t^i u$ for $i = 1, 2, \ldots, m-1$. Then the following system of equations is obtained:

$$\partial_t u_i = \Lambda u_{i+1}; \quad 0 \leq i \leq m-1$$
$$\partial_t u_{m-1} = \Lambda^{-m+1}(\partial_t^m u)$$

where it is understood that $\partial_t^m u$ should be expressed in terms of $u_0, \ldots, u_{m-1}$ by means of the original system. This system is first order in a sense which is hopefully obvious intuitively and which can be made precise using the theory of pseudodifferential operators. This is not a system of differential equations, due to the nonlocality of Λ, but rather a system of pseudodifferential equations. Suppose we write it as $\partial_t v + A^i(v)\partial_i v + B(v) = 0$, where v is an abbreviation for $u_0, \ldots, u_{m-1}$. The idea is now to find a pseudodifferential operator $A^0(v)$ of order zero depending on the unknown so that $A^0(v)A^i(v)$ is symmetric for all v and i and A^0 has suitable positivity properties. If an A^0 of this kind can be found, then multiplying the equation by it gives something which looks like a pseudodifferential analogue of a symmetric hyperbolic system. This symmetrization of the reduced equation can in fact be carried out in the case of strictly hyperbolic systems and the resulting first order pseudodifferential equation be treated by a generalization of the methods used for symmetric hyperbolic systems [89], [90]. This gives an existence theorem, but does not directly give information about the domain of dependence. This information can be obtained afterwards by a different method [88].

The algebra involved in symmetrizing the reduction of a strictly hyperbolic system has not been detailed. Instead we will present some general information about symmetrization of first order systems using pseudodifferential operators. Consider then the equation

$$\partial_t u + A^i(u)\partial_i u = 0$$

If there exists a function $A^0(u)$ such that A^0 is positive definite and $A^0 A^i$ is symmetric for each i then multiplying the equation by $A^0(u)$ gives the equation:

$$A^0(u)\partial_t u + A^0(u)A^i(u)\partial_i u = 0$$

which is symmetric hyperbolic. This can be generalized as follows. Each differential operator has a symbol which is polynomial in ξ. In the theory of pseudodifferential operators, operators are associated to symbols which no longer have a polynomial dependence on ξ. Within this theory the general statement (which must of course be limited in order to become strictly true) is that any algebra which can be done with symbols can be mimicked by operators. In terms of symbols the problem of symmetrization which has been stated above can be formulated as follows. Given symbols $\xi_i a^i(x, u)$, which are matrix-valued functions, find a positive definite symmetric real matrix-valued function $a^0(x, u)$ such that $a^0(x, u)\xi_i a^i(x, u)$ is symmetric. When the problem has been formulated in this way a generalization becomes obvious. Why not allow a^0 to depend on ξ? In the context of pseudodifferential operators this can be done. In order for this to be useful a^0 must be of order zero, which means that it must be bounded in ξ for each fixed (x, u) and that its derivatives with respect to ξ and x satisfy similar conditions which will not be given here. In fact it is enough to achieve the symmetrization of the symbol for ξ of unit length, since the ξ dependence of the procedure is essentially homogeneous in ξ anyway. It is also important that it be possible to choose a^0 in such a way that it depends smoothly on ξ, since the theory of pseudodifferential operators requires sufficiently smooth symbols. To prevent confusion in the notation, a^0 will from now on be denoted by r while we write $a(x, \xi, u) = a^i(x, u)\xi_i$.

A criterion which ensures that a first order system of PDE can be symmetrized in the generalized sense just discussed is that of strong hyperbolicity [63]. It is supposed that all characteristics are real, that they are of constant multiplicity, and that the symbol is everywhere diagonalizable. Furthermore, it is assumed that this diagonalization can be carried out in a way which depends smoothly on x, ξ and u. This means that there is a symbol $b(x, \xi, u)$ which satifies

$$b^{-1}(x, \xi, u)a(x, \xi, u)b(x, \xi, u) = d(x, \xi, u)$$

where d is diagonal. Let $r(x, \xi, u) = [b(x, \xi, u)b^T(x, \xi, u)]^{-1}$ where the superscript T denotes the transpose. Then r has the desired properties. For $r = b^{-1}(b^{-1})^T$ is symmetric and positive definite while $ra = (b^{-1})^T d b^{-1}$ is also symmetric. The one part of this criterion which is not purely algebraic is the smoothness condition. It is, however, often not hard to check once it is known how to verify the other conditions.

Another related notion of hyperbolicity is Leray hyperbolicity [66]. It allows operators with principal parts of different orders in the same system. We will discuss this in terms of using powers of the operator Λ to adjust the orders of the equations. This is somewhat different from the approach of Leray but is closely related to it. Suppose we have a system of PDE with unknown u. Let us split the vector u into a sequence of vectors $u_1, \ldots, u_L$. Then the differential equation which we may write schematically as $P(u) = 0$ can be rewritten as $P_j(u_i) = 0$, where i and j run from 1 to L. Let $v_i = \Lambda^{s(i)} u_i$ multiply the equations $P_j(u_i) = 0$ with $\Lambda^{t(j)}$. The result is a system of equations of the form $Q_j(v_i) = 0$, where $Q_j(v_i) = \Lambda^{t(j)} P_j(\Lambda^{-s(i)} v_i)$. The order of Q_j in its dependence on v_i is the order of P_j in its dependence on u_i plus $t(j) - s(i)$. Now we would like to choose $t(j)$ and $s(i)$ in such a way that all operators Q_i have the same order (say one) in their dependence on the corresponding variable v_i and lower order (say zero) in their dependence on v_j for $i \neq j$. Adding the same amount to all indices simultaneously is irrelevant. The indices are only determined up to a common additive constant. The system is Leray hyperbolic if this can be achieved by suitable choices of the decomposition and the indices, if the operators P_i corresponding to the given decomposition, considered with respect to their dependence on u_i, are strictly hyperbolic, and if the characteristics of these strictly hyperbolic operators satisfy a certain condition. (We say more on this condition later.) It is now plausible that this decomposition can be combined with the symmetrization of strictly hyperbolic systems to obtain a system of first order pseudodifferential equations which admit a pseudodifferential symmetrization. We do however stress that as far as we know the only place where the existence proof for Leray hyperbolic systems is written down in the literature is in the original lecture notes of Leray [66] and that the details of the plausibility argument presented here have not been worked out. It has the advantage that it gives an intuitive interpretation of the meaning of the indices $s(i)$ and $t(j)$. All the components of v in the solution will have the same degree of differentiability. If this is H^k then the variables u_i will have differentiability $H^{k-s(i)}$.

We now comment on the condition on the characteristics mentioned in the discussion of Leray hyperbolicity. For a strictly hyperbolic system there is a notion of spacelike covectors analogous to that for symmetric hyperbolic systems. The position of these spacelike covectors is closely related to the position of the characteristics. The required condition is that there should be vectors which are simultaneously spacelike for the L strictly hyperbolic systems occurring in the definition.

These general ideas concerning Leray hyperbolic systems will now be illustrated by the example of the Einstein-dust system. This example was first treated by Choquet-Bruhat [17]. A discussion of this and other examples can also be found in the books of Lichnerowicz [67], [68]. We will have more to say about this system in Sects. 4.2 and 5.4. The Einstein-dust equations are

$$G_{\alpha\beta} = 8\pi\rho U_\alpha U_\beta$$

$$\nabla_\alpha(\rho U^\alpha) = 0$$
$$U^\alpha \nabla_\alpha U^\beta = 0$$

The problem with these equations from the point of view of symmetric hyperbolic systems is that while the equation for the evolution of ρ contains derivatives of U^α the equation for the evolution of U^α does not contain ρ This could be got around if the derivatives of U^α in the evolution equation for ρ were considered as lower order terms. This is only possible if the first derivatives of U^α are considered on the same footing as ρ. The adjustment of orders involved in the Leray theory, combined with an extra device, allows this to be achieved. In order that everything be consistent we expect the order of differentiability of the metric to be one greater than that of U^α. For the Christoffel symbols occur in the evolution equations for U^α. Combining these two things means that the density should be two times less differentiable than the metric. This creates problems with the Einstein equations. For these are essentially (and the harmonically reduced equations are precisely) non-linear wave equations for the metric. The solution of a system of this kind is only one degree more differentiable than the right hand side. This does not fit, since the density occurs on the right hand side. The extra device consists is differentiating the Einstein equations once more in the direction U^α and then substituting the evolution equation for ρ into the result. This gives the equation:

$$U^\gamma \nabla_\gamma G_{\alpha\beta} = -8\pi\rho U_\alpha U_\beta \nabla_\gamma U^\gamma$$

Note that the right hand side of this contains no derivatives of ρ and so is not worse than the right hand side of the undifferentiated equation. On the other hand, the left hand side is, in harmonic coordinates, a third order hyperbolic equation in $g_{\alpha\beta}$ and the solution of an equation of this type has (by the Leray theory) two more degrees of differentiability than the right hand side.

After this intuitive discussion of the Einstein-dust system, let us show how it is related to the choice of indices needed to make the Leray hyperbolicity of the system manifest. The first thing which needs to be done is to decide which equations should be considered as the evolution equations for which variables. This has already been done in the above discussion. The relative orders of differentiability we have discussed suggest that we choose $s(g) = 1$, $s(U) = 2$ and $s(\rho) = 3$. This equalizes the expected differentiability of the different variables. In the new variables the evolution equations for g, U and ρ are of order four, three and four respectively. To equalize the orders we can choose $t(g) = 1$, $t(U) = 2$ and $t(\rho) = 1$. The blocks into which the system must be split are not just three but fifteen, one for each component. The indices $s(i)$ and $t(j)$ are here chosen the same for each component of one geometrical object. (Here we have ignored the difficulty that because of the normalization condition $U_\alpha U^\alpha = -1$ the variables U^α are not all independent.)

4 Reductions

In Sect. 2 we saw an example of a hyperbolic reduction for Einstein's field equations which is, at least in the vacuum case, sufficient to obtain local existence results and to demonstrate that the local evolution is dominated by the light cone structure and the associated concept of the domain of dependence. Thus, besides the existence problem, it settles the conceptual question whether the evolution process determined by Einstein's equations is consistent with the basic tenets of the theory.

Nevertheless, there are various reasons for considering other types of reductions. Different matter fields may require different treatments, various physical or geometrical considerations may require reductions satisfying certain side conditions, the desire to control the gauge for an arbitrarily long time may motivate the search for new gauge conditions. In particular, in recent years various systems of hyperbolic equations deduced from Einstein's equations have been put forward with the aim of providing 'good' systems for numerical calculations (cf. [1], [4], [14], [22], [38], [43], [56]). Since it is difficult to judge the relative efficiency of such systems by a few abstract arguments, detailed numerical calculations are needed to test them and it still remains to be seen which of these systems will serve the intended purpose best.

A general discussion of the problem of finding reductions which are useful in numerical or analytical studies should also include systems combining hyperbolic with elliptic equations (cf. [25], [77], [24] for analytic treatments of such situations) or with systems of still other types. For simplicity we will restrict ourselves to purely hyperbolic reductions. But even then a general discussion does not yet appear feasible. Ideally, one would like to exhibit a kind of 'hyperbolic skeleton' of the Einstein equations and a complete characterization of the freedom to fix the gauge from which all hyperbolic reductions should be derivable. Instead, there are at present various different methods available which have been invented to serve specific needs.

Therefore, we will present some of these methods without striving for completeness. There are various different boundary value problems of interest for Einstein's equations and tomorrow a new question may lead to a new solution of the reduction problem. Our aim is rather to illustrate the enormous richness of possibilities to adapt the equations to various geometrical and physical situations, and to comment on some of the new features which may be observed. Apart from the illustrative purpose it should not be forgotten, though, that each of the reductions outlined here implies, if worked out in detail, the local existence of solutions satisfying certain side conditions.

We begin by recalling in general terms the steps to be considered in a reduction procedure. It should be noted that these steps are not independent of each other, and that the same is true of the following considerations.

As a first step we need to *choose a representation of the field equations.* Above we used the standard representation in which Einstein's equations are written as a system of second order for the metric coefficients $g_{\mu\nu}$ and we also considered the ADM equations with the first and second fundamental forms of the time slicing as the basic variables. Later we shall also employ representations involving equations of third order in the metric field, in which the metric is expressed in terms of an orthonormal frame field.

Contrary to what has been claimed by too ambitious or suggested by ambiguous formulations, there is no way to get hyperbolic evolution equations without fixing a gauge. Hyperbolicity implies uniqueness in the PDE sense (in contrast to the notion of geometric uniqueness used for the Einstein equations, cf. Sect. 5.2). It is necessary to make *a choice of precisely four gauge source functions.*

Although in the proof of an existence result one has to fix a coordinate system, it is of interest to note that there are choices of gauge conditions which render the reduced equations hyperbolic for any fixed choice of coordinates. In Sect. 2 we considered the functions $F^{\nu}(x^{\mu})$ as gauge source functions determining the coordinates. Since these functions can be chosen arbitrarily and since the principal part of the reduced equations (2.39) will not be changed if they depend on the metric coefficients, they can be assumed to be of the form $F^{\nu} = F^{\nu}(x^{\mu}, g_{\lambda\rho})$.

The following choice is particularly interesting. Let an affine connection $\bar{\Gamma}_{\nu}{}^{\mu}{}_{\lambda}$ be given on the manifold $M = \mathbb{R} \times S$ on which we want to construct our solution. Since the difference $\Gamma_{\nu}{}^{\mu}{}_{\lambda} - \bar{\Gamma}_{\nu}{}^{\mu}{}_{\lambda}$, where the first term denotes the connection of the prospective solution $g_{\mu\nu}$, defines a tensor field, the requirement that the equation $\Gamma^{\mu} = g^{\nu\lambda}\,\bar{\Gamma}_{\nu}{}^{\mu}{}_{\lambda}$ be satisfied, has an invariant meaning. This suggests a way to impose a gauge condition which removes the freedom to perform diffeomorphisms while leaving all the freedom to perform coordinate transformations. In particular, if we assume the connection $\bar{\Gamma}_{\nu}{}^{\mu}{}_{\lambda}$ to be the Levi–Civita connection of a metric $\bar{g}_{\mu\nu}$ on M, the condition $F^{\mu} = g^{\nu\lambda}\,\bar{\Gamma}_{\nu}{}^{\mu}{}_{\lambda}$ in equation (2.39) corresponds to the requirement that the identity map of M onto itself is a harmonic map under which $\bar{g}$ is pulled back to g (cf. also [37], [38]).

Again, we do not have a good overview of all the possibilities to impose gauge conditions. There exist conditions which work well with quite different representations of the field equations. Examples are given by the choice of gauge source functions $F^{\nu}(x^{\mu})$ considered above, which work with suitably chosen gauge conditions for the frame field also in the frame formalism ([36]), or by the gauge, considered also below, in which the shift β^{a} and the function $q = \log(\alpha\, h^{-\sigma})$, with $h = \det(h_{ab})$ and $\sigma = \text{const.} > 0$, are prescribed as gauge source functions (cf. [4], [22], [38] for reductions based on $\sigma = \frac{1}{2}$ and different representations of the Einstein equations). There are other gauge conditions which only work for specific representations like some of the ones we shall consider in the context of the frame formalism. For some gauge conditions, like the ones employing the gauge source functions $F^{\nu}(x^{\mu})$, the universal applicability has been shown (cf. our argument in Sect. 2) or is easy to see. For others, like the ones using the gauge source function q above, it has apparently never been shown.

The gauge problem does not admit a 'universal solution' which works in all possible situations of interest. In fact, choosing the gauge is related to some of the most complicated questions of constructing general solutions to Einstein's equations. Often one would like to find a system of coordinates which covers the complete domain of existence of a solution arising from given initial data. If there existed such coordinates x^μ, we could, in principle, characterize them in terms of the associated gauge source functions $F^\nu(x^\mu)$ considered in Sect. 2. However, the domain of validity, in particular the 'lifetime', of a coordinate system depends on the data, the equations, the type of gauge condition, as well as on the given gauge source functions. In practice, gauge source functions which ensure that the coordinates exist globally, if there exist any at all, have to be identified in the course of constructing the solution.

Having implemented the gauge conditions, we have to *find a hyperbolic system of reduced equations* from our representation of the Einstein equations. As we shall see, there are often various possibilities and the final choice will depend on the desired application. With the reduced equations at hand we have to *arrange initial data* which satisfy the constraints and are consistent with the gauge conditions. While the second point usually poses no problem, the first point involves solving elliptic equations and requires a seperate discussion.

The reduced equations define together with the initial data a well-posed initial value problem and we can show, by using standard techniques of the theory of partial differential equations (cf. Sect. 3.1), the *existence of solutions to the reduced problem*, work out differentiability properties of solutions etc. or start the numerical evolution.

As the final step one needs to show that the *constraints and gauge conditions are preserved* by the evolution defined by the reduced equations. One may wonder whether there exists a universal argument which tells us that this will be the case for any system of hyperbolic reduced equations deduced from the Einstein equations. The example of the spinor equations discussed below shows that this cannot be true without restrictions on the matter fields. However, even if we ask this question about reduced problems for the Einstein vacuum field equations, the answer seems not to be known. The standard method here is to use the reduced equations, as in Sect. 2, together with some differential identities, to show that the 'constraint quantities' satisfy a certain subsidiary system which allows us to argue that these quantities vanish.

If the gauge condition used in our reduced equations has been shown to be universally applicable and if there already exists another hyperbolic reduction using these gauge conditions and applying to the same geometric and physical situation (choice of matter model), it can be argued, invoking the uniqueness property of initial value problems for hyperbolic equations, that our reduced equations must preserve the constraints and the gauge conditions.

In the examples which follow we shall not always produce the complete reduction argument. Often we shall only exhibit a symmetric hyperbolic system of reduced equations and remark on some of its properties.

4.1 Hyperbolic Systems from the ADM Equations

In the article [11] (cf. also [86]) the authors derived a system from equations (2.10), (2.11), (2.22) and (2.23) which seems to be numerically distinctively better behaved than any other system derived so far from the ADM equations. However, at present there appears to be no clear understanding as to why this should be so. The symmetric hyperbolic system to be discussed below was found in the course of an attempt to understand whether the system considered in [11] is in any sense related to a hyperbolic system. Since other hyperbolic systems related to the equations in [11] have been discussed in [2], [44] there is now a large family of such systems available.

In the following we shall consider the fields

$$\beta^a, \quad q = \log(\alpha\, h^{-\sigma}), \tag{4.1}$$

with $h = \det(h_{ab})$, $\sigma = \text{const.} > 0$, as the gauge source functions. The density h will be used to rescale the 3-metric and the trace-free part of the extrinsic curvature to obtain the densities $\tilde{h}_{ab} = h^{-\frac{1}{3}}\, h_{ab}$, $\tilde{\chi}_{ab} = h^{-\frac{1}{3}}\,(\chi_{ab} - \frac{1}{3}\,\chi\, h_{ab})$. For simplicity we shall refer to $\tilde{h}_{ab}$ as to the conformal metric or, if no confusion can arise, simply as to the metric. The following equations are derived from the ADM equations by using the standard rules for conformal rescalings. The occurrence of some of the terms in the following equations find their proper explanation in the general calculus for densities but we shall not discuss this here any further.

The unknowns in our equations will be the fields

$$\tilde{h}_{ab}, \quad \eta = \log\, h, \quad \eta_a = \tilde{D}_a \log\, h, \quad \chi = h^{ab}\chi_{ab}, \tag{4.2}$$

$$\hat{\gamma}_a = \tilde{\gamma}_a - (\frac{1}{6} + \sigma)\,\tilde{D}_a \log\, h, \quad \tilde{\chi}_{ab}, \quad \tilde{h}_{abc} = \tilde{h}_{ab,c}, \tag{4.3}$$

where $\tilde{\gamma}_a$ is defined for the metric $\tilde{h}_{ab}$ in analogy to (2.32). Note that the power of the scaling factor h has been chosen such that we have $\tilde{h} = \det(\tilde{h}_{ab}) = 1$, whence $0 = \tilde{h}_{,c} = \tilde{h}\,\tilde{h}^{ab}\,\tilde{h}_{ab,c} = \tilde{h}^{ab}\,\tilde{h}_{abc}$. We denote by $\tilde{D}$ the covariant derivative, by $\tilde{\gamma}_a{}^b{}_c$ the connection coefficients, and by $\tilde{R}$ the Ricci scalar which are defined for the conformal metric $\tilde{h}_{ab}$ by the standard rules. In all expressions involving quantities carrying a tilde any index operations are performed with the metric $\tilde{h}_{ab}$.

Our equations are obtained as follows (cf. also [11]). From (2.22) we get by a direct calculation

$$\partial_t \tilde{h}_{ab} - \tilde{h}_{ab,c}\,\beta^c = 2\,\tilde{h}_{c(a}\,\beta^c{}_{,b)} + \frac{2}{3}\,\tilde{h}_{ab}\,\beta^c{}_{,c} + 2\,\alpha\,\tilde{\chi}_{ab}, \tag{4.4}$$

$$\partial_t \eta - \eta_{,c}\,\beta^c = 2\,\beta^c{}_{,c} + 2\,\alpha\,\chi. \tag{4.5}$$

Taking derivatives on both sides of the second equation gives

$$\partial_t \eta_a - \eta_{a,c}\,\beta^c = 2\,\alpha\,\tilde{D}_a\,\chi + 2\,\chi\,(\sigma\,\eta_a + \tilde{D}_a\,q) + \eta_c\,\beta^c{}_{,a} + 2\,\beta^c{}_{,ca}. \tag{4.6}$$

Equation (2.23) implies together with (2.22) that

$$\partial_t \chi - \chi_{,c}\,\beta^c = D_c\,D^c\,\alpha - \alpha\,((1-a)\,R + a\,R + \chi^2 - 3\,\lambda + \frac{\kappa}{2}\,(h^{ab}\,T_{ab} - 3\,\rho)),$$

with an arbitrary real number a. To replace R in the second term on the right hand side, we use the Hamiltonian constraint (2.10), to replace R in the third term we use the transformation law of the Ricci scalar under conformal rescalings and the expression of the Ricci scalar $\tilde{R}$ in terms of the conformal quantities, which give

$$R = h^{-\frac{1}{3}}\left\{\tilde{D}^c\,(\tilde{\gamma}_c - \frac{2}{3}\,\tilde{D}_c\,\log\,h) - \tilde{\gamma}_c{}^c{}_b\,\tilde{\gamma}^b + \right.$$
$$\left. + \tilde{h}^{ab}\,\tilde{\gamma}_a{}^c{}_d\,\tilde{\gamma}_c{}^d{}_b - \frac{1}{18}\,\tilde{D}_a\,\log\,h\,\tilde{D}^a\,\log\,h\right\}.$$

Thus we get

$$h^{\frac{1}{3}}\,\frac{1}{\alpha}(\partial_t\,\chi - \chi_{,c}\,\beta^c) = -a\,\tilde{D}^a\,\hat{\gamma}_a + (\sigma + a\,(\frac{1}{2} - \sigma))\,\tilde{D}^a\,\eta_a \tag{4.7}$$
$$+\tilde{D}^a\,\tilde{D}_a q + a\,(\tilde{\gamma}_c{}^c{}_d\,\tilde{\gamma}^d - \tilde{h}^{ab}\,\tilde{\gamma}_a{}^c{}_d\,\tilde{\gamma}_c{}^d{}_b)$$
$$+(\frac{1}{18}\,a + \frac{1}{6}\,\sigma + \sigma^2)\,\eta_a\,\eta^a + (\frac{1}{6} + 2\,\sigma)\,\eta^a\,\tilde{D}_a q + \tilde{D}^a q\,\tilde{D}_a q$$
$$-h^{\frac{1}{3}}\,\{(1-a)\,\tilde{\chi}_{ab}\,\tilde{\chi}^{ab} + \frac{1}{3}\,(1+2\,a)\,\chi^2 - (1+2\,a)\,\lambda + \frac{\kappa}{2}\,\{h^{ab}\,T_{ab} + (1-2\,a)\,\rho)\}.$$

Using the equations above, the definition of $\hat{\gamma}_a$, and the momentum constraint (2.11) as well as its expression in terms of the conformal quantities,

$$\tilde{D}^c\,\tilde{\chi}_{ca} - \frac{2}{3}\,\tilde{D}_a\,\chi + \frac{1}{2}\,\tilde{\chi}_{ca}\,\tilde{D}^c\,\log\,h + \frac{1}{6}\,\chi\,\tilde{D}_a\,\log\,h = \kappa\,T_{\mu\nu}\,n^\mu\,h^\nu{}_a,$$

we obtain, with arbitrary real number c, by a direct calculation the equation

$$\frac{1}{\alpha}\,\{\partial_t\,\hat{\gamma}_a - \hat{\gamma}_{a,c}\,\beta^c\} = -c\,\tilde{D}^c\,\tilde{\chi}_{ca} + 2\,(\frac{1}{2} + \frac{c}{3} - \sigma)\,\tilde{D}_a\,\chi \tag{4.8}$$
$$+(2+c)\,\kappa\,T_{\mu\nu}\,n^\mu\,h^\nu{}_a + 2\,\tilde{\chi}_a{}^c\,\{\tilde{\gamma}_c - (\frac{1}{2} + \frac{c}{4} - 2\,\sigma)\,\eta_c + \tilde{D}_c\,q\}$$

$$-\chi\,((\frac{1}{3}+2\,\sigma)\,(\sigma\,\eta_a+\tilde{D}_a\,q)+\frac{c}{6}\,\eta_a)-2\,\tilde{\chi}^{cd}\,\tilde{\gamma}_c{}^b{}_d\,\tilde{h}_{ab}$$

$$+\frac{1}{\alpha}\left\{h^{cd}\,h_{ac,b}\,\beta^b{}_{,d}-D^{(c}\,\beta^{d)}\,(h_{ab}\,\gamma_c{}^b{}_d+h_{cb}\,\gamma_d{}^b{}_a)-h^{cd}{}_{,b}\,h_{ac,d}\,\beta^b\right.$$

$$\left.-(\frac{1}{2}+\sigma)\,\eta_b\,\beta^b{}_{,a}+h^{cd}\,(h_{bc}\,\beta^b{}_{,a}+h_{ab}\,\beta^b{}_{,c})_{,d}-(1+2\,\sigma)\,\beta^b{}_{,ba}\right\}.$$

For the rescaled trace free part of the extrinsic curvature we get from (2.23) the equation

$$\partial_t\,\tilde{\chi}_{ab}-\tilde{\chi}_{ab,c}\,\beta^c-2\,\tilde{\chi}_{c(a}\,\beta^c{}_{,b)}-\frac{2}{3}\,\tilde{\chi}_{ab}\,\beta^c{}_{,c}$$

$$=-\alpha\,h^{-\frac{1}{3}}\,(R_{ab}-\frac{1}{3}\,R\,h_{ab})+h^{-\frac{1}{3}}\,(D_a\,D_b\,\alpha-\frac{1}{3}\,D_c\,D^c\,\alpha\,h_{ab})$$

$$-\alpha\,(\chi\,\tilde{\chi}_{ab}-2\,\tilde{h}^{cd}\,\tilde{\chi}_{ac}\,\tilde{\chi}_{bd}).$$

To express the equation in terms of the conformal quantities we use the transformation law of the Ricci tensor under conformal rescalings and the expression of the conformal Ricci tensor in terms of the connection coefficients to get

$$R_{ab}-\frac{1}{3}\,R\,h_{ab}=-\frac{1}{2}\,\tilde{h}^{cd}\,\tilde{h}_{ab,cd}+\tilde{D}_{(a}\,\tilde{\gamma}_{b)}-\frac{1}{3}\,\tilde{h}_{ab}\,\tilde{D}_c\,\tilde{\gamma}^c$$

$$+\tilde{\gamma}_c{}^d{}_a\,\tilde{h}_{ed}\,\tilde{h}^{cf}\,\tilde{\gamma}_f{}^e{}_b+2\,\tilde{\gamma}_d{}^e{}_c\,\tilde{h}^{df}\,\tilde{h}_{e(a}\,\tilde{\gamma}_{b)}{}^c{}_f+\frac{1}{3}\,\tilde{h}_{ab}\,(\tilde{\gamma}_c{}^c{}_d\,\tilde{\gamma}^d-\tilde{h}^{ef}\,\tilde{\gamma}_e{}^c{}_d\,\tilde{\gamma}_c{}^d{}_f)$$

$$+\frac{1}{36}\,(\tilde{D}_a\,\log\,h\,\tilde{D}_b\,\log\,h-\frac{1}{3}\,\tilde{h}_{ab}\,\tilde{h}^{cd}\,\tilde{D}_c\,\log\,h\,\tilde{D}_d\,\log\,h).$$

Thus we obtain

$$h^{\frac{1}{3}}\,\frac{1}{\alpha}\,\{\partial_t\,\tilde{\chi}_{ab}-\tilde{\chi}_{ab,c}\,\beta^c\}=\frac{1}{2}\,\tilde{h}^{cd}\,\tilde{h}_{ab,cd}-\tilde{D}_{(a}\hat{\gamma}_{b)}+\frac{1}{3}\,\tilde{h}_{ab}\,\tilde{D}^c\,\hat{\gamma}_c \tag{4.9}$$

$$+\tilde{D}_a\,\tilde{D}_bq-\frac{1}{3}\,\tilde{h}_{ab}\,\tilde{D}^c\,\tilde{D}_cq+\tilde{D}_aq\,\tilde{D}_bq-\frac{1}{3}\,h_{ab}\,\tilde{D}^cq\tilde{D}_cq$$

$$+(2\,\sigma-\frac{1}{3})\,(\eta_{(a}\,\tilde{D}_{b)}q-\frac{1}{3}\,h_{ab}\,\eta^c\,\tilde{D}_cq)+(\sigma^2-\frac{1}{3}\,\sigma-\frac{1}{36})\,(\eta_a\,\eta_b-\frac{1}{3}\,h_{ab}\,\eta_c\,\eta^c)$$

$$-\tilde{\gamma}_c{}^d{}_a\,\tilde{h}_{ed}\,\tilde{h}^{cf}\,\tilde{\gamma}_f{}^e{}_b-2\,\tilde{\gamma}_d{}^e{}_c\,\tilde{h}^{df}\,\tilde{h}_{e(a}\,\tilde{\gamma}_{b)}{}^c{}_f-\frac{1}{3}\,\tilde{h}_{ab}\,(\tilde{\gamma}_c{}^c{}_e\,\tilde{\gamma}^e-\tilde{h}^{cd}\,\tilde{\gamma}_c{}^e{}_f\,\tilde{\gamma}_e{}^f{}_d)$$

$$+2\,\tilde{\chi}_{c(a}\,S^c{}_{,b)}+\frac{2}{3}\,\tilde{\chi}_{ab}\,S^c{}_{,c}-h^{\frac{1}{3}}\,(\chi\,\tilde{\chi}_{ab}-2\,\tilde{h}^{cd}\,\tilde{\chi}_{ac}\,\tilde{\chi}_{bd}).$$

In all the equations expressions like $\tilde{D}_a\tilde{\gamma}_b$ etc. are defined by the expressions which would hold if $\tilde{\gamma}_b$ denoted a tensor field. Finally, we get from (4.4) that

$$\partial_t\,\tilde{h}_{abc}-\tilde{h}_{abc,d}\,\beta^d=2\,\alpha\,\tilde{D}_c\,\tilde{\chi}_{ab} \tag{4.10}$$

$$+2\,\tilde{\chi}_{ab}\,\alpha\,(\sigma\,\eta_c + \tilde{D}_c q) + 4\,\alpha\,\tilde{\gamma}_c{}^d{}_{(a}\,\tilde{\chi}_{b)d} + \tilde{h}_{abd}\,\beta^d{}_{,c}$$

$$+2\,\beta^d{}_{(,a}\,\tilde{h}_{b)dc} + 2\,\tilde{h}_{d(b}\,\beta^d{}_{,a)c} + \frac{2}{3}\,(\tilde{h}_{abc}\,\beta^d{}_{,d} + \tilde{h}_{ab}\,\beta^d{}_{,dc}).$$

If $c > 0$, the system (4.4), (4.5), (4.6), (4.7), (4.8), (4.9), (4.10) is symmetric hyperbolic. This can be seen as follows. (i) We choose real numbers e, f satisfying

$$e > 0, \quad f > 0, \quad f\,a = -2\,(\frac{1}{2}+\frac{c}{3}-\sigma), \quad e = \frac{\sigma}{2}\,f-(\frac{1}{2}-\sigma)\,(\frac{1}{2}+\frac{c}{3}-\sigma), \tag{4.11}$$

and multiply some of the equations by overall factors c, e, f, α^{-1}, to obtain them in the form (writing out only their principal parts here)

$$\partial_t\,\tilde{h}_{ab} - \tilde{h}_{ab,c}\,\beta^c = \ldots, \tag{4.12}$$

$$\partial_t\,\eta - \eta_{,c}\,\beta^c = \ldots, \tag{4.13}$$

$$\frac{e}{\alpha}\,\{\partial_t\,\eta_a - \eta_{a,c}\,\beta^c\} = e\,\{2\,\tilde{D}_a\,\chi + \ldots\}, \tag{4.14}$$

$$h^{\frac{1}{3}}\,\frac{f}{\alpha}\,\{\partial_t\,\chi - \chi_{,c}\,\beta^c\} = f\,\{-a\,\tilde{D}^a\,\hat{\gamma}_a + 2\,\frac{e}{f}\,\tilde{D}^a\,\eta_a + \ldots\}, \tag{4.15}$$

$$\frac{1}{\alpha}\,\{\partial_t\,\hat{\gamma}_a - \hat{\gamma}_{a,c}\,\beta^c\} = -c\,\tilde{D}^c\,\tilde{\chi}_{ca} - f\,a\,\tilde{D}_a\,\chi + \ldots, \tag{4.16}$$

$$h^{\frac{1}{3}}\,\frac{c}{\alpha}\,\{\partial_t\,\tilde{\chi}_{ab} - \tilde{\chi}_{ab,c}\,\beta^c\} = c\,\{\frac{1}{2}\,\tilde{h}^{cd}\,\tilde{h}_{ab,cd} - \tilde{D}_{(a}\,\hat{\gamma}_{b)} + \frac{1}{3}\,\tilde{h}_{ab}\,\tilde{D}^c\,\hat{\gamma}_c + \ldots\}, \tag{4.17}$$

$$\frac{c}{4\,\alpha}\{\partial_t\,\tilde{h}_{abc} - \tilde{h}_{abc,d}\,\beta^d\} = c\,\{\frac{1}{2}\,\tilde{D}_c\,\tilde{\chi}_{ab} + \ldots\}. \tag{4.18}$$

(ii) We contract both sides of (4.14) and (4.16) with $\tilde{h}^{ba}$, both sides of (4.17) with $\tilde{h}^{a(c}\,\tilde{h}^{d)b}$, both sides of (4.18) with $\tilde{h}^{a(d}\,\tilde{h}^{e)b}\,\tilde{h}^{fc}$, and add on the right hand side of the equation obtained in this way from (4.16) a term of the form $\frac{c}{3}\,\tilde{h}^{ba}\,\tilde{h}^{cd}\,\tilde{D}_a\tilde{\chi}_{cd}$, which vanishes identically but whose formal occurence makes the symmetry manifest.

The range in which the coefficients a, c, e, f, σ are consistent with (4.11) can be seen by considering the following cases:

$$i) \quad a = 0 \leftrightarrow \sigma = \frac{1}{2} + \frac{c}{3}.$$

In this case we have $e = (\frac{1}{4} + \frac{c}{6})\, f$ and we can choose $c, f > 0$ arbitrarily. If $a \neq 0$ we have $f = -\frac{2}{a}\,(\frac{1}{2} + \frac{c}{3} - \sigma)$ and $e = \{\sigma\,\frac{a-1}{a} - \frac{1}{2}\}\,(\frac{1}{2} + \frac{c}{3} - \sigma)$, which gives the following restrictions

$$ii) \quad a < 0, \qquad \frac{|a|}{2\,(|a|+1)} < \sigma < \frac{1}{2} + \frac{c}{3},$$

$$iii) \qquad 0 < a \leq 1, \quad \frac{1}{2} + \frac{c}{3} < \sigma,$$

$$iv) \quad 1 < a < \frac{3 + 2\,c}{2\,c}, \quad \frac{1}{2} + \frac{c}{3} < \sigma < \frac{a}{2\,(a-1)}.$$

We cannot have $e > 0$, $f > 0$ with $a \geq \frac{3+2\,c}{2\,c}$.

It is of interest to study the characteristics of the system. Equations (4.12) and (4.13) contribute factors $n(\xi) = n^\mu\,\xi_\mu$ to the characteristic polynomial. To find the other characteristics we analyse for which $\xi_\mu \neq 0$ the following linear system of equations, defined by the principal symbol map, admits non-trivial solutions. We set $\xi^a = \tilde{h}^{ab}\,\xi_b$.

$$n(\xi)\,\eta_a = 2\,\xi_a\,\chi,$$

$$h^{\frac{1}{3}}\,n(\xi)\,\chi = -a\,\xi^a\,\hat{\gamma}_a + 2\,\frac{e}{f}\,\xi^a\,\eta_a,$$

$$n(\xi)\,\hat{\gamma}_a = -c\,\xi^b\,\tilde{\chi}_{ba} - f\,a\,\xi_a\,\chi,$$

$$h^{\frac{1}{3}}\,n(\xi)\,\tilde{\chi}_{ab} = \frac{1}{2}\,\tilde{h}_{abc}\,\xi^c - \xi_{(a}\,\hat{\gamma}_{b)} + \frac{1}{3}\,\tilde{h}_{ab}\,\xi^c\,\hat{\gamma}_c,$$

$$n(\xi)\,\tilde{h}_{abc} = 2\,\xi_c\,\tilde{\chi}_{ab}.$$

The condition $n(\xi) = 0$ implies $\tilde{\chi}_{ab} = 0$ and $\chi = 0$ but there remains a fifteen-parameter freedom to choose the remaining unknowns. If

$$n(\xi) \neq 0, \tag{4.19}$$

we derive from the equations above the further equations (writing $g(\xi,\xi) = g^{\mu\nu}\,\xi_\mu\,\xi_\nu$)

$$h^{\frac{1}{3}}\,g(\xi,\xi)\,\tilde{\chi}_{ab} = n(\xi)\,(\xi_{(a}\,\hat{\gamma}_{b)} - \frac{1}{3}\,\tilde{h}_{ab}\,\xi^c\,\hat{\gamma}_c),$$

whence

$$\{g(\xi,\xi) + \frac{c}{2}\,h^{cd}\,\xi_c\,\xi_d\}\,\hat{\gamma}_a = -(\frac{c}{6}\,h^{-\frac{1}{3}}\,\xi^c\,\hat{\gamma}_c + \frac{g(\xi,\xi)}{n(\xi)}\,f\,a\,\chi)\,\xi_a.$$

The latter equation implies

$$\{g(\xi,\xi) + \frac{2\,c}{3}\,h^{cd}\,\xi_c\,\xi_d\}\,\xi^a\,\hat{\gamma}_a + \frac{g(\xi,\xi)}{n(\xi)}\,\xi^c\,\xi_c\,f\,a\,\chi = 0, \tag{4.20}$$

and, with p^a such that $p^a \, \xi_a = 0$,

$$\{g(\xi,\xi) + \frac{c}{2} \, h^{cd} \, \xi_c \, \xi_d\} \, p^a \, \hat{\gamma}_a = 0,$$

Finally we get

$$\{g(\xi,\xi) + (a-1)\,(1-2\,\sigma)\, h^{cd} \, \xi_c \, \xi_d\} \, \chi - a \, h^{-\frac{1}{3}} \, n(\xi) \, \xi^c \, \hat{\gamma}_c = 0. \qquad (4.21)$$

Equations (4.20) and (4.21), are of the form $A \, u = 0$ with the unknown u the transpose of $(\chi, \xi^i \, \hat{\gamma}_i)$ and the matrix A satisfying

$$\det A = \{g(\xi,\xi) + \frac{2\,c}{3}\,(1-a)\, h^{cd} \, \xi_c \, \xi_d\} \, \{g(\xi,\xi) + (2\,\sigma - 1)\, h^{cd} \, \xi_c \, \xi_d\}.$$

If

$$g(\xi,\xi) = 0, \qquad (4.22)$$

it follows from the first of these equations that $\hat{\gamma}_a = 0$. If $a \neq 1$, $\sigma \neq \frac{1}{2}$ there remains the two-parameter freedom to choose $\tilde{\chi}_{ab}$ with $\xi^a \, \tilde{\chi}_{ab} = 0$. If $a = 1$ or $\sigma = \frac{1}{2}$ there remains the three-parameter freedom to choose χ, $\tilde{\chi}_{ab}$ satisfying $\xi^a \, \tilde{\chi}_{ab} = -\frac{f\,a}{c} \, \xi_b \, \chi$. If $g(\xi,\xi) \neq 0$ the field $\tilde{\chi}_{ij}$ is known once $\hat{\gamma}_a$ has been determined. If

$$g(\xi,\xi) + \frac{c}{2} \, h^{cd} \, \xi_c \, \xi_d = 0, \qquad (4.23)$$

there is a two-parameter freedom to choose $p^i \, \hat{\gamma}_i$ as above. This is the only freedom unless $a = \frac{1}{4}$ or $\sigma = \frac{1}{2} + \frac{c}{4}$, conditions which exclude each other. Further characteristics are given by the equations

$$g(\xi,\xi) + \frac{2\,c}{3}\,(1-a)\, h^{cd} \, \xi_c \, \xi_d = 0, \qquad (4.24)$$

$$g(\xi,\xi) + (2\,\sigma - 1)\, h^{cd} \, \xi_c \, \xi_d = 0. \qquad (4.25)$$

The 'physical' characteristics correspond of course to (4.22), the two-parameter freedom pointed out in the case $a \neq 1$, $\sigma \neq \frac{1}{2}$ corresponding to the two polarization states of gravitational waves and the additional freedom in the other cases corresponding essentially to a gauge freedom. The timelike characteristics corresponding to (4.19) occur because of the transition from the system of second order to a system of first order. These characteristics are neither 'physical' nor 'harmful'. Characteristics corresponding to (4.23) are spacelike, while the nature of the characteristics corresponding to (4.24) and (4.25) depends on the constants a and σ. It can happen that one is spacelike while the other is timelike or both are spacelike.

We note here that it is possible to obtain by similar procedures reduced equations, for somewhat more complicated unknowns, which have only characteristics which are timelike or null [44].

Though the number $c > 0$ can be chosen arbitrarily (suitably adjusting the others) it is not possible to perform the limit $c \to 0$ while keeping the symmetric hyperbolicity of the system, however, the equations in [11] can be considered as limit of equations which are algebraically equivalent to our systems. We note also that it is not possible to perform a regular limit $\sigma \to 0$ which would make lapse and shift the gauge source functions. In the limit as $c \to 0$ (which we can perform if we do not insist on hyperbolicity) all characteristics, with the exception of the gauge dependent characteristics corresponding to (4.25), become non-spacelike.

We finally remark on our gauge conditions. Under certain assumptions the gauge (4.1) coincides with the gauge with harmonic time coordinate and prescribed shift. From (2.7) follows the general relation

$$\partial_t \alpha - \alpha_{,a}\, \beta^a = \alpha^2\, \chi - \alpha^3\, \Gamma^0.$$

Equation (4.5), written as an equations for h, entails together with (4.1) the equation

$$\partial_t \alpha - \alpha_{,a}\, \beta^a = 2\,\sigma\,\alpha^2\, \chi + \alpha\,(2\,\sigma\,\beta^a{}_{,a} + \partial_t q - q_{,a}\beta^a).$$

Thus the time coordinate t is harmonic in our gauge, i.e. $\Gamma^0 = 0$, if $\sigma = \frac{1}{2}$ and $\partial_t q - q_{,a}\beta^a = -\beta^a{}_{,a}$. In more general situations the expression for Γ^0 implied by the equations above contains information about the solution and admits no direct conclusion about time-harmonicity in terms of α and β^a.

It follows from complete reductions based on the gauge source function q (cf. e.g [38], [44]) that on solutions of Einstein's equations it is possible to achieve the corresponding gauge close to some initial surface. However, it has apparently never been shown that for prescribed $\sigma > 0$, β^a and given spacelike hypersurface of some arbitrary Lorentz manifold, coordinates can be constructed which realize these gauge source functions. It would be useful to have a proof of the universal applicability of this type of gauge and information about its general behaviour.

4.2 The Einstein–Euler System

In the following we shall discuss the Einstein–Euler equations, i.e. Einstein's equation coupled to the Euler equation for a simple perfect fluid. Its hyperbolicity has been studied by various authors (cf. [17], [68], [76]). Though the system is also important in the cosmological context, our main concern in analysing the system here is to control the evolution of compact perfect fluid bodies, which are considered as models for 'gaseous stars'. In this situation arises, besides the need to cast the equations into hyperbolic form, the side condition to control the evolution of the timelike boundary along which the Einstein–Euler equations go over into the Einstein vacuum field equations.

The analysis of the evolution of the fields in the neighbourhood of this boundary poses the basic difficulty in the discussion of compact fluid bodies. In the case of spherical symmetry, this problem was overcome in [61]. If in more general situations the coordinates used in the hyperbolic reduction are governed e.g. by wave equations, as is the case in the harmonic gauge, there appears to be no way to control the motion of this boundary in these coordinates. In [39] a system of equations has been derived from the Einstein–Euler system which combines hyperbolicity with the Lagrangian description of the flow, so that the spatial coordinates are constant along the flow lines. Thus the location of the body is known in these coordinates. In the following we shall discuss this system and derive the subsidiary system, which was not given in [39].

The Basic Equations We shall use a frame formalism in which the information on the metric is expressed in terms of an orthonormal frame $\{e_k\}_{k=0,\dots,3}$ and all fields, with the possible exception of the frame itself, are given in this frame. To make the formalism easily comparable with the spin frame formalism which will be used later, we shall use a signature such that $g_{ik} \equiv g(e_i, e_k) = \mathrm{diag}(1,-1,-1,-1)$. Let ∇ denote the the Levi–Civita connection of $g_{\mu\nu}$. The basic unknowns of our representation of the Einstein equations are given by

$$e^\mu{}_k, \quad \Gamma_i{}^j{}_k, \quad C^i{}_{jkl}, \quad \text{matter variables},$$

where $e^\mu{}_k = \langle e_k, x^\mu \rangle$ are the coefficients of the frame field in some coordinates x^μ, $\Gamma_i{}^j{}_k$ are the connection coefficients, defined by $\nabla_i\, e_k = \Gamma_i{}^j{}_k\, e_j$ and satisfying $\Gamma_i{}^j{}_k\, g_{jl} + \Gamma_i{}^j{}_l\, g_{jk} = 0$, and $C^i{}_{jkl}$ denotes the conformal Weyl tensor in the frame e_k. The latter is obtained from the decomposition

$$R_{ijkl} = C_{ijkl} + \{g_{i[k}\, S_{l]j} - g_{j[k}\, S_{l]i}\}, \tag{4.26}$$

of the curvature tensor

$$R^i{}_{jpq} = e_p(\Gamma_q{}^i{}_j) - e_q(\Gamma_p{}^i{}_j) - \Gamma_k{}^i{}_j\,(\Gamma_p{}^k{}_q - \Gamma_q{}^k{}_p) \tag{4.27}$$
$$+\Gamma_p{}^i{}_k\,\Gamma_q{}^k{}_j - \Gamma_q{}^i{}_k\,\Gamma_p{}^k{}_j,$$

where we also set $S_{ij} = R_{ij} - \frac{1}{6}\, g_{ij}\, R$, with R_{ij} and R denoting the Ricci tensor and the Ricci scalar. We shall need the notation

$$T_i{}^k{}_j\, e_k = -[e_i, e_j] + (\Gamma_i{}^l{}_j - \Gamma_j{}^l{}_i)\, e_l,$$

$$\Delta^i{}_{jkl} = R^i{}_{jkl} - C^i{}_{jkl} - g^i{}_{[k}\, S_{l]j} + g_{j[k}\, S_{l]}{}^i,$$

with $R^i{}_{jkl}$ understood as being given by (4.27). Furthermore we set

$$F_{jkl} = \nabla_i\, F^i{}_{jkl}, \quad \text{with} \quad F^i{}_{jkl} = C^i{}_{jkl} - g^i{}_{[k}\, S_{l]j}.$$

In the equations above we take account of the Einstein–Euler equations in the form

$$S_{ik} = \kappa\,(T_{ik} - \frac{1}{3}\, g_{jk}\, T), \tag{4.28}$$

with an energy-momentum tensor of a simple perfect fluid

$$T_{ik} = (\rho + p)\, U_i\, U_k - p\, g_{ik}. \tag{4.29}$$

Here ρ is the total energy density and p the pressure, as measured by an observer moving with the fluid, and U denotes the (future directed) flow vector field, which satisfies $U_i\, U^i = 1$.

We shall need the decomposition

$$\nabla^j\, T_{jk} = q_k + q\, U_k, \tag{4.30}$$

and the field

$$J_{jk} = \nabla_{[j}\, q_{k]}, \tag{4.31}$$

with

$$q = U^i\, \nabla_i\, \rho + (\rho + p)\, \nabla_i\, U^i, \tag{4.32}$$

$$q_k = (\rho + p)\, U^i\, \nabla_i\, U_k + \{U_k\, U^i\, \nabla_i - \nabla_k\}\, p. \tag{4.33}$$

We assume that the fluid is simple, i.e. it consists of only one class of particles, and denote by n, s, T the number density of particles, the entropy per particle, and the absolute temperature as measured by an observer moving with the fluid. We shall assume the first law of equilibrium thermodynamics which has the familiar form $d\,e = -p\,d\,v + T\,d\,s$ in terms of the volume $v = \frac{1}{n}$ and the energy $e = \frac{\rho}{n}$ per particle. In terms of the variables above, we have

$$d\,\rho = \frac{\rho + p}{n}\, d\,n + n\,T\,d\,s. \tag{4.34}$$

We assume an equation of state given in the form

$$\rho = f(n, s), \tag{4.35}$$

with some suitable non-negative function f of the number density of particles and the entropy per particle. Using this in (4.34), we obtain

$$p = n\,\frac{\partial\,\rho}{\partial\,n} - \rho, \quad T = \frac{1}{n}\,\frac{\partial\,\rho}{\partial\,s}, \tag{4.36}$$

as well as the speed of sound ν, given by

$$\nu^2 \equiv (\frac{\partial\,p}{\partial\,\rho})_s = \frac{n}{\rho + p}\,\frac{\partial\,p}{\partial\,n}, \tag{4.37}$$

as known functions of n and s. We require that the specific enthalpy and the speed of sound are positive, i. e.

$$\frac{\rho + p}{n} > 0, \quad \frac{\partial\,p}{\partial\,n} = n\,\frac{\partial^2\,\rho}{\partial\,n^2} > 0. \tag{4.38}$$

We assume the law of particle conservation

$$U^i \nabla_i n + n \nabla_i U^i = 0. \tag{4.39}$$

It implies together with $q = 0$ and (4.34) that the flow is adiabatic, i.e. the entropy per particle is conserved along the flow lines,

$$U^i \nabla_i s = 0. \tag{4.40}$$

The case of an isentropic flow, where the entropy is constant in space and time, is of some interest. In this case the equation of state can be given in the form

$$p = h(\rho) \tag{4.41}$$

with some suitable function h. As a special subcase we shall consider pressure free matter ('dust'), where $h \equiv 0$.

We note that (4.35), (4.36), (4.37), (4.39), and (4.40) imply

$$U^i \nabla_i p = -(\rho + p) \nu^2 \nabla_i U^i,$$

from which we get

$$q_k = -\nabla_k p + (\rho + p) \{U^i \nabla_i U_k - \nu^2 U_k \nabla_i U^i\}. \tag{4.42}$$

Finally, (4.40) implies the equation

$$\mathcal{L}_U s_k = 0, \tag{4.43}$$

for $s_k = \nabla_k s$, where $\mathcal{L}_U$ denotes the Lie derivative in the direction of U.

The Einstein–Euler equations are given in our representation by the equations $z = 0$, $q = 0$, (4.35), (4.36), (4.37), (4.39), and (4.40), where we denote by z the vector-valued quantity

$$z = (T_i{}^k{}_j,\ \Delta^i{}_{jkl},\ F_{jkl},\ q_k). \tag{4.44}$$

which we shall refer to (as well as to each of its components) as a 'zero quantity'. These equations entail furthermore $J_{jk} = 0$ and (4.43). After making a suitable choice of gauge conditions we shall extract hyperbolic evolution equations from this highly overdetermined system .

Decomposition of Unknowns and Equations From now on we shall assume that

$$e_0 = U,$$

so that we have $U^i = \delta^i{}_0$. For the further discussion of the equations we decompose the unknowns and the equations. With the vector field U we associate 'spatial' tensor fields, i.e. tensor fields $T_{i_1,\dots,i_p}$ satisfying

$$T_{i_1,\dots,i_l,\dots,i_p} U^{i_l} = 0, \quad l = 1, \cdots, p.$$

The subspaces orthogonal to U inherit the metric $h_{ij} = g_{ij} - U_i\,U_j$, and $h_i{}^j$ (indices being raised and lowered with g_{ij}) is the orthogonal projector onto these subspaces.

We shall have to consider the projections of various tensor fields with respect to U and its orthogonal subspaces. For a given tensor any contraction with U will be denoted by replacing the corresponding index by U and the projection with respect to $h_i{}^j$ will be indicated by a prime, so that for a tensor field T_{ijk} we write e.g.

$$T'_{iUk} = T_{mpq}\,h_i{}^m\,U^p\,h_k{}^q,$$

etc. Denoting by ϵ_{ijkl} the totally antisymmetric tensor field with $\epsilon_{0123} = 1$ and setting $\epsilon_{jkl} = \epsilon'_{Ujkl}$, we have the decomposition

$$\epsilon_{ijkl} = 2\,(U_{[i}\,\epsilon_{j]kl} - \epsilon_{ij[k}\,U_{l]}),$$

and the relations

$$\epsilon^{jkl}\,\epsilon_{jpq} = -2\,\epsilon\,h^k{}_{[p}\,h^l{}_{q]}, \quad \epsilon^{jkl}\,\epsilon_{jkq} = -2\,\epsilon\,h^l{}_q.$$

Denoting by $C^*_{ijkl} = \frac{1}{2}\,C_{ijpq}\,\epsilon_{kl}{}^{pq}$ the dual of the conformal Weyl tensor, its U-electric and U-magnetic parts are given by by $E_{jl} = C'_{UjUl}$, $B_{jl} = C^{*'}_{UjUl}$ respectively. With the notation $l_{jk} = h_{jk} - U_j\,U_k$, we get the decompositions

$$C_{ijkl} = 2\,(l_{j[k}\,E_{l]i} - l_{i[k}\,E_{l]j}) - 2\,(U_{[k}\,B_{l]p}\,\epsilon^p{}_{ij} + U_{[i}\,B_{j]p}\,\epsilon^p{}_{kl}), \tag{4.45}$$

$$C^*_{ijkl} = 2\,U_{[i}\,E_{j]p}\,\epsilon^p{}_{kl} - 4\,E_{p[i}\,\epsilon_{j]}{}^p{}_{[k}\,U_{l]} - 4\,U_{[i}\,B_{j][k}\,U_{l]} - B_{pq}\,\epsilon^p{}_{ij}\,\epsilon^q{}_{kl}. \tag{4.46}$$

We set

$$a^i = U^k\,\nabla_k\,U^i, \quad \chi_{ij} = h_i{}^k\,\nabla_k\,U_j, \quad \chi = h^{ij}\,\chi_{ij},$$

so that we have

$$\nabla_j\,U^i = U_j\,a^i + \chi_j{}^i, \quad a^i = h_j{}^i\,\Gamma_0{}^j{}_0, \quad \chi_{ij} = -h_i{}^k\,h_j{}^l\,\Gamma_k{}^0{}_l.$$

Since U is not required to be hypersurface orthogonal, the field χ_{ij} will in general not be symmetric. If the tensor field T is spatial, i.e. $T = T'$, we define its spatial covariant derivative by $\mathcal{D}\,T = (\nabla\,T)'$. i.e.

$$\mathcal{D}_i\,T_{i_1,\ldots,i_p} = \nabla_j\,T_{j_1,\ldots,j_p}\,h_i{}^j\,h_{i_1}{}^{j_1}\ldots h_{i_p}{}^{j_p}.$$

It follows that

$$\mathcal{D}_i\,h_{jk} = 0, \quad \mathcal{D}_i\,\epsilon_{jkl} = 0.$$

Equation (4.40) then implies $s_k = \mathcal{D}_k\,s$.

To decompose the equations, we observe the relations

$$q = \frac{2}{\kappa}\,h^{ij}\,F'_{iUj}, \quad q_k = -\frac{2}{\kappa}\,(h^{ij}\,F'_{ijk} - F'_{UkU}),$$

and set

$$P_i = F'_{UiU}$$

$$= \mathcal{D}^j \{E_{ji} - \kappa\,(\frac{1}{3}\,\rho + \frac{1}{2}\,p)\,h_{ji}\} + \frac{1}{2}\,\kappa\,(\rho + p)\,a_i + 2\,\chi^{kl}\,\epsilon^j\,{}_{l(i}\,B_{k)j},$$

$$Q_j = -\frac{1}{2}\,\epsilon_j\,{}^{kl}\,F'_{Ukl}$$

$$= \mathcal{D}^k\,B_{kj} + \epsilon_j\,{}^{kl}\,(2\,\chi^i\,{}_k - \chi_k\,{}^i)\,E_{il} + \kappa\,(\rho + p)\chi_{kl}\,\epsilon_j\,{}^{kl},$$

$$P_{ij} = \{F'_{(i|U|j)} - \frac{1}{3}\,h_{ij}\,h^{kl}\,F'_{kUl}\} = \mathcal{L}_U\,E_{ij} + \mathcal{D}_k\,B_{l(i}\,\epsilon_{j)}\,{}^{kl}$$

$$-2\,a_k\,\epsilon^{kl}\,{}_{(i}\,B_{j)l} - 3\,\chi_{(i}\,{}^l\,E_{j)l} - 2\,\chi^l\,{}_{(i}\,E_{j)l} + h_{ij}\,\chi^{kl}\,E_{kl} + 2\,\chi\,E_{ij}$$

$$+\frac{\kappa}{2}\,(\rho + p)\,(\chi_{(ij)} - \frac{1}{3}\,\chi\,h_{ij}),$$

$$Q_{kl} = \frac{1}{2}\,\epsilon_{(k}\,{}^{ij}\,F'_{l)ij} = \mathcal{L}_U\,B_{kl} - \mathcal{D}_i\,E_{j(k}\,\epsilon_{l)}\,{}^{ij}$$

$$+2\,a_i\,\epsilon^{ij}\,{}_{(k}\,E_{l)j} - \chi^i\,{}_{(k}\,B_{l)i} - 2\,\chi_{(k}\,{}^i\,B_{l)i} + \chi\,B_{kl} - \chi_{ij}\,B_{pq}\,\epsilon^{pi}\,{}_{(k}\,\epsilon^{jq}\,{}_{l)}.$$

Then we find the splitting

$$F_{jkl} = 2\,U_j\,P_{[k}\,U_{l]} + h_{j[k}\,P_{l]} + Q_i\,(U_j\,\epsilon^i\,{}_{kl} - \epsilon^i\,{}_{j[k}\,U_{l]}) \tag{4.47}$$

$$-2\,P_{j[k}\,U_{l]} - Q_{ji}\epsilon^i\,{}_{kl} - \frac{1}{2}\,\kappa\,h_{j[k}\,q_{l]} - \frac{1}{3}\,\kappa\,q\,h_{j[k}\,U_{l]}.$$

The Reduced System i) In the case of pressure free matter the equation $q_k = 0$ tells us that e_0 is geodesic and we can assume the frame field to be parallelly transported in the direction of e_0. Furthermore, we can assume the coordinates x^α, $\alpha = 1, 2, 3$ to be constant on the flow lines of e_0 and the coordinate $t = x^0$ to be a parameter on the integral curves of e_0. These conditions are equivalent to

$$\Gamma_0\,{}^j\,{}_k = 0, \qquad e^\mu\,{}_0 = U^\mu = \delta^\mu\,{}_0,$$

which, in turn, imply together with the requirement $p = 0$ the relation $q_k = 0$ if $\rho + p > 0$. The unknowns to be determined are given by

$$u = (e^\mu\,{}_a,\ \Gamma_b\,{}^i\,{}_j,\ E_{ij},\ B_{ij},\ \rho),$$

where $a, b = 1, 2, 3$. The reduced system is given by

$$T_0\,{}^j\,{}_a = 0, \quad \Delta^i_{j0a} = 0, \quad P_{ij} = 0, \quad Q_{kl} = 0, \quad q = 0. \tag{4.48}$$

ii) In the general case we cannot assume the frame to be parallelly transported because the evolution of U is governed by the Euler equations. We assume the vector fields e_a, $a = 1, 2, 3$, to be Fermi transported in the direction of U and the coordinates to be chosen as before such that

$$\Gamma_0{}^a{}_b = 0, \qquad e^\mu{}_0 = U^\mu = \delta^\mu{}_0.$$

Our unknowns are given by

$$u = (e^\mu{}_a,\ \Gamma_0{}^0{}_a,\ \Gamma_a{}^k{}_l,\ E_{ij},\ B_{kl},\ \rho,\ n,\ s,\ s_k),$$

and the reduced system by

$$T_0{}^j{}_a = 0, \quad \Delta^c{}_{b0a} = 0, \tag{4.49}$$

$$\nu^2\, \Delta^c{}_{0ac} - \frac{1}{\rho + p}\, J'_{Ua} = 0, \quad \nu^2\, (\Delta^0{}_{a0b} + \frac{1}{\rho + p}\, J'_{ab}) = 0,$$

$$P_{ij} = 0, \quad Q_{kl} = 0, \quad q = 0, \tag{4.50}$$

$$\mathcal{L}_U\, n = -n\, \chi, \quad \mathcal{L}_U\, s = 0, \quad \mathcal{L}_U\, s_k = 0,$$

where it is assumed that the functions p, ν^2, α, β are determined from the equation of state (4.35) according to (4.36), (4.37), etc. and $\rho + p > 0$.

Since the functions ρ, s, n are then determined as functions of the coordinates x^μ, we remark that the relation $\rho(x^\mu) = f(n(x^\mu), s(x^\mu))$ is satisfied as a consequence of the equations for ρ, s, and n, and the relation (4.36), since $q = 0$ implies $\partial_t\, \rho = \frac{d}{dt}\, f(n, s)$. Furthermore, we note that in our formalism $\mathcal{L}_U\, s_k = \partial_t\, s_k - (\Gamma_0{}^j{}_k - \Gamma_k{}^j{}_0)\, s_j = e^\mu{}_k\, \partial_t\, s_\mu$ etc.

When we solve the equations $P_{ij} = 0$ and $Q_{kl} = 0$, the symmetry of the fields E_{ij}, B_{kl} has to be made explicit. The trace-free condition then follows as a consequence of the equations and the fact that the initial data are trace-free. With this understanding it is easy to see that the system (4.48) is symmetric hyperbolic, the remaining equations only containing derivatives in the direction of U. To see that the system consisting of (4.49) and (4.50) is also symmetric hyperbolic, we write out some of the equations explicitly (taking the opportunity to correct some misprints in [39]). It follows directly from the definition that

$$J_{kj} = (\rho + p)\, \{U^i\, (\nabla_k\, \nabla_i\, U_j - \nabla_j\, \nabla_i\, U_k) - \nu^2\, U_j\, \nabla_k\, \nabla_i\, U^i \tag{4.51}$$

$$+\nu^2\, U_k\, \nabla_j\, \nabla_i\, U^i - \nu^2\, \nabla_l\, U^l\, (\nabla_k\, U_j - \nabla_j\, U_k) + \nabla_k\, U^i\, \nabla_i\, U_j - \nabla_j\, U^i\, \nabla_i\, U_k$$

$$+\epsilon\, \frac{\rho + p}{\nu^2}\, (\frac{\partial^2\, p}{\partial\, \rho^2})_s\, \nabla_l\, U^l\, (U_k\, U^i\, \nabla_i\, U_j - U_j\, U^i\, \nabla_i\, U_k)\Big\}$$

$$+(\alpha\, U_k\, \nabla_i\, U^i - \beta\, U^i\, \nabla_i\, U_k)\, s_j - (\alpha\, U_j\, \nabla_i\, U^i - \beta\, U^i\, \nabla_i\, U_j)\, s_k,$$

where we set

$$\alpha = (\rho + p)\,\frac{\partial\,\nu^2}{\partial\,s} - (1 + \frac{n}{\nu^2}\,\frac{\partial\,\nu^2}{\partial\,n})\,\frac{\partial\,p}{\partial\,s} + \nu^2\,n\,T, \quad \beta = n\,T - \frac{1}{\nu^2}\,\frac{\partial\,p}{\partial\,s}.$$

In particular,

$$-\frac{1}{\rho + p}\,J'_{Ua} = e_0(\Gamma_0{}^0{}_a) - \nu^2\,e_a(\Gamma_c{}^c{}_0) + \Gamma_0{}^0{}_c\,(\Gamma_a{}^c{}_0 - \Gamma_0{}^c{}_a) \tag{4.52}$$

$$+ \left(\frac{\rho + p}{\nu^2}\,(\frac{\partial^2\,p}{\partial\,\rho^2})_s - \nu^2\right)\,\Gamma_c{}^c{}_0\,\Gamma_0{}^0{}_a - \frac{\alpha}{\rho + p}\,\Gamma_c{}^c{}_0\,s_a,$$

and

$$-\frac{1}{\rho + p}\,J'_{ab} = e_a(\Gamma_0{}^0{}_b) - e_b(\Gamma_0{}^0{}_a) - \Gamma_0{}^0{}_c\,(\Gamma_a{}^c{}_b - \Gamma_b{}^c{}_a) \tag{4.53}$$

$$-\nu^2\,\Gamma_c{}^c{}_0\,(\Gamma_a{}^0{}_b - \Gamma_b{}^0{}_a)\} - \frac{\beta}{\rho + p}\,(\Gamma_0{}^0{}_a\,s_b - \Gamma_0{}^0{}_b\,s_a).$$

From this follows that the last two equations of (4.49) are given by

$$\partial_t\,\Gamma_0{}^0{}_a - \nu^2\,e_c(\Gamma_a{}^c{}_0) = -\Gamma_0{}^0{}_c\,\Gamma_a{}^c{}_0 \tag{4.54}$$

$$- \left(\frac{\rho + p}{\nu^2}\,(\frac{\partial^2\,p}{\partial\,\rho^2})_s - \nu^2\right)\,\Gamma_c{}^c{}_0\,\Gamma_0{}^0{}_a + \frac{\alpha}{\rho + p}\,\Gamma_c{}^c{}_0\,s_a$$

$$+\nu^2\,\left(\Gamma_k{}^c{}_0\,(\Gamma_a{}^k{}_c - \Gamma_c{}^k{}_a) - \Gamma_a{}^c{}_k\,\Gamma_c{}^k{}_0 + \Gamma_c{}^c{}_k\,\Gamma_a{}^k{}_0 + R^c{}_{0ac}\right),$$

$$\nu^2\,\partial_t\,\Gamma_a{}^0{}_b - \nu^2\,e_b(\Gamma_0{}^0{}_a) = \nu^2\,\Big(\Gamma_k{}^0{}_b\,(\Gamma_0{}^k{}_a - \Gamma_a{}^k{}_0) \tag{4.55}$$

$$-\Gamma_0{}^0{}_c\,\Gamma_a{}^c{}_b + R^0{}_{b0a} + \Gamma_0{}^0{}_c\,(\Gamma_a{}^c{}_b - \Gamma_b{}^c{}_a)$$

$$+\nu^2\,\Gamma_c{}^c{}_0\,(\Gamma_a{}^0{}_b - \Gamma_b{}^0{}_a) + \frac{\beta}{\rho + p}\,(\Gamma_0{}^0{}_a\,s_b - \Gamma_0{}^0{}_b\,s_a)\Big).$$

The remaining equations of (4.49) and (4.50) (besides $P_{ij} = 0$, $Q_{kl} = 0$) again only contain derivatives in the direction of U.

The Derivation of the Subsidiary Equations We have to show that any solution to the reduced equations which satisfies the constraints on an initial hypersurface, i.e. for which $z = 0$ on the initial hypersurface, will satisfy $z = 0$ in the domain of dependence of the initial hypersurface with respect to the metric supplied by the solution. For this purpose we will derive a system of partial differential equations for those components of z which do not vanish already because of the reduced equations and the gauge conditions.

We begin by deriving equations for F_{jkl}. There exist two different expressions for $F_{kl} \equiv \nabla^j\,F_{jkl}$. From the definition of F_{jkl} and from the symmetries of the tensor field involved follows

$$F_{kl} = \nabla^j \nabla^i F_{ijkl} = \nabla^j \nabla^i C_{ijkl} - \nabla^j \nabla_{[k} S_{l]j}$$

$$= -R^p{}_{[i}{}^{ij} C_{j]pkl} + C_{ijp[l} R^p{}_{k]}{}^{ij} + \frac{1}{2} T_i{}^p{}_j \nabla_p C^{ij}{}_{kl}$$

$$+\nabla_{[k} \nabla^j S_{l]j} - R^p{}_{[l}{}^j{}_{k]} S_{pj} - R^p{}_j{}^j{}_{[k} S_{l]p} + \nabla_p S^j{}_{[l} T_{k]}{}^p{}_j,$$

where we took into account that we do not know at this stage whether the connection coefficients $\Gamma_i{}^j{}_k$ supplied by the solution define a torsion free connection. From the reduced field equations, the definition of the zero quantities, and the symmetries of C_{ijkl}, it follows that

$$F_{kl} = -\Delta^p{}_{[i}{}^{ij} C_{j]pkl} + C_{ijp[l} \Delta^p{}_{k]}{}^{ij} + \frac{1}{2} T_i{}^p{}_j \nabla_p C^{ij}{}_{kl} \tag{4.56}$$

$$+\kappa J_{kl} + \Delta^p{}_{[l}{}^j{}_{k]} S_{pj} - \Delta^p{}_j{}^j{}_{[k} S_{l]p} = N(z),$$

where $N(z)$ is, as in the following, a generic symbol for a smooth function (which may change from equation to equation) of the zero quantities which satisfies $N(0) = 0$.

On the other hand, because of the reduced equations, equation (4.47) takes the form

$$F_{jkl} = 2\, U_j\, P_{[k}\, U_{l]} + h_{j[k}\, P_{l]} + Q_i\, (U_j\, \epsilon^i{}_{kl} - \epsilon^i{}_{j[k}\, U_{l]}) - \frac{1}{2}\, \kappa\, h_{j[k}\, q_{l]}. \tag{4.57}$$

Contracting with ∇^j, decomposing the resulting expression into F'_{kU}, F'_{kl} and equating with the corresponding expressions obtained from (4.56), we arrive at equations of the form

$$\mathcal{L}_U\, P_k + \frac{1}{2}\, \epsilon_k{}^{ij}\, \mathcal{D}_i\, Q_j = N(z) \tag{4.58}$$

$$\mathcal{L}_U\, Q_k - \frac{1}{2}\, \epsilon_k{}^{ij}\, \mathcal{D}_i\, P_j = N(z). \tag{4.59}$$

The connection defined by the $\Gamma_i{}^j{}_k$ and the associated torsion and curvature tensors satisfy the first Bianchi identity

$$\sum_{(jkl)} \nabla_j T_k{}^i{}_l = \sum_{(jkl)} (R^i{}_{jkl} + T_j{}^m{}_k T_l{}^i{}_m),$$

where $\sum_{(jkl)}$ denotes the sum over the cyclic permutation of the indices jkl. Setting here $j = 0$, observing that the symmetries of $C^i{}_{jkl}$, S_{kl} imply $\sum_{(jkl)} R^i{}_{jkl} = \sum_{(jkl)} \Delta^i{}_{jkl}$, and taking into account the reduced equations, we get from this an equation of the form

$$e_0(T_a{}^i{}_b) = N(z). \tag{4.60}$$

To derive equations for $\Delta^i{}_{jkl}$, we use the second Bianchi identity

$$\sum_{(jkl)} \nabla_j R^i{}_{mkl} = -\sum_{(jkl)} R^i{}_{mnj} T_k{}^n{}_l. \tag{4.61}$$

We write

$$R^i{}_{jkl} = \Delta^i{}_{jkl} + C^i{}_{jkl} + E^i{}_{jkl} + G^i{}_{jkl},$$

with $E^i{}_{jkl} = g^i{}_{[k} S^*_{l]j} - g_{j[k} S^*_{l]}{}^i$, $G^i{}_{jkl} = \frac{1}{2} S g^i{}_{[k} g_{l]j}$ and $S^*_{jk} = S_{jk} - \frac{1}{4} S g_{jk}$. Using the well known facts that the left and right duals of C_{ijkl} and G_{ijkl} are equal while the left dual of E_{ijkl} differs from its right dual by a sign, and using the reduced equations, we get

$$\sum_{(jkl)} \nabla_j R^i{}_{mkl} = \sum_{(jkl)} \nabla_j \Delta^i{}_{mkl}$$

$$+\frac{1}{2} \epsilon_{jkl}{}^p (\nabla_q C^q{}_{prs} - \nabla_q E^q{}_{prs} + \nabla_q G^q{}_{prs}) \epsilon_m{}^{irs}$$

$$= \sum_{(jkl)} \nabla_j \Delta^i{}_{mkl} + \frac{1}{2} \epsilon_{jkl}{}^p (F_{prs} + \kappa g_{pr} q_s) \epsilon_m{}^{irs},$$

whence, by (4.57) and (4.61),

$$\sum_{(jkl)} \nabla_j \Delta^i{}_{mkl} = N(z). \tag{4.62}$$

i) In the case of pressure free matter we get from (4.62), by setting $j = 0$ and using (4.48), an equation of the form

$$e_0(\Delta^i{}_{kab}) = N(z). \tag{4.63}$$

Since, as remarked earlier, $q_k = 0$ by our gauge conditions, the system of equations consisting of (4.58), (4.59), (4.60), and (4.63) constitutes the desired 'subsidiary system' for the zero quantities in the pressure free case.

ii) Using (4.49), we get in the general case by the analogous procedure only an equation of the form

$$e_0(\Delta^a{}_{bcd}) = N(z). \tag{4.64}$$

By the antisymmetry of J'_{ab} we know already that

$$\Delta^0{}_{a0b} + \Delta^0{}_{b0a} = 0.$$

We can express the equations for $\Delta^0{}_{[b|0|c]}$, $\Delta^0{}_{abc}$. in terms of the quantities

$$\Delta_a = \frac{1}{2} \epsilon_a{}^{bc} \Delta^0{}_{b0c}, \quad \Delta^*_{ab} = \frac{1}{2} \epsilon_{(a}{}^{cd} \Delta^0{}_{b)cd}, \quad \Delta^*_a = \frac{1}{2} \Delta^c{}_{0ac}, \tag{4.65}$$

because

$$\Delta^0{}_{[b|0|c]} = -\Delta_a\, \epsilon^a{}_{bc}, \quad \Delta^0{}_{abc} = -\Delta^*_{ad}\, \epsilon_{bc}{}^d + 2\, h_{a[b}\, \Delta^*_{c]}.$$

From (4.62) we get

$$e_0\,(\Delta^0{}_{abc}) + e_c\,(\Delta^0{}_{a0b}) + e_b\,(\Delta^0{}_{ac0}) = N(z),$$

which implies for the quantities (4.65) equations

$$2\, e_0\,(\Delta^*_a) - \epsilon_a{}^{bc}\, e_b\,(\Delta_c) = N(z),$$

and

$$e_0\,(\Delta^*_{ab}) - e_{(a}\,(\Delta_{b)}) + h_{ab}\, h^{cd}\, e_c\,(\Delta_d) = N(z).$$

Since we have

$$h^{ab}\, e_a\,(\Delta_b) = -\frac{1}{2}\, \epsilon^{abc}\, e_a\,(\frac{1}{\rho+p}\, J'_{bc}) = -\frac{1}{2}\, \frac{1}{\rho+p}\, \epsilon^{ijk}\, \nabla_i\, \nabla_j\, q_k + N(z) = N(z),$$

we can write the second equation in the form

$$e_0\,(\Delta^*_{ab}) - e_{(a}\,(\Delta_{b)}) = N(z).$$

From (4.62) we get furthermore

$$e_a\,(\Delta^0{}_{bcd}) + e_d\,(\Delta^0{}_{bac}) + e_c\,(\Delta^0{}_{bda}) = N(z),$$

which implies in terms of the quantities (4.65) an equation of the form

$$h^{ab}\, e_a\,(\Delta^*_{bc}) - \epsilon_c{}^{ab}\, e_a\,(\Delta^*_b).$$

By a direct calculation we derive from (4.31) the equation

$$2\, \mathcal{L}_U\, J'_{ij} = 4\, \mathcal{D}_{[i}\, J'_{U j]} - h_i{}^p\, h_j{}^q\, (\Delta^l{}_{npq} + \Delta^l{}_{qnp} + \Delta^l{}_{pqn})\, U^n\, q_l$$
$$-h_i{}^p\, h_j{}^q\, (T_l{}^n{}_p\, \nabla_n\, q_q + T_q{}^n{}_l\, \nabla_n\, q_p + T_p{}^n{}_q\, \nabla_n\, q_l)\, U^l$$
$$-2\, a_i\, J'_{Uj} + 2\, a_j\, J'_{Ui},$$

which can be rewritten by (4.49) in the form

$$(\rho+p)\, \{e_0\,(\Delta_a) + 2\, \nu^2\, \epsilon_a{}^{bc}\, e_b\,(\Delta^*_c)\} = N(z). \tag{4.66}$$

From the equations above we obtain the system

$$(\rho+p)\, \{2\, \nu^2\, e_0\,(\Delta^*_a) - \nu^2\, \epsilon_a{}^{bc}\, e_b\,(\Delta_c)\} = N(z), \tag{4.67}$$

$$(\rho+p)\, \{e_0\,(\Delta_a) + \nu^2\, \epsilon_a{}^{bc}\, e_b\,(\Delta^*_c) + \nu^2\, h^{bc}\, e_b\,(\Delta^*_{ca})\} = N(z), \tag{4.68}$$

$$c_{ab}\,(\rho+p)\, \{\nu^2\, e_0\,(\Delta^*_{ab}) - \nu^2\, e_{(a}\,(\Delta_{b)})\} = N(z), \tag{4.69}$$

where $c_{ab} = 1$ if $a = b$ and $c_{ab} = 2$ if $a \neq b$. Finally, we obtain from (4.31), (4.49) the equation

$$\mathcal{L}_U \, q_a = 2 \, \nu^2 \, (\rho + p) \, \Delta^c{}_{0ac}. \tag{4.70}$$

Equations (4.58), (4.59), (4.60), (4.64), (4.67), (4.68), (4.69), and (4.70) constitute the subsidiary equations for the zero quantities in the general case.

We note here that in the reduced system it has not been built in explicitly that the energy-momentum tensor has vanishing divergence (cf. equation (4.30)). While we assumed the equation $q = 0$ as part of the reduced equations, we verify the vanishing of the quantity q_k by deriving the subsidiary equations and using the uniqueness property for these equations.

In the present formalism the gauge conditions are taken care of by the explicit form of some of the unknowns, however the list of constraints is much longer than in the previous discussions. We shall not try to demonstrate that the constraints are preserved in the specific case of 'floating fluid balls'.. Though the construction of data for fluid balls of compact support which are embedded in asymptotically vacuum data has been shown and their smoothness properties near the boundary have been discussed [70], the evolution in time of these data and the precise smoothness properties of the fields near and possible jumps travelling along the boundary have not been worked out yet. However, without a precise understanding of the behaviour of the solution near the boundary the conservation of the constraints cannot be demonstrated.

Our reduced system also found applications in cosmological context where the fluid is spread out, with $\rho + p > 0$, over the time slices (cf. [31], [79]). In this case the desired conclusion follows from the fact that the subsidiary systems are symmetric hyperbolic, have right hand sides of the form $N(z)$, and the characteristics of the reduced system and the subsidiary system are as follows (where we use only the frame components $\xi_k = \xi_\mu \, e^\mu{}_k$ of the covector ξ).

(i) In the case of pressure free matter the characteristic polynomial of the reduced system is of the form

$$c \, (\xi_0)^K \, (\xi_0^2 + \frac{1}{4} \, h^{ab} \, \xi_a \, \xi_b)^L \, (g^{\mu\nu} \, \xi_\mu \, \xi_\nu)^N,$$

with positive integers K, L, N and constant factor c, while the characteristic polynomial of the subsidiary system only contains the first two factors. Thus the characteristics of the subsidiary system are timelike with respect to $g_{\mu\nu}$ (cf. also the remarks in Sect. 2).

(ii) In the general case the characteristic polynomial of the reduced equations is of the form

$$c \, (\xi_0)^K \, (\xi_0^2 + \frac{1}{4} \, h^{ab} \, \xi_a \, \xi_b)^L \, (\xi_0^2 + \nu^2 \, h^{cd} \, \xi_c \, \xi_d)^M \, (g^{\mu\nu} \, \xi_\mu \, \xi_\nu)^N, \tag{4.71}$$

with positive integers K, L, M, N and constant factor c, while the characteristic polynomial of the subsidiary system is generated by powers of the first three factors.

We note that the equations (4.54) and (4.55) contribute to the principal symbol the third factor which corresponds to the sound cone pertaining to the fluid. If one wants to ensure that the sound does not travel with superluminal speed one has to require the equation of state (4.35) to be such that $\nu \leq 1$. For the question of existence and uniqueness of solutions there is no need to impose such a condition, but if $\nu > 1$, the domain of dependence with respect to $g_{\mu\nu}$, as it has been defined in Sect. 2, cannot be shown any longer, by the arguments given in Sect. 2, to be also a domain of uniqueness. In any case, it follows from the characteristic polynomials that in a domain where the solution of the reduced system is unique according to those arguments, the constraints will be satisfied if they hold on the initial hypersurface.

We note that the system simplifies considerably in the isentropic case. In the reduced system the function $\rho+p$ then neither occurs in the principal part nor in a denominator. In the case of the subsidiary system a more detailed discussion is required to understand the consequences of the occurrence of the various factors $\rho + p$.

In our procedure the fluid equations serve two purposes, they determine the motion of the fluid as well as the evolution of the frame. If we set $\kappa = 0$ in all equations the fluid equations decouple from the geometric equations and we obtain a new hyperbolic reduction of the vacuum field equations. In this procedure any exotic 'equation of state' may be prescribed as long as it ensures a useful, long-lived gauge.

If the initial data for the Einstein–Euler equations are such that U, ρ, p, n, s, and the equation of state can be smoothly extended through the boundaries of the fluid balls, this suggests using the 'extended fluid' to control the evolution of the gauge in the vacuum part of the solution near the boundary.

4.3 The Initial Boundary Value Problem

In the previous section we studied a problem involving a distinguished timelike hypersurface. Its evolution in time was determined by a physical process. There are also important problems where the Einstein equations are solved near timelike hypersurfaces which are prescribed for practical reasons, e.g. to perform numerical calculations on finite grids. The underlying initial boundary value problem for Einstein's field equations, where initial data are prescribed on a (spacelike) hypersurface S and boundary data a (timelike) boundary T which intersect at a 2-surface $\Sigma = T \cap S$, has been analysed in detail in the article [42]. The solution to this problem requires a hyperbolic reduction which needs to satisfy, beyond the conditions discussed at the beginning of this section, certain side conditions. In the following we want to comment on those aspects of the work in [42] which illustrate the flexibility of the field equations in performing reductions and on certain characteristics of the reduced system. For the full analysis of the initial boundary value problem we refer to [42].

Since we are dealing with a problem for equations which are essentially hyperbolic, the problem can be localized. In suitably adapted coordinates x^μ, defined on some neighbourhood of a point $p \in \Sigma$, the manifold M on which the solution is to be determined will then be given in the form $M = \{x \in \mathbb{R}^4 | x^0 \geq 0, x^3 \geq 0\}$, the initial hypersurface by $S = \{x \in M | x^0 = 0\}$ and the boundary by $T = \{x \in M | x^3 = 0\}$. Clearly, we will have to prescribe Cauchy data on S as before, we will have to analyse which kind of boundary data are admitted by the equations, and on the edge $\Sigma = \{x \in M | x^0 = 0,\ x^3 = 0\}$ the data will have to satisfy some consistency conditions, as is always the case in initial boundary value problems.

Maximally Dissipative Initial Boundary Value Problems We have seen in Sect. 3.1 that energy estimates provide a basic tool for obtaining results about the existence and uniqueness of solutions to symmetric hyperbolic systems. To explain the side conditions which have to be satisfied in a hyperbolic reduction of an initial boundary value problem for Einstein's field equations, we consider what will happen if we try to obtain energy estimates in the present situation. Assume that we are given on M in the coordinates x^μ a linear symmetric hyperbolic system of the form

$$A^\mu \partial_\mu u = B\, u + f(x), \tag{4.72}$$

for an $\mathbb{R}^N$-valued unknown u. The matrices $A^\mu = A^\mu(x)$, $\mu = 0, 1, 2, 3$, are smooth functions on M which take values in the set of symmetric $N \times N$-matrices, there exists a 1-form ξ_μ such that $A^\mu\, \xi_\mu$ is positive definite, $B = B(x)$ is a smooth matrix-valued function and $f(x)$ a smooth $\mathbb{R}^N$-valued function on M. For convenience we assume that the positivity condition is satisfied with $\xi_\mu = \delta^0{}_\mu$.

If we assume that u vanishes for large positive values of x^α, $\alpha = 1, 2, 3$, and if the relation

$$\partial_\mu({}^t u\, A^\mu\, u) = {}^t u\, K\, u + 2\, {}^t u\, f \quad \text{with} \quad K = B + {}^t B + \partial_\mu\, A^\mu,$$

implied by (4.72), is integrated over a set $M_\tau = \{x \in M | 0 \leq x^0 \leq \tau\}$, defined by some number $\tau \geq 0$, we obtain the relation

$$\int_{S_\tau} {}^t u\, A^0\, u\, dS = \int_S {}^t u\, A^0\, u\, dS + \int_{M_\tau} \{{}^t u\, K\, u + 2\, {}^t u\, f\}\, dV + \int_{T_\tau} {}^t u\, A^3\, u\, dS,$$

involving boundary integrals over $S_\tau = \{x \in M | x^0 = \tau\}$ and $T_\tau = \{x \in M | 0 \leq x^0 \leq \tau, x^3 = 0\}$. Obviously, the structure of the *normal matrix* A^3 plays a prominent role here. If the last term on the right hand side is non-positive, we can use the equation above to obtain energy estimates for proving the existence and uniqueness of solutions.

By this (and certain considerations which will become clear when we have set up our reduced system) we are led to consider the following *maximally dissipative boundary value problem*.

We choose $g \in C^\infty(S, \mathbb{R}^N)$ and require as initial condition $u(x) = g(x)$ for $x \in S$. We choose a smooth map Q of T into the set of linear subspaces of $\mathbb{R}^N$ and require as boundary condition $u(x) \in Q(x)$ for $x \in T$. The type of map Q admitted here is restricted by the following assumptions.

(i) The set T is a characteristic of (4.72) of constant multiplicity, i.e.

$$\dim(\ker A^3(x)) = \text{const.} > 0, \quad x \in T.$$

(ii) The map Q is chosen such as to ensure the desired non-positivity

$${}^t u\, A^3(x)\, u \leq 0, \quad u \in Q(x), \quad x \in T.$$

(iii) The subspace $Q(x)$, $x \in T$, is a maximal with (ii), i.e. the dimension of $Q(x)$ is equal to number of non-positive eigenvalues of A^3 counting multiplicity.

The specification of Q can be expressed in terms of linear equations. Since A^3 is symmetric, we can assume, possibly after a transformation of the dependent variable, that at a given point $x \in T$

$$A^3 = \kappa \begin{bmatrix} -I_j & 0 & 0 \\ 0 & 0_k & 0 \\ 0 & 0 & I_l \end{bmatrix}, \quad \kappa > 0,$$

where I_j is a $j \times j$ unit matrix, 0_k is a $k \times k$ zero matrix etc. and $j + k + l = N$. Writing $u = {}^t(a, b, c) \in \mathbb{R}^j \times \mathbb{R}^k \times \mathbb{R}^l$ we find that at x the linear subspaces admitted as values of Q are neccessarily given by equations of the form $0 = c - H\, a$ where $H = H(x)$ is a $l \times j$ matrix satisfying

$$-{}^t a\, a + {}^t a\, {}^t H\, H\, a \leq 0, \quad a \in \mathbb{R}^j, \quad \text{i.e.} \quad {}^t H\, H \leq I_j.$$

We note that there is no freedom to prescribe data for the component b of u associated with the kernel of A^3. More specifically, if $A^3 \equiv 0$ on T, energy estimates are obtained without imposing conditions on T and the solutions are determined uniquely by the initial condition on S. By subtracting a suitable smooth function from u and redefining the function f, we can convert the homogeneous problem above to an inhomogeneous problem and vice versa. Inhomogeneous maximal dissipative boundary conditions are of the form

$$q = c - H\, a, \tag{4.73}$$

with $q = q(x)$, $x \in T$, a given $\mathbb{R}^l$-valued function representing the free boundary data on T.

Maximally dissipative boundary value problems as outlined above have been worked out in detail in [74], [83] for the linear case and in [48] [84] for quasi-linear problems (see also these articles for further references). If we want to make use of this theory to analyse the initial boundary value problem for Einstein's field equations we will have to solve two problems which go beyond what is known from the standard Cauchy problem.

(i) We will have to find a reduction, involving a symmetric hyperbolic system, which gives us sufficient information on the normal matrix so that we can control the conditions above.

(ii) To demonstrate the preservation of the constraints we will have to discuss an initial boundary value problem for the subsidiary system. This should be such as to admit a uniqueness proof. Moreover, there is the problem of getting sufficient control on the solution near T. While the initial data for the reduced system will of course be arranged such that the constraints are satisfied on the initial hypersurface S, it will a priori not be clear that we will have sufficient information on the behaviour of the solution to the reduced equations and on the data on T in order to conclude that the constraints will be satisfied on T.

The choice of representation of the field equations, of the gauge conditions and the gauge source functions, and, in particular, the choice of the reduced equations will largely be dominated by the second problem.

The Representation of the Einstein Equations In [42] the initial boundary value problem for Einstein's vacuum field equations was analysed in terms of the equations

$$T_i{}^k{}_j = 0, \quad \Delta^i{}_{jkl} = 0, \quad F_{jkl} = 0, \tag{4.74}$$

of the previous section with everywhere vanishing energy-momentum tensor. We shall use these equations together with the conventions and notation introduced in the previous section.

The Gauge Conditions The gauge, which we assume here for simplicity extends to all of M, has been chosen as follows. On the initial hypersurface $x^0 = 0$ and x^α, $\alpha = 1,2,3$, are coordinates with $x^3 = 0$ on Σ and $x^3 > 0$ elsewhere. The timelike unit vector field e_0 on M is tangent to T, orthogonal to the 2-surfaces $S_c = \{x^3 = c = \text{const.} > 0\}$ in S, and it points towards M on $S \cap U$. The coordinates x^μ satisfy $e^\mu{}_0 = e_0(x^\mu) = \delta^\mu{}_0$ on M and the sets $T_c = \{x^3 = c\}$ are smooth timelike hypersurfaces of M with $T_0 = T$. The unit vector field e_3 is normal to the hypersurfaces T_c and points towards M on T. The vector fields e_A, $A = 1,2$, are tangent to $T_c \cap S$ and such that they form with e_0, e_3 a smooth orthonormal frame field on S. On the hypersurfaces T_c these fields are Fermi transported in the direction of e_0 with respect to the Levi–Civita connection D defined by the metric induced on T_c. The e_k form a smooth orthonormal frame field on U. We refer this type of gauge as an 'adapted gauge'. Notice that it leaves a freedom to choose the timelike vector field e_0 on $M \setminus S$.

In analysing the initial boundary value problem it will be necessary to distinguish between interior equations on the submanifolds S, T, T_c, Σ, S_c. Since our frame is adapted to these submanifolds, this can be done by distinguishing four groups of indices. They are given, together with the values they take, as follows

$$a, c, d, e, f = 0, 1, 2; \quad i, j, k, l, m, n = 0, 1, 2, 3;$$
$$p, q, r, s, t = 1, 2, 3; \quad A, B, C, D = 1, 2.$$

We assume the summation convention for each group.

By our conditions the frame coefficients $e^\mu{}_k$ satisfy

$$e^\mu{}_0 = \delta^\mu{}_0, \quad e^3{}_a = 0, \quad e^3{}_3 > 0 \quad \text{on} \quad M, \tag{4.75}$$

while the frame connection coefficients satisfy

$$\Gamma_0{}^A{}_B = 0. \tag{4.76}$$

The fields e_a satisfy on T_c the equations

$$D_{e_0}\, e_0 = \Gamma_0{}^A{}_0\, e_A, \quad D_{e_0}\, e_A = -g_{AB}\, \Gamma_0{}^B{}_0\, e_0. \tag{4.77}$$

Thus, given the hypersurfaces T_c, the evolution of the coordinates x^α, $\alpha = 0, 1, 2$, and the frame vector fields e_a off S is governed by the coefficients $\Gamma_0{}^A{}_0$.

Another part of the connection coefficients defines the intrinsic connection D on T_c, since $D_a\, e_c = D_{e_a}\, e_c = \Gamma_a{}^b{}_c\, e_b$. The remaining connection coefficients, given by

$$\chi_{ab} = g(\nabla_{e_a} e_3, e_b) = \Gamma_a{}^j{}_3\, g_{jb} = \Gamma_a{}^3{}_b = \Gamma_{(a}{}^3{}_{b)}, \tag{4.78}$$

define the second fundamental form of the hypersurfaces T_c in the frame e_a. In the reduced equations, the symmetry of χ_{ab} has to be taken into account explicitly. A special role is played by mean extrinsic curvature

$$\chi \equiv g^{ab}\, \chi_{ab} = g^{jk}\, \Gamma_j{}^3{}_k = \nabla_\mu\, e^\mu{}_3, \tag{4.79}$$

since it can be regarded as the quantity controlling the evolution of the hypersurfaces T_c and thus of the coordinate x^3.

We now choose two smooth functions $F^A \in C^\infty(M)$ as gauge source functions. These will occur explicitly in the reduced equations and will play the role of connection coefficients for the solution, namely $F^A = \Gamma_0{}^A{}_0$. Furthermore we will choose a function $f \in C^\infty(M)$ which will play the role of the mean extrinsic curvature on the hypersurfaces T_c. Here the interpretation is somewhat more complicated. On T the function $\chi = f|_T$ must be regarded as the free datum which, together with certain data on Σ, indirectly specifies the boundary T. However, for $x^3 > 0$ the function f plays the role of a gauge source function which determines the gauge dependent hypersurface T_c, $c > 0$. It is a remarkable feature of the Codazzi equations that they admit this freedom while at the same time implying hyperbolic equations.

This example clearly shows the importance of the freedom to dispose of the gauge source functions. While we could choose $F^A = 0$ locally (cf. the remarks in [42] about certain subtleties arising here), we need the full freedom to make use of the gauge source functions f, since otherwise we could only handle restricted types of boundaries.

The Reduced Equations Using the gauge conditions above, we extract from (4.74) the following reduced system for those components of the unknowns $e^{\mu}{}_{k}$, $\Gamma_{i}{}^{j}{}_{k}$, E_{ij}, B_{kl} which are not determined already by the gauge conditions and the chosen gauge source functions. Where it has not already been done explicitly, it is understood that in the following equations the connection coefficients $\Gamma_{0}{}^{A}{}_{0}$ and χ are replaced by the gauge source functions F^{A} and f respectively. The torsion free condition gives

$$0 = -T_{0}{}^{k}{}_{p}\, e^{\mu}{}_{k} = \partial_{t}\, e^{\mu}{}_{p} - (\Gamma_{0}{}^{q}{}_{p} - \Gamma_{p}{}^{q}{}_{0})\, e^{\mu}{}_{q} - \Gamma_{0}{}^{0}{}_{p}\, \delta^{\mu}{}_{0}. \tag{4.80}$$

The Gauss equations with respect to T_c provide the equations

$$0 = \Delta^{B}{}_{00A} = e_{0}(\Gamma_{A}{}^{B}{}_{0}) - e_{A}(F^{B}) + \Gamma_{C}{}^{B}{}_{0}\,\Gamma_{A}{}^{C}{}_{0} \tag{4.81}$$

$$-\Gamma_{A}{}^{B}{}_{C}\, F^{C} + F^{B}\, F^{C}\, g_{AC} + \chi_{0}{}^{B}\, \chi_{A0} - \chi_{A}{}^{B}\, \chi_{00} - C^{B}{}_{00A},$$

$$0 = \Delta^{B}{}_{C0A} = e_{0}(\Gamma_{A}{}^{B}{}_{C}) + F^{B}\,\Gamma_{A}{}^{0}{}_{C} + \Gamma_{A}{}^{B}{}_{0}\, F^{D}\, g_{CD} \tag{4.82}$$

$$+\Gamma_{D}{}^{B}{}_{C}\,\Gamma_{A}{}^{D}{}_{0} + \chi_{0}{}^{B}\,\chi_{AC} - \chi_{A}{}^{B}\,\chi_{0C} - C^{B}{}_{C0A}.$$

Codazzi's equations with respect to T_c imply

$$0 = g^{ab}\,\Delta^{3}{}_{ab1} = D_{0}\,\chi_{01} - D_{1}\,\chi_{11} - D_{2}\,\chi_{12} - D_{1}(f), \tag{4.83}$$

$$0 = g^{ab}\,\Delta^{3}{}_{ab2} = D_{0}\,\chi_{02} - D_{1}\,\chi_{12} - D_{2}\,\chi_{22} - D_{2}(f), \tag{4.84}$$

$$0 = \Delta^{3}{}_{101} = D_{0}\,\chi_{11} - D_{1}\,\chi_{01} - C^{3}{}_{101}, \tag{4.85}$$

$$0 = \Delta^{3}{}_{201} + \Delta^{3}{}_{102} = 2\,D_{0}\,\chi_{12} - D_{1}\,\chi_{02} - D_{2}\,\chi_{01} - C^{3}{}_{201} - C^{3}{}_{102}, \tag{4.86}$$

$$0 = \Delta^{3}{}_{202} = D_{0}\,\chi_{22} - D_{2}\,\chi_{02} - C^{3}{}_{202}, \tag{4.87}$$

where it is understood that the component χ_{00}, which appears only in undifferentiated form, is given by $\chi_{00} = \chi_{11} + \chi_{22} + f$. The remaining Ricci identities give

$$0 = \Delta^{A}{}_{B03} = e_{0}(\Gamma_{3}{}^{A}{}_{B}) + F^{A}\,\Gamma_{3}{}^{0}{}_{B} + \Gamma_{3}{}^{A}{}_{0}\,F^{C}\,g_{BC} + \Gamma_{C}{}^{A}{}_{B}\,\Gamma_{3}{}^{C}{}_{0} \tag{4.88}$$

$$+\Gamma_{3}{}^{A}{}_{B}\,\Gamma_{3}{}^{3}{}_{0} + \chi_{0}{}^{A}\,\Gamma_{3}{}^{3}{}_{B} - \Gamma_{3}{}^{A}{}_{3}\,\chi_{0B} - \Gamma_{C}{}^{A}{}_{B}\,\chi_{0}{}^{C} - C^{A}{}_{B03},$$

$$0 = \Delta^{A}{}_{003} = e_{0}(\Gamma_{3}{}^{A}{}_{0}) - e_{3}(F^{A}) + \chi_{0}{}^{A}\,\Gamma_{3}{}^{3}{}_{0} - \Gamma_{3}{}^{A}{}_{B}\,F^{B} + \Gamma_{B}{}^{A}{}_{0}\,\Gamma_{3}{}^{B}{}_{0} \tag{4.89}$$

$$+\Gamma_3{}^A{}_0\,\Gamma_3{}^3{}_0 - \Gamma_B{}^A{}_0\,\chi_0{}^B - \Gamma_3{}^3{}_B\,g^{BA}\,\chi_{00} - F^A\,\chi_{00} - C^A{}_{003},$$

$$0 = \Delta^3{}_{A03} + \Delta^3{}_{03A} = e_0(\Gamma_3{}^3{}_A) - e_A(\Gamma_3{}^3{}_0) \qquad (4.90)$$
$$+\Gamma_3{}^3{}_0\,F^B\,g_{BA} + \Gamma_3{}^3{}_C\,\Gamma_A{}^C{}_0,$$

$$0 = g^{ab}\,\Delta^3{}_{ab3} = e_0(\Gamma_3{}^3{}_0) + g^{AB}\,e_A(\Gamma_3{}^3{}_B) - e_3(f) \qquad (4.91)$$
$$-g^{ab}\,\Gamma_3{}^3{}_k\Gamma_b{}^k{}_a + g^{ab}\,\Gamma_b{}^3{}_k\Gamma_3{}^k{}_a + g^{ab}\,\Gamma_m{}^3{}_a(\Gamma_3{}^m{}_b - \Gamma_b{}^m{}_3).$$

In the previous section we saw how to extract a symmetric hyperbolic system from the Bianchi identities. However, for reasons given below, we shall not choose that system here. Instead we choose the 'boundary adapted system'

$$\begin{aligned} P_{11} - P_{22} &= 0 & Q_{11} - Q_{22} &= 0 \\ 2\,P_{12} &= 0 & 2\,Q_{12} &= 0 \\ P_{11} + P_{22} &= 0 & Q_{11} + Q_{22} &= 0 \\ P_{13} &= \tfrac{1}{2}Q_2 & Q_{13} &= -\tfrac{1}{2}P_2 \\ P_{23} &= -\tfrac{1}{2}Q_1 & Q_{23} &= \tfrac{1}{2}P_1, \end{aligned} \qquad (4.92)$$

written as a system for the unknown vector u which is the transpose of

$$((E_-, 2E_{12}, E_+, E_{13}, E_{23}), (B_-, 2B_{12}, B_+, B_{13}, B_{23})).$$

Here $E_\pm = E_{11} \pm E_{22}$, and $B_\pm = B_{11} \pm B_{22}$ and it is understood that the relations $g^{ij}\,E_{ij} = 0$ and $g^{ij}\,B_{ij} = 0$ are used everywhere to replace the fields E_{33} and B_{33} by our unknowns. Written out explicitly, this system takes the form $(\mathbf{I}^\mu + \mathbf{A}^\mu)\,\partial_\mu\,u = b$, with

$$\mathbf{I}^\mu = \begin{bmatrix} I^\mu & 0 \\ 0 & I^\mu \end{bmatrix}, \qquad \mathbf{A}^\mu = \begin{bmatrix} 0 & A^\mu \\ {}^T A^\mu & 0 \end{bmatrix},$$

where

$$I^\mu = \delta^\mu{}_0 \begin{bmatrix} 1&0&0&0&0 \\ 0&1&0&0&0 \\ 0&0&1&0&0 \\ 0&0&0&1&0 \\ 0&0&0&0&1 \end{bmatrix}, \qquad A^\mu = \begin{bmatrix} 0 & -e^\mu{}_3 & 0 & e^\mu{}_2 & e^\mu{}_1 \\ e^\mu{}_3 & 0 & 0 & -e^\mu{}_1 & e^\mu{}_2 \\ 0 & 0 & 0 & e^\mu{}_2 & -e^\mu{}_1 \\ -e^\mu{}_2 & e^\mu{}_1 & -e^\mu{}_2 & 0 & 0 \\ -e^\mu{}_1 & -e^\mu{}_2 & e^\mu{}_1 & 0 & 0 \end{bmatrix}. \qquad (4.93)$$

The reduced system consisting of (4.80) to (4.92), is symmetric hyperbolic. However, beyond that the choice of this particular system was motivated by the following specific features.

(i) The theory of maximally dissipative initial value problems applies to our reduced equations. In equations (4.80) to (4.91) the derivative ∂_{x^3}, which by our gauge conditions occurs only with the directional derivative e_3, is applied to the gauge source functions but not to the unknowns, while (4.93) shows that we have perfect control on the non-trivial part of the normal matrix arising from (4.92). Our discussion of maximally dissipative initial value problems, which led to (4.73), and the form of the matrices (4.93) suggest that we can prescribe besides the datum χ, which characterizes the boundary, precisely two free functions as boundary data on T. This is confirmed by the detailed discussion in [42], though in general a number of technical details have to be taken care of .

(ii) If instead of (4.92) we had chosen the system $P_{ij} = 0$, $Q_{kl} = 0$ as equations for the electric and magnetic part of the conformal Weyl tensor, the theory of maximally dissipative initial value problems would also have applied. We would, however, have come to the conclusion that besides the mean extrinsic curvature four functions could be prescribed freely on T. This apparent contradiction is resolved when one tries to show the preservation of the constraints, i.e. that those equations contained in (4.74) are satisfied which are not already solved because of the gauge conditions and the reduced equations. In the case of the reduced equations above this can be shown for the following reason. The subsidiary system splits in this case into a hierachy of symmetric hyperbolic subsystems with the following property. The first subsystem has vanishing normal matrix on T. This implies under suitable assumptions on the domain of the solution to the reduced equations that all unknows in this subsystem must vanish, because the data on S are of course arranged such that all constraints are satisfied. Furthermore, it follows for any subsystem in the hierarchy that its normal matrix vanishes if the unknowns of all previous subsystems in the hierarchy vanish. From this the desired conclusion follows in a finite number of steps. If we had considered instead the system $P_{ij} = 0$, $Q_{kl} = 0$, the discussion whether the constraints are preserved would have become quite complicated and would have led us in the end to the conclusion that only two functions are really free on T while the others are subject to restrictions determined by the evolution properties of the reduced system.

We end our discussion of the initial boundary value problem with an observation about the characteristics of the reduced system. Equations (4.80) to (4.91) contribute a factor of the form

$$\xi_0^K \, (\xi_0^2 - \xi_1^2 - \xi_2^2)^L \, (2\,\xi_0^2 - \xi_1^2 - \xi_2^2)^M,$$

to the characteristic polynomial (using again only the frame components of the covector ξ). The corresponding characteristics are timelike or null with respect to $g_{\mu\nu}$. However, the subsystem (4.92) contributes a factor

$$\xi_0^2 \, (\xi_0^2 - \xi_1^2 - \xi_2^2)^2 \, (\xi_0^2 - 2\,\xi_1^2 - 2\,\xi_2^2 - \xi_3^2)^2.$$

If we denote by σ^j the 1-forms dual to the vector fields e_k, so that $< \sigma^j, e_k >$ $= \delta^j{}_k$ and $g_{\mu\nu} = \sigma^0_\mu\,\sigma^0_\nu - \sigma^1_\mu\,\sigma^1_\nu - \sigma^2_\mu\,\sigma^2_\nu - \sigma^3_\mu\,\sigma^3_\nu$, the characteristics associated with the third factor in the polynomial above can be described as the null hypersurfaces with respect to the metric

$$k_{\mu\nu} = \sigma^0_\mu \sigma^0_\nu - \frac{1}{2}\sigma^1_\mu \sigma^1_\nu - \frac{1}{2}\sigma^2_\mu \sigma^2_\nu - \sigma^3_\mu \sigma^3_\nu. \tag{4.94}$$

4.4 The Einstein–Dirac System

Apparently, not much has been shown so far about the existence of solutions to the Einstein–Dirac system for general data. The initial value problem for this system was considered in [9], but no existence theorem for the evolution equations was proved there. J. Isenberg has suggested to us that it should be possible to show the well-posedness of the equations by formulating the equations as a system of wave equations in a way similar to what was done for the Cauchy problem for classical supergravity in [8]. However, this idea has not been worked out in the literature.

We shall indicate here how to obtain symmetric hyperbolic evolution equations from the Einstein–Dirac system. To avoid lengthy calculations, we shall not discuss the complete reduction procedure but only use this system to illustrate certain questions arising in the reduction.

We shall write the equations in terms of the 2-component spin frame formalism, which may be thought of as the spinor version of the frame formalism used in the previous sections. The fields and the equations will be expressed in terms of a spin frame $\{\iota_a\}_{a=0,1}$, which is normalized with respect to the antisymmetric bilinear form ϵ in the sense that it satisfies $\epsilon_{ab} = \epsilon(\iota_a, \iota_b)$ with $\epsilon_{01} = 1$. The associated double null frame is given by $e_{aa'} = \iota_a \, \bar{\iota}_{a'}$, it satisfies $\bar{e}_{aa'} = e_{aa'}$ and $g(e_{aa'}, e_{bb'}) = \epsilon_{ab} \, \epsilon_{a'b'}$

All spinor fields (with the possible exception of the basic spin frame itself and the vector fields $e_{aa'}$) will be given with respect to the spin frame above and we shall use ϵ_{ab} and ϵ^{ab}, defined by the requirement $\epsilon_{ab} \, \epsilon^{cb} = \delta_a{}^c$ (the Kronecker symbol), to move indices according to the rule $\omega^a = \epsilon^{ab} \, \omega_b$, $\omega_a = \omega^b \, \epsilon_{ba}$.

We use the covariant derivative operator ∇ acting on spinors, which is derived from the Levi–Civita connection of g and satisfies $\nabla \, \epsilon_{ab} = 0$, to define connection coefficients $\Gamma_{aa'bc} = \Gamma_{aa'(bc)}$ by

$$\nabla_{aa'} \, \iota_b = \nabla_{e_{aa'}} \, \iota_b = \Gamma_{aa'}{}^c{}_b \, \iota_c.$$

For any spinor field ω^a we have

$$(\nabla_{cc'} \, \nabla_{dd'} - \nabla_{dd'} \, \nabla_{cc'}) \, \omega^a = -R^a{}_{bcc'dd'} \, \omega^b - T_{cc'}{}^{ee'}{}_{dd'} \, \nabla_{ee'} \omega^a,$$

with vanishing torsion

$$0 = T_{bb'}{}^{dd'}{}_{cc'} \, e_{dd'} = \nabla_{bb'} \, e_{cc'} - \nabla_{cc'} \, e_{bb'} - [e_{bb'}, e_{cc'}] \tag{4.95}$$

$$= \Gamma_{bb'}{}^d{}_c \, e_{dc'} + \bar{\Gamma}_{bb'}{}^{d'}{}_{c'} \, e_{cd'} - \Gamma_{cc'}{}^d{}_b \, e_{db'} - \bar{\Gamma}_{cc'}{}^{d'}{}_{b'} \, e_{bd'} - [e_{bb'}, e_{cc'}],$$

and curvature spinor field

$$R_{abcc'dd'} = e_{dd'}(\Gamma_{cc'ab}) - e_{cc'}(\Gamma_{dd'ab}) + \Gamma_{dd'ae}\,\Gamma_{cc'}{}^{e}{}_{b} \tag{4.96}$$
$$+\Gamma_{ed'ab}\,\Gamma_{cc'}{}^{e}{}_{d} - \Gamma_{cc'ae}\,\Gamma_{dd'}{}^{e}{}_{b} - \Gamma_{ec'ab}\,\Gamma_{dd'}{}^{e}{}_{c}$$
$$+\Gamma_{de'ab}\,\bar{\Gamma}_{cc'}{}^{e'}{}_{d'} - \Gamma_{ce'ab}\,\bar{\Gamma}_{dd'}{}^{e'}{}_{c'} - T_{cc'}{}^{ee'}{}_{dd'}\,\Gamma_{ee'ab}.$$

The latter has the decomposition

$$R_{abcc'dd'} = -\Psi_{abcd}\,\epsilon_{c'd'} - \Phi_{abc'd'}\,\epsilon_{cd} + \Lambda\,\epsilon_{c'd'}\,(\epsilon_{bd}\,\epsilon_{ac} + \epsilon_{ad}\,\epsilon_{bc}), \tag{4.97}$$

into the conformal Weyl spinor field $\Psi_{abcd} = \Psi_{(abcd)}$ as well as the Ricci spinor $\Phi_{aba'b'} = \Phi_{(ab)(a'b')} = \bar{\Phi}_{aba'b'}$ and the scalar Λ, which allow us to represent the Ricci tensor in the form

$$R_{aa'bb'} = 2\,\Phi_{aba'b'} + 6\,\Lambda\,\epsilon_{ab}\,\epsilon_{a'b'}.$$

The Bianchi identity reads

$$\nabla^{f}{}_{a'}\,\Psi_{abcf} = \nabla_{(a}{}^{f'}\,\Phi_{bc)a'f'}. \tag{4.98}$$

The Field Equations The Einstein–Dirac system is specified (cf. [72]) by a pair of 2-spinor fields ϕ_a, $\chi_{a'}$ satisying the Dirac equations

$$\nabla^{a}{}_{a'}\phi_a = \mu\,\chi_{a'}, \qquad \nabla_{a}{}^{a'}\chi_{a'} = \mu\,\phi_a, \tag{4.99}$$

with a real constant μ, and the Einstein equations with energy-momentum tensor

$$T_{aa'bb'} = \frac{i\,k}{2}\,\{\phi_a\,\nabla_{bb'}\bar{\phi}_{a'} - \bar{\phi}_{a'}\,\nabla_{bb'}\phi_{a'} + \phi_b\,\nabla_{aa'}\bar{\phi}_{b'} - \bar{\phi}_{b'}\,\nabla_{aa'}\phi_b \tag{4.100}$$
$$-\bar{\chi}_a\,\nabla_{bb'}\chi_{a'} + \chi_{a'}\,\nabla_{bb'}\bar{\chi}_a - \bar{\chi}_b\,\nabla_{aa'}\chi_{b'} + \chi_{b'}\,\nabla_{aa'}\bar{\chi}_b\}.$$

The Einstein equations then take the form

$$\Lambda = -\frac{i\,k\,\kappa\,\mu}{3}\,(\phi_a\bar{\chi}^a - \bar{\phi}_{a'}\chi^{a'}), \tag{4.101}$$

$$\Phi_{aba'b'} = \frac{i\,k\,\kappa}{2}\,\{\phi_{(a}\,\nabla_{b)(a'}\bar{\phi}_{b')} - \bar{\phi}_{(a'}\,\nabla_{b')(a}\phi_{b)} \tag{4.102}$$
$$-\bar{\chi}_{(a}\,\nabla_{b)(a'}\chi_{b')} + \chi_{(a'}\,\nabla_{b')(a}\bar{\chi}_{b)}\}.$$

The discussion of this system is complicated by the fact that the Dirac equations are of first order while derivatives of the spinor fields also appear on the right hand side of (4.102). Consequently, the right hand side of the Bianchi identity (4.98) is given by an expression involving the derivatives of the spinor fields from zeroth to second order. Therefore we need to derive equations for these quantities as well.

By taking derivatives of the Dirac equations and commuting derivatives, we obtain

$$\nabla^a{}_{a'}\nabla_{bb'}\phi_a = -R^h{}_a{}^a{}_{a'bb'}\,\phi_h + \mu\,\nabla_{bb'}\chi_{a'} \tag{4.103}$$

$$\nabla^a{}_{a'}\nabla_{cc'}\nabla_{bb'}\phi_a = -\nabla_{cc'}R^h{}_a{}^a{}_{a'bb'}\,\phi_h \tag{4.104}$$

$$-R^h{}_a{}^a{}_{a'bb'}\,\nabla_{cc'}\phi_h + \mu\,\nabla_{bb'}\nabla_{cc'}\chi_{a'} - R^h{}_b{}^a{}_{a'cc'}\,\nabla_{hb'}\phi_a$$

$$-\bar{R}^{h'}{}_{b'}{}^a{}_{a'\,cc'}\,\nabla_{bh'}\phi_a - R^h{}_a{}^a{}_{a'cc'}\,\nabla_{bb'}\phi_h$$

and similar equations for the derivatives of $\chi_{a'}$. It is important here that the curvature quantity

$$R^h{}_a{}^a{}_{a'bb'} = -\Phi^h{}_{da'b'} - 3\,\Lambda\,\epsilon_{a'b'}\epsilon_d{}^h$$

which occurs in these equations does not contain the conformal Weyl spinor. We can use (4.101) and (4.102) to express $R^h{}_a{}^a{}_{a'bb'}$ and its derivative in (4.103) and (4.104) in terms of the spinor fields and their derivatives to obtain a complete system of equations for ϕ_a, $\nabla_{bb'}\phi_a$, $\nabla_{cc'}\nabla_{bb'}\phi_a$ and the corresponding fields derived from $\chi_{a'}$.

These fields are not quite independent of each other. If we define symmetric fields $\phi_{aca'}$, $\phi_{abca'b'}$, $\chi_{aa'c'}$, $\chi_{aba'b'c'}$ by setting

$$\phi_{ac}{}^{a'} = \nabla_{(a}{}^{a'}\phi_{c)}, \qquad \phi_{abc}{}^{a'b'} = \nabla_{(a}{}^{(a'}\nabla_b{}^{b')}\phi_{c)}$$

$$\chi_a{}^{a'c'} = \nabla_a{}^{(a'}\chi^{c')}, \qquad \chi_{ab}{}^{a'b'c'} = \nabla_{(a}{}^{(a'}\nabla_{b)}{}^{b'}\chi^{c')},$$

we get from the Dirac equations

$$\nabla_{aa'}\phi_b = \phi_{aba'} - \frac{\mu}{2}\,\epsilon_{ab}\,\chi_{a'},$$

$$\nabla_{cc'}\nabla_{bb'}\phi_a = \phi_{abcb'c'} - \frac{1}{2}\,\epsilon_{b'c'}\,\Psi_{abch}\,\phi^h + \frac{2}{3}\,\epsilon_{c(a}\,\Phi_{b)hb'c'}\,\phi^h$$

$$+2\,\Lambda\,\phi_{(a}\epsilon_{b)c}\,\epsilon_{b'c'} + \frac{2}{3}\,\mu\,\epsilon_{a(b}\,\chi_{c)b'c'} - \frac{1}{2}\,\mu^2\,\phi_a\,\epsilon_{bc}\,\epsilon_{b'c'},$$

and similar relations for the derivatives of $\chi_{a'}$. From these, the equations above, and (4.98), (4.101) and (4.102) we can derive equations of the form

$$\nabla^a{}_{a'}\phi_{abb'} = M^1_{ba'b'}, \quad \nabla^a{}_{a'}\phi_{abcb'c'} = M^2_{bca'b'c'}, \tag{4.105}$$

$$\nabla_a{}^{a'}\chi_{ba'b'} = N^1_{abb'}, \quad \nabla_a{}^{a'}\chi_{bca'b'c'} = N^2_{abcb'c'}, \tag{4.106}$$

where $M^1_{ba'b'}$, $N^1_{abb'}$ denote functions of ϕ_a, $\chi_{a'}$, $\phi_{aca'}$, $\chi_{aa'c'}$, while $M^2_{bca'b'c'}$, $N^2_{abcb'c'}$ depend in addition on $\phi_{abca'b'}$, $\chi_{aba'b'c'}$, and Ψ_{abcd}. Note that this introduces (or rather makes explicit) further non-linearities.

Hyperbolic Equations from the Einstein–Dirac System Our field equations for the unknowns

$$e^{\mu}{}_{k},\ \ \Gamma_{aa'bc},\ \ \Psi_{abcd},\ \ \phi_{a},\ \ \phi_{abb'},\ \ \phi_{abcb'c'},\ \ \chi_{a'},\ \ \chi_{ba'b'},\ \ \chi_{bca'b'c'},$$

are now given by (4.95) and (4.97) (with the left hand side understood as being given by (4.96)), (4.98), (4.99), (4.105) and (4.106). Here Einstein's equations (4.101) and (4.102) are used to express quantities derived from the Ricci tensor in terms of the spinor fields and their derivatives.

When we try to deduce a hyperbolic reduced system from these equations, the first two equations, which determine the gauge dependent quantities, will require the choice of a gauge, while we expect the remaining equations, which are tensorial, to contain subsystems which are hyperbolic irrespective of any gauge. This is indeed the case and there are, due to the fact that most of the equations are overdetermined, various possibilites to extract such systems.

In [36] sytems of spinor equations have been considered which are built from systems of the type

$$\nabla^{b}{}_{a'}\,\psi_{b\beta} = F_{a'\beta}(x^{\mu}, \psi_{c\gamma}),$$

or their complex conjugates, where β denotes a multi-index of some sort. If the components corresponding to different values of the indices b, β are independent of each other the equations

$$-\nabla^{b}{}_{0'}\,\psi_{b\beta} = -F_{0'\beta}$$

$$\nabla^{b}{}_{1'}\,\psi_{b\beta} = F_{1'\beta}$$

form a symmetric hyperbolic system. Equations (4.99) are thus symmetric hyperbolic as they stand. If symmetries are present which relate the index b to a group of unprimed spinor indices comprised by β, the equations to be extracted are slightly different. For instance, we obtain from (4.98) a symmetric hyperbolic system (regarding all fields besides the Weyl spinor field as given) of form

$$\nabla^{f}{}_{1'}\,\Psi_{000f} = \dots,$$

$$\nabla^{f}{}_{1'}\,\Psi_{ab1f} - \nabla^{f}{}_{0'}\,\Psi_{ab0f} = \dots,$$

$$-\nabla^{f}{}_{0'}\,\Psi_{111f} = \dots$$

where the symmetry $\Psi_{abcd} = \Psi_{(abcd)}$ is assumed explicitly so that there are five complex unknown functions. Equations (4.105) and (4.106) can be dealt with similarly.

To compare the characteristics of the system above with previous hyperbolic systems extracted from the Bianchi identity, we set

$$e_0 = \frac{1}{\sqrt{2}}(e_{00'} + e_{11'}), \quad e_1 = \frac{1}{\sqrt{2}}(e_{01'} + e_{10'}),$$

$$e_2 = \frac{-i}{\sqrt{2}}(e_{01'} - e_{10'}), \quad e_3 = \frac{1}{\sqrt{2}}(e_{00'} - e_{11'}).$$

Then the characteristic polynomial is given up to a positive constant factor by

$$\xi_\mu \, e^\mu{}_0 \, k^{\rho\nu} \, \xi_\rho \, \xi_\nu \, g^{\lambda\sigma} \, \xi_\lambda \, \xi_\sigma, \tag{4.107}$$

which contains the degenerate quadratic form

$$k^{\mu\nu} = 2 \, e^\mu{}_0 \, e^\nu{}_0 - e^\mu{}_1 \, e^\nu{}_1 - e^\mu{}_2 \, e^\nu{}_2.$$

The cone $\{k^{\rho\nu} \, \xi_\rho \, \xi_\nu = 0\}$ is the product of a 2-dimensional cone in the plane $\{\xi_\mu \, e^\mu{}_3 = 0\}$ with the real line so that its set of generators is diffeomorphic to $S^1 \times \mathbb{R}$. The associated characteristics are timelike. The special role played here by the vector field e_3 allows us to adapt the system to situations containing a distinguished direction.

Another method to extract symmetric hyperbolic equations from spinor equations has been discussed in [37]. It is based on the space-spinor formalism in which an arbitrary normalized timelike vector field is used to express all spinor fields and spinor equations in terms of fields and equations containing only unprimed indices. If the fields and equations are then decomposed into their irreducible parts (a direct, though somewhat lengthy algebraic procedure), the equations almost automatically decompose into symmetric hyperbolic propagation equations and constraints.

For simplicity we choose the timelike vector field to be

$$\sqrt{2} \, e_0 = \tau^{aa'} \, e_{aa'} \quad \text{with} \quad \tau_{aa'} = \epsilon_0{}^a \, \epsilon_{0'}{}^{a'} + \epsilon_1{}^a \, \epsilon_{1'}{}^{a'}.$$

Since $\tau_{aa'} \, \tau^{ba'} = \epsilon_a{}^b$ etc., maps generalizing the map $\omega_{a'} \rightarrow \tau_a{}^{a'} \, \omega_{a'}$ to spinors of arbitrary valence are bijective and allow us to obtain faithful representations of all spinor relations in terms of unprimed spinors. Writing $\nabla_{ab} = \tau_b{}^{a'} \, \nabla_{aa'} = \frac{1}{2} \, \epsilon_{ab} \, P + \mathcal{D}_{ab}$, we obtain a representation of the covariant derivative operator in terms of the directional derivative operators $P = \tau^{aa'} \, \nabla_{aa'}$, $\mathcal{D}_{ab} = \tau_{(b}{}^{a'} \, \nabla_{a)a'}$ acting in the direction of e_0 and in directions orthogonal to e_0 respectively. In particular, (4.98) splits under the operations indicated above into 'constraints'

$$\mathcal{D}^{fg} \, \Psi_{abfg} = \ldots,$$

and 'evolution equations'

$$P \, \Psi_{abcd} - 2 \, \mathcal{D}_{(a}{}^f \, \Psi_{bcd)f} = \ldots$$

If the latter are multiplied by the binomial coefficients $\binom{4}{a+b+c+d}$, they are seen to be symmetric hyperbolic. The characteristic poynomial of this system is again of the form

$$\xi_\mu \, e^\mu{}_0 \, k^{\rho\nu} \, \xi_\rho \, \xi_\nu \, g^{\lambda\sigma} \, \xi_\lambda \, \xi_\sigma,$$

however, since there is no privileged spacelike direction singled out here, we have a non-degenerate quadratic form

$$k^{\mu\nu} = (1+c)\, e^{\mu}{}_{0}\, e^{\nu}{}_{0} - e^{\mu}{}_{1}\, e^{\nu}{}_{1} - e^{\mu}{}_{2}\, e^{\nu}{}_{2} - e^{\mu}{}_{3}\, e^{\nu}{}_{3}, \tag{4.108}$$

with some constant $c > 0$, as in the case of the system used in the case of the Einstein–Euler equations.

By the method outlined above symmetric hyperbolic systems are also obtained for equations (4.105) and (4.106). For instance, from the first of equations (4.105) we obtain for $\phi_{abc} = \tau_c{}^{b'}\, \phi_{abb'} = \phi_{(ab)c}$ an equation of the form $\nabla^a{}_d\, \phi_{abc} = \ldots$, where we only indicate the principal part. Using the decomposition

$$\phi_{abc} = \phi^*_{abc} - \frac{2}{3}\, \epsilon_{c(a}\, \phi^*_{b)} \quad \text{with} \quad \phi^*_{abc} = \phi_{(abc)}, \quad \phi^*_b = \phi_{fb}{}^f,$$

and the decomposition of ∇_{ab}, we get a system of the form

$$P\, \phi^*_a - \frac{2}{3}\, \mathcal{D}^b{}_a\, \phi^*_b + 2\, \mathcal{D}^{bc}\, \phi^*_{abc} = 2\, \nabla^{bc}\, \phi_{bac} = \ldots,$$

$$\left(\frac{3}{a+b+c}\right) \left\{ P\, \phi^*_{abc} - 2\, \mathcal{D}^d{}_{(a}\, \phi^*_{bc)d} - \frac{2}{3}\, \mathcal{D}_{(ab}\, \phi^*_{c)} \right\}$$

$$= -2 \left(\frac{3}{a+b+c}\right) \nabla^f{}_{(a}\, \phi_{|f|bc)} = \ldots$$

which is symmetric hyperbolic.

It is well known that equations for spinor fields of spin $\frac{m}{2}$, $m > 2$ give rise to consistency conditions (cf. [72]). For instance, the equation

$$\nabla^a{}_{a'}\, \phi_{abcb'c'} = H_{bca'b'c'}, \tag{4.109}$$

where we consider the right hand side as given, implies the relation

$$\phi_{abcb'c'}\, \Psi^{abc}{}_d + 4\, \phi_{abcd'(b'}\, \Phi^{abd'}{}_{c')} = \nabla^{aa'}\, H_{ada'b'c'},$$

which reduces e.g. in the case of vanishing right hand side to a particular relation between the background curvature and the unknown spinor field. Depending on the type of equation and the background space-time, such consistency conditions may forbid the existence of any solution at all. Nevertheless, equation (4.109) implies a symmetric hyperbolic system for which the existence of solutions is no problem. Difficulties will arise if one wants to show that the constraints implied by (4.109) are preserved.

This example emphasizes the need to show the preservation of the constraints. Because in our case the right hand sides of the equations are given by very specific functions of the unknowns themselves, we can expect to obtain useful subsidiary equations.

There are again various methods to obtain hyperbolic equations for the gauge-dependent frame and connection coefficients, which depend in particular on the choice of gauge conditions. In [36], [37] the coordinates and the frame field have been subject to wave equations (nonlinear in the case of the frame field). Here we shall indicate a gauge considered in [38] (which can, of course, also be implemented in the frame formalism considered in the previous sections).

We shall denote by T a 'time flow vector field' and by x^μ coordinates on some neighbourhood of an initial hypersurface S. We assume T to be transverse to S, the 'time coordinate' $t \equiv x^0$ to vanish on S, and the relation $< dx^\mu, T >= \delta^\mu{}_0$ to hold on the neighbourhood such that we can write $T = \partial_t$.

The frame $e_{aa'}$ is chosen such that the timelike vector field $\tau^{aa'} e_{aa'}$ is orthogonal to S. Using the expansion

$$e_{aa'} = \frac{1}{2}\tau_{aa'}\tau^{cc'} e_{cc'} - \tau^b{}_{a'} e_{ab} \quad \text{with} \quad e_{ab} = \tau_{(b}{}^{a'} e_{a)a'},$$

we can write

$$T = \alpha\,\tau^{cc'} e_{cc'} + \beta^{cc'} e_{cc'} = \alpha\,\tau^{cc'} e_{cc'} + \beta^{ab} e_{ab},$$

with

$$\tau^{cc'} \beta_{cc'} = 0, \qquad \beta_{ab} = \tau_{(a}{}^{a'} \beta_{b)a'}. \tag{4.110}$$

Thus the evolution of the coordinates off S is determined by the fields $\alpha \neq 0$ and $\beta^{aa'}$ and we can write

$$\tau^{cc'} e_{cc'} = \frac{1}{\alpha}(\partial_t - \beta^{ab} e_{ab}). \tag{4.111}$$

Since we have $\nabla_T \iota_c = \Gamma^b{}_c \iota_b$, the evolution of the frame is determined by the functions

$$\Gamma_{bc} = T^{aa'} \Gamma_{aa'bc}.$$

and we can write

$$\tau^{aa'} \Gamma_{aa'bc} = \frac{1}{\alpha}(\Gamma_{bc} - \beta^{ae} \Gamma_{aebc}), \tag{4.112}$$

with $\Gamma_{aebc} = \tau_{(e}{}^{a'} \Gamma_{a)a'bc}$.

We now consider the fields $\alpha = \alpha(x^\mu) > 0$, $\beta^{ab} = \beta^{ab}(x^\mu)$ (together four real functions) as 'coordinate gauge source functions' and the field $\Gamma_{bc} = \Gamma_{bc}(x^\mu)$ (six real functions) as 'frame gauge source functions'. This is feasible, because given these functions, we can find smooth coordinates and a frame fields close to an initial hypersurface such that the given functions assume the meaning given to them above.

The gauge conditions are then expressed by the requirement that the right hand sides of (4.111) and (4.112) are given in terms of the gauge source functions and e_{ab} and Γ_{aebc}. Thus it remains to obtain evolution equations for $e^{\mu}{}_{ab}$ and Γ_{aebc}.

Reading the quantity $e^{\mu}{}_{aa'}$ for fixed index μ as the expression of the differential of x^{μ} in the frame $e_{aa'}$, we can write (4.95) in the form

$$\nabla_{aa'} e^{\mu}{}_{bb'} - \nabla_{bb'} e^{\mu}{}_{aa'} = 0.$$

Contracting this equation with $T^{aa'}$ and $\tau_c{}^{b'}$ and symmetrizing, we obtain the equation

$$0 = \nabla_T\, e^{\mu}{}_{cb} - e^{\mu}{}_{b'(b} \nabla_T\, \tau_{c)}{}^{b'} - e^{\mu}{}_{aa'} \nabla_{bc} T^{aa'},$$

which can be rewritten in the form

$$\partial_t\, e^{\mu}{}_{ab} = \ldots,$$

where the right hand side can be expressed in terms of the gauge source functions and their derivatives and the unknowns. By using (4.97) with (4.96) on the left hand side, we can derive in a similar way an equation

$$\partial_t\, \Gamma_{aebc} = \ldots,$$

with the right hand side again being given in terms of the gauge source functions and their derivatives and the unknowns.

Thus we obtain symmetric hyperbolic reduced equations for all unknowns except those given by the left hand sides of the gauge conditions (4.111) and (4.112). Our procedure applies of course to various other sytems. Our choice of gauge is of interest because of the direct relation between the gauge source functions and the evolution of the gauge. The causal nature of the evolution can be controlled explicitly because the formalism allows us to calculate the value of the norm $g(T,T) = 2\,\alpha^2 + \beta_{ab}\,\beta^{ab}$. This may prove useful if it is desired to control the effect of the choice of gauge source functions on the long time evolution of the gauge in numerical calculations of space-times.

4.5 Remarks on the Structure of the Characteristic Set

We have seen that for certain reduced systems there occur besides the 'physical' characteristics, given by null hypersurfaces, also characteristics which are timelike or spacelike with respect to the metric $g_{\mu\nu}$. Timelike characteristics, which usually occur if a system of first order is deduced from a system of second order, are usually harmless and of no physical significance. The spacelike characteristics, which are partly due to the choice of gauge condition and partly due to the use made of the constraints, have no physical significance either. Though they are associated with non-causal propagation, there is a priori nothing bad about them and it rather depends on the applications one wants to make whether they are harmful or not.

In the characteristic polynomial (4.71) of the reduced equations for the Einstein–Euler system there appears a factor $\xi_0^2 + \frac{1}{4} h^{ab} \xi_a \xi_b$ which corresponds to timelike characteristics. In vacuum the corresponding cone has of course no physical meaning, since in general there is no preferred timelike vector field available. In the perfect fluid case there is a distinguished timelike vector field present. This, and perhaps the symmetry of the inner cone with respect to the fluid vector, has led some people to speculate on the physical significance of that cone [31]. However, in some of the later examples of hyperbolic equations deduced from the Bianchi identity, which could also be used in the fluid case, the structure of the characteristics is drastically different from the one observed in the fluid case (cf. (4.94) and also the degenerate cone arising in (4.107)). In particular, the factor above does not occur in their characteristic polynomials. The large arbitrariness in extracting hyperbolic equations, which arises from different use made of the constraints implied by the Bianchi identity, suggests that in the case of the Einstein–Euler system the only 'physical characteristics' are those associated with the fluid vector, the null cone of $g_{\mu\nu}$, and the sound cone.

The null cone of the metric (4.94) touches the null cone of the metric g in the directions of $\pm e_3$ but it is spacelike in all other directions. Thus all null hypersurfaces of it are spacelike or null for $g_{\mu\nu}$. Such a cone has the effect that the 'domain of uniqueness' defined by the techniques discussed in Sect. 3.1 may decrease. However, as we have seen, is does not prevent us from proving useful results.

There is also no reason to assume that the additional characteristics necessarily create problems in numerical calculations. In situations where the maximal slicing condition can be used, the occurrence of spacelike characteristics which are related, as in our examples, in a rigid way with the metric should be innocuous. Also, numerical calculations based on equations with inner characteristic cones as observed above have been performed without difficulties ([35], [55]).

5 Local Evolution

5.1 Local Existence Theorems for the Einstein Equations

The purpose of this section is to present a local existence theorem for the Einstein vacuum equations. By (abstract) vacuum initial data we mean a three-dimensional manifold S together with a Riemannian metric h_{ab} and a symmetric tensor χ_{ab} on S which satisfy the vacuum constraints (see Sect. 2). A corresponding solution of the vacuum Einstein equations is a Lorentzian metric $g_{\alpha\beta}$ on a four-dimensional manifold M and an embedding ϕ of S into M such that h_{ab} and χ_{ab} coincide with the pull-backs via ϕ of the induced metric on $\phi(S)$ and the second fundamental form of that manifold respectively and the Einstein tensor of $g_{\alpha\beta}$ vanishes. If $\phi(S)$ is a Cauchy surface for the space-time $(M, g_{\alpha\beta})$ then this space-time is said to be a Cauchy development of the data (S, h_{ab}, χ_{ab}). The basic local existence theorem for the vacuum Einstein equations says that every vacuum initial data set has at least one Cauchy development. In fact to make this precise it is necessary to fix the differentiability properties which are assumed for the data and demanded of the solution. For instance, the result holds if the differentiability class for both data and solutions is taken to be C^∞. Note that there is no need to require any further conditions on the spatial dependence of the data.

The proof of local existence will now be outlined. We follow essentially the original method of [16] except for the fact that we reduce second order equations to first order symmetric hyperbolic systems and that we use harmonic mappings rather than harmonic coordinates. The use of harmonic mappings, as discussed in Sect. 4, allows us to work globally in space even if the manifold S cannot be covered by a single chart. Using harmonic coordinates it would be necessary to construct solutions local in space and time and then piece them together. Choose a fixed Lorentz metric on $S \times \mathbb{R}$, for instance the metric product of the metric h_{ab} with $-dt^2$. This comparison metric will be denoted by $\bar{g}_{\alpha\beta}$. The idea is to look for a solution $g_{\alpha\beta}$ on an open subset U of $\mathbb{R} \times S$ such that the identity is a harmonic map from $(U, g_{\alpha\beta})$ to $(U, \bar{g}_{\alpha\beta})$. This is a condition which is defined in a global invariant way. Its expression in local coordinates is $g^{\beta\gamma}(\Gamma^\alpha_{\beta\gamma} - \bar{\Gamma}^\alpha_{\beta\gamma}) = 0$ where $\Gamma^\alpha_{\beta\gamma}$ and $\bar{\Gamma}^\alpha_{\beta\gamma}$ are the Christoffel symbols of $g_{\alpha\beta}$ and $\bar{g}_{\alpha\beta}$ respectively. In the terminology of Sect. 2.4 this means that we choose $g^{\beta\gamma}\bar{\Gamma}^\alpha_{\beta\gamma}$ as a gauge source function.

Next consider the question of reduction of nonlinear wave equations to symmetric hyperbolic form. This is done as follows. Let $g^{\alpha\beta}(t, x, u)$ be functions of (t, x, u) which for each fixed value of (t, x, u) make up a symmetric matrix of Lorentz signature and consider an equation of the form:

$$g^{\alpha\beta}\partial_\alpha\partial_\beta u + F(t, x, u, Du) = 0 \tag{5.1}$$

This has been formulated in a local way but a corresponding class of equations can be defined in the case that the unknown u is a section of a fibre bundle. As in the above treatment of symmetric hyperbolic equations on a manifold, consideration will be restricted to the case of sections of a vector bundle V. Choose a fixed connection on V. Then the class of equations to be considered is obtained by replacing the partial derivatives in the above equation by covariant derivatives defined by the given connection. Let $u_\alpha = \nabla_\alpha u$. Then the equation (5.1) can be written as:

$$\begin{aligned}
-g^{00}\nabla_0 u_0 - 2g^{0a}\nabla_a u_0 &= g^{ab}\nabla_a u_b + F(t, x, u_0, u_a) \\
g^{ab}\nabla_0 u_a &= g^{ab}\nabla_a u_0 + K^b \\
\nabla_0 u &= u_0
\end{aligned}$$

Here K^b is a term involving the curvature of the connection which is of order zero in the unknowns of the system. This is a symmetric hyperbolic system for the unknowns u and u_α. Since u is allowed to be a section of a vector bundle we are dealing with a system of equations. However the functions $g^{\alpha\beta}$ must be scalars. The appropriate initial data for the second order equation consists of the values of u and $\partial_t u$ on the initial hypersurface. From these the values of the functions u_α on the initial hypersurface may be determined. Thus an initial data set for the symmetric hyperbolic system is obtained. It satisfies the additional constraint equation $\nabla_\alpha u = u_\alpha$. Applying the existence theory for symmetric hyperbolic systems gives a solution (u, u_α). To show that the function u obtained in this way is a solution of the original second order equation it is necessary to show that the constraint equation is satisfied everwhere. That $\nabla_0 u = u_0$ follows directly from the first order system. It also follows from the first order system that $\nabla_0(u_a - \nabla_a u) = 0$. Since this is a first order homogeneous ODE for $u_a - \nabla_a u$, the vanishing of the latter quantity for $t = 0$ implies its vanishing everywhere.

In the case of the vacuum Einstein equations the bundle V can be taken to be the bundle of symmetric covariant second rank tensors. The connection can be chosen to be the Levi–Civita connection defined by $\bar{g}_{\alpha\beta}$. This is only one possible choice but note that it is important that this connection does not depend on the unknown in the equations, in this case the metric $g_{\alpha\beta}$.

Now a proof of local in time existence for the vacuum Einstein equations will be presented. Let h_{ab} and χ_{ab} be the initial data. In Sect. 2 it was shown that the vacuum Einstein equations reduce to a system of nonlinear wave equations when harmonic coordinates, or the generalization involving gauge source functions, are used. As was already indicated in that section, there is no loss of generality in imposing this condition locally in time. If there exists a development of particular initial data then there exists a diffeomorphism ϕ of a neighbourhood of the initial hypersurface such that the pull-back of the metric with the given diffeomorphism satisfies the harmonic condition with respect to $\bar{g}_{\alpha\beta}$. In fact ϕ can be chosen to satisfy some additional conditions. The harmonic condition is equivalent to a nonlinear wave equation for ϕ. Solving the local in time Cauchy problem for this wave equation provides the desired diffeomorphism. The existence theory for this Cauchy problem follows from that for symmetric hyperbolic systems by the reduction to first order already presented. The initial data for ϕ will be specified as follows. It is the identity on the initial hypersurface as is the contraction of its derivative with the normal vector with respect to $\bar{g}_{\alpha\beta}$. Since the vector $\partial/\partial t$ is the unit normal vector to S with respect to $\bar{g}_{\alpha\beta}$ it will also have this property with respect to $g_{\alpha\beta}$.

A local solution of the Einstein equations corresponding to prescribed initial data can be obtained as follows. Let h_{ab} and χ_{ab} denote the components of the tensors making up the initial data in a local chart as above. A set of initial data for the harmonically reduced vacuum Einstein equations consists of values for the whole metric $g_{\alpha\beta}$ and its time derivative on the initial hypersurface. A data set of this kind can be constructed from h_{ab} and χ_{ab} as follows. (The following equations are expressed in local coordinates, but their invariant meaning should be clear.)

$$
\begin{aligned}
&g_{ab} = h_{ab}, \quad g_{0a} = 0, \quad g_{00} = -1 \\
&\partial_t g_{ab} = 2\chi_{ab}, \quad \partial_t g_{0a} = h^{bc}(h_{ab,c} - (1/2)h_{bc,a}) - h_{ab}\bar{\Gamma}^b_{cd}h^{cd}, \\
&\partial_t g_{00} = -2h^{ab}\chi_{ab}
\end{aligned}
$$

These data are chosen in such a way that the harmonic condition is satisfied on the initial hypersurface.

Corresponding to the initial data for the reduced equations there is a unique local solution of these equations. It remains to show that it is actually a solution of the Einstein equations provided the initial data h_{ab} and χ_{ab} satisfy the Einstein constraint equations. In order to do this it suffices to show that the harmonic conditions are satisfied everywhere since under those conditions the reduced equations are equivalent to the Einstein equations. Let $\Delta^\alpha = g^{\beta\gamma}(\Gamma^\alpha_{\beta\gamma} - \bar{\Gamma}^\alpha_{\beta\gamma})$. That the harmonic conditions are satisfied can be verified using the fact that the quantities Δ^α satisfy a linear homogeneous system of wave equations. By uniqueness for this system the Δ^α vanish provided the initial data Δ^α and $\partial_t\Delta^\alpha$ vanish on the initial hypersurface. The first of these was built into the construction of the data for the reduced equations. The second is a consequence of the combination of the reduced equations with the Einstein constraints.

5.2 Uniqueness

The argument of the last section gives an existence proof for solutions of the vacuum Einstein equations, local in time. It does not immediately say anything about uniqueness of the space-time constructed. The solution of the reduced equations is unique, as a consequence of the uniqueness theorem for solutions of symmetric hyperbolic systems. However the freedom to do diffeomorphisms has not yet been explored. In fact it is straightforward to obtain a statement of uniqueness of the solution corresponding to given abstract initial data, up to diffeomorphism. Suppose two solutions g_1 and g_2 with the same initial data are given. Choose a reference metric $\bar{g}$ as before and determine diffeomorphisms ϕ_1 and ϕ_2 such that the identity is a harmonic map with respect to the pairs $(\bar{g}, g_1)$ and $(\bar{g}, g_2)$ respectively. The transformed metrics satisfy the same system of reduced equations with the same initial data and thus must coincide on a neighbourhood of the initial hypersurface. Thus g_1 and $(\phi_2 \circ \phi_1^{-1})^* g_2$ coincide on a neighbourhood of the initial hypersurface. This statement is often referred to as 'geometric uniqueness'.

For hyperbolic equations it is in general hard to prove theorems about global existence of solutions due to the possibility of the formation of singularities. On the other hand, it is possible to prove global uniqueness theorems. Proving global uniqueness for the Einstein equations is more difficult due to difficulties with controlling the freedom to do diffeomorphisms. In this context we use the word 'global' to mean not just applying to a subset of a given solution, but to whole solutions having suitable intrinsic properties. The strategy which has just been used to prove local existence using harmonic maps cannot be applied directly. For doing so would require a global existence theorem for harmonic maps and that we do not have in general.

A global uniqeness theorem was proved by abstract means by Choquet-Bruhat and Geroch [19], who introduced the notion of the maximal Cauchy development. If initial data for the Einstein equations coupled to some matter fields on a manifold S are prescribed, a *development* of the data is a solution of the Einstein-matter system on a manifold M together with an embedding ϕ of S into a M which induces the correct initial data and for which the image of S is a Cauchy surface. Another development with a solution of the Einstein-matter equations on a manifold M' and an embedding ϕ' is called an *extension* of the first if there is a diffeomorphism ψ from M to an open subset U of M' which maps the given metric on M onto the restriction of the metric on M' to U and also maps the matter fields on M to those on U obtained by restriction from M'. If we are dealing with the vacuum Einstein equations then the requirement on the matter fields is absent. In [19] the following theorem was proved for the vacuum case:

Theorem Let S be an initial data set. Then there exists a development M of S which is an extension of every other development of S. This development is unique up to isometry.

The development whose existence and uniqueness is asserted by this theorem is called the *maximal Cauchy development* of the initial data set. Uniqueness up to isometry means the following. If we have two developments of the same data given by embeddings ϕ and ϕ' of S into manifolds M and M' respectively then there exists a diffeomorphism $\psi : M \to M'$ which is an isometry and satisfies $\phi' = \phi \circ \psi$. The proof of this theorem does not depend strongly on the vacuum assumption. One potential problem in extending it to certain matter fields is gauge freedom in those fields. This requires the concept of extension to be defined in a slightly different way. For instance in the case of gauge fields it is necessary to consider not only diffeomorphisms of the base manifold, but also automorphisms of the principal bundle entering into the definition of the theory. We do not expect that this leads to any essential difficulty, but in any concrete example one should pay attention to this point.

The proof of this theorem applies directly to Zorn's lemma and it is an open question whether it is possible to remove the use of the axiom of choice from the argument. The maximal Cauchy development is often very useful in formulating certain arguments. However it remains very abstract and gives the subjective impression of being difficult to pin down.

It should be emphasized that, despite its global aspects, it would be misleading to consider the above theorem as a global existence theorem for the Einstein equations in any sense. A comparison with ordinary differential equations may help to make this clear. If an ordinary differential equation for a function $u(t)$ is given (with smooth coefficients) then the standard local existence theorem for ordinary differential equations says that given an initial value u_0 there exists a $T > 0$ and a unique solution $u(t)$ on the interval $(-T, T)$ with $u(0) = u_0$. Now one can ask for the longest time interval $(-T_1, T_2)$ on which a solution of this kind exists for a fixed initial value u_0. This is called the maximal interval of existence and has a similar status to that of the maximal Cauchy development. The fact that the maximal interval of existence is well-defined says nothing about the question whether the solution exists globally or not, which is the question whether T_1 and T_2 are infinite or not. The existence of the maximal Cauchy development says nothing about the global properties of the solution obtained in this way, whereas global existence of the solution of an ordinary differential equation does mean that the solution has a certain global property, namely that it exists for an infinitely long time.

5.3 Cauchy Stability

Cauchy stability of the initial value problem for the Einstein equations is the statement that, in an appropriate sense, the solution of the Einstein equations depends continuously on the initial data. This continuity statement has two parts, which can be stated intuitively in the following way. Firstly, if a solution corresponding to one initial data set is defined on a suitable closed region, then the solution corresponding to any initial data set close enough to the original one will be defined on the same region. Closeness is defined in terms of Sobolev norms. Care is needed with the interpretation of the phrase 'the same region' due to the diffeomorphism invariance. To make it precise, something has to be said about how regions of the different spacetimes involved are to be compared with each other. Secondly, the solution defined on this common region depends continuously on the initial data, where continuity is again defined in terms of Sobolev norms. In non-compact situations it is appropriate to use local Sobolev norms for this, i.e. the Sobolev norms of restrictions of a function to compact sets.

Rather than make this precise in general we will restrict to one case where the formulation of the statement is relatively simple, but which is still general enough to give a good idea of the basic concepts. Consider initial data sets for the Einstein equations on a compact manifold S. For definiteness let us restrict to the vacuum case. As has been discussed above, solutions can be constructed by using the harmonically reduced equations. One step in this process is to associate to geometric data (h_{ab}, χ_{ab}) full data $(g_{\alpha\beta}, \partial_t g_{\alpha\beta})$. For each of these pairs let us choose the topology of the Sobolev space $H^s(S)$ for the first member and that of $H^{s-1}(S)$ for the second. Then standard properties of Sobolev spaces show that if s is sufficiently large the mapping from geometric data to full data is continuous. Suppose now that we have one particular solution of the Einstein vacuum equations with data on S. The corresponding solution of the harmonically reduced equations exists on some region of the form $S \times [-T, T]$. If s is sufficiently large then there exists an open neighbourhood of the given data in $H^s(S) \times H^{s-1}(S)$ such that for any data in this neighbourhood there exists a corresponding solution in $H^s(S\times[-T,T])$. Moreover the mapping from data to solutions defined on this neighbourhood is continuous (in fact differentiable). This has been proved by Choquet-Bruhat [18].

The theorem concerning a compact initial hypersurface can also be modified to give a local statement of the following type. Let initial data for the Einstein equations be given on some manifold S and suppose that a corresponding solution is given on a manifold M. There is a neighbourhood U of S where harmonic reduction is possible globally. This identifies U with an open subset of $S \times \mathbb{R}$. This contains a set of the form $V \times [-T,T]$ (for some open subset V of S and some $T > 0$) which contains any given point of the initial hypersurface. If we cut off the the initial data for the harmonically reduced equations and use the domain of dependence, we can use the above statement for a compact initial hypersurface to get continuous dependence on initial data for a possibly smaller set $V' \times [-T',T']$. Summing up, each point sufficiently close to the initial hypersurface has a neighbourhood W_1 with compact closure such that there is an open subset W_2 of the initial hypersurface with compact closure such that the following properties hold. If the restriction of an initial data set for the Einstein equations is sufficiently close to that of the original data set in $H^s(W_2)$ then there exists a corresponding solution of class H^s on a neighbourhood of W_1. Moreover, the resulting mapping from $H^s(W_2)$ to $H^s(W_1)$ is continuous.

5.4 Matter Models

To specify a matter model in general relativity three elements are required. The first is a set of tensors (or perhaps other geometrical objects) on space-time which describe the matter fields. The second is the equations of motion which are to be satisifed by these fields. The third is the expression for the energy-momentum tensor in terms of the matter fields which is to be used to couple the matter to the Einstein equations. Note that in general both the matter field equations and the expression for the energy-momentum tensor involve the space-time metric. Thus it is impossible to consider matter in isolation from the space-time metric. In solving the Cauchy problem it is necessary to deal with the coupled system consisting of the Einstein equations and the equations of motion for the matter fields.

There are two broad classes of matter models which are considered in general relativity, the field theoretical and phenomenological matter models. The distinction between these is not sharply defined but is useful in order to structure the different models. The intuitive idea is that the field theoretic matter models correspond to a fundamental description while the phenomenological models represent an effective description of matter which may be useful in certain situations. Within the context of classical general relativity, which is the context of this article, the pretension of the field theoretic matter models to be more fundamental is not well founded since on a fundamental level the quantum mechanical nature of matter should be taken into account.

Before going further, a general remark on the Einstein-matter equations is in order. Suppose that in any given coordinate system the matter equations can be written in symmetric hyperbolic form in terms of a variable u. Consider the system obtained by coupling the harmonically reduced Einstein equations, written in first order symmetric hyperbolic form, to the sytem for u. If the coupling is only by terms of order zero then the combined system is symmetric hyperbolic and a local existence theorem for the reduced Einstein-matter system is obtained. The condition for this to happen is that the equations for the matter fields contain at most first derivatives of the metric (in practice the Christoffel symbols) and that the energy-momentum tensor contains no derivatives of u. When these conditions are satisfied, local existence for the Einstein-matter equations (not just the reduced equations) can be proved using the same strategy as we presented in the vacuum case. The fact, which should hold for any physically reasonable matter model, that the energy-momentum tensor is divergence free as a consequence of the matter field equations, can be is used derive the equation which allows it to be proved that the harmonic condition propagates.

It would be unreasonable to try and describe here all the matter models which have ever been used in general relativity. We will, however, attempt to give a sufficiently wide variety of examples to illustrate most of the important features to be expected in general. We start with the field theoretic models.

The simplest case is where the matter field is a single real-valued function ϕ. The equations of motion are:

$$\nabla^\alpha \nabla_\alpha \phi = m^2 \phi + V'(\phi)$$

Here m is a constant and V is a smooth function which is $O(\phi^3)$ for ϕ close to zero. A typical example would be $V(\phi) = \phi^4$. The energy-momentum tensor is:

$$T_{\alpha\beta} = \nabla_\alpha \phi \nabla_\beta \phi - [(1/2)(\nabla^\gamma \phi \nabla_\gamma \phi) + m^2 \phi^2 + 2V(\phi)] g_{\alpha\beta}$$

The equation for ϕ is a nonlinear wave equation and so may be reduced to a symmetric hyperbolic system. When it is coupled to the harmonically reduced Einstein equations via the energy-momentum tensor above and the whole system reduced to first order there is no coupling in the principal part. As mentioned above this is enough to allow a local existence theorem to be proved. Note that the splitting off of the nonlinear term V and the sign condition following from the form m^2 of the coefficient in the linear term are irrelevant for the local well-posedness of the equations. On the one hand they are motivated by considerations of the physical interpretation of the equations. On the other hand they have an important influence on the global behaviour of solutions. This matter model is often referred to as 'the scalar field' , although when used without qualification this often means the special case $m = 0$, $V = 0$.

The scalar field can be thought of as a mapping into the real line. It can be generalized by considering mappings into a Riemannian manifold N. An equivalent of the massless scalar field with vanishing potential is the nonlinear σ-model or wave map as it is often known to physicists and mathematicians respectively. It is a mapping from space-time into a manifold N with Riemannian metric h called the target manifold. The field equations and energy-momentum tensor have a coordinate-invariant meaning but we will content ourselves with giving the expressions in coodinate systems on M and N. The field equations are

$$\nabla_\alpha \nabla^\alpha \phi^A + \Gamma^A_{BC}(\phi) \nabla_\alpha \phi^B \nabla^\alpha \phi^C = 0$$

where Γ^A_{BC} are the Christoffel symbols of h in some coordinate system. The energy-momentum tensor is:

$$T_{\alpha\beta} = h_{AB}[\nabla_\alpha \phi^A \nabla_\beta \phi^B - (1/2)(\nabla_\gamma \phi^A \nabla^\gamma \phi^B) g_{\alpha\beta}]$$

The special case where N is the complex plane with the flat Euclidean metric corresponds to the complex scalar field. In contrast to the case of the scalar field, the wave map does not allow the addition of a mass term or a potential term in any obvious way. If N has the structure of a vector space there is an obvious way of defining a mass term and further structure on N may lead to natural ways of defining a potential. These features occur in the case of Higgs fields. It may be noted that the wave maps considered here, which are sometimes also called hyperbolic harmonic maps, are related mathematically to the harmonic gauge discussed in Sect. 4. The role of the connection $\bar{\Gamma}$ in Sect. 4 is played here by the Levi–Civita connection of the target manifold.

One of the most familiar matter models in general relativity is the Maxwell field. This is described by an antisymmetric tensor $F_{\alpha\beta}$. The equations of motion for the source-free Maxwell field are $\nabla_\alpha F^{\alpha\beta} = 0$ and $\nabla_\alpha F_{\beta\gamma} + \nabla_\beta F_{\gamma\alpha} + \nabla_\gamma F_{\alpha\beta} = 0$ and the energy-momentum tensor is

$$T_{\alpha\beta} = F_\alpha{}^\gamma F_{\beta\gamma} - (1/4) F^{\gamma\delta} F_{\gamma\delta} g_{\alpha\beta}$$

The second set of Maxwell equations can be solved locally by writing $F_{\alpha\beta} = \nabla_\alpha A_\beta - \nabla_\beta A_\alpha$ for a potential A_α. Then the other equations can be regarded as second order equations for A_α. Note, however, that if space-time has a non-trivial topology then it may be impossible to find a global potential which reproduces a given field $F_{\alpha\beta}$. In the same way that the scalar field can be generalized to get wave maps, the Maxwell field can be generalized to get Yang–Mills fields. We will not give the global description of these fields involving principal fibre bundles but only give expressions in local coordinates and a local gauge. The model is defined by the choice of a Lie algebra (which we describe via a basis) and a positive definite quadratic form on the Lie algebra with components h^{IJ} in this basis. Let C^I_{JK} be the structure constants in this basis. The basic matter field is a one-form A^I_α with values in the Lie algebra and the field strength is defined by

$$F^I_{\alpha\beta} = \nabla_\alpha A^I_\beta - \nabla_\beta A^I_\alpha + C^I_{JK} A^J_\alpha A^K_\beta$$

The field equations are

$$\nabla_\alpha F^{I\alpha\beta} + C^I_{JK} A^J_\alpha F^{K\alpha\beta} = 0$$

In the special case where the Lie algebra is one-dimensional (and hence Abelian) the Yang–Mills equations reduce to the Maxwell equations. Note however that the Yang–Mills field cannot be described by the field strength alone. The description in terms of a potential is indispensible.

A complication which arises when studying the initial value problem for the Yang–Mills or Einstein–Yang–Mills systems is that of gauge invariance. Although the potential is required it is not uniquely determined. Gauge transformations of the form:

$$A^I_\alpha \mapsto A^I_\alpha + (g^{-1}\nabla_\alpha g)^I$$

leave $F^I_{\alpha\beta}$ invariant. Here g is a function taking values in a Lie group with the given Lie algebra and the expression $g^{-1}\nabla g$ can naturally be identified with a Lie-algebra-valued one-form. Fields related by a gauge transformation describe the same physical system. The ambiguity here is similar to that of the ambiguity of different coordinate systems in the case of the Einstein equations. It can be solved in an analogous way by the use of the Lorentz gauge. This is similar to the harmonic coordinate condition and reduces the Yang–Mills equations on any background to a system of nonlinear wave equations which can, if desired, be reduced to a symmetric hyperbolic system. Combining harmonic coordinates and Lorentz gauge produces a reduced Einstein–Yang–Mills system which can be handled by the same sort of techniques as the reduced vacuum Einstein equations. Of course there are a number of steps which have to be checked, such as the propagation of Lorentz gauge. All these comments apply equally well to the Einstein–Yang–Mills–Higgs system obtained by coupling the Yang–Mills field to a Higgs field. (Now gauge transformations for the Higgs field must also be specified.)

A field theoretic matter model whose Cauchy problem does not fit easily into the above framework is that given by the Dirac equation. It has been discussed in Sect. 4.4.

Probably the best known phenomenological matter model in general relativity is the perfect fluid. This has already been discussed at some length in Sect. 4.2. Here we mention some complementary aspects. Recall that the basic matter fields are the energy density ρ, a non-negative real-valued function, and the four-velocity U^α, a unit timelike vector field. The energy-momentum tensor is given by

$$T^{\alpha\beta} = (\rho + p)U^\alpha U^\beta + p g^{\alpha\beta}$$

in the signature used here, namely $(-,+,+,+)$. The pressure p is given in the isentropic case in terms of an equation of state $p = h(\rho)$. The usual assumptions on this equation of state is that it is a non-negative continuous function defined on an interval $[\rho_0, \infty)$ which is positive for $\rho > 0$. Also it should be C^1 for $\rho > 0$ with positive derivative $h'(\rho) = dp/d\rho$ there. The equations of motion of the fluid, the Euler equations, are given by the condition that the energy-momentum tensor should be divergence-free. The Euler equations can be written as a well-posed symmetric hyperbolic system in terms of the basic variables provided we restrict to cases where $\rho \geq C > 0$ for some $C > 0$. Here the condition $h' > 0$ is crucial. In the study of spatially homogeneous cosmological models the equation of state $p = k\rho$ with $k < 0$ is sometimes considered. We emphasize that a fluid with an equation of state of this kind cannot be expected to have a well-posed initial value problem. This has to do with the fact that the speed of sound, which is the square root of h', is imaginary in that case. In Sect. 5.5 we prove a related but simpler result, namely that in special relativity the Euler equations with this equation of state, linearized about a constant state, have an ill-posed initial value problem. The system obtained by coupling the Euler equations to the harmonically reduced Einstein equations can be written as a symmetric hyperbolic system in the case that the Euler equations can be written symmetric hyperbolic in terms of the basic variables, as stated above.

In the case where the density is everywhere positive, writing the Euler equations in symmetric hyperbolic form is not trivial. One approach is to take ρ and the spatial components U^i of the velocity as variables and to express U^0 in terms of the U^i and the metric via the normalization condition $U^\alpha U_\alpha = -1$ (see [87]). Another possibility (see [15]) is to consider the Euler equations as evolution equations for ρ and U^α and treat the normalization condition as a constraint, whose propagation must be demonstrated. The treatments above are limited to the isentropic Euler equations. For a fluid which is not isentropic the conservation equation for the energy-momentum tensor must be supplemented by the equation of conservation of entropy $u^\alpha \nabla_\alpha s = 0$. The equation of state can then be written in the form $p = h(\rho, s)$ or, equivalently, in the form $p = f(n, s)$, where n is the number density of particles. We are not aware that a local existence theorem non-isentropic Euler equations has been proved by a generalization of the method for the isentropic case just outlined, although there is no reason to suppose that it cannot be done, provided the condition $\partial h/\partial \rho > 0$ is satisfied. As we saw in Sect. 4.2, the Einstein equations coupled to the non-isentropic Euler equations can be brought into symmetric hyperbolic form by introducing additional variables in a suitable way, following [39]. The Cauchy problem for the general (i.e. not necessarily isentropic) Euler equations coupled to the Einstein equations had much earlier been solved by other means by Choquet-Bruhat using the theory of Leray hyperbolic systems [17].

If it is desired to show the existence of solutions of the Einstein–Euler system representing dynamical fluid bodies (such as oscillating stars) then problems arise. Either the density must become zero somewhere, in which case the Euler equations as written in the usual variables fail to be symmetric hyperbolic there, or at least the density must come arbitrarily close to zero at infinity, which is almost as bad. Treating this situation as a pure initial value problem is only possible under restrictive circumstances [76] and the general problem is still open. It would seem more promising to try to use the theory of initial boundary value problems, explicitly taking account of the boundary of the fluid. So far this has only been achieved in the spherically symmetric case with $\rho_0 > 0$ [61].

One case of a fluid which is frequently considered in general relativity is dust. This is defined by the condition that the presure should be identically zero. Since in that case $h' = 0$ the straightforward method of writing the fluid equations as a symmetric hyperbolic system does not work. The symmetric hyperbolic system of [39] discussed in Sect. 4.2 also covers the dust case as does the existence theorem of Choquet-Bruhat [17] using a Leray hyperbolic system.

The next phenomenological matter model we will consider comes from kinetic theory. It does not quite fit into the framework we have used so far to describe matter models since the fundamental matter field is a non-negative function f on the cotangent bundle of space-time. (Often the case of particles with a fixed mass is considered in which case it is defined on the subset of the cotangent bundle defined by the condition $g^{\alpha\beta}p_\alpha p_\beta = -1$, known as the mass shell.) The idea is that the matter consists of particles which are described statistically with respect to their position and momentum. The function f represents the density of particles. The geodesic flow of the space-time metric defines a vector field (Liouville vector field) on the cotangent bundle. Call it L. This flow describes the evolution of individual test particles. The equation of motion for the particles is $Lf = Q(f)$ where $Q(f)$ is an integral expression which is quadratic in its argument. This is the Boltzmann equation. It describes collisions between the particles in a statistical way. The case $Q = 0$ is the collisionless case, where the equation $Lf = 0$ obtained is often called the Vlasov equation. The energy-momentum tensor is defined by:

$$T^{\alpha\beta} = \int f p^\alpha p^\beta d\omega(p)$$

where $d\omega(p)$ represents a natural measure on the cotangent space or mass shell, depending on the case being considered.

The coupled Einstein–Boltzmann system in harmonic coodinates cannot be a hyperbolic system is any usual sense for the simple reason that it is not even a system of differential equations, due to the integrals occurring. Nevertheless, the techniques used to prove existence and uniqueness for hyperbolic equations can be adapted to prove local existence for the Einstein–Boltzmann system [7]. In contrast to the case of the perfect fluid nothing particular happens when the energy density vanishes so that there is no problem in describing isolated concentrations of matter. The Einstein–Vlasov system can be used to describe globular clusters and galaxies. For the analogue of this in Newtonian theory see [13]. Literature on the relativistic case can be found in [85].

Another kind of phenomenological matter model is elasticity theory. Apart from its abstract interest, self-gravitating relativistic elasticity is of interest for describing the solid crust of neutron stars. As in the case of a perfect fluid the field equations are equivalent to the equation that the divergence of the energy-momentum tensor is zero. What is different is the nature of the matter variables and the way they enter into the definition of the energy-momentum tensor. This is complicated and will not be treated here. The Cauchy problem for the Einstein equations coupled to elasticity theory has been discussed by Choquet-Bruhat and Lamoureux–Brousse [21]. A local existence theorem for the equations of relativistic elasticity has been proved by Pichon [73] for data belonging to Gevrey classes. These are classes of functions which are more special that C^∞ functions in that the growth of their Taylor coefficients is limited. However they do not have the property of analytic functions, that fixing the function on a small open set determines it everywhere.

In non-relativistic physics, the Euler equations are an approximation to the Navier–Stokes equations where viscosity and heat conduction are neglected. The Navier–Stokes equations are dissipative with no limit to the speed at which effects can propagate. Mathematically this has the effect that the equations are parabolic with no finite domain of dependence. It is problematic to find an analogue of the Navier–Stokes equations in general relativity which takes account of the effects of diffusion and heat conduction. One possibility is to start from the Boltzmann equation, which does have a finite domain of dependence and try to do an expansion in the limit where the collision term is large. This is analogous to the Hilbert and Chapman–Enskog expansions in non-relativistic physics. The first attempts to do this led to equations which probably have no well-posed Cauchy problem (Landau–Lifschitz and Eckart models). Hiscock and Lindblom [53] have shown that the linearization of these equations about an equilibrium state have solutions which grow at arbitrarily large exponential rates. In response to this other classes of models were developed where the fluid equations are symmetric hyperbolic. (For information on this see [46]). These models do have a well-posed Cauchy problem and the main difficulty seems to be to decide between the many possible models. Since the variables used to formulate the symmetric hyperbolic system for the fluid are not differentiated in forming the energy-momentum tensor, the system obtained by coupling these fluids to the Einstein equations can be written in symmetric hyperbolic form.

More general matter models can be obtained from those already mentioned by combining different types of matter field. For instance there is the charged scalar field which combines a scalar field and a Maxwell field. This often produces no extra difficulties at all for the local in time Cauchy problem due to the fact, mentioned above, that the system obtained by taking two symmetric hyperbolic systems together is also symmetric hyperbolic, provided the coupling between the two systems is via terms of order zero. It is also routine to allow charged particles in a kinetic model, obtaining the Einstein–Maxwell–Boltzmann system [7]. Charged fluids are more complicated. The local in time Cauchy problem has been treated in two cases, those of zero conductivity and infinite conductivity. The latter model is also known as magnetohydrodynamics. It has been shown to be have a well-posed initial value problem only for data in Gevrey classes [67], [68].

5.5 An Example of an Ill-Posed Initial Value Problem

It may be hard to appreciate the significance of a system of equations having a well-posed initial value problem since most examples which come up in practice do have this property. In this section we present an ill-posed example which is close to examples which relativists are familiar with. Consider the special relativistic Euler equations with equation of state $p = k\rho$ where $k < 0$. A special solution is given by constant density and zero spatial velocity. Now consider the equations obtained by linearizing the Euler equations about this background solution. The unknowns in the linearized system will be denoted by adding a tilde to the corresponding unknowns in the nonlinear system. For simplicity we take the background density to be unity. The linearized equations are:

$$\begin{aligned}\partial_t\tilde{\rho} &= -2(1+k)\partial_a\tilde{u}^a \\ \partial_t\tilde{u}^a &= -\frac{k}{2(1+k)}\delta^{ab}\partial_b\tilde{\rho}\end{aligned}$$

It will be shown that given any $T > 0$ there exist periodic initial data $(\tilde{\rho}_0, \tilde{u}_0^a)$ of class C^∞ such that it is not true that there is a unique corresponding solution, periodic in the space coordinates, on the time interval $[0, T]$. The condition of periodicity here does not play an essential role. It is adopted for convenience. Instead of thinking of smooth periodic functions on $\mathbb{R}^3$ with can equally well think in terms of smooth functions on a torus T^3. The space of smooth functions on T^3 can be made into a topological vector space X in a standard way. Convergence of functions in the sense of this topology means uniform convergence of the functions and their derivatives of all orders. In a similar way the space of smooth functions on $[0, T] \times T^3$ can be made into a topological vector space Y. These are Fréchet spaces [81]. The unknown in the linearized Euler equations can be thought of as an element of X^4 and the data as an element of Y^4. Let Z be the closed linear subspace of X^4 consisting of solutions of the linearized Euler equations. Consider the linear mapping $L : Z \to Y^4$ defined by restricting solutions to $t = 0$. It is continuous with respect to the relevant topologies. If existence and uniqueness held for all initial data then this linear map would be invertible. Hence, by the open mapping theorem [81], it would have continuous inverse. What this means concretely is that given any sequence $(\tilde{\rho}_{0,n}, \tilde{u}^a_{0,n})$ which is uniformly bounded together with each of its derivatives, the corresponding sequence of solutions (which exists by assumption) must also be bounded together with each of each of its derivatives. Thus to prove the desired theorem it is enough to exhibit a uniformly bounded sequence of initial data and a corresponding sequence of solutions which is not uniformly bounded. This can be done explicitly as follows:

$$\begin{aligned}
\tilde{\rho}_{0,n} &= \sin nx \\
\tilde{u}^a_{0,n} &= 0 \\
\tilde{\rho}_n &= \sin nx \cosh(n\sqrt{-k}t) \\
\tilde{u}^1_n &= \frac{\sqrt{-k}n}{2(1+k)} \sin nx \sinh(n\sqrt{-k}t) \\
\tilde{u}^2_n &= \tilde{u}^3_n = 0
\end{aligned}$$

From these explicit formulae we get an idea what is going wrong. Fourier modes of increasing frequencies of the initial data grow at increasing exponential rates. In the case of a fluid where the equation of state has a positive value of k the hyperbolic functions in the above formulae are replaced by trigonometric ones and the problem does not arise.

In fact in the above example the density perturbation satisfies a second order equation which is elliptic. After a rescaling of the time coordinate it reduces to the Laplace equation. The computation which has just been done should be compared with the remarks on the Cauchy problem for the Laplace equation on p. 229 of [27], Vol. 2. The corresponding example for the Laplace equation goes back to Hadamard [49].

Solutions of the Einstein equations coupled to a fluid with an equation of state of the type considered in this section have been considered in the context of inflationary models [10], [93]. While this is unproblematic for spatially homogeneous models, the above ill-posedness result suggests strongly that this kind of model cannot give reasonable results in the inhomogeneous case.

Some general comments on well-posedness and stability will now be made. Suppose a solution of a system of evolution equations is given. Assume for simplicity that this solution is time-independent, although a similar discussion could be carried out more generally. The solution is called stable if in order to ensure that a solution stays close to the original solution to any desired accuracy, it is enough to require it to be sufficiently close at one time. Closeness is measured in some appropriate norm. Well-posedness has no influence on stability in this sense. Already for ordinary differential equations with smooth coefficients, which always have a well-posed initial value problem, stability does not in general hold. Solutions of the linearized sytem about the given solution can grow exponentially. However there is a constant k such that no linearized solution can grow faster than Ce^{kt}. This is a rather general feature of well-posed evolution equations. For instance any linear symmetric hyperbolic system allows an exponential bound with some constant k independent of the solution. If no such bound is possible, then it is said that there is a violent instability. (Cf. [69], Sect. 4.4 for this terminology.) The presence of a violent instability is closely related to ill-posedness, as can be seen in the above example.

5.6 Symmetries

If an initial data set for the Einstein-matter equations possesses symmetries, then we can expect these to be inherited by the corresponding solutions. This will be discussed here in the case of the vacuum Einstein equations. There is nothing in the argument which obviously makes essential use of the vacuum condition and it should extend to reasonable types of matter. It makes use of the maximal Cauchy development and so any restrictions on the matter model which might come up there would appear again in the present context.

The following only covers symmetries of spacetime which leave a given Cauchy surface invariant. It is based on group actions rather than Killing vectors. A more extensive discussion of symmetries of spacetime and their relations to the Cauchy problem can be found in Sect. 2.1 of [26].

By a symmetry of an initial data set (S, h_{ab}, χ_{ab}) for the vacuum Einstein equations we mean a diffeomorphism $\psi : S \to S$ which leaves h_{ab} and χ_{ab} invariant. Let ϕ be the embedding of S into its maximal Cauchy development. Then $\bar{\phi} = \phi \circ \psi$ also satisfies the properties of the embedding in the definition of the maximal Cauchy devlopment. Hence, by uniqueness up to isometry, there exists an isometry $\bar{\psi}$ of the maximal Cauchy development onto itself such that $\bar{\psi} \circ \phi = \bar{\phi}$. This means that $\bar{\psi} \circ \phi = \phi \circ \psi$. Thus $\bar{\psi}$ is an isometry of M whose restriction to $\phi(S)$ is equal to ψ. We see that a symmetry of the initial data extends to a symmetry of the solution. Next we wish to show that this extension is unique. Since a general theorem of Lorentzian (or Riemannian) geometry says that two isometries which agree on a open set agree everywhere it suffices to show that any two isometries $\bar{\psi}$ with the properties described agree on a neighbourhood of $\phi(S)$. Let p be a point of $\phi(S)$. A neighbourhood of p can be covered with Gauss coordinates based on $\phi(S)$. An isometry preserves geodesics and orthogonality. Hence if, when expressed in Gauss coordinates, it is the identity for $t = 0$ it must be the identity everywhere. This completes the proof of the uniqueness of $\bar{\psi}$.

Now consider the situation where a Lie group G acts on S in such a way that each transformation ψ_g of S corresponding to an element of the group is a symmetry of the initial data. Let H be the isometry group of the maximal Cauchy development and H_S the group of all isometries of the maximal Cauchy development which leave $\phi(S)$ invariant. The group H_S is a closed subgroup of the Lie group H and thus is itself a Lie group. Each

ψ_g is the restriction of a unique element $\bar{\psi}_g$ of H_S. Using uniqueness again we must have $\bar{\psi}_{gh} = \bar{\psi}_g\bar{\psi}_h$ for all elements g and h of G. Thus we obtain a homomorphism from G to H_S. This shows that there exists an action of the group G on M by isometries which extends the action on $\phi(S)$ arising from the original action on S by means of the identification using ϕ. However this argument does not show that the resulting action of G is smooth. To show this consider first the group H_I of all symmetries of the initial data. It is a closed subgroup of the isometry group of h and therefore has the structure of a Lie group. The above considerations show that restriction defines an isomorphism of groups from H_S to H_I. (We identify S with $\phi(S)$ here.) If we knew that this mapping was continuous a general theorem on Lie groups [92] would show that it is also an isomorphism of Lie groups. The continuity can be seen by noting that the topology of an isometry group coming from its Lie group structure coincides with the compact open topology [62]. The continuity of the restriction mapping in the compact open topology follows immediately from the definitions. We conclude that as Lie groups H_S and H_I can be identified.

Now we come back to the action of G. The action of G on initial data is a smooth mapping $G \times S \to S$. It is the composition of a smooth homomorphism from G to H_I with the action of H_I on S. By the comments of the last paragraph this can be identified with the composition of a smooth homomorphism from G to H_S with the action of H_S. In this way we obtain a smooth action of G on M which leaves S invariant and restricts to the original action on the initial data. It is the action of G which we previously considered.

6 Outlook

This article is intended to be an informative tour through its subject, rather than an exhaustive account. The latter would in any case be impossible in an article of this length, given the amount of literature which now exists. The aim of this section is to mention a few of the important things which have been left out, and to direct the reader to useful sources of information concerning these. A good starting point is the review article of Choquet-Bruhat and York [23] which is still very useful. (See also [34].) There is an extensive treatment of the constraints in reference [23]. For a selection of newer results on the constraints, see e.g. [5], [6], [57], [58], [59] and [20].

The most obvious omission of the present article is the lack of statements on the global Cauchy problem. A review with pointers to further sources can be found in [78]. This material is too recent to be discussed in [23] and a lot has happened since then. It is natural that once some of the basic local questions had been solved attention turned to global issues. The latter are now central to present research on the Cauchy problem. Most of the existing results concern spacetimes with high symmetry, although in the meantime there are also a few theorems on space-times without symmetries which are small but finite perturbations of space-times of special types. This is an area which is developing vigorously at the moment. Often the statements obtained about global existence of solutions are accompanied by information on the global qualitative properties of the solutions. A discussion of the asymptotic behaviour of a particular class of solutions and its relevance for the modelling of certain physical systems can be found in [40] and [41].

In the introduction we mentioned the possible applications of ideas connected with the Cauchy problem to analytical and numerical approximations. Up to now progress on establishing an effective interaction between theoretical developments and the applications of approximate methods to concrete physical problems such as the generation of gravitational waves has been limited. For instance, little has been done on the question of proving theorems on analytical approximations in general relativity since [28], [75], and [77]. There is no shortage of things to be done. For instance one tempting goal would be a precise formulation and justification of the quadrupole formula.

As for the link to numerical relativity, the discussions in Sect. 4, apart from their interest for purely analytical reasons, could potentially be exploited for improving numerical codes. Many new hyperbolic reductions have been suggested recently with the aim of providing equations which would ensure a stable time evolution. We have added a few more. We also pointed out various different gauge conditions. Their usefulness for stable long-time numerical calculations still has to be explored. We have seen that different representations of the field equations and different formalisms allow us to employ different gauge conditions. The possibilities of controlling the lifetime of a gauge by a judicious choice of gauge source functions have neither been investigated analytically nor numerically in a systematic way.

One of the main interests in the analysis of the initial boundary problem lies in the fact that many approaches to numerical relativity require the introduction of timelike boundary hypersurfaces which reduce the calculations to spatially finite grids. A good analytical understanding of the initial boundary value problem does not guarantee the stability of long-time evolutions for problems with timelike boundaries, but the latter are likely to fail without a proper understanding of the analytical background. In this respect we also consider the analytical investigations of the conformal Einstein equations (cf. [40], [41]) as an opportunity for numerical relativity. They allow us in principle to calculate entire spacetimes, including their asymptotic behaviour, on finite grids without the need to introduce artificial boundaries in the physical spacetime.

Another place where the theory of the Cauchy problem could have something to contribute is the application of numerical methods developed for systems of conservation laws (or, more generally, systems of balance laws) to the Einstein equations, a procedure which has been popular recently. In a certain sense this means accepting an analogy between the Einstein equations and the Euler equations. At the moment this amounts to no more than the fact that both systems are quasi-linear hyperbolic and can be formulated as systems of balance laws. Unlike the Euler equations, the Einstein equations do not appear to admit a preferred formulation of this kind. There are different alternatives and no known criterion for choosing between them. It is very tempting to import the vast amount of knowledge which has been accumulated concerning the numerical solution of the Euler equations into general relativity. On the other hand it is not at all clear how many of these techniques are really advantageous for, say, the vacuum Einstein equations. One might hope that analytical theory could throw some light on these questions.

There can be little doubt that increased cooperation between people working on analytical and numerical aspects of the evolution of solutions of the Einstein equations would lead to many new insights. As both fields progress cases where a rewarding collaboration would be possible are bound to present themselves. It suffices for someone to show the initiative required to profit from this situation.

An open problem which has been touched on already is that of the existence of solutions of the Einstein–Euler system describing fluid bodies. This kind of free boundary problem is poorly understood even in classical physics, although there has been significant progress recently [95]. The central importance of this becomes clear when it is borne in mind that the usual applications of gravitational theory in astrophysics (except for cosmology) concern self-gravitating *bodies.*

We hope to have succeeded in showing in this article that the study of the Cauchy problem for the Einstein equations is a field which presents a variety of fascinating challenges. Perhaps, with luck, it will stimulate others to help tackle them.

Acknowledgements During the preparation of this article we have benefitted from discussions with many people. We are grateful to all of them. We thank in particular Yvonne Choquet-Bruhat, Oliver Henkel, Jim Isenberg, Satyanad Kichenassamy and Bernd Schmidt.

References

1. Abrahams, A., Anderson, A., Choquet-Bruhat, Y., York, J. (1997) Geometrical hyperbolic systems for general relativity and gauge theories. Class. Quantum Grav. 14, A9–A22.
2. Alcubierre, M., Brügmann, B., Miller, M., Suen, W.-M. (1999) A Conformal Hyperbolic Formulation of the Einstein Equations Preprint gr-qc/9903030.
3. Alinhac, S., Gérard, P. (1991) Opérateurs pseudodifferentiels et théorème de Nash-Moser. InterEditions, Paris.
4. Anderson, A., York, J. W. (1999) Fixing Einstein's equations. Phys. Rev. Lett. 82, 4384–4387.

5. Andersson, L., Chruściel, P. T., Friedrich, H. (1992) On the regularity of solutions to the Yamabe equation and the existence of smooth hyperboloidal initial data for Einstein's field equations. Commun. Math. Phys. 149, 587–612.
6. Andersson, L., Chruściel, P. T.(1994) On hyperboloidal Cauchy data for the vacuum Einstein equations and obstructions to the smoothness of Scri. Commun. Math. Phys. 161, 533–568.
7. Bancel, D., Choquet-Bruhat, Y. (1973) Existence, uniqueness and local stability for the Einstein–Maxwell–Boltzmann system. Commun. Math. Phys. 33, 83–96.
8. Bao, D., Choquet-Bruhat, Y., Isenberg, J., Yasskin, P. B. (1985) The well-posedness of (N=1) classical supergravity. J. Math. Phys. 26, 329–333.
9. Bao, D., Isenberg, J., Yasskin, P. B. (1985) The dynamics of the Einstein–Dirac system I. A principal bundle formulation of the theory and its canonical analysis. Ann. Phys. (NY) 164, 103–171.
10. Barrow, J. D. (1987) Cosmic no hair theorems and inflation. Phys. Lett. B 187, 12–16.
11. Baumgarte, T. W., Shapiro, S. L. (1998) Numerical integration of Einstein's field equations. Phys. Rev. D 59, 024007.
12. Besse, A. (1987) Einstein Manifolds. Springer, New York.
13. Binney, J., Tremaine, S. (1987) Galactic dynamics. Princeton University Press, Princeton.
14. Bona, C., Masso, J. (1992) Hyperbolic Evolution System for Numerical Relativity. Phys. Rev. Lett. 68, 1097–1099.
15. Brauer, U. (1995) Singularitäten in relativistischen Materiemodellen. PhD Thesis, University of Potsdam.
16. Bruhat, Y. (1952) Théorème d'existence pour certains systèmes d'équations aux dérivées partielles non linéaires. Acta Mathematica 88, 141–225.
17. Choquet-Bruhat, Y. (1958) Théorème d'existence en mécanique des fluides relativistes. Bull. Soc. Math. France 86, 155–175.
18. Choquet-Bruhat, Y. (1974) The stability of the solutions of nonlinear hyperbolic equations on a manifold. Russ. Math. Surveys 29, 327–335.
19. Choquet-Bruhat, Y., Geroch, R. (1969) Global aspects of the Cauchy problem in general relativity. Commun. Math. Phys. 14, 329–335.
20. Choquet-Bruhat, Y., Isenberg, J., York, J. (1999) Einstein Constraints on Asymptotically Euclidean Manifolds. In preparation.
21. Choquet-Bruhat, Y., Lamoureux-Brousse, L. (1973) Sur les équations de l'élasticité relativiste. C. R. Acad. Sci. Paris 276, 1317–1320.
22. Choquet-Bruhat, Y.,Ruggeri, T. (1983) Hyperbolicity of the $3+1$ system System of Einstein Equations. Commun. Math. Phys. 89, 269–275.
23. Choquet-Bruhat, Y., York, J. (1980) The Cauchy problem. In: Held, A. (Ed.) General Relativity and Gravitation, Vol. I. Plenum, New York, 99–172.
24. Choquet-Bruhat, Y., York, J. (1996) Mixed elliptic and hyperbolic systems for the Einstein equations. Preprint gr-qc/9601030.
25. Christodoulou, D., Klainerman, S. (1993) The global nonlinear stability of the Minkowski space. Princeton University Press, Princeton.
26. Chruściel, P. T. (1991) On uniqueness in the large of solutions of Einstein's equations (Strong Cosmic Censorship). Proceedings of the CMA Vol. 27. Australian National University.
27. Courant, R., Hilbert, D. (1962) Methods of Mathematical Physics. Wiley, New York.

28. Damour, T., Schmidt, B. G. (1990) Reliability of perturbation theory in general relativity. J. Math. Phys. 31, 2441-2453.
29. Dieudonné, J. (1969) Foundations of modern analysis. Academic Press, New York.
30. Dieudonné, J. (1974) Treatise on Analysis IV. Academic Press, New York.
31. Elst, H. van, Ellis, G. F. R. (1998) Causal propagation of geometrical fields in relativistic cosmology Phys. Rev. D 59, 024013.
32. Evans, L. C. (1998) Partial differential equations. American Mathematical Society, Providence.
33. Fischer, A. E., Marsden, J. E. (1972) The Einstein Equations of Evolution – A Geometric Approach. J. Math. Phys. 13, 546–568.
34. Fischer, A. E., Marsden, J. E. (1979) The initial value problem and the dynamical formulation of general relativity. In: Hawking, S. W., Israel, W. (Eds.) General Relativity: an Einstein Centenary Survey. Cambridge University Press, Cambridge.
35. Frauendiener, J. (1998) Numerical treatment of the hyperboloidal initial value problem for the vacuum Einstein equations. II. The evolution equations. Phys. Rev. D 58, 064004.
36. Friedrich, H. (1985) On the Hyperbolicity of Einstein's and Other Gauge Field Equations. Commun. Math. Phys. 100, 525–543.
37. Friedrich, H. (1991) On the global existence and the asymptotic behaviour of solutions to the Einstein–Maxwell–Yang–Mills equations. J. Diff. Geom. 34, 275–345.
38. Friedrich, H. (1996) Hyperbolic reductions for Einstein's equations. Class. Quantum Grav. 13, 1451–1469.
39. Friedrich, H. (1998) Evolution equations for gravitating ideal fluid bodies in general relativity. Phys. Rev. D 57, 2317–2322.
40. Friedrich, H. (1998) Einstein Equations and Conformal Structure. In: S. A. Hugget et al (Eds.) The Geometric Universe: Science, Geometry, and the Work of Roger Penrose. Oxford University Press, Oxford, 81-98.
41. Friedrich, H. (1998) Einstein's Equations and Geometric Asymptotics. In: N. Dadhich, J. Narlikar (Eds.) Gravitation and Relativity at the Turn of the Millennium. Inter-University Centre for Astronomy and Astrophysics, Pune, India. gr-qc/9804009.
42. Friedrich, H., Nagy, G. (1999) The initial boundary value problem for Einstein's vacuum field equations. Commun. Math. Phys. 201, 619–655.
43. Frittelli, S., Reula, O. (1996) First-order symmetric-hyperbolic Einstein equations with arbitrary fixed gauge. Phys. Rev. Lett. 76, 4667–4670.
44. Frittelli, S., Reula, O. (1999) Well-posed forms of the 3 + 1 conformally-decomposed Einstein equations. Preprint gr-qc/9904048.
45. Garabedian, P. R. (1964) Partial Differential Equations. Wiley, New York.
46. Geroch, R. (1991) Causal theories of dissipative relativistic fluids. Ann. Phys. 207, 394–416.
47. Griffiths, P., Harris, J. (1978) Principles of algebraic geometry. Wiley, New York.
48. Guès, O. (1990) Problème mixte hyperbolique quasi-linéaire caractéristique. Commun. Part. Diff. Eq. 15, 595–645.
49. Hadamard, J. (1952) Lectures on Cauchy's problem in linear partial differential equations. Dover, New York.

50. Hamilton, R. S. (1982) The inverse function theorem of Nash and Moser. Bull. Amer. Math. Soc. 7, 65–222.
51. Hartman, P. (1982) Ordinary differential equations. Birkhäuser, Boston.
52. Hawking, S. W., Ellis, G. F. R. (1973) The large scale structure of space-time. Cambridge University Press, Cambridge.
53. Hiscock, W. A., Lindblom, L. (1985) Generic instabilities in first-order dissipative relativistic fluid theories. Phys. Rev. D 31, 725–733.
54. Hörmander, L. (1983) The Analysis of Linear Partial Differential Operators I. Springer, Berlin.
55. Hübner, P. (1999) A Scheme to Numerically Evolve Data for the Conformal Einstein Equation. AEI Preprint 105, gr-qc/9903088.
56. Iriondo, M. S., Leguizamón, E. O., Reula, O. A. (1998) On the Dynamics of Einstein's Equations in the Ashtekar Formulation. Adv. Theor. Math. Phys. 2, 1075–1103.
57. Isenberg, J. (1995) Constant mean curvature solutions of the Einstein constraint equations on closed manifolds. Class. Quantum Grav. 12, 2249–2274.
58. Isenberg, J., Moncrief, V. (1996) A set of nonconstant mean curvature solutions of the Einstein constraint equations on closed manifolds. Class. Quantum Grav., 13, 1819–1847.
59. Kánnár, J. (1996) Hyperboloidal initial data for the vacuum Einstein equations with cosmological constant. Class. Quantum Grav. 13, 3075–3084.
60. Kato, T. (1975) The Cauchy problem for quasi-linear symmetric hyperbolic systems. Arch. Rat. Mech. Anal. 58, 181–205.
61. Kind, S., Ehlers, J. (1993) Initial boundary value problem for the spherically symmetric Einstein equations for a perfect fluid. Class. Quantum Grav. 18, 2123–2136.
62. Kobayashi, S. (1972) Transformation groups in differential geometry. Springer, Berlin.
63. Kreiss, H.-O., Lorenz, J. (1989) Initial boundary value problems and the Navier-Stokes equations. Academic Press, New York.
64. Kazdan, J. L. (1981) Another Proof of Bianchi's Identity in Riemannian Geometry. Proc. Amer. Math. Soc. 81, 341–342.
65. Lanczos, C. (1922) Ein vereinfachendes Koordinatensystem für die Einsteinschen Gravitationsgleichungen. Phys. Z. 23, 537–539.
66. Leray, J. (1953) Hyperbolic differential equations. Lecture Notes, Princeton.
67. Lichnerowicz, A. (1967) Relativistic hydrodynamics and magnetohydrodynamics. Benjamin, New York.
68. Lichnerowicz, A. (1994) Magnetohydrodynamic waves and shock waves in curved spacetime. Kluwer, Dordrecht.
69. Majda, A. (1984) Compressible fluid flow and systems of conservation laws in several space variables. Springer, Berlin.
70. Nagy, G. (1999) Initial data for ideal fluid bodies. In preparation
71. Palais, R. (1968) Foundations of global nonlinear analysis. Benjamin, New York.
72. Penrose, R., Rindler, W. (1995) Spinors and space-time. Cambridge University Press, Cambridge.
73. Pichon, G. (1966) Théorèmes d'existence pour les équations des milieux élastiques. J. Math. Pures Appl. 45, 395–409.
74. Rauch, J. (1985) Symmetric positive systems with boundary characteristics of constant multiplicity. Trans. Amer. Math. Soc. 291, 167–187.

75. Rendall, A. D. (1992) On the definition of post-newtonian approximations. Proc. R. Soc. Lond. A 438, 341–360.
76. Rendall, A. D. (1992) The initial value problem for a class of general relativistic fluid bodies. J. Math. Phys. 33, 1047–1053.
77. Rendall, A. D. (1995) The Newtonian Limit for Asymptotically Flat Solutions of the Vlasov-Einstein System. Commun. Math. Phys. 163, 89–112.
78. Rendall, A. D. (1998) Local and global existence theorems for the Einstein equations. Living Reviews, article 1998-4. http://www.livingreviews.org
79. Reula, O. (1999) Exponential Decay for Small Non-Linear Perturbations of Expanding Flat Homogeneous Cosmologies. AEI Preprint 108, gr-qc/9902006.
80. Rudin, W. (1987) Real and complex analysis. McGraw-Hill, New York.
81. Rudin, W. (1991) Functional analysis. McGraw-Hill, New York.
82. Schoen, R., Yau, S. T. (1979) On the proof of the positive mass conjecture in general relativity. Commun. Math. Phys. 65, 45–76.
83. Secchi, P. (1996) The initial boundary value problem for linear symmetric hyperbolic systems with characteristic boundary of constant multiplicity. Diff. Int. Eqs. 9, 671–700.
84. Secchi, P. (1996) Well-Posedness of Characteristic Symmetric Hyperbolic Systems. Arch. Rational Mech. Anal. 134, 155–197.
85. Shapiro, s. L, Teukolsky, S. A. (1985) Relativistic stellar dynamics on the computer I. Motivation and numerical method. Astrophys. J. 298, 34–57.
86. Shibata,M., Nakamura, T. (1995) Evolution of three-dimensional gravitational waves: Harmonic slicing case. Phys. Rev. D 10, 5428–5444.
87. Stewart, J. (1990) Advanced general relativity. Cambridge University Press, Cambridge.
88. Taylor, M. E. (1981) Pseudodifferential operators. Princeton University Press, Princeton.
89. Taylor, M. E. (1991) Pseudodifferential operators and nonlinear PDE. Birkhäuser, Boston.
90. Taylor, M. E. (1996) Partial Differential Equations III. Nonlinear Equations. Springer, Berlin.
91. Treves, F. (1975) Basic Linear Partial Differential Equations. Academic Press, New York.
92. Varadarajan, V. S. (1984) Lie groups, Lie algebras and their representations. Springer, Berlin.
93. Wainwright, J., Ellis, G.F.R. (eds.) Dynamical Systems in Cosmology. Cambridge University Press, Cambridge.
94. Witten, E. (1981) A new proof of the positive energy theorem. Commun. Math. Phys. 80, 381–402.
95. Wu, S. (1999) Well-posedness in Sobolev spaces of the full water wave problem. J. Amer. Math. Soc. 12, 445–495.

Post-Newtonian Gravitational Radiation

Luc Blanchet

Département d'Astrophysique Relativiste et de Cosmologie,
Centre National de la Recherche Scientifique (UMR 8629),
Observatoire de Paris, 92195 Meudon Cedex, France

1 Introduction

1.1 On Approximation Methods in General Relativity

Let us declare that the most important *devoir* of any physical theory is to draw firm predictions for the outcome of laboratory experiments and astronomical observations. Unfortunately, the devoir is quite difficult to fulfill in the case of general relativity, essentially because of the complexity of the Einstein field equations, to which only few exact solutions are known. For instance, it is impossible to settle the exact prediction of this theory when there are no symmetry in the problem (as is the case in the problem of the gravitational dynamics of separated bodies). Therefore, one is often obliged, in general relativity, to resort to approximation methods.

It is beyond question that approximation methods do work in general relativity. Some of the great successes of this theory were in fact obtained using approximation methods. We have particularly in mind the test by Taylor and collaborators [1–3] regarding the orbital decay of the binary pulsar PSR 1913+16, which is in agreement to within 0.35% with the general-relativistic *post-Newtonian* prediction. However, a generic problem with approximation methods (especially in general relativity) is that it is non trivial to define a clear framework within which the approximation method is mathematically well-defined, and such that the results of successive approximations could be considered as *theorems* following some precise (physical and/or technical) assumptions. Even more difficult is the problem of the relation between the approximation method and the *exact* theory. In this context one can ask: What is the mathematical nature of the approximation series (convergent, asymptotic, ...)? What its "reliability" is (i.e., does the approximation series come from the Taylor expansion of a family of exact solutions)? Does the approximate solution satisfy some "exact" boundary conditions (for instance the no-incoming radiation condition)?

Since the problem of theoretical prediction in general relativity is complex, let us distinguish several approaches (and ways of thinking) to it, and illustrate them with the example of the prediction for the binary pulsar. First we may consider what could be called the "physical" approach, in which one analyses the relative importance of each physical phenomena at work by using crude numerical estimates, and where one uses only the lowest-order approximation, relating if necessary the local physical quantities to observables by means of balance equations (perhaps not well defined in terms of basic theoretical concepts). The physical approach to the problem of the binary pulsar is well illustrated by Thorne in his beautiful Les Houches review [4] (see also the round table discussion moderated by Ashtekar [5]): one derives the loss of energy by gravitational radiation from the (Newtonian) quadrupole formula applied formally to point-particles, assumed to be test-masses though they are really self-gravitating, and one argues "physically" that the effect comes from the variation of the Newtonian binding energy in the center-of-mass frame – indeed, on physical grounds, what else could this be (since we expect the rest masses won't vary)? The physical approach yields the correct result for the rate of decrease of the period of the binary pulsar. Of course, thinking physically is extremely useful, and indispensable in a preliminary stage, but certainly it should be completed by a solid study of the connection to the mathematical structure of the theory. Such a study would *a posteriori* demote the physical approach to the status of "heuristic" approach. On the other hand, the physical approach may fall short in some situations requiring a sophisticated mathematical modelling (like in the problem of the dynamics of singularities), where one is often obliged to follow one's mathematical rather than physical insight.

A second approach, that we shall qualify as "rigorous", has been advocated mainly by Jürgen Ehlers (see, e.g., [6]). It consists of looking for a high level of mathematical rigor, within the exact theory if possible, and otherwise using an approximation scheme that we shall be able to relate to the exact theory. This does not mean that we will be so much wrapped up by mathematical rigor as to forget about physics. Simply, in the rigorous approach, the prediction for the outcome of an experiment should follow mathematically from first theoretical principles. Clearly this approach is the one we should ideally adhere to. As an example, within the rigorous approach, one was not permitted, by the end of the seventies, to apply the standard quadrupole formula to the binary pulsar. Indeed, as pointed out by Ehlers *et al* [7], it was not clear that gravitational radiation reaction on a self-gravitating system implies the standard quadrupole formula for the energy flux, notably because computing the radiation reaction demands *a priori* three non-linear iterations of the field equations [8], which were not fully available at that time. Ehlers and collaborators [7] remarked also that the exact results concerning the structure of the field at infinity (nota bly the asymptotic shear of null geodesics whose variation determines the flux of radiation) were not connected to the actual dynamics of the binary.

Maybe the most notable result of the rigorous approach concerns the relation between the exact theory and the approximation methods. In the case of the post-Newtonian approximation (limit $c \to \infty$), Jürgen Ehlers has provided with his frame theory [9–11] a conceptual framework in which the post-Newtonian approximation can be clearly formulated (among other purposes). This theory unifies the theories of Newton and Einstein into a single generally covariant theory, with a parameter $1/c$ taking the value zero in the case of Newton and being the inverse of the speed of light in the case of Einstein. Within the frame theory not only does one understand the limit relation of Einstein's theory to Newton's, but one explains why it is legitimate when describing the predictions of general relativity to use the common-sense language of Newton (for instance thinking that the trajectories of particles in an appropriately defined coordinate system take place in some Euclidean space, and viewing the coordinate velocities as being defined with respect to absolute time). It was shown by Lottermoser [12] that the constraint equations of the (Hamiltonian formulation of the) Ehlers frame theory admit solutions with a well-defined post-Newtonian limit. Further in the spirit of the rigorous approach, we quote the work of Rendall [13] on the definition of the post-Newtonian approximation, and the link to the post-Newtonian equations used in practical computations. (See also [14,15] for an attempt at showing, using restrictive assumptions, that the post-Newtonian series is asymptotic.)

The important remarks of Jürgen Ehlers *et al* [7] on the applicability of the quadrupole formula to the binary pulsar stimulated research to settle down this question with (al least) acceptable mathematical rigor. The question was finally answered positively by Damour and collaborators [16–19], who obtained in algebraically closed form the general-relativistic equations of motion of two compact objects, up to the requisite 5/2 post-Newtonian order (2.5PN order or $1/c^5$) where the gravitational radiation reaction force appears. This extended to 2.5PN order the work at 1PN of Lorentz and Droste [20], and Einstein, Infeld and Hoffmann [21]. The net result is that the dynamics of the binary pulsar as predicted by (post-Newtonian) general relativity is in full agreement both with the prediction of the quadrupole formula, as derived earlier within the "physical" approach, and with the observations by Taylor *et al* (see [22] for discussion).

Motivated by the success of the theoretical prediction in the case of the binary pulsar [16–19,22], we shall try to follow in this article the spirit of the "rigorous" approach of Jürgen Ehlers, notably in the way it emphasizes the mathematical proof, but we shall also differ from it by a systematic use of approximation methods. This slightly different approach recognizes from the start that in certain difficult problems, it is impossible to derive a physical result all the way through the exact theory without any gap, so that one must proceed with approximations. *But*, in this approach, one implements a mathematically well-defined framework for the approximation method, and within this framework one proves theorems that (ideally) guarantee the correctness of the theoretical prediction to be compared with experiments. Because the comparison with experiments is the only thing which matters *in fine* for a pragmatist, we qualify this third approach as "pragmatic".

In this article we describe the pragmatic approach to the problem of gravitational radiation emitted by a general isolated source, based on the rigorous post-Minkowskian iteration of the field outside the source [23], and on the general connection of the exterior field to the field inside a slowly-moving source [24,25]. Note that for this particular problem the pragmatic approach is akin to the rigorous one in that it permits to establish some results on the connection between approximate and exact methods. For instance it was proved by Damour and Schmidt [26] (see also [27,28]) that the post-Minkowskian algorithm generates an asymptotic approximation to exact solutions, and it was shown [29] that the solution satisfies to any order in the post-Minkowskian expansion a rigorous definition of asymptotic flatness at future null infinity. However it remains a challenge to analyse in the manner of the rigorous approach the relation to exact theory of the whole formalism of [23–25,29].

By combining the latter post-Minkowskian approximation and a post-Newtonian expansion inside the system, it was proved (within this framework of approximations) that the quadrupole formula for slowly-moving, weakly-stressed and self-gravitating systems is correct, even including post-Newtonian corrections [30]; and *idem* for the radiation reaction forces acting locally inside the system, and for the associated balance equations [31,32]. These results answered positively Ehlers' remarks [7] in the case of slowly-moving extended (fluid) systems. However we are also interested in this article to the application to binary systems of compact objects modelled by point-masses. Indeed the latter sources of radiation are likely to be detected by future gravitational-wave experiments, and thus concern the pragmatist. We shall see how one can address the problem in this case. (When specialized to point-mass binaries, the results on radiation reaction [31,32] are in agreement with sepa rate work of Iyer and Will [33,34].) For other articles on the problem of gravitational radiation from general and binary point-mass sources, see [35–39].

1.2 Field Equations and the No-Incoming-Radiation Condition

The problem is to find the solutions, in the form of analytic approximations, of the Einstein field equations in $\mathbb{R}^4$,

$$R^{\mu\nu} - \frac{1}{2}g^{\mu\nu}R = \frac{8\pi G}{c^4}T^{\mu\nu} , \tag{1}$$

and thus also of their consequence, the equations of motion of the matter source, $\nabla_\nu T^{\mu\nu} = 0$. Throughout this work we assume the existence and unicity of a global harmonic (or de Donder) coordinate system. This means that we can choose the gauge condition

$$\partial_\nu h^{\mu\nu} = 0 \; ; \qquad h^{\mu\nu} \equiv \sqrt{-g}g^{\mu\nu} - \eta^{\mu\nu} , \tag{2}$$

where g and $g^{\mu\nu}$ denote the determinant and inverse of the covariant metric $g_{\mu\nu}$, and where $\eta^{\mu\nu}$ is an auxiliary flat metric [i.e. $\eta^{\mu\nu} = \mathrm{diag}(-1,1,1,1) = \eta_{\mu\nu}$]. The Einstein field equations (1) can then be replaced by the *relaxed* equations

$$\Box h^{\mu\nu} = \frac{16\pi G}{c^4}\tau^{\mu\nu} \,, \tag{3}$$

where the box operator is the flat d'Alembertian, $\Box \equiv \Box_\eta = \eta^{\mu\nu}\partial_\mu\partial_\nu$, and where the source term is the sum of a matter part and a gravitational part,

$$\tau^{\mu\nu} \equiv |g|T^{\mu\nu} + \frac{c^4}{16\pi G}\Lambda^{\mu\nu} \,. \tag{4}$$

In harmonic coordinates the field equations take the form of simple wave equations, but whose source term is actually a complicated functional of the gravitational field $h^{\mu\nu}$; notably the gravitational part depends on $h^{\mu\nu}$ and its first and second space-time derivatives:

$$\begin{aligned}\Lambda^{\mu\nu} =& - h^{\rho\sigma}\partial^2_{\rho\sigma}h^{\mu\nu} + \partial_\rho h^{\mu\sigma}\partial_\sigma h^{\nu\rho} + \frac{1}{2}g^{\mu\nu}g_{\rho\sigma}\partial_\lambda h^{\rho\tau}\partial_\tau h^{\sigma\lambda} \\ & - g^{\mu\rho}g_{\sigma\tau}\partial_\lambda h^{\nu\tau}\partial_\rho h^{\sigma\lambda} - g^{\nu\rho}g_{\sigma\tau}\partial_\lambda h^{\mu\tau}\partial_\rho h^{\sigma\lambda} + g_{\rho\sigma}g^{\lambda\tau}\partial_\lambda h^{\mu\rho}\partial_\tau h^{\nu\sigma} \\ & + \frac{1}{8}(2g^{\mu\rho}g^{\nu\sigma} - g^{\mu\nu}g^{\rho\sigma})(2g_{\lambda\tau}g_{\epsilon\pi} - g_{\tau\epsilon}g_{\lambda\pi})\partial_\rho h^{\lambda\pi}\partial_\sigma h^{\tau\epsilon} \,.\end{aligned} \tag{5}$$

The point is that $\Lambda^{\mu\nu}$ is at least quadratic in h, so the relaxed field equations (3) are very naturally amenable to a perturbative non-linear expansion. As an immediate consequence of the gauge condition (2), the right side of the relaxed equations is conserved in the usual sense, and this is equivalent to the equations of motion of matter:

$$\partial_\nu \tau^{\mu\nu} = 0 \quad \Leftrightarrow \quad \nabla_\nu T^{\mu\nu} = 0 \,. \tag{6}$$

We refer to $\tau^{\mu\nu}$ as the total stress-energy pseudo-tensor of the matter and gravitational fields in harmonic coordinates. Since the harmonic coordinate condition is Lorentz covariant, $\tau^{\mu\nu}$ is a tensor with respect to Lorentz transformations (but of course not with respect to general diffeomorphisms).

In order to select the physically sensible solution of the field equations in the case of a bounded system, one must choose some boundary conditions at infinity, i.e. the famous no-incoming radiation condition, which ensures that the system is truly isolated (no radiating sources located at infinity). In principle the no-incoming radiation condition is to be formulated at past null infinity $\mathcal{J}^-$. Here, we shall simplify the formulation by taking advantage of the presence of the Minkowski background $\eta_{\mu\nu}$ to define the no-incoming radiation condition with respect to the Minkowskian past null infinity $\mathcal{J}^-_{\mathrm{M}}$. Of course, this does not make sense in the exact theory where only exists the metric $g_{\mu\nu}$ and where the metric $\eta_{\mu\nu}$ is fictituous, but within approximate (post-Minkowskian) methods it is legitimate to view the gravitational field as propagating on the flat background $\eta_{\mu\nu}$, since $\eta_{\mu\nu}$ does exist at any finite order of approximation.

We formulate the no-incoming radiation condition in such a way that it suppresses any homogeneous, regular in $\mathbb{R}^4$, solution of the d'Alembertian equation $\Box h = 0$. We have at our disposal the Kirchhoff formula which expresses $h(\mathbf{x}', t')$ in terms of values of $h(\mathbf{x}, t)$ and its derivatives on a sphere centered on $\mathbf{x}'$ with radius $\rho \equiv |\mathbf{x}' - \mathbf{x}|$ and at retarded time $t \equiv t' - \rho/c$:

$$h(\mathbf{x}', t') = \iint \frac{d\Omega}{4\pi} \left[\frac{\partial}{\partial \rho}(\rho h) + \frac{1}{c}\frac{\partial}{\partial t}(\rho h) \right](\mathbf{x}, t) \tag{7}$$

where $d\Omega$ is the solid angle spanned by the unit direction $(\mathbf{x} - \mathbf{x}')/\rho$. From the Kirchhoff formula we obtain the no-incoming radiation condition as a limit at $\mathcal{J}^-_{\mathrm{M}}$, that is $r \to +\infty$ with $t + r/c =$const (where $r = |\mathbf{x}|$). In fact we obtain two conditions: the main one,

$$\lim_{\substack{r \to +\infty \\ t + r/c = \text{const}}} \left[\frac{\partial}{\partial r}(r h^{\mu\nu}) + \frac{1}{c}\frac{\partial}{\partial t}(r h^{\mu\nu}) \right](\mathbf{x}, t) = 0 \; , \tag{8}$$

and an auxiliary condition, that $r\partial_\lambda h^{\mu\nu}$ should be bounded at $\mathcal{J}^-_{\mathrm{M}}$, coming from the fact that ρ in the Kirchhoff formula (7) differs from r [we have $\rho = r - \mathbf{x}'.\mathbf{n} + O(1/r)$ where $\mathbf{n} = \mathbf{x}/r$].

In fact, we adopt in this article a much more restrictive condition of no-incoming radiation, namely that the field is stationary before some finite instant $-\mathcal{T}$ in the past:

$$t \leq -\mathcal{T} \Rightarrow \frac{\partial}{\partial t}[h^{\mu\nu}(\mathbf{x}, t)] = 0 \; . \tag{9}$$

In addition we assume that before $-\mathcal{T}$ the field $h^{\mu\nu}(\mathbf{x})$ is of order $O(1/r)$ when $r \to +\infty$. These restrictive conditions are imposed for technical reasons following [23], since they allow constructing rigorously (and proving theorems about) the metric outside some time-like world tube $r \equiv |\mathbf{x}| > \mathcal{R}$. We shall assume that the region $r > \mathcal{R}$ represents the exterior of an actual compact-support system with constant radius $d < \mathcal{R}$ [i.e. d is the maximal radius of the adherence of the compact support of $T^{\mu\nu}(\mathbf{x}, t)$, for any time t].

Now if $h^{\mu\nu}$ satisfies for instance (9), so does the pseudo-tensor $\tau^{\mu\nu}$ built on it, and then it is clear that the retarded integral of $\tau^{\mu\nu}$ satisfies itself the same condition. Therefore one infers that the unique solution of the Einstein equation (3) satisfying the condition (9) is

$$h^{\mu\nu} = \frac{16\pi G}{c^4} \Box^{-1}_R \tau^{\mu\nu} \; , \tag{10}$$

where the retarded integral takes the standard form

$$(\Box^{-1}_R \tau)(\mathbf{x}, t) \equiv -\frac{1}{4\pi} \int \frac{d^3\mathbf{x}'}{|\mathbf{x} - \mathbf{x}'|} \tau\left(\mathbf{x}', t - |\mathbf{x} - \mathbf{x}'|/c\right) \; . \tag{11}$$

Notice that since $\tau^{\mu\nu}$ depends on h and its derivatives, the equation (10) is to be viewed rather as an integro-differential equation equivalent to the Einstein equation (3) with no-incoming radiation.

1.3 Method and General Physical Picture

We want to describe an isolated system, for instance a "two-body system", in Einstein's theory. We expect (though this is not proved) that initial data sets $g_{\mu\nu}$, $\partial_t g_{\mu\nu}$, ρ, $\mathbf{v}$ satisfying the constraint equations on the space-like hypersurface $t = t_0$ exist, and that this determines a unique solution of the field equations for any time t, which approaches in the case of two bodies a "scattering state" when $t \to -\infty$, in which the bodies move on unbound (hyperbolic-like) orbits. We assume that the space-times generated by such data admit a past null infinity $\mathcal{J}^-$ (or, if one uses approximate methods, $\mathcal{J}^-_{\mathrm{M}}$) with no incoming radiation. (Note that in a situation with initial scattering the field might not satisfy the rigorous definitions of asymptotic flatness at $\mathcal{J}^-$; see [40–43].) The point to make is that in this class of space-times there is no degree of freedom for the gravitational field (we could consider other situations where the motion is influenced by incoming radiation).

Both our technical assumptions of compact support for the matter source (with constant radius d) and stationarity before the time $-\mathcal{T}$ contradict our expectation that a two-body system follows an unbound orbit in the remote past. We do not solve this conflict but argue as follows: (i) these technical assumptions permit to derive rigorously some results, for instance the expression [given by (52) with (56) below] of the far-field of an isolated past-stationary system; (ii) it is clear that these results do not depend on the constant radius d, and furthermore we check that they admit in the "scattering" situation a well-defined limit when $-\mathcal{T} \to -\infty$; (iii) this makes us confident that the results are actually valid for a more realistic class of physical systems which become unbound in the past and are never stationary (and, even, one can give *a posteriori* conditions under which the limit $-\mathcal{T} \to -\infty$ exists for a general system at some order of approximation).

Suppose that the system is "slowly-moving" [in the sense of (12) below], so that we can compute the field inside its compact support by means of a post-Newtonian method, say $h^{\mu\nu}_{\mathrm{in}} \equiv \overline{h}^{\mu\nu}$ where the overbar refers to the *formal* post-Newtonian series. The post-Newtonian iteration (say, for hydrodynamics) is not yet defined to all orders in $1/c$, but many terms are known: see the works of Lorentz and Droste [20], Einstein, Infeld and Hoffmann [21], Fock [44], Chandrasekhar and collaborators [45–47], Ehlers and followers [48–54], and many other authors [55–58,30,24].

On the other hand, outside the isolated system, the field is weak everywhere and it satisfies the vacuum equations. Therefore, the equations can be solved conjointly by means of a weak-field or post-Minkowskian expansion ($G \to 0$), and, for each coefficient of G^n in the latter expansion, by means of a multipole expansion (valid because we are outside). The general

Multipolar-post-Minkowskian (MPM) metric was constructed in [23,29] as a functional of two sets of "multipole moments" $M_L(t)$ and $S_L(t)$ which were left arbitrary at this stage (i.e. not connected to the source). The idea of combining the post-Minkowskian and multipole expansions comes from the works of Bonnor [59] and Thorne [60]. We denote by $h^{\mu\nu}_{\rm ext} \equiv \mathcal{M}(h^{\mu\nu})$ the exterior solution, where $\mathcal{M}$ stands for the multipole expansion (as it will turn out, the post-Minkowskian expansion appears in this formalism to be somewhat less fundamental than the multipole expansion).

The key assumption is that the two expansions $h^{\mu\nu}_{\rm in} = \overline{h}^{\mu\nu}$ and $h^{\mu\nu}_{\rm ext} = \mathcal{M}(h^{\mu\nu})$ should match in a region of common validity for both the post-Newtonian and multipole expansions. Here is where our physical restriction to slow motion plays a crucial role, because such an overlap region exists (this is the so-called exterior near-zone) if and only if the system is slowly-moving. The matching is a variant of the well-known method of matching of asymptotic expansions, very useful in gravitational radiation theory [61–65,30,66,24]. It consists of decomposing the inner solution into multipole moments (valid in the outside), re-expanding the exterior solution in the near zone ($r/c \to 0$), and equating term by term the two resulting expansion series. From the requirement of matching we obtain in [25], and review in Sects. 2 and 3 below, the general formula for the multipole expansion $\mathcal{M}(h^{\mu\nu})$ in terms of the "source" multipole moments (notably a mass-type moment I_L and a current-type J_L), given as functionals of the *post-Newtonian* expansion of the pseudo-tensor, i.e. $\overline{\tau}^{\mu\nu}$. [The previous moments M_L and S_L (referred below to as "canonical") are deduced from the source moments after a suitable coordinate transformation.] In addition the matching equation determines the radiation reaction contributions in the inner post-Newtonian metric [67,68,32].

To obtain the source multipole moments in terms of basic source parameters (mass density, pressure), it remains to replace $\overline{\tau}^{\mu\nu}$ by the result of an explicit post-Newtonian iteration of the inner field. This was done to 1PN order in [30,66], then to 2PN order in [24], and the general formulas obtained in [25] permit recovering these results. See Sect. 6. On the other hand, if one needs the equations of motion of the source, simply one inserts the post-Newtonian metric into the conservation law $\partial_\nu \overline{\tau}^{\mu\nu} = 0$. (Note that we are speaking of the equations of motion, which take for instance the form of Euler-type equations with many relativistic corrections, but not of the *solutions* of these equations, which are typically impossible to obtain analytically.)

From the harmonic coordinates, one can perform to all post-Minkowskian orders [29] a coordinate transformation to some radiative coordinates such that the metric admits a far-field expansion in powers of the inverse of the distance R (without the powers of $\ln R$ which plague the harmonic coordinates). Considering the leading order $1/R$ one compares the exterior metric, which is parametrized by the source moments (connected to the source via the matching equation), to the metric defined with "radiative" multipole moments, say

U_L and V_L. This gives U_L and V_L in terms of the source moments, notably I_L and J_L, and *a fortiori* of the source parameters. This solves, within approximate methods, the problem of the relation between the far field and the source. The radiative moments have been obtained with increasing precision reaching now 3PN [69–71], as reviewed in Sect. 5.

The previous scheme is developed for a general description of matter, however restricted to be smooth (we have in mind a general "hydrodynamical" $T^{\mu\nu}$). Thus the scheme *a priori* excludes the presence of singularities (no "point-particles" or black holes), but this is a serious limitation regarding the application to compact objects like neutron stars, which can adequately be approximated by point-masses when studying their dynamics. Fortunately, the formalism *is* applicable to a singular $T^{\mu\nu}$ involving Dirac measures, at the price of a further ansatz, that the infinite self-field of point-masses can be regularized in a certain way. By implementing consistently the regularization we obtain the multipole moments and the radiation field of a system of two point-masses at 2.5PN order [72,73], as well as their equations of motion at the same order in the form of ordinary differential equations [74] (the result agrees with previous works [16–19]); see Sect. 7.

2 Multipole Decomposition

In this section we construct the multipole expansion $\mathcal{M}(h^{\mu\nu}) \equiv h^{\mu\nu}_{\text{ext}}$ of the gravitational field outside an isolated system, supposed to be at once self-gravitating and slowly-moving. By slowly-moving we mean that the typical current and stress densities are small with respect to the energy density, in the sense that

$$\max\left\{\left|\frac{T^{0i}}{T^{00}}\right|, \left|\frac{T^{ij}}{T^{00}}\right|^{1/2}\right\} = O\left(\frac{1}{c}\right) , \tag{12}$$

where $1/c$ denotes (slightly abusively) the small post-Newtonian parameter. The point about (12) is that the ratio between the size of the source d and a typical wavelength of the gravitational radiation is of order $d/\lambda = O(1/c)$. Thus the domain of validity of the post-Newtonian expansion covers the source: it is given by $r < b$ where the radius b can be chosen so that $d < b = O(\lambda/c)$.

2.1 The Matching Equation

The construction of the multipole expansion is based on several technical assumptions, the crucial one being that of the consistency of the asymptotic matching between the exterior and interior fields of the isolated system. In some cases the assumptions can be proved from the properties of the exterior field $h^{\mu\nu}_{\text{ext}}$ as obtained in [23] by means of a post-Minkowskian algorithm. However, since our assumptions are free of any reference to the post-Minkowskian expansion, we prefer to state them more generally, without invoking the existence of such an approximation (refer to [25] for the full detailed assumptions). In many cases the assumptions have been explicitly verified at some low post-Newtonian orders [30,66,24,73].

The field h (skipping space-time indices), solution in $\mathbb{R}^4$ of the relaxed field equations and the no-incoming radiation condition, is given as the retarded integral (10). We now *assume* that outside the isolated system, say, in the region $r > \mathcal{R}$ where $\mathcal{R}$ is a constant radius strictly larger than d, we have $h = \mathcal{M}(h)$ where $\mathcal{M}(h)$ denotes the multipole expansion of h, a solution of the *vacuum* field equations in $\mathbb{R}^4$ deprived from the spatial origin $r = 0$, and admitting a spherical-harmonics expansion of a certain structure (see below). Thus, in $\mathbb{R} \times \mathbb{R}^3_*$ where $\mathbb{R}^3_* \equiv \mathbb{R}^3 - \{\mathbf{0}\}$,

$$\partial_\nu \mathcal{M}(h^{\mu\nu}) = 0 \; , \tag{13a}$$

$$\Box \mathcal{M}(h^{\mu\nu}) = \mathcal{M}(\Lambda^{\mu\nu}) \; . \tag{13b}$$

The source term $\mathcal{M}(\Lambda)$ is obtained from inserting $\mathcal{M}(h)$ in place of h into (5), i.e. $\mathcal{M}(\Lambda) \equiv \Lambda(\mathcal{M}(h))$. [Since the matter tensor has a compact support, $\mathcal{M}(T) = 0$ so that $\mathcal{M}(\tau) = \frac{c^4}{16\pi G}\mathcal{M}(\Lambda)$.] Of course, inside the source (when $r \leq d$), the true solution h differs from the vacuum solution $\mathcal{M}(h)$, the latter becoming in fact singular at the origin ($r = 0$). We assume that the spherical-harmonics expansion of $\mathcal{M}(h)$ in $\mathbb{R} \times \mathbb{R}^3_*$ reads

$$\mathcal{M}(h)(\mathbf{x}, t) = \sum_{a \leq N} \hat{n}_L r^a (\ln r)^p {}_L F_{a,p}(t) + R_N(\mathbf{x}, t) \; . \tag{14}$$

This expression is valid for any $N \in \mathbb{N}$. The powers of r are positive or negative, $a \in \mathbb{Z}$, and we have $a \leq N$ (the negative powers of r show that the multipole expansion is singular at $r = 0$). For ease of notation we indicate only the summation over a, but there are two other summations involved: one over the powers $p \in \mathbb{N}$ of the logarithms, and one over the order of multipolarity $l \in \mathbb{N}$. The summations are considered only in the sense of formal series, as we do not control the mathematical nature of the series. The factor $\hat{n}_L$ is a product of l unit vectors, $n_L \equiv n^L \equiv n^{i_1}...n^{i_l}$, where $L \equiv i_1...i_l$ is a multi-index with l indices, on which the symmetric and trace-free (STF) projection is applied: $\hat{n}_L \equiv \mathrm{STF}[n_L]$. The decomposition in terms of STF tensors $\hat{n}_L(\theta, \varphi)$ is equivalent to the decomposition in usual spherical harmonics. The functions ${}_L F_{a,p}(t)$ are smooth (C^∞) functions of time, which become constant when $t \leq -\mathcal{T}$ because of our assumption (9). [Of course, the ${}_L F_{a,p}$'s depend also on c: ${}_L F_{a,p}(t, c)$.] Finally the function $R_N(\mathbf{x}, t)$ is defined by continuity throughout $\mathbb{R}^4$. Its two essential properties are $R_N \in C^N(\mathbb{R}^4)$ and $R_N = O(r^N)$ when $r \to 0$ with fixed t. In addition R_N is zero before the time $-\mathcal{T}$. Though the function $R_N(\mathbf{x}, t)$ is given "globally" (as is the multipole expansion), it represents a small remainder $O(r^N)$ in the expansion of $\mathcal{M}(h)$ when $r \to 0$, which is to be identified with the "near-zone" expansion of the field outside the source. It is convenient to introduce a special notation for the formal near-zone expansion (valid to any order N):

$$\overline{\mathcal{M}(h)}(\mathbf{x},t) = \sum \hat{n}_L r^a (\ln r)^p {}_L F_{a,p}(t) \ , \tag{15}$$

where the summation is to be understood in the sense of formal series. [Note that (14) and (15) are written for the field variable $\mathcal{M}(h)$, but it is easy to check that the same type of structure holds also for the source term $\mathcal{M}(\Lambda)$.]

Our justification of the assumed structure (14) is that it has been *proved* to hold for metrics in the class of Multipolar-post-Minkowskian (MPM) metrics considered in [23], i.e. formal series $h_{\text{ext}} = \sum G^n h_n$ which satisfy the vacuum equations, are stationary in the past, and depend on a *finite* set of independent multipole moments. More precisely, from the Theorem 4.1 in [23], the general MPM metric h_{ext}, that we identify in this paper with $\mathcal{M}(h^{\mu\nu})$, is such that the property (14) holds for the h_n's to any order n, with the only difference that to any finite order n the integers a, p, l vary into some finite ranges, namely $a_{\text{min}}(n) \leq a \leq N$, $0 \leq p \leq n-1$ and $0 \leq l \leq l_{\text{max}}(n)$, with $a_{\text{min}}(n) \to -\infty$ and $l_{\text{max}}(n) \to +\infty$ when $n \to +\infty$. The functions ${}_L F_{a,p}$ and the remainder R_N in (14) should therefore be viewed as post-Minkowskian series $\sum G^n {}_L F_{a,p,n}$ and $\sum G^n R_{N,n}$. What we have done in writing (14) and (15) is to assume that one can legitimately consider such formal post-Minkowskian series. Note that because the general MPM metric represents the most general solution of the field equations outside the source (Theorem 4.2 in [23]), it is quite appropriate to identify the general multipole expansion $\mathcal{M}(h)$ with the MPM metric h_{ext}. Actually we shall justify this assumption in Sect. 5 by recovering from $\mathcal{M}(h)$, step by step in the post-Minkowskian expansion, the MPM metric h_{ext}. Because the properties are proved in [23] for any n, and because we consider the formal post-Minkowskian sum, we see that (14)-(15), viewed as if it were "exact", constitutes a quite natural assumption. In particular we have assumed in (14)-(15) that the multipolar series involves an infinite number of independent multipoles. In summary, we give to the properties (14)-(15) a scope larger than the one of MPM expansions (maybe they could be proved for exact solutions), at the price of counting them among our basic assumptions.

The multipole expansion $\mathcal{M}(h)$ is a mathematical solution of the vacuum equations in $\mathbb{R} \times \mathbb{R}^3_*$, but whose "multipole moments" (the functions ${}_L F_{a,p}$) are not determined in terms of the source parameters. When the isolated system is slowly moving in the sense of (12), there exists an overlapping region between the domains of validity of the post-Newtonian expansion: the "near-zone" $r < b$, where $d < b = O(\lambda/c)$, and of the multipole expansion: the exterior zone $r > \mathcal{R}$. For this to be true it suffices to choose $\mathcal{R}$, which is restricted only to be strictly larger than d, such that $d < \mathcal{R} < b$. We assume that the field h given by (10) admits in the near-zone a formal post-Newtonian expansion, $h = \overline{h}$ when $r < b$. On the other hand, recall that $h = \mathcal{M}(h)$ when $r > \mathcal{R}$. Matching the two asymptotic expansions $\overline{h}$ and $\mathcal{M}(h)$ in the "matching" region $\mathcal{R} < r < b$ means that the (formal) double series obtained by considering the multipole expansion of (all the coefficients of) the post-Newtonian expansion $\overline{h}$ is *identical* to the double series obtained by taking the near-zone expansion of the multipole expansion. [We use the same overbar notation for the post-Newtonian and near-zone expansions because the near-zone expansion $(r/c \to 0)$ of the exterior multipolar field is mathematically equivalent to the expansion when $c \to \infty$ with fixed multipole moments.] The resulting matching equation reads

$$\overline{\mathcal{M}(h)} = \mathcal{M}(\overline{h}) \; . \tag{16}$$

This equation should be true term by term, after both sides of the equation are re-arranged as series corresponding to the same expansion parameter. Though looking quite reasonable (if the theory makes sense), the matching equation cannot be justified presently with full generality; however up to 2PN order it was shown to determine a unique solution valid everywhere inside and outside the source [30,66,24]. The matching assumption complements the framework of MPM approximations [23], by giving physical "pith" to the arbitrary multipole moments used in the construction of MPM metrics (see Sect. 4).

2.2 The Field in Terms of Multipole Moments

Let us consider the relaxed vacuum Einstein equation (13b), whose source term $\mathcal{M}(\Lambda)$, according to our assumptions, owns the structure (14) [recall that (14) applies to $\mathcal{M}(h)$ as well as $\mathcal{M}(\Lambda)$]. We obtain a *particular* solution of this equation (in $\mathbb{R} \times \mathbb{R}^3_*$) as follows. First we multiply each term composing $\mathcal{M}(\Lambda)$ in (14) by a factor $(r/r_0)^B$, where B is a complex number and r_0 a constant with the dimension of a length. For each term we can choose the real part of B large enough so that the term becomes regular when $r \to 0$, and then we can apply the retarded integral (11). The resulting B-dependent retarded integral is known to be analytically continuable for any $B \in \mathbb{C}$ except at integer values including in general the value of interest $B = 0$. Furthermore one can show that the finite part (in short $\mathrm{FP}_{B=0}$) of this integral, defined to be the coefficient of the zeroth power of B in the expansion when $B \to 0$, is a retarded solution of the corresponding wave equation. In the case of a regular term in (14) such as the remainder R_N, this solution simply reduces to the retarded integral. Summing all these solutions, corresponding to all the separate terms in (14), we thereby obtain as a particular solution of (13b) the object $\mathrm{FP}_{B=0}\,\Box^{-1}_R[(r/r_0)^B\mathcal{M}(\Lambda)]$. This is basically the method employed in [23] to solve the vacuum field equations in the post-Minkowskian approximation.

Now all the problem is to find *the* homogeneous solution to be added to the latter particular solution in order that the multipole expansion $\mathcal{M}(h)$ matches with the post-Newtonian expansion $\overline{h}$, solution within the source of the field equation (3) [or, rather, (10)]. Finding this homogeneous solution means finding the general consequence of the matching equation (16). The result [24,25] is that the multipole expansion $h^{\mu\nu}$ satisfying the Einstein equation (10) together with the matching equation (16) reads

$$\mathcal{M}(h^{\mu\nu}) = \mathrm{FP}_{B=0}\,\Box_R^{-1}[(r/r_0)^B \mathcal{M}(\Lambda^{\mu\nu})] - \frac{4G}{c^4}\sum_{l=0}^{+\infty}\frac{(-)^l}{l!}\partial_L\left\{\frac{1}{r}\mathcal{H}_L^{\mu\nu}(t-r/c)\right\} \tag{17}$$

where the first term is the previous particular solution, and where the second term is a retarded solution of the source-free (homogeneous) wave equation, whose "multipole moments" are given explicitly by ($u \equiv t - r/c$)

$$\mathcal{H}_L^{\mu\nu}(u) = \mathrm{FP}_{B=0}\int d^3\mathbf{x}\; |\mathbf{x}/r_0|^B x_L\, \overline{\tau}^{\mu\nu}(\mathbf{x},u)\;. \tag{18}$$

Here $\overline{\tau}^{\mu\nu}$ denotes the post-Newtonian expansion of the stress-energy pseudo-tensor $\tau^{\mu\nu}$ appearing in the right side of (10). In (17) and (18) we denote $L = i_1 \dots i_l$ and $\partial_L \equiv \partial_{i_1} \dots \partial_{i_l}$, $x_L \equiv x_{i_1} \dots x_{i_l}$.

It is important that the multipole moments (18) are found to depend on the *post-Newtonian* expansion $\overline{\tau}^{\mu\nu}$ of the pseudo-tensor, and not of $\tau^{\mu\nu}$ itself, as this is precisely where our assumption of matching to the inner post-Newtonian field comes in. The formula is *a priori* valid only in the case of a slowly-moving source; it is *a priori* true only after insertion of a definite post-Newtonian expansion of the pseudo-tensor, where in particular all the retardations have been expanded when $c \to \infty$ [the formulas (17)-(18) assume implicitly that one can effectively construct such a post-Newtonian expansion].

Like in the first term of (17), the moments (18) are endowed with a finite part operation defined by complex analytic continuation in B. Notice however that the two finite part operations in the first term of (17) and in (18) act quite differently. In the first term of (17) the analytic continuation serves at regularizing the singularity of the multipole expansion at the spatial *origin* $r = 0$. Since the pseudo-tensor is smooth inside the source, there is no need in the moments (18) to regularize the field near the origin; still the finite part is essential because it applies to the bound of the integral at *infinity* ($|\mathbf{x}| \to \infty$). Otherwise the integral would be (*a priori*) divergent at infinity, because of the presence of the factor $x_L = O(r^l)$ in the integrand, and the fact that the pseudo-tensor $\overline{\tau}^{\mu\nu}$ is non-compact supported. The two finite parts present in the two separate terms of (17) involve the same arbitrary constant r_0, but this constant can be readily checked to cancel out between the two terms [i.e. the differentiation of $\mathcal{M}(h^{\mu\nu})$ with respect to r_0 yields zero].

The formulas (17)-(18) were first obtained (in STF form) up to the 2PN order in [24] by performing explicitly the matching. This showed in particular that the matching equation (16) is correct to 2PN order. Then the proof valid to any post-Newtonian order, but at the price of *assuming* (16) to all orders, was given in Sect. 3 of [25] (see also Appendix A of [25] for an alternative proof). The crucial step in the proof is to remark that the finite part of the integral of $\overline{\mathcal{M}(\Lambda)}$ over the *whole* space $\mathbb{R}^3$ is identically zero by analytic continuation:

$$\mathrm{FP}_{B=0} \int_{\mathbb{R}^3} d^3\mathbf{x}\; |\mathbf{x}/r_0|^B x_L \overline{\mathcal{M}(\Lambda)}(\mathbf{x}, u) = 0\;. \tag{19}$$

This follows from the fact that $\overline{\mathcal{M}(\Lambda)}$ can be written as a formal series of the type (15). Using (15) it is easy to reduce the computation of the integral (19) to that of the elementary radial integral $\int_0^{+\infty} d|\mathbf{x}||\mathbf{x}|^{B+2+l+a}$ (since the powers of the logarithm can be obtained by repeatedly differentiating with respect to B). The latter radial integral can be split into a "near-zone" integral, extending from zero to radius $\mathcal{R}$, and a "far-zone" integral, extending from $\mathcal{R}$ to infinity (actually any finite non-zero radius fits instead of $\mathcal{R}$). When the real part of B is a large enough positive number, the value of the near-zone integral is $\mathcal{R}^{B+3+l+a}/(B+3+l+a)$, while when the real part of B is a large *negative* number, the far-zone integral reads the opposite, $-\mathcal{R}^{B+3+l+a}/(B+3+l+a)$. Both obtained values represent the unique analytic continuations of the near-zone and far-zone integrals for any $B \in \mathbb{C}$ except $-3-l-a$. The complete integral $\int_0^{+\infty} d|\mathbf{x}||\mathbf{x}|^{B+2+l+a}$ is defined as the sum of the analytic continuations of the near-zone and far-zone integrals, and is therefore identically zero ($\forall B \in \mathbb{C}$); this proves (19).

One may ask why the whole integration over $\mathbb{R}^3$ contributes to the multipole moment (18) – a somewhat paradoxical fact because the integrand is in the form of a post-Newtonian expansion, and is thus expected to be physically valid (i.e. to give accurate results) only in the near zone. This fact is possible thanks to the technical identity (19) which enables us to transform a near-zone integration into a complete $\mathbb{R}^3$-integration (refer to [25] for details).

2.3 Equivalence with the Will–Wiseman Multipole Expansion

Recently a different expression of the multipole decomposition, with correlatively a different expression of the multipole moments, was obtained by Will and Wiseman [75], extending previous work of Epstein and Wagoner [76] and Thorne [60]. Basically, the multipole moments in [75] are defined by an integral extending over a ball of *finite* radius $\mathcal{R}$ (essentially the same $\mathcal{R}$ as here), and thus do not require any regularization of the bound at infinity. By contrast, our multipole moments (18) involve an integration over the whole $\mathbb{R}^3$, which is allowed thanks to the analytic continuation [leading to the identity (19)]. Let us outline the proof of the equivalence between the Will–Wiseman formalism [75] and the present one [24,25].

Will and Wiseman [75] find, instead of (17)-(18),

$$\mathcal{M}(h^{\mu\nu}) = \Box_R^{-1}[\mathcal{M}(\Lambda^{\mu\nu})]_{|\mathcal{R}} - \frac{4G}{c^4}\sum_{l=0}^{+\infty}\frac{(-)^l}{l!}\partial_L\left\{\frac{1}{r}\mathcal{W}_L^{\mu\nu}(t-r/c)\right\} . \quad (20)$$

The first term is given by the retarded integral (11) acting on $\mathcal{M}(\Lambda)$, but *truntated*, as indicated by the subscript $\mathcal{R}$, to extend only in the "far zone": $|\mathbf{x}'| > \mathcal{R}$ in the notation (11). Thus, the near-zone part of the retarded integral, which contains the source, is removed, and there is no problem with the singularity of the multipole expansion at the origin. Then, the multipole moments $\mathcal{W}_L$ are given by an integral extending over the "near zone" only:

$$\mathcal{W}_L^{\mu\nu}(u) = \int_{|\mathbf{x}|<\mathcal{R}} d^3\mathbf{x}\; x_L\, \overline{\tau}^{\mu\nu}(\mathbf{x},u) . \quad (21)$$

The integral being compact-supported is well-defined. The multipole moments $\mathcal{W}_L$ look technically more simple than ours given by (18). On the other hand, practically speaking, the analytic continuation in (18) permits deriving many closed-form formulas to be used in applications [72,77]. Of course, one is free to choose any definition of the multipole moments as far as it is used in a consistent manner.

We compute the difference between the moments $\mathcal{H}_L$ and $\mathcal{W}_L$. For the comparison we split $\mathcal{H}_L$ into far-zone and near-zone integrals corresponding to the radius $\mathcal{R}$. Since the analytic continuation factor in $\mathcal{H}_L$ deals only with the bound at infinity, it can be removed from the near-zone integral, which is then clearly seen to agree with $\mathcal{W}_L$. So the difference $\mathcal{H}_L - \mathcal{W}_L$ is given by the far-zone integral:

$$\mathcal{H}_L(u) - \mathcal{W}_L(u) = \mathrm{FP}_{B=0}\int_{|\mathbf{x}|>\mathcal{R}} d^3\mathbf{x}\; |\mathbf{x}/r_0|^B x_L \overline{\tau}(\mathbf{x},u) . \quad (22)$$

Next we transform the integrand. Successively we write $\overline{\tau} = \mathcal{M}(\overline{\tau})$ because we are in the far zone; $\mathcal{M}(\overline{\tau}) = \overline{\mathcal{M}(\tau)}$ from the matching equation (16); and $\overline{\mathcal{M}(\tau)} = \frac{c^4}{16\pi G}\overline{\mathcal{M}(\Lambda)}$ because T has a compact support. At this stage, the technical identity (19) allows one to transform the far-zone integration into a near zone integration (changing simply the overall sign in front of the integral). So,

$$\mathcal{H}_L(u) - \mathcal{W}_L(u) = -\frac{c^4}{16\pi G}\mathrm{FP}_{B=0}\int_{|\mathbf{x}|<\mathcal{R}} d^3\mathbf{x}\; |\mathbf{x}/r_0|^B x_L \overline{\mathcal{M}(\Lambda)}(\mathbf{x},u) . \quad (23)$$

It is straightforward to check that the right side of this equation, when summed up over all multipolarities l, accounts exactly for the near-zone part that was removed from the retarded integral of $\mathcal{M}(\Lambda)$ [first term in (20)], so that the "complete" retarded integral as given by the first term in (17) is exactly reconstituted. In conclusion the two formalisms [24,25] and [75] are equivalent.

3 Source Multipole Moments

Quite naturally our source multipole moments will be closely related to the $\mathcal{H}_L$'s obtained in (18). However, before giving a precise definition, we need to find the equivalent of the multipole decomposition (17)-(18) in terms of symmetric and trace-free (STF) tensors, and we must reduce the number of independent tensors by imposing the harmonic gauge condition (13a). This leads to the definition of a "linearized" metric associated with the multipole expansion $\mathcal{M}(h)$, and parametrized by six sets of STF source multipole moments.

3.1 Multipole Expansion in Symmetric Trace-Free Form

The moments $\mathcal{H}_L$ given by (18) are non-trace-free because x_L owns all its traces (i.e. $\delta_{i_l i_{l-1}} x_L = \mathbf{x}^2 x_{L-2}$, where $L-2 = i_1 ... i_{l-2}$). Instead of $\mathcal{H}_L$, there are certain advantages in using STF multipole moments: indeed the STF moments are uniquely defined, and they often yield simpler computations in practice. It is not difficult, using STF techniques, to obtain the multipole decomposition equivalent to (17)-(18) but expressed in terms of STF tensors. We find

$$\mathcal{M}(h^{\mu\nu}) = \mathrm{FP}_{B=0} \Box_R^{-1}[(r/r_0)^B \mathcal{M}(\Lambda^{\mu\nu})] - \frac{4G}{c^4} \sum_{l=0}^{+\infty} \frac{(-)^l}{l!} \partial_L \left\{ \frac{1}{r} \mathcal{F}_L^{\mu\nu}(t - r/c) \right\} \tag{24}$$

where the STF multipole moments are given by [25]

$$\mathcal{F}_L^{\mu\nu}(u) = \mathrm{FP}_{B=0} \int d^3\mathbf{x}\ |\mathbf{x}/r_0|^B \hat{x}_L \int_{-1}^{1} dz\ \delta_l(z) \overline{\tau}^{\mu\nu}(\mathbf{x}, u + z|\mathbf{x}|/c)\ . \tag{25}$$

The notation for a STF product of vectors is $\hat{x}_L \equiv \mathrm{STF}(x_L)$ (such that $\hat{x}_L$ is symmetric in L and $\delta_{i_l i_{l-1}} \hat{x}_L = 0$; for instance $\hat{x}_{ij} = x_i x_j - \frac{1}{3}\delta_{ij}\mathbf{x}^2$). As we see, the STF moments (25) involve an extra integration, over the variable z, with respect to the non-STF ones (18). The weighting function associated with the z-integration reads, for any l,

$$\delta_l(z) = \frac{(2l+1)!!}{2^{l+1} l!}(1 - z^2)^l\ ; \qquad \int_{-1}^{1} dz\ \delta_l(z) = 1\ . \tag{26}$$

In the limit of large l the weighting function tends toward the Dirac delta measure (hence its name): $\lim_{l\to\infty} \delta_l = \delta$. Remark that since (25) is valid only in the post-Newtonian approximation, the z-integration is to be expressed as a post-Newtonian series. Here is the relevant formula [30]:

$$\int_{-1}^{1} dz\ \delta_l(z)\overline{\tau}(\mathbf{x}, u + z|\mathbf{x}|/c) = \sum_{k=0}^{\infty} \frac{(2l+1)!!}{2^k k!(2l+2k+1)!!} \left(\frac{|\mathbf{x}|}{c}\frac{\partial}{\partial u}\right)^{2k} \overline{\tau}(\mathbf{x}, u) \ . \tag{27}$$

In the limiting case of linearized gravity, one can neglect the first term in (24), and the pseudo-tensor $\overline{\tau}^{\mu\nu}$ in (25) can be replaced by the matter stress-energy tensor $T^{\mu\nu}$ (we have $\overline{T}^{\mu\nu} = T^{\mu\nu}$ inside the slowly-moving source). Since $T^{\mu\nu}$ has a compact support the finite part prescription can be removed, and we recover the known multipole decomposition corresponding to a compact-support source (see the appendix B in [30]).

3.2 Linearized Approximation to the Exterior Field

Up to now we have solved the *relaxed* field equation (10) in the exterior zone, with result the multipole decomposition (24)-(25). In this section we further impose the harmonic gauge condition (13a), and from this we find a solution of the linearized vacuum equation, appearing as the first approximation in a post-Minkowskian expansion of the multipole expansion $\mathcal{M}(h)$.

Let us give a notation to the first term in (24):

$$u^{\mu\nu} \equiv \mathrm{FP}_{B=0}\, \Box_R^{-1}[(r/r_0)^B \mathcal{M}(\Lambda^{\mu\nu})] \ . \tag{28}$$

Applying on (24) the condition $\partial_\nu \mathcal{M}(h^{\mu\nu}) = 0$, we find that the divergence $w^\mu \equiv \partial_\nu u^{\mu\nu}$ is equal to a retarded solution of the source-free wave equation, given by

$$w^\mu = \frac{4G}{c^4}\partial_\nu \left(\sum_{l=0}^{+\infty} \frac{(-)^l}{l!} \partial_L \left\{ \frac{1}{r} \mathcal{F}_L^{\mu\nu}(t - r/c) \right\} \right) \ . \tag{29}$$

Now, associated to any w^μ of this type, there exists some $v^{\mu\nu}$ which is like w^μ a retarded solution of the source-free wave equation, $\Box(v^{\mu\nu}) = 0$, and furthermore whose divergence is the opposite of w^μ, $\partial_\nu v^{\mu\nu} = -w^\mu$. We refer to [23,70] for the explicit formulas allowing the "algorithmic" construction of $v^{\mu\nu}$ once we know w^μ. For definiteness, we adopt the formulas (2.12) in [70], which represent themselves a slight modification of the earlier formulas (4.13) in [23] (see also the appendix B in [25]).

With $v^{\mu\nu}$ at our disposal we define what constitutes the linearized approximation to the exterior metric, say $Gh_1^{\mu\nu}$ where we factorize out G in front of the metric in order to emphasize its linear character:

$$Gh_1^{\mu\nu} \equiv -\frac{4G}{c^4} \sum_{l=0}^{+\infty} \frac{(-)^l}{l!} \partial_L \left\{ \frac{1}{r} \mathcal{F}_L^{\mu\nu}(t - r/c) \right\} - v^{\mu\nu} \ . \tag{30}$$

The linearized metric satisfies the linearized vacuum equations in harmonic gauge: $\Box h_1^{\mu\nu} = 0$ since both terms in (30) satisfy the source-free wave equation, and $\partial_\nu h_1^{\mu\nu} = 0$ thanks to (29) and $\partial_\nu v^{\mu\nu} = -w^\mu$. Using the definition (30) one can re-write the multipole expansion of the exterior field as

$$\mathcal{M}(h^{\mu\nu}) = G h_1^{\mu\nu} + u^{\mu\nu} + v^{\mu\nu} \,. \tag{31}$$

Quite naturally the $u^{\mu\nu}$ and $v^{\mu\nu}$ will represent the *non-linear* corrections to be added to the "linearized" metric $Gh_1^{\mu\nu}$ in order to reconstruct the complete exterior metric (see Sect. 4).

Since $h_1^{\mu\nu}$ satisfies $\Box h_1^{\mu\nu} = 0 = \partial_\nu h_1^{\mu\nu}$, there is a unique way to decompose it into the sum of a "canonical" metric introduced by Thorne [60] (see also [23]) plus a linearized gauge transformation,

$$h_1^{\mu\nu} = h_{\mathrm{can}1}^{\mu\nu} + \partial^\mu \varphi_1^\nu + \partial^\nu \varphi_1^\mu - \eta^{\mu\nu}\partial_\lambda \varphi_1^\lambda \,. \tag{32}$$

The canonical linearized metric is defined by

$$h_{\mathrm{can}1}^{00} = -\frac{4}{c^2}\sum_{l\geq 0}\frac{(-)^l}{l!}\partial_L\left(\frac{1}{r}I_L(u)\right) \,, \tag{33a}$$

$$h_{\mathrm{can}1}^{0i} = \frac{4}{c^3}\sum_{l\geq 1}\frac{(-)^l}{l!}\left\{\partial_{L-1}\left(\frac{1}{r}I^{(1)}_{iL-1}(u)\right) + \frac{l}{l+1}\varepsilon_{iab}\partial_{aL-1}\left(\frac{1}{r}J_{bL-1}(u)\right)\right\} \,, \tag{33b}$$

$$h_{\mathrm{can}1}^{ij} = -\frac{4}{c^4}\sum_{l\geq 2}\frac{(-)^l}{l!}\left\{\partial_{L-2}\left(\frac{1}{r}I^{(2)}_{ijL-2}(u)\right) + \frac{2l}{l+1}\partial_{aL-2}\left(\frac{1}{r}\varepsilon_{ab(i}J^{(1)}_{j)bL-2}(u)\right)\right\} \,, \tag{33c}$$

where the I_L's and J_L's are two sets of functions of the retarded time $u = t-r/c$ [the subscript (n) indicates n time derivatives], and which are STF with respect to all their indices $L = i_1 \dots i_l$ (the symmetrization is denoted with parenthesis). As for the gauge vector φ_1^μ, it satisfies $\Box\varphi_1^\mu = 0$ and depends in a way similar to (33) on four other sets of STF functions of u, denoted W_L, X_L, Y_L and Z_L (one type of function for each component of the vector). See [25] for the expression of $\varphi_1^\mu = \varphi_1^\mu[W_L, X_L, Y_L, Z_L]$.

3.3 Derivation of the Source Multipole Moments

The two sets of multipole moments I_L and J_L parametrizing the metric (33) constitute our definitions for respectively the mass-type and current-type multipole moments of the source. Actually, there are also the moments W_L,

X_L, Y_L, Z_L, and we refer collectively to $\{I_L, J_L, W_L, X_L, Y_L, Z_L\}$ as the set of six *source* multipole moments.

With (32) it is easily seen (because $\Box\varphi_1^\mu = 0$) that the gauge condition $\partial_\nu h_1^{\mu\nu} = 0$ imposes no condition on the source moments except the conservation laws appropriate to the gravitational monopole I (having $l = 0$) and dipoles I_i, J_i ($l = 1$): namely,

$$I^{(1)} = 0 \; ; \qquad I_i^{(2)} = 0 \; ; \qquad J_i^{(1)} = 0 \; . \tag{34}$$

The mass monopole I and current dipole J_i are thus constant, and agree respectively with the ADM mass and total angular momentum of the isolated system (later we shall denote the ADM mass by $M \equiv I$). According to (34) the mass dipole I_i is a linear function of time, but since we assumed that the metric is stationary in the past, I_i is in fact also constant, and equal to the (ADM) center of mass position.

The expressions of I_L and J_L (as well as of the other moments W_L, X_L, Y_L, Z_L) come directly from (30) with (32)-(33) and the result of the matching, which is personified by the formula (25). To simplify the notation we define

$$\Sigma \equiv \frac{\overline{\tau}^{00} + \overline{\tau}^{ii}}{c^2} \; , \tag{35a}$$

$$\Sigma_i \equiv \frac{\overline{\tau}^{0i}}{c} \; , \tag{35b}$$

$$\Sigma_{ij} \equiv \overline{\tau}^{ij} \; , \tag{35c}$$

(where $\overline{\tau}^{ii} \equiv \delta_{ij}\overline{\tau}^{ij}$). The result is [25]

$$I_L(u) = \mathrm{FP}_{B=0} \int d^3\mathbf{x}\; |\mathbf{x}/r_0|^B \int_{-1}^{1} dz \Big\{ \delta_l \hat{x}_L \Sigma - \frac{4(2l+1)}{c^2(l+1)(2l+3)} \delta_{l+1} \hat{x}_{iL} \partial_t \Sigma_i + \frac{2(2l+1)}{c^4(l+1)(l+2)(2l+5)} \delta_{l+2} \hat{x}_{ijL} \partial_t^2 \Sigma_{ij} \Big\} (\mathbf{x}, u + z|\mathbf{x}|/c) \; , \tag{36a}$$

$$J_L(u) = \varepsilon_{ab<i_l} \mathrm{FP}_{B=0} \int d^3\mathbf{x}\; |\mathbf{x}/r_0|^B \int_{-1}^{1} dz \Big\{ \delta_l \hat{x}_{L-1>a} \Sigma_b - \frac{2l+1}{c^2(l+2)(2l+3)} \delta_{l+1} \hat{x}_{L-1>ac} \partial_t \Sigma_{bc} \Big\} (\mathbf{x}, u + z|\mathbf{x}|/c) \; , \tag{36b}$$

(<> refers to the STF projection). In a sense these expressions are *exact*, since they are formally valid up to any post-Newtonian order. [See (68)-(69) below for explicit formulas at 2PN.]

By replacing $\overline{\tau}^{\mu\nu}$ in (36) by the compact-support matter tensor $T^{\mu\nu}$ we recover the expressions of the multipole moments worked out in linearized gravity by Damour and Iyer [78] (see also [79]). On the other hand the formulas (36) contain the results obtained by explicit implementation ("order by order") of the matching up to the 2PN order [24].

4 Post-Minkowskian Approximation

In linearized gravity, the source multipole moments represent also the moments which are "measured" at infinity, using an array of detectors surrounding the source. However, in the non-linear theory, the gravitational source $\Lambda^{\mu\nu}$ cannot be neglected and the first term in (24) plays a crucial role, notably it implies that the measured multipole moments at infinity differ from the source moments. Thus, we must now supplement the formulas of the source multipole moments (36) by the study of the "non-linear" term $u^{\mu\nu} \equiv \mathrm{FP}_{B=0}\Box_R^{-1}[(r/r_0)^B \mathcal{M}(\Lambda^{\mu\nu})]$ in (24). For this purpose we develop following [23] a post-Minkowskian approximation for the exterior vacuum metric.

4.1 Multipolar Post-Minkowskian Iteration of the Exterior Field

The work started already with the formulas (31)-(33), where we expressed the exterior multipolar metric $h^{\mu\nu}_{\mathrm{ext}} \equiv \mathcal{M}(h^{\mu\nu})$ as the sum of the "linearized" metric $Gh_1^{\mu\nu}$ and the "non-linear" corrections $u^{\mu\nu}$, given by (28), and $v^{\mu\nu}$, algorithmically constructed from $w^\mu = \partial_\nu u^{\mu\nu}$ [see (29)]. The linearized metric is a functional of the source multipole moments: $h_1 = h_1[I, J, W, X, Y, Z]$. We regard G as the book-keeping parameter for the post-Minkowskian series, and consider that Gh_1 is purely of first order in G, and thus that h_1 itself is purely of zeroth order. Of course we know from the previous section that this is untrue, because the source multipole moments depend on G; supposing $h_1 = O(G^0)$ is simply a convention allowing the systematic implementation of the post-Minkowskian iteration.

Here we check that the non-linear corrections $u^{\mu\nu}$ and $v^{\mu\nu}$ in (31) generate the whole post-Minkowskian algorithm of [23]. The detail demanding attention is how the post-Minkowskian expansions of $u^{\mu\nu}$ and $v^{\mu\nu}$ are related to a splitting of the gravitational source $\Lambda^{\mu\nu}$ into successive non-linear terms. Let us pose, with obvious notation,

$$\Lambda^{\mu\nu} = N^{\mu\nu}[h,h] + M^{\mu\nu}[h,h,h] + O(h^4) \ , \tag{37}$$

where, from the exact formula (5), the quadratic-order piece reads (all indices being lowered with the Minkowski metric, and h denoting $\eta^{\rho\sigma}h_{\rho\sigma}$):

$$\begin{aligned} N^{\mu\nu}[h,h] = &- h^{\rho\sigma}\partial^2_{\rho\sigma}h^{\mu\nu} + \frac{1}{2}\partial^\mu h_{\rho\sigma}\partial^\nu h^{\rho\sigma} - \frac{1}{4}\partial^\mu h \partial^\nu h \\ &- \partial^\mu h_{\rho\sigma}\partial^\rho h^{\nu\sigma} - \partial^\nu h_{\rho\sigma}\partial^\rho h^{\mu\sigma} + \partial_\sigma h^{\mu\rho}(\partial^\sigma h^\nu_\rho + \partial_\rho h^{\nu\sigma}) \\ &+ \eta^{\mu\nu}\left[-\frac{1}{4}\partial_\lambda h_{\rho\sigma}\partial^\lambda h^{\rho\sigma} + \frac{1}{8}\partial_\rho h \partial^\rho h + \frac{1}{2}\partial_\rho h_{\sigma\lambda}\partial^\sigma h^{\rho\lambda}\right] \ , \end{aligned} \tag{38}$$

and where the cubic-order piece $M[h,h,h]$ and all higher-order terms can be obtained in a straightforward way.

First, reasoning *ad absurdio*, we prove (see [25] for details) that both u and v indeed represent non-linear corrections to the linearized metric since they start at order G^2: $u = G^2 u_2 + O(G^3)$ and $v = G^2 v_2 + O(G^3)$. Next we obtain explicitly u_2 by substituting the linearized metric h_1 into (38) and applying the finite part of the retarded integral, i.e.

$$u_2^{\mu\nu} = \mathrm{FP}_{B=0}\Box_R^{-1}\left\{(r/r_0)^B N^{\mu\nu}[h_1, h_1]\right\} . \tag{39}$$

In this way we have a particular solution of the wave equation in $I\!R \times I\!R^3_*$, $\Box u_2 = N[h_1, h_1]$. From u_2 one deduces v_2 by the same "algorithmic" equations as used when deducing v from u [see after (29)]. Then $\Box v_2 = 0$ and the sum $u_2 + v_2$ is divergenceless, so we can solve the quadratic-order vacuum equations in harmonic coordinates by posing

$$h_2^{\mu\nu} = u_2^{\mu\nu} + v_2^{\mu\nu} . \tag{40}$$

With this definition it is clear that the multipole expansion (31) reads to quadratic order:

$$\mathcal{M}(h^{\mu\nu}) = G h_1^{\mu\nu} + G^2 h_2^{\mu\nu} + O(G^3) . \tag{41}$$

Continuing in this fashion to the next order we find successively

$$u_3^{\mu\nu} = \mathrm{FP}_{B=0}\Box_R^{-1}\left\{(r/r_0)^B\Big(M^{\mu\nu}[h_1, h_1, h_1] + N^{\mu\nu}[h_1, h_2] + N^{\mu\nu}[h_2, h_1]\Big)\right\}; \tag{42a}$$

$$h_3^{\mu\nu} = u_3^{\mu\nu} + v_3^{\mu\nu}; \tag{42b}$$

$$\mathcal{M}(h^{\mu\nu}) = G h_1^{\mu\nu} + G^2 h_2^{\mu\nu} + G^3 h_3^{\mu\nu} + O(G^4) . \tag{42c}$$

This process continues *ad infinitum*. The latter post-Minkowskian algorithm is exactly the one proposed in [23] (see also Sect. 2 of [70]). That is, starting from $h_1[I, J, W, X, Y, Z]$ given by (32)-(33), one generates the infinite post-Minkowskian (MPM) series of [23], solving the vacuum (harmonic-coordinate) Einstein equations in $I\!R \times I\!R^3_*$, and this formal series happens to be equal, term by term in G, to the *general* multipole decomposition of $h^{\mu\nu}$ given by (24). For any n, we have $h_n^{\mu\nu} = u_n^{\mu\nu} + v_n^{\mu\nu}$, and

$$\mathcal{M}(h^{\mu\nu}) = \sum_{n=1}^{+\infty} G^n h_n^{\mu\nu} . \tag{43}$$

This result is perfectly consistent with the fact that the MPM algorithm generates the most *general* solution of the field equations in $I\!R \times I\!R^3_*$. Furthermore, the latter post-Minkowskian approximation is known [26] to be reliable (existence of a one-parameter family of exact solutions whose Taylor expansion when $G \to 0$ reproduces the approximation) – an interesting result which indicates that the multipole decomposition $\mathcal{M}(h)$ given by (24)-(25) might be proved within a context of exact solutions.

Recall that the source multipole moments I_L, J_L, W_L, X_L, Y_L, Z_L entering the linearized metric h_1 at the basis of the post-Minkowskian algorithm are given by formulas like (36). Thus, in the present formalism, the source moments, including formally all post-Newtonian corrections [and all possible powers of G] as contained in (36), serve as "seeds" for the post-Minkowskian iteration of the exterior field, which as it stands leads to all possible non-linear interactions between the moments. As we can imagine, rapidly the formalism becomes extremely complicated when going to higher and higher post-Minkowskian and/or post-Newtonian approximations. Most likely the complexity is not due to the formalism but reflects the complexity of the field equations. It is probably impossible to find a different formalism in which things would be much simpler (except if one restricts to a particular type of source).

4.2 The "Canonical" Multipole Moments

The previous post-Minkowskian algorithm started with h_1, a functional of *six* types of source multipole moments, I_L and J_L entering the "canonical" linearized metric $h_{\mathrm{can}1}$ given by (33), and W_L, X_L, Y_L, Z_L parametrizing the gauge vector φ_1 in (32). All these moments deserve their name of source moments, but clearly the moments W_L, X_L, Y_L and Z_L do not play a physical role at the level of the linearized approximation, as they simply parametrize a linear gauge transformation. But because the theory is covariant with respect to (non-linear) diffeomorphisms and not merely to linear gauge transformations, these moments do contribute to physical quantities at the non-linear level.

In practice, the presence of the moments W_L, X_L, Y_L, Z_L complicates the post-Minkowskian iteration. Fortunately one can take advantage of the fact (proved in [23]) that it is always possible to parametrize the vacuum metric by means of two and only two types of multipole moments M_L and S_L (different from I_L and J_L). The metric is then obtained by the same post-Minkowskian algorithm as in (39)-(43), but starting with the "canonical" linearized metric $h_{\mathrm{can}1}[M,S]$ instead of $h_1[I,J,W,X,Y,Z]$. The resulting non-linear metric h_{can} is isometric to our exterior metric $h_{\mathrm{ext}} \equiv \mathcal{M}(h)$, provided that the moments M_L and S_L are given in terms of the source moments $I_L, J_L, \ldots, Z_L$ by some specific relations

$$M_L = M_L[I,J,W,X,Y,Z] \;, \tag{44a}$$

$$S_L = S_L[I,J,W,X,Y,Z] \;. \tag{44b}$$

The two coordinate systems in which h_{can} and h_{ext} are defined satisfy the harmonic gauge condition in the exterior zone, but (probably) only the one associated with h_{ext} meshes with the harmonic coordinates in the interior zone. With the notation (32) the coordinate change reads $\delta x^\mu = G\varphi_1^\mu +$ non-linear corrections. We shall refer to the moments M_L and S_L as the mass-type and current-type *canonical* multipole moments. Of course, since at the linearized approximation the only "physical" moments are I_L and J_L, we have

$$M_L = I_L + O(G) \; , \tag{45a}$$

$$S_L = J_L + O(G) \; , \tag{45b}$$

where $O(G)$ denotes the post-Minkowskian corrections. Furthermore, it can be shown [73] that in terms of a post-Newtonian expansion the difference between both types of moments is very small: 2.5PN order, i.e.

$$M_L = I_L + O\left(\frac{1}{c^5}\right) \tag{46}$$

[note that $M = M_{\text{ADM}} = I$]. Thus, from (46), the canonical moments are only "slightly" different from the source moments. Their usefulness is merely practical – in general they are used in place of the source moments to simplify a computation.

4.3 Retarded Integral of a Multipolar Extended Source

The previous post-Minkowskian algorithm has only theoretical interest unless we supply it with some *explicit* formulas for the computation of the coefficients h_n. Happily for us pragmatists, such formulas exist, and can be found in a rather elegant way thanks to the process of analytic continuation. Basically we need the retarded integral of an extended (non-compact-support) source with a definite multipolarity l. Here we present three exemplifying formulas; see the appendices A in [70] and [71] for more discussion.

Very often we meet a wave equation whose source term is of the type $\hat{n}_L F(t-r/c)/r^k$, where $\hat{n}_L$ has multipolarity l and F denotes a certain product of multipole moments. [Clearly, the near-zone expansion of such a term is of the form (15).] When the power k is such that $3 \leq k \leq l+2$ (this excludes the scalar case $l=0$), we obtain the solution of the wave equation as [23,68]

$$\text{FP}_{B=0}\Box_R^{-1}\left[(r/r_0)^B \frac{\hat{n}_L}{r^k} F(t-r/c)\right] = -\frac{(k-3)!(l+2-k)!}{(l+k-2)!}\hat{n}_L \times \sum_{j=0}^{k-3} 2^{k-3-j} \frac{(l+j)!}{j!(l-j)!} \frac{F^{(k-3-j)}(t-r/c)}{c^{k-3-j}\, r^{j+1}} \; . \tag{47}$$

As we see the (finite part of the) retarded integral depends in this case on the values of the extended source at the *same* retarded time $t - r/c$ (for simplicity we use the same notation for the source and field points). But it is well known (see e.g. [80,81]) that this feature is exceptional; in most cases the retarded integral depends on the whole integrated past of the source. A chief example of such a "hereditary" character is the case with $k = 2$ in the previous example, for which we find [68,69]

$$\Box_R^{-1}\left[\frac{\hat{n}_L}{r^2}F(t-r/c)\right] = -\frac{\hat{n}_L}{r}\int_{-\infty}^{ct-r} ds F(s/c) Q_l\left(\frac{ct-s}{r}\right) \tag{48}$$

where Q_l denotes the Legendre function of the second kind, related to the usual Legendre polynomial P_l by the formula

$$Q_l(x) = \frac{1}{2}P_l(x)\ln\left(\frac{x+1}{x-1}\right) - \sum_{j=1}^{l}\frac{1}{j}P_{l-j}(x)P_{j-1}(x) \; . \tag{49}$$

Since the retarded integral (48) is in fact convergent when $r \to 0$, we have removed the factor $(r/r_0)^B$ and finite part prescription. When the source term itself is given by a "hereditary" expression such as the right side of (48), we get a more complicated but still manageable formula, for instance [71]

$$\Box_R^{-1}\left[\frac{\hat{n}_L}{r^2}\int_{-\infty}^{ct-r} ds F(s/c) Q_p\left(\frac{ct-s}{r}\right)\right] = \frac{c\hat{n}_L}{r}\int_{-\infty}^{ct-r} ds F^{(-1)}(s/c) R_{lp}\left(\frac{ct-s}{r}\right) \tag{50}$$

where $F^{(-1)}$ denotes that anti-derivative of F which is zero in the past [from (9) we have restricted F to be zero in the past], and where

$$R_{lp}(x) = Q_l(x)\int_1^x dy\, Q_p(y)\frac{dP_l}{dy}(y) + P_l(x)\int_x^{+\infty} dy\, Q_p(y)\frac{dQ_l}{dy}(y) \; . \tag{51}$$

Like in (48) we do not need a finite part operation. The function R_{lp} is well-defined thanks to the behaviour of the Legendre function at infinity: $Q_l(x) \sim 1/x^{l+1}$ when $x \to \infty$.

The formulas (48)-(51) are needed to investigate the so-called tails of gravitational waves appearing at quadratic non-linear order, and even the tails generated by the tails themselves ("tails of tails") which arise at cubic order [69,71]. (These formulas do not show a dependence on the constant r_0, but other formulas do.)

5 Radiative Multipole Moments

In Sect. 2 we introduced the *definition* of a set of multipole moments $\{I_L, J_L, W_L, X_L, Y_L, Z_L\}$ for the isolated source, and in Sect. 3 we showed that the exterior field, and in particular the asymptotic field therein, is actually a complicated non-linear functional of the latter moments. Therefore, to define some source multipole moments is not sufficient by itself; this must be completed by a study of the relation between the adopted definition and some convenient far-field observables. The same is true of other definitions of source moments in different formalisms, such as in the Dixon local description of extended bodies [82–84], which should be completed by a connection to the far-zone gravitational field, for instance along the line proposed by [85,86] in the case of the Dixon moments. In the present formalism, the connection rests on the relation between the so-called *radiative* multipole moments, denoted U_L and V_L, and the source moments I_L, J_L,..., Z_L [in fact, for simplicity's sake, we prefer using the two moments M_L and S_L instead of the more basic six source moments].

5.1 Definition and General Structure

The radiative moments U_L (mass-type) and V_L (current-type) are the coefficients of the multipolar decomposition of the leading $1/R$ part of the transverse-tracefree (TT) projection of the radiation field in radiative coordinates $(T,\mathbf{X})$ (with $R = |\mathbf{X}|$ the radial distance to the source). Radiative coordinates are such that the metric coefficients admit an expansion when $R \to \infty$ in powers of $1/R$ (no logarithms of R). In radiative coordinates the retarded time $T - R/c$ is light-like, or becomes asymptotically light-like when $R \to \infty$. By *definition*,

$$h_{ij}^{TT}(\mathbf{X},T) = \frac{4G}{c^2 R}\mathcal{P}_{ijab}(\mathbf{N})\sum_{l\geq 2}\frac{1}{c^l l!}\Big\{N_{L-2}U_{abL-2} - \frac{2l}{c(l+1)}N_{cL-2}\varepsilon_{cd(a}V_{b)dL-2}\Big\} + O\left(\frac{1}{R^2}\right) , \quad (52)$$

where $N_i = X^i/R$, $N_{L-2} = N_{i_1}\dots N_{i_{l-2}}$, $N_{cL-2} = N_c N_{L-2}$, and the TT *algebraic* projector reads $\mathcal{P}_{ijab} = (\delta_{ia} - N_i N_a)(\delta_{jb} - N_j N_b) - \frac{1}{2}(\delta_{ij} - N_i N_j)(\delta_{ab} - N_a N_b)$. The radiative moments U_L and V_L depend on $T - R/c$; from (52) they are defined $\forall\, l \geq 2$. The radiative-coordinate retarded time differs from the corresponding harmonic-coordinate time by the well-known logarithmic deviation of light cones,

$$T - \frac{R}{c} = t - \frac{r}{c} - \frac{2GM}{c^3}\ln\left(\frac{r}{r_0}\right) + O(G^2) , \quad (53)$$

where we have introduced in the logarithm the same constant r_0 as in (39) (this corresponds simply to a choice of the origin of time in the far zone).

Now from the post-Minkowskian algorithm of Sect. 3, it is clear that the radiative moments U_L and V_L can be obtained to any post-Minkowskian order in principle, in the form of a non-linear series in the source or equivalently the canonical multipole moments M_L and S_L. The practical detail (worked out in [29]) is to determine the transformation between harmonic and radiative coordinates, generalizing (53) to any post-Minkowskian order. The structure of e.g. the mass-type radiative moment is

$$U_L = M_L^{(l)} + \sum_{n=2}^{+\infty} \frac{G^{n-1}}{c^{3(n-1)+2k}} X_{nL} \, . \tag{54}$$

The first term comes from the fact that the radiative moment reduces at the linearized approximation to the (lth time derivative of the) source or canonical moment. The second term represents the series of non-linear corrections, each of them is given by a certain X_{nL} which is a n-linear functional of derivatives of multipole moments M_L or S_L. Furthermore we know from e.g. (48) and (50) that each new non-linear iteration (which always involves a retarded integral) brings *a priori* a new "hereditary" integration with respect to the previous approximation. So we expect that X_{nL} is of the form ($U \equiv T - R/c$)

$$X_{nL}(U) = \sum \int_{-\infty}^{U} du_1 \dots \int_{-\infty}^{U} du_n \mathcal{Z}_n(U, u_1, \dots, u_n) M_{L_1}^{(a_1)}(u_1) \dots S_{L_n}^{(a_n)}(u_n) \tag{55}$$

where $\mathcal{Z}_n$ denotes a certain kernel depending on time variables $U, u_1, \dots, u_n$, and where the sum refers to all possibilities of coupling together the n moments. [See (56) below for examples of kernels $\mathcal{Z}_2$ and $\mathcal{Z}_3$.] A useful information is obtained from imposing that $\mathcal{Z}_n$ be dimensionless; this yields the powers of G and $1/c$ in front of each non-linear term in (54), where k is the number of contractions among the indices present on the n moments (the current moments carrying their associated Levi-Civita symbol).

As an example of application of (54) let us suppose that one is interested in the 3PN or $1/c^6$ approximation. From (54) we have $3(n-1)+2k = 6$, and we deduce that the only possibility is $n = 3$ (cubic non-linearity) and $k = 0$ (no contractions between the moments). From this we infer immediately that the only possible multipole interaction at that order is between two mass monopoles and a multipole, i.e. $M \times M \times M_L$. This corresponds to the "tails of tails" computed explicitly in (56) below.

5.2 The Radiative Quadrupole Moment to 3PN Order

To implement the formula (54) a tedious computation is to be done, following in details the post-Minkowskian algorithm of Sect. 4 augmented by explicit formulas such as (47)-(51), and changing the coordinates from harmonic to radiative according to the prescription in [29]. Here we present the result of the computation of the mass-type radiative quadrupole ($l = 2$) up to the 3PN order:

$$
\begin{aligned}
U_{ij}(U) &= M^{(2)}_{ij}(U) + 2\frac{GM}{c^3}\int_0^{+\infty} d\tau M^{(4)}_{ij}(U-\tau)\left[\ln\left(\frac{c\tau}{2r_0}\right)+\frac{11}{12}\right] \\
&+\frac{G}{c^5}\left\{-\frac{2}{7}\int_0^{+\infty} d\tau\left[M^{(3)}_{a<i}M^{(3)}_{j>a}\right](U-\tau) - \frac{2}{7}M^{(3)}_{a<i}M^{(2)}_{j>a}\right. \\
&\left.-\frac{5}{7}M^{(4)}_{a<i}M^{(1)}_{j>a} + \frac{1}{7}M^{(5)}_{a<i}M_{j>a} + \frac{1}{3}\varepsilon_{ab<i}M^{(4)}_{j>a}S_b\right\} \\
&+2\left(\frac{GM}{c^3}\right)^2\int_0^{+\infty} d\tau M^{(5)}_{ij}(U-\tau)\left[\ln^2\left(\frac{c\tau}{2r_0}\right)+\frac{57}{70}\ln\left(\frac{c\tau}{2r_0}\right)+\frac{124627}{44100}\right] \\
&+O\left(\frac{1}{c^7}\right) . \qquad (56)
\end{aligned}
$$

Recall that in this formula the moment M_{ij} is the canonical moment which agrees with the source moment I_{ij} up to a 2.5PN term [see (46)], and that the source moment I_{ij} itself is given in terms of the pseudo-tensor of the source by (36a). See also the formulas (68)-(69) below for a more explicit expression of the source moment at the 2PN order [of course, to be consistent, one should use (56) conjointly with 3PN expressions of the source moments].

The "Newtonian" term in (56) corresponds to the quadrupole formalism. Next, there is a quadratic non-linear correction with multipole interaction $M \times M_{ij}$ representing the dominant effect of tails (scattering of linear waves off the space-time curvature generated by the mass M). This correction, computed in [69], is of order $1/c^3$ or 1.5PN and has the form of a hereditary integral with logarithmic kernel. The constant 11/12 depends on the coordinate system chosen to cover the source, here the harmonic coordinates; for instance the constant would be 17/12 in Schwarzschild-like coordinates [87,88]. The next correction, of order $1/c^5$ or 2.5PN, is constituted by quadratic interactions between two mass quadrupoles, and between a mass quadrupole and a constant current dipole [70]. This term contains a hereditary integral, of a type different from the tail integral, which is due to the gravitational radiation generated by the stress-energy distribution of linear waves [89–91,69]. Sometimes this integral is referred to as the non-linear memory integral because it corresponds to the contribution of gravitons in the so-called linear memory effect [92]. The non-linear memory integral can easily be found by using the effective stress-energy tensor of gravitational waves in place of the right side of (3); it follows also from rigorous studies of the field at future null infinity [93,94]. Finally, at 3PN order in (56) appears the dominant cubic non-linear correction, corresponding to the interaction $M \times M \times M_{ij}$ and associated with the tails of tails of gravitational waves [71].

5.3 Tail Contributions in the Total Energy Flux

Observable quantities at infinity are expressible in terms of the radiative mass and current multipole moments. For instance the total gravitational-wave power emitted in all spatial directions (total gravitational flux or "luminosity" $\mathcal{L}$) is given by the positive-definite multipolar series

$$\mathcal{L} = \sum_{l=2}^{+\infty} \frac{G}{c^{2l+1}} \left\{ \frac{(l+1)(l+2)}{l(l-1)l!(2l+1)!!} U_L^{(1)} U_L^{(1)} + \frac{4l(l+2)}{c^2(l-1)(l+1)!(2l+1)!!} V_L^{(1)} V_L^{(1)} \right\} . \quad (57)$$

In the case of inspiralling compact binaries (a most prominent source of gravitational waves) the rate of inspiral is fixed by the flux $\mathcal{L}$, which is therefore a crucial quantity to predict. Excitingly enough, we know that $\mathcal{L}$ should be predicted to 3PN order for detection and analysis of inspiralling binaries in future experiments [95,96].

To 3PN order we can use the relation (56) giving the 3PN radiative quadrupole moment. Here we concentrate our attention on tails and tails of tails. The dominant tail contribution at 1.5PN order yields correspondingly a contribution in the total flux (with $U = T - R/c$):

$$\mathcal{L}_{\text{tail}} = \frac{4G^2M}{5c^8} I_{ij}^{(3)}(U) \int_0^{+\infty} d\tau I_{ij}^{(5)}(U-\tau) \left[\ln\left(\frac{c\tau}{2r_0}\right) + \frac{11}{12} \right] . \quad (58)$$

Since we are interested in the dominant tail we have replaced using (46) the canonical mass quadrupole by the source quadrupole. Similarly there are some tail contributions due to the mass octupole, current quadrupole and all higher-order multipoles, but these are correlatively of higher post-Newtonian order [see the factors $1/c$ in (57)]. It has been shown [68] that the work done by the dominant "hereditary" contribution in the radiation reaction force within the source – which arises at 4PN order in the equations of motion – agrees exactly with (58).

Next, because $\mathcal{L}$ is made of squares of (derivatives of) radiative moments, it contains a term with the square of the tail integral at 1.5PN. This term arises at the 3PN relative order and reads

$$\mathcal{L}_{(\text{tail})^2} = \frac{4G^3M^2}{5c^{11}} \left(\int_0^{+\infty} d\tau I_{ij}^{(5)}(U-\tau) \left[\ln\left(\frac{c\tau}{2r_0}\right) + \frac{11}{12} \right] \right)^2 . \quad (59)$$

Finally, there is also the direct 3PN contribution of tails of tails in (56):

$$\mathcal{L}_{\text{tail(tail)}} = \frac{4G^3M^2}{5c^{11}} I_{ij}^{(3)}(U) \int_0^{+\infty} d\tau I_{ij}^{(6)}(U-\tau) \times \left[\ln^2\left(\frac{c\tau}{2r_0}\right) + \frac{57}{70} \ln\left(\frac{c\tau}{2r_0}\right) + \frac{124627}{44100} \right] . \quad (60)$$

By a control of all the hereditary integrals in $\mathcal{L}$ up to 3PN we have checked [71] that the terms (59)-(60) do exist. The two contributions (59) and (60) appear somewhat on the same footing – of course both should be taken into account in practical computations. Note that in a physical situation where the emission of radiation stops after a certain date, in the sense that the source multipole moments become constant after this date (assuming a consistent matter model which would do this at a given post-Newtonian order), the only contribution to $\mathcal{L}$ which survives after the end of emission is the 3PN tail-square contribution (59).

6 Post-Newtonian Approximation

In Sects. 2 and 3 we have reasoned upon the formal post-Newtonian expansion $\overline{h}^{\mu\nu}$ of the near-zone field to obtain the source multipole moments as functionals of the post-Newtonian pseudo-tensor $\overline{\tau}^{\mu\nu}$. We have also considered in Sects. 4 and 5 the formal expansion $c \to \infty$ of the radiation field when holding the multipole moments fixed. Clearly missing in this scheme is an *explicit* algorithm for the computation of $\overline{h}^{\mu\nu}$ in the near zone. No such algorithm (say, in the spirit of the post-Minkowskian algorithm in Sect. 4) is known presently, but a lot is known on the first few post-Newtonian iterations [20,21,44–58,30,24].

The main difficulty in setting up a post-Newtonian algorithm is the appearance at some post-Newtonian order of divergent Poisson-like integrals. This comes from the fact that the post-Newtonian expansion is actually a near-zone expansion [44], which is valid only in the region where $r = O(\lambda/c)$, and that such an expansion blows up when taking formally the limit $r \to +\infty$. For instance, Rendall [13] has shown that the post-Newtonian solution cannot be asymptotically flat starting at the 2PN or 3PN level, depending on the gauge. This is clear from the structure of the exterior near-zone expansion (15), which involves many positive powers of the radial distance r. Thus, one is not allowed in general to consider the limit $r \to +\infty$. In consequence, using the Poisson integral for solving a Poisson equation with non-compact-support source at a given post-Newtonian order is *a priori* meaningless. Indeed the Poisson integral not only extends over the near-zone but also over the regions at infinity. This means that the Poisson integral does not constitute the correct solution of the Poisson equation in this context. However, to the lowest post-Newtonian orders it works; for instance it was shown by Kerlick [50,51] and Caporali [52] that the post-Newtonian iteration (including the suggestion by Ehlers [48,49] of an improvement with respect to previous work [55]) is well-defined up to the 2.5PN order where radiation reaction terms appear, but that some divergent integrals show up at the 3PN order.

Another difficulty is that the post-Newtonian approximation is in a sense not self-supporting, because it necessitates information coming from outside its own domain of validity. Of course we have in mind the boundary conditions at infinity which determine the radiation reaction in the source's local equations of motion. Again, to the lowest post-Newtonian orders one can circumvent this difficulty by considering *retarded* integrals that are formally expanded when $c \to \infty$ as series of "instantaneous" Poisson-like integrals [55]. However, this procedure becomes incorrect at the 4PN order, not to mention the problem of divergencies, because the near-zone field (as well as the source's dynamics) ceases to be given by an instantaneous functional of the source parameters, due to the appearance of "tail-transported" hereditary integrals modifying the lowest-order radiation reaction damping [68,32].

Let us advocate here that the cure of the latter difficulty (and perhaps of all difficulties) is the matching equation (16). Indeed suppose that one knows a particular solution of the Poisson equation at some post-Newtonian order. This solution might be in the form of some "finite part" of a Poisson integral. The correct post-Newtonian solution will be the sum of this particular solution and of a homogeneous solution satisfying the Laplace equation, namely a harmonic solution, regular at the origin, which can always be written in the form $\sum A_L \hat{x}_L$, for some unknown constant tensors A_L. The homogeneous solution is associated with radiation reaction effects. Now the matching equation states that the multipole expansion of the post-Newtonian solution agrees with the near-zone expansion of the exterior field (which has been computed beforehand in Sect. 4). The multipole expansion of the known particular solution can be obtained by a standard method, and the multipole expansion of the homogeneous solution is simply itself, i.e. $\mathcal{M}(\sum A_L \hat{x}_L) = \sum A_L \hat{x}_L$. Therefore, we see that the matching equation determines in principle the homogeneous solution (i.e. all the unknown tensors A_L), and since the exterior field satisfies relevant boundary conditions at infinity, the A_L's should correspond to the radiation reaction on a truly isolated system. See [67,68,31,32] for implementation of this method to determine the radiation reaction force to 4PN order (1.5PN relative order).

6.1 The Inner Metric to 2.5PN Order

Going to high post-Newtonian orders can become prohibitive because of the rapid proliferation of terms. Typically any allowed term (compatible dimension, correct index structure) does appear with a definite non-zero coefficient in front. However, high post-Newtonian orders can be manageable if one chooses some appropriate matter variables, and if one avoids expanding systematically the retardations due to the speed of propagation of gravity. Often it is sufficient, and clearer, to present a result in terms of matter variables still containing some c's, and perhaps also in terms of some convenient retarded potentials (being clear that any retardation going to an order higher than the prescribed post-Newtonian order of the calculation is irrelevant). See for instance (65) and (68)-(69) below. Anyway, only in a final stage, when a result to the prescribed order is in hands, should we introduce the more basic matter variables (e.g. the coordinate mass density) and perform all necessary retardations. Then of course one does not escape to a profusion of terms, but at least we have been able to carry the post-Newtonian iteration using some reasonably simple expressions.

The matter variables are chosen [30,24] in a way consistent with our earlier definitions (35), i.e.

$$\sigma \equiv \frac{T^{00}+T^{ii}}{c^2} ; \tag{61a}$$

$$\sigma_i \equiv \frac{T^{0i}}{c} ; \tag{61b}$$

$$\sigma_{ij} \equiv T^{ij} . \tag{61c}$$

To 2.5PN order one defines some *retarded* potentials V, V_i, $\hat{W}_{ij}$, $\hat{X}$ and $\hat{R}_i$, with V and V_i looking like some retarded versions of the Newtonian and gravitomagnetic potentials, and $\hat{W}_{ij}$ being associated with the matter and gravitational-field stresses:

$$V \equiv \Box_R^{-1}\{-4\pi G\sigma\} , \tag{62a}$$

$$V_i \equiv \Box_R^{-1}\{-4\pi G\sigma_i\} , \tag{62b}$$

$$\hat{W}_{ij} \equiv \Box_R^{-1}\{-4\pi G(\sigma_{ij}-\delta_{ij}\sigma_{kk}) - \partial_i V\partial_j V\} , \tag{62c}$$

$$\hat{R}_i \equiv \Box_R^{-1}\left\{-4\pi G(V\sigma_i - V_i\sigma) - 2\partial_k V\partial_i V_k - \frac{3}{2}\partial_t V\partial_i V\right\} , \tag{62d}$$

$$\hat{X} \equiv \Box_R^{-1}\left\{-4\pi G V\sigma_{ii} + 2V_i\partial_t\partial_i V + V\partial_t^2 V + \frac{3}{2}(\partial_t V)^2 - 2\partial_i V_j\partial_j V_i + \hat{W}_{ij}\partial_{ij}^2 V\right\} , \tag{62e}$$

where $\Box_R^{-1}$ denotes the retarded integral (11). All these potentials but V and V_i have a spatially non-compact support. The highest non-linearity entering them is cubic; it appears in the last term of $\hat{X}$.

Based on the latter potentials one can show [24,74] that the inner metric to order 2.5PN (in harmonic coordinates, $\partial_\nu(\sqrt{-g}g^{\mu\nu})=0$) takes the form

$$g_{00} = -1 + \frac{2}{c^2}V - \frac{2}{c^4}V^2 + \frac{8}{c^6}\left[\hat{X} + V_iV_i + \frac{V^3}{6}\right] + O\left(\frac{1}{c^8}\right) , \tag{63a}$$

$$g_{0i} = -\frac{4}{c^3}V_i - \frac{8}{c^5}\hat{R}_i + O\left(\frac{1}{c^7}\right) , \tag{63b}$$

$$g_{ij} = \delta_{ij}\left(1 + \frac{2}{c^2}V + \frac{2}{c^4}V^2\right) + \frac{4}{c^4}\hat{W}_{ij} + O\left(\frac{1}{c^6}\right) , \tag{63c}$$

(writing $\overline{g}_{\mu\nu}$ would be more consistent with the notation of Sect. 2). With this form, we believe, the computational problems encountered in applications are conveniently divided into the specific problems associated with the computation of the various potentials (62), which constitute in this approach some appropriate computational "blocks" (having of course no physical signification separately). By expanding all powers of $1/c$ present into the matter densities (61) and into the retardations of the potentials (62), we find that the metric (63) becomes extremely complicated, as it really is (see e.g. [46,47,50,51]).

Because of our use of retarded potentials, the metric (63) involves explicitly only even post-Newtonian terms (using the post-Newtonian terminology that even terms correspond to even powers of $1/c$ in the equations of motion). We have checked [24] that the *odd* post-Newtonian terms (responsible for radiation reaction), contained in (63) via the expansion of retardations, match, in the sense of the equation (16), to the exterior metric satisfying the no-incoming radiation condition (9).

The harmonic gauge condition implies some differential equations to be satisfied by the previous potentials. To 2.5PN order we find

$$\partial_t \left\{ V + \frac{1}{c^2}\left[\frac{1}{2}\hat{W}_{ii} + 2V^2\right]\right\} + \partial_i \left\{ V_i + \frac{2}{c^2}\left[\hat{R}_i + VV_i\right]\right\} = O\left(\frac{1}{c^4}\right) , \tag{64a}$$

$$\partial_t V_i + \partial_j \left\{ \hat{W}_{ij} - \frac{1}{2}\delta_{ij}\hat{W}_{kk} \right\} = O\left(\frac{1}{c^2}\right) , \tag{64b}$$

where $\hat{W}_{ii} \equiv \delta_{ij}\hat{W}_{ij}$. These equations are in turn equivalent to the equation of continuity and the equation of motion for the matter system,

$$\partial_t\sigma + \partial_i\sigma_i = \frac{1}{c^2}\left(\partial_t\sigma_{ii} - \sigma\partial_t V\right) + O\left(\frac{1}{c^4}\right) , \tag{65a}$$

$$\partial_t\sigma_i + \partial_j\sigma_{ij} = \sigma\partial_i V + O\left(\frac{1}{c^2}\right) . \tag{65b}$$

Note that the precision is 1PN for the equation of continuity but only Newtonian for the equation of motion.

6.2 The Mass-Type Source Moment to 2.5PN Order

From the 2.5PN metric (63) we obtain the pseudo-tensor $\overline{\tau}$ and the auxiliary quantities (35), that we replace into the formulas (36) to obtain the 2.5PN source multipole moments. Recall that the z-integration in the moments is to be carried out using the formula (27). Let us first see how this works at the 1PN order.

We need Σ to 1PN order and Σ_i to Newtonian order. The latter quantity reduces to the matter part, $\Sigma_i = \sigma_i + O(1/c^2)$, and the former one reads after a simple transformation

$$\Sigma = \sigma - \frac{1}{2\pi G c^2}\Delta(V^2) + O\left(\frac{1}{c^4}\right) . \tag{66}$$

The substitution into the moments I_L given by (36a) leads to

$$I_L = \mathrm{FP}_{B=0}\int d^3\mathbf{x}\ |\mathbf{x}/r_0|^B\Big\{\hat{x}_L\sigma - \frac{\hat{x}_L}{2\pi G c^2}\Delta(V^2) + \frac{|\mathbf{x}|^2\hat{x}_L}{2c^2(2l+3)}\partial_t^2\sigma - \frac{4(2l+1)\hat{x}_{iL}}{c^2(l+1)(2l+3)}\partial_t\sigma_i\Big\} + O\left(\frac{1}{c^4}\right) . \tag{67}$$

The integrand is non-compact-supported because of the contribution of the second term, and accordingly we keep the regularization factor $|\mathbf{x}/r_0|^B$ and finite part operation. But let us operate by parts the second term, using the fact that $|\mathbf{x}|^B\hat{x}_L\Delta(V^2) - \Delta(|\mathbf{x}|^B\hat{x}_L)V^2 = \partial_i\{|\mathbf{x}|^B\hat{x}_L\partial_i(V^2) - \partial_i(|\mathbf{x}|^B\hat{x}_L)V^2\}$ is a pure divergence. When the real part of B is a large *negative* number, we see thanks to the Gauss theorem that the latter divergence will not contribute to the moment, therefore by the unicity of the analytic continuation it will always yield zero contribution. Thus, using $\Delta\hat{x}_L = 0$, we can replace $|\mathbf{x}|^B\hat{x}_L\Delta(V^2)$ in the second term of (67) by $\Delta(|\mathbf{x}|^B\hat{x}_L)V^2 = B(B+l+1)|\mathbf{x}|^{B-2}\hat{x}_L V^2$, and because of the explicit factor B we see that the second term can be non-zero only in the case where the factor B multiplies an integral owning a simple pole $\sim 1/B$ due to the integration bound $|\mathbf{x}| \to \infty$. Expressing V^2 (to Newtonian order) in terms of source points $\mathbf{z}_1$ and $\mathbf{z}_2$, we obtain the integral $\int d^3\mathbf{x}\ |\mathbf{x}|^{B-2}\hat{x}_L|\mathbf{x}-\mathbf{z}_1|^{-1}|\mathbf{x}-\mathbf{z}_2|^{-1}$. When $|\mathbf{x}| \to \infty$ each $|\mathbf{x}-\mathbf{z}_{1,2}|^{-1}$ can be expanded as a series of $\hat{n}_{L_{1,2}}|\mathbf{x}|^{-l_{1,2}-1}$; then performing the angular integration shows that the sum of "multipolarities" $l+l_1+l_2$ is necessarily an even integer. When this is realized the remaining radial integral reads $\int d|\mathbf{x}|\ |\mathbf{x}|^{B+l-l_1-l_2-2}$ which develops a pole only when $l-l_1-l_2-2 = -1$. But that is incompatible with the previous finding. Thus the second term in (67) is identically zero, and we end up simply with a compact-support expression on which we no longer need to implement the finite part,

$$I_L = \int d^3\mathbf{x}\Big\{\hat{x}_L\sigma + \frac{|\mathbf{x}|^2\hat{x}_L}{2c^2(2l+3)}\partial_t^2\sigma - \frac{4(2l+1)\hat{x}_{iL}}{c^2(l+1)(2l+3)}\partial_t\sigma_i\Big\} + O\left(\frac{1}{c^4}\right) . \tag{68}$$

This expression was first obtained in [30] using a different method valid at 1PN order. Here we have recovered the same expression from the formula (36a) valid to any post-Newtonian order [24,25].

Only starting at the 2PN order does the mass multipole moment have a non-compact support (so the finite part becomes crucial at this order). By a detailed computation in [24] we arrive at the following 2PN (or rather 2.5PN) expression:

$$
\begin{aligned}
I_L(t) = \mathrm{FP}_{B=0} \int d^3\mathbf{x}\, |\mathbf{x}/r_0|^B \Bigg\{ & \hat{x}_L \left[\sigma + \frac{4}{c^4}\sigma_{ii} V\right] + \frac{|\mathbf{x}|^2 \hat{x}_L}{2c^2(2\ell+3)} \partial_t^2 \sigma \\
& + \frac{|\mathbf{x}|^4 \hat{x}_L}{8c^4(2\ell+3)(2\ell+5)} \partial_t^4 \sigma - \frac{2(2\ell+1)|\mathbf{x}|^2 \hat{x}_{iL}}{c^4(\ell+1)(2\ell+3)(2\ell+5)} \partial_t^3 \sigma_i \\
& + \frac{2(2\ell+1)\hat{x}_{ijL}}{c^4(\ell+1)(\ell+2)(2\ell+5)} \partial_t^2 \left[\sigma_{ij} + \frac{1}{4\pi G}\partial_i V \partial_j V\right] \\
& + \frac{\hat{x}_L}{\pi G c^4}\left[-\hat{W}_{ij}\partial^2_{ij} V - 2V_i \partial_t \partial_i V + 2\partial_i V_j \partial_j V_i - \frac{3}{2}(\partial_t V)^2 - V\partial_t^2 V\right] \\
& - \frac{4(2\ell+1)\hat{x}_{iL}}{c^2(\ell+1)(2\ell+3)} \partial_t \Bigg[\left(1 + \frac{4V}{c^2}\right)\sigma_i \\
& + \frac{1}{\pi G c^2}\left(\partial_k V[\partial_i V_k - \partial_k V_i] + \frac{3}{4}\partial_t V \partial_i V\right)\Bigg]\Bigg\} + O\left(\frac{1}{c^6}\right) . \qquad (69)
\end{aligned}
$$

Recall that the canonical moment M_L differs from the source moment I_L at precisely the 2.5PN order [see (46)].

7 Point-Particles

So far the post-Newtonian formalism has been developed for *smooth* (i.e. C^∞) matter distributions. As such, the source multipole moments (36) become ill-defined in the presence of singularities. We now argue that the formalism is in fact also applicable to singular sources (notably point-particles described by Dirac measures) provided that we add to our other basic assumptions a certain method for removing the infinite self-field of point-masses. Our main motivation is the inspiralling compact binary – a system of two compact objects (neutron stars or black holes) which can be described with great precision by two point-particles moving on a circular orbit, and whose orbital phase evolution should be computed prior to gravitational-wave detection with relative 3PN precision [95,96].

For this application we restrict ourselves to two point-masses m_1 and m_2 (constant Schwarzschild masses). The trajectories are $\mathbf{y}_1(t)$ and $\mathbf{y}_2(t)$ and the coordinate velocities $\mathbf{v}_{1,2} = d\mathbf{y}_{1,2}/dt$; we pose $v^\mu_{1,2} = (c, \mathbf{v}_{1,2})$. The symbol $1 \leftrightarrow 2$ means the same term but with the labels of the two particles exchanged. A model for the stress-energy tensor of point-masses (say, at 2PN order) is

$$
T^{\mu\nu}_{\text{point-mass}}(\mathbf{x}, t) = \mu_1(t) v_1^\mu(t) v_1^\nu(t) \delta[\mathbf{x} - \mathbf{y}_1(t)] + 1 \leftrightarrow 2 \; ; \qquad (70a)
$$

$$
\mu_1(t) \equiv \frac{m_1}{\sqrt{(g g_{\rho\sigma})_1 \frac{v_1^\rho v_1^\sigma}{c^2}}} \, , \qquad (70b)
$$

where δ denotes the three-dimensional Dirac measure, and $g_{\mu\nu}$ the metric coefficients in harmonic coordinates ($g \equiv \det g_{\mu\nu}$). The notation $(gg_{\mu\nu})_1$ means the value at the location of particle 1. However, due to the presence of the Dirac measure at particles 1 and 2, the metric coefficients will be singular at 1 and 2. Therefore, we must supplement the model (70) by a method of "regularization" able to give a sense to the ill-defined limit at 1 or 2. *A priori* the choice of one or another regularization constitutes a fully-qualified element of the model of point-particles. In the following we systematically employ the Hadamard regularization, based on the Hadamard "partie finie" of a divergent integral [97,98].

Let us discuss an example. The "Newtonian" potential U, defined by $U = \Delta^{-1}(-4\pi G\sigma)$, where σ is given by (61a) [we have $V = U + O(1/c^2)$], follows from (70a) as

$$U = \frac{G\mu_1}{r_1}\left[1 + \frac{v_1^2}{c^2}\right] + 1 \leftrightarrow 2 \, , \tag{71}$$

where $r_1 = |\mathbf{x} - \mathbf{y}_1|$. To Newtonian order $U = Gm_1/r_1 + O(1/c^2) + 1 \leftrightarrow 2$. We compute U at the 1PN order: from (70b) we deduce at this order $\mu_1/m_1 = 1 - (U)_1/c^2 + v_1^2/2c^2 + O(1/c^4)$, which involves U itself taken at point 1, but of course this does not make sense because U is singular at 1 and 2. Now, after applying the Hadamard regularization (described below), we obtain unambiguously the standard Newtonian result $(U)_1 = Gm_2/r_{12} + O(1/c^2)$, where $r_{12} = |\mathbf{y}_1 - \mathbf{y}_2|$, that we insert back into μ_1. So, U at 1PN, and its regularized value at 1, read

$$U = \frac{Gm_1}{r_1}\left(1 + \frac{1}{c^2}\left[-\frac{Gm_2}{r_{12}} + \frac{3}{2}v_1^2\right]\right) + O\left(\frac{1}{c^4}\right) + 1 \leftrightarrow 2 \, , \tag{72a}$$

$$(U)_1 = \frac{Gm_2}{r_{12}}\left(1 + \frac{1}{c^2}\left[-\frac{Gm_1}{r_{12}} + \frac{3}{2}v_2^2\right]\right) + O\left(\frac{1}{c^4}\right) . \tag{72b}$$

7.1 Hadamard Partie Finie Regularization

We consider the class of functions of the field point $\mathbf{x}$ which are smooth on $\mathbb{R}^3$ except at the location of the two source points $\mathbf{y}_{1,2}$, around which the functions admit some power-like expansions in the radial distance $r_1 = |\mathbf{x} - \mathbf{y}_1|$, with fixed spatial direction $\mathbf{n}_1 = (\mathbf{x} - \mathbf{y}_1)/r_1$ (and idem for 2). Thus, for any $F(\mathbf{x})$ in this class, we have

$$F = \sum_a r_1^a f_{1(a)}(\mathbf{n}_1) \qquad (\text{when } r_1 \to 0) \; ; \tag{73a}$$

$$F = \sum_a r_2^a f_{2(a)}(\mathbf{n}_2) \qquad (\text{when } r_2 \to 0) \; , \tag{73b}$$

where the summation index a ranges over values in $\mathbb{Z}$ bounded from below, $a \geq -a_0$ (we do not need to be more specific), and where the coefficients of the various powers of $r_{1,2}$ depend on the spatial directions $\mathbf{n}_{1,2}$. In (73) we do not write the remainders for the expansions because we don't need them; simply, we regard the expansions (73) as listings of the various coefficients $f_{1(a)}$ and $f_{2(a)}$. We assume also that the functions F in this class decrease sufficiently rapidly when $|\mathbf{x}| \to \infty$, so that all integrals we consider are convergent at infinity.

The integral $\int d^3\mathbf{x}F$ is in general divergent because of the singular behaviour of F near $\mathbf{y}_{1,2}$, but we can compute its partie finie (Pf) in the sense of Hadamard [97,98]. Let us consider two volumes surrounding the two singularities, of the form $r_1 \leq s\rho_1(\mathbf{n}_1)$ (and similarly for 2), where s measures the size of the volume and ρ_1 gives its shape as a function of the direction $\mathbf{n}_1$ ($\rho_1 = 1$ in the case of a spherical ball). Using (73) it is easy to determine the expansion when $s \to 0$ of the integral extending on $\mathbb{R}^3$ deprived from the two previous volumes, and then to subtract from the integral all the divergent terms when $s \to 0$ in the latter expansion. The Hadamard partie finie is defined to be the limit when $s \to 0$ of what remains. As it turns out, the result can be advantageously re-expressed in terms of an integral on $\mathbb{R}^3$ deprived from two *spherical* balls ($\rho_{1,2} = 1$), at the price of introducing two constants $s_{1,2}$ which depend on the shape of the two regularizing volumes originally considered. With full generality the Hadamard partie finie of the divergent integral reads

$$\begin{aligned}\mathrm{Pf}\int d^3\mathbf{x}\ F \equiv \lim_{s\to 0}\Bigg\{&\int_{\substack{r_1>s\\ r_2>s}} d^3\mathbf{x}\ F \\ &+ \sum_{a+3\leq -1}\frac{s^{a+3}}{a+3}\int d\Omega_1 f_{1(a)} + \ln\left(\frac{s}{s_1}\right)\int d\Omega_1 f_{1(-3)} + 1 \leftrightarrow 2\Bigg\} \qquad (74)\end{aligned}$$

where s_1 is given by

$$\ln s_1 = \frac{\int d\Omega_1 f_{1(-3)} \ln\rho_1}{\int d\Omega_1 f_{1(-3)}} \ . \qquad (75)$$

Because of the two arbitrary constants $s_{1,2}$ the Hadamard partie finie is ambiguous, and one could think *a priori* that there is no point about defining a divergent integral by means of an ambiguous expression. Actually the point is that we control the origin of these constants: they come from the coefficients of $1/r^3_{1,2}$ in the expansions of F, which generate logarithmic terms in the integral. As we shall see the constants $s_{1,2}$ do not appear in the post-Newtonian metric up to the 2.5PN order (they are expected to appear only at 3PN order).

We can also give a meaning to the value of the function F at the location of particle 1 for instance, by taking the average over all directions $\mathbf{n}_1$ of the coefficient of the zeroth power of r_1 in (73a), namely

$$(F)_1 \equiv \int \frac{d\Omega_1}{4\pi} f_{1(0)} \ . \tag{76}$$

We refer also to the definition (76) as the Hadamard partie finie (of the function F at 1) because this definition is closely related to the definition (74) of the Hadamard partie finie of a divergent integral. To see this, apply (74) to the case where the function F is actually a gradient, $F = \partial_i G$, where G satisfies (73) [it is then clear that F itself satisfies (73)]. We find

$$\mathrm{Pf} \int d^3\mathbf{x}\ \partial_i G = -4\pi (n_1^i r_1^2 G)_1 - 4\pi (n_2^i r_2^2 G)_2 \tag{77}$$

where in the right side the values at 1 and 2 are taken in the sense of the Hadamard partie finie (76). This nice connection between the Hadamard partie finie of a divergent integral and that of a singular function is clearly understood from applying the Gauss theorem on two surfaces $r_{1,2} = s$ surrounding the singularities (there is no dependence on the constants $s_{1,2}$).

7.2 Multipole Moments of Point-Mass Binaries

To compute the source moments (36) of two point-particles we insert (70) in place of the stress-energy tensor $T^{\mu\nu}$ of a continuous source, and we pick up the Hadamard partie finie [in the sense of (74)] of all integrals. This *ansatz* reads

$$(I_L)_{\mathrm{point-mass}} = \mathrm{Pf} \left\{ I_L[T^{\mu\nu}_{\mathrm{point-mass}}] \right\} \ ; \tag{78a}$$

$$(J_L)_{\mathrm{point-mass}} = \mathrm{Pf} \left\{ J_L[T^{\mu\nu}_{\mathrm{point-mass}}] \right\} \ . \tag{78b}$$

As we have seen in (69), the source multipole moments involve at high PN order many (non-compact-support) non-linear contributions which can be expressed in terms of retarded potentials such as V. The paradigm of such non-linear contributions is a term involving the quadratic product of two (derivatives of) potentials V, say $\partial V \partial V$, or, neglecting $O(1/c^2)$ corrections, $\partial U \partial U$. To Newtonian order U is given by $Gm_1/r_1 + Gm_2/r_2$ and it is easily checked that this paradigmatic term can be written as a certain derivative operator, say $\partial\partial$, acting on the elementary integral (assuming for simplicity $l = 2$)

$$Y_{ij}(\mathbf{y}_1, \mathbf{y}_2) \equiv -\frac{1}{2\pi} \mathrm{FP}_{B=0} \int d^3\mathbf{x}\ |\mathbf{x}/r_0|^B \frac{\hat{x}_{ij}}{r_1 r_2} \ . \tag{79}$$

We see that the integral would be divergent at infinity without the finite part operation. However, it is perfectly well-behaved near 1 and 2 where there is no need of a regularization. The integral (79) can be evaluated in various ways; the net result is [24,72]

$$Y_{ij} = \frac{r_{12}}{3}\left[y_1^{<ij>} + y_1^{<i}y_2^{j>} + y_2^{<ij>}\right] \tag{80}$$

where $< ij >\equiv \mathrm{STF}(ij)$. Starting at 3PN order we meet some elementary integrals which need the regularization at 1 or 2 in addition to involving the finite part at infinity. An example is

$$Z_{ij}(\mathbf{y}_1) \equiv -\frac{1}{2\pi}\mathrm{Pf}\left\{\mathrm{FP}_{B=0}\int d^3\mathbf{x}\ |\mathbf{x}/r_0|^B \frac{\hat{x}_{ij}}{r_1^3}\right\}. \tag{81}$$

To obtain this integral one splits it into a near-zone integral extending over the domain $r_1 < \mathcal{R}_1$ (say), and a far-zone integral extending over $\mathcal{R}_1 < r_1$. The Hadamard regularization at 1 applies only to the near-zone integral, while the finite part at $B = 0$ is needed only for the far-zone integral. The result, found to be independent of the radius $\mathcal{R}_1$, reads [77]

$$Z_{ij} = \left[2\ln\left(\frac{s_1}{r_0}\right) + \frac{16}{15}\right] y_1^{<ij>}. \tag{82}$$

In this case we find an explicit dependence on both the constants r_0 due to the finite part at infinity, and s_1 due to the Hadamard partie finie near 1 [see (74)]. However these constants do not enter the multipole moments before the 3PN order (collaboration with Iyer and Joguet [77]).

A long computation, done in [72], yields the mass-type quadrupole moment at the 2PN order fully reduced in the case of two point-masses moving on a circular orbit. The method is to start from (69) (issued from [24]) and to employ notably the elementary integral (79)-(80) (see also [72] for the treatment of a cubically non-linear term). An equivalent result has been obtained by Will and Wiseman using their formalism [75]. In a mass-centered frame the moment is of the form

$$I_{ij} = \mu\left(A\,\hat{y}_{ij} + B\,\frac{\hat{v}_{ij}}{\omega^2}\right) + O\left(\frac{1}{c^5}\right), \tag{83}$$

where $y_i = y_1^i - y_2^i$ and $v_i = v_1^i - v_2^i$, where ω denotes the binary's Newtonian orbital frequency [$\omega^2 = Gm/r_{12}^3$ with $m = m_1+m_2$], and where $\mu = m_1m_2/m$ is the reduced mass. The point is to obtain the coefficients A and B developed to 2PN order in terms of the post-Newtonian parameter $\gamma = Gm/r_{12}c^2$, where we recall that r_{12} is the distance between the two particles in harmonic coordinates. Untill 2PN we find some definite polynomials in the mass ratio $\nu = \mu/m$ (such that $0 < \nu \leq 1/4$):

$$A = 1 + \gamma\left[-\frac{1}{42} - \frac{13}{14}\nu\right] + \gamma^2\left[-\frac{461}{1512} - \frac{18395}{1512}\nu - \frac{241}{1512}\nu^2\right], \tag{84a}$$

$$B = \gamma\left[\frac{11}{21} - \frac{11}{7}\nu\right] + \gamma^2\left[\frac{1607}{378} - \frac{1681}{378}\nu + \frac{229}{378}\nu^2\right]. \tag{84b}$$

The 2PN mass quadrupole moment (83)-(84) is part of a program aiming at computing the orbital phase evolution of inspiralling compact binaries to high post-Newtonian order (see Sect. 7.4). First-order black-hole perturbations, valid in the test-mass limit $\nu \to 0$ for one body, have already achieved the very high 5.5PN order [87,99–101]. Recovering the result of black-hole perturbations in this limit constitutes an important check of the overall formalism. For the moment it passed the check to 2.5PN order [72,73]; this is quite satisfactory regarding the many differences between the present approach and the black-hole perturbation method.

7.3 Equations of Motion of Compact Binaries

The equations of motion of two point-masses play a crucial role in accounting for the observed dynamics of the binary pulsar PSR1913+16 [1–3,22], and constitute an important part of the program concerning inspiralling compact binaries. The motivation for investigating rigorously the equations of motion came in part from the salubrious criticizing remarks of Jürgen Ehlers *et al* [7]. Four different approaches have succeeded in obtaining the equations of motion of point-mass binaries complete up to the 2.5PN order (dominant order of radiation reaction): the "post-Minkowskian" approach of Damour, Deruelle and colleagues [16–19]; the "Hamiltonian" approach of Schäfer and predecessors [102,103,57,58] ; the "extended-body" approach of Kopejkin *et al* [104,105]; and the "post-Newtonian" approach of Blanchet, Faye and Ponsot [74]. The four approaches yield mutually agreeing results.

The post-Newtonian approach [74] consists of (i) inserting the point-mass stress-energy tensor (70) into the 2.5PN metric in harmonic coordinates given by (63); (ii) curing systematically the self-field divergences of point-masses using the Hadamard regularization; and (iii) substituting the regularized metric into the standard geodesic equations. For convenience we write the geodesic equation of the particle 1 in the Newtonian-like form

$$\frac{d\mathcal{P}_1^i}{dt} = \mathcal{F}_1^i \tag{85}$$

where the (specific) linear momentum $\mathcal{P}_1^i$ and force $\mathcal{F}_1^i$ are given by

$$\mathcal{P}_1^i = c\left(\frac{v_1^\mu g_{i\mu}}{\sqrt{-g_{\rho\sigma}v_1^\rho v_1^\sigma}}\right)_1 \; ; \qquad \mathcal{F}_1^i = \frac{c}{2}\left(\frac{v_1^\mu v_1^\nu \partial_i g_{\mu\nu}}{\sqrt{-g_{\rho\sigma}v_1^\rho v_1^\sigma}}\right)_1 . \tag{86}$$

Crucial in this method, the quantities are evaluated at the location of particle 1 according to the rule (76). All the potentials (62) and their gradients are evaluated in a way similar to our computation of U in (72), and then inserted into (85)-(86). We "order-reduce" the result, i.e. we replace each acceleration, consistently with the approximation, by its equivalent in terms of the positions and velocities as given by the (lower-order) equations of motion. After simplication we find, in agreement with other methods,

$$
\begin{aligned}
\frac{dv_1^i}{dt} = &- \frac{Gm_2}{r_{12}^2} n_{12}^i + \frac{Gm_2}{r_{12}^2 c^2} \Bigg\{ v_{12}^i \left[4(n_{12}v_1) - 3(n_{12}v_2)\right] \\
&+ n_{12}^i \left[-v_1^2 - 2v_2^2 + 4(v_1v_2) + \frac{3}{2}(n_{12}v_2)^2 + 5\frac{Gm_1}{r_{12}} + 4\frac{Gm_2}{r_{12}} \right] \Bigg\} \\
&+ \frac{Gm_2}{r_{12}^2 c^4} n_{12}^i \Bigg\{ \Big[-2v_2^4 + 4v_2^2(v_1v_2) - 2(v_1v_2)^2 \\
&+ \frac{3}{2}v_1^2(n_{12}v_2)^2 + \frac{9}{2}v_2^2(n_{12}v_2)^2 - 6(v_1v_2)(n_{12}v_2)^2 - \frac{15}{8}(n_{12}v_2)^4 \Big] \\
&+ \frac{Gm_1}{r_{12}} \Big[-\frac{15}{4}v_1^2 + \frac{5}{4}v_2^2 - \frac{5}{2}(v_1v_2) \\
&+ \frac{39}{2}(n_{12}v_1)^2 - 39(n_{12}v_1)(n_{12}v_2) + \frac{17}{2}(n_{12}v_2)^2 \Big] \\
&+ \frac{Gm_2}{r_{12}} \left[4v_2^2 - 8(v_1v_2) + 2(n_{12}v_1)^2 - 4(n_{12}v_1)(n_{12}v_2) - 6(n_{12}v_2)^2 \right] \\
&+ \frac{G^2}{r_{12}^2} \left[-\frac{57}{4}m_1^2 - 9m_2^2 - \frac{69}{2}m_1m_2 \right] \Bigg\} \\
&+ \frac{Gm_2}{r_{12}^2 c^4} v_{12}^i \Bigg\{ v_1^2(n_{12}v_2) + 4v_2^2(n_{12}v_1) - 5v_2^2(n_{12}v_2) - 4(v_1v_2)(n_{12}v_1) \\
&+ 4(v_1v_2)(n_{12}v_2) - 6(n_{12}v_1)(n_{12}v_2)^2 + \frac{9}{2}(n_{12}v_2)^3 \\
&+ \frac{Gm_1}{r_{12}} \left[-\frac{63}{4}(n_{12}v_1) + \frac{55}{4}(n_{12}v_2) \right] + \frac{Gm_2}{r_{12}} \left[-2(n_{12}v_1) - 2(n_{12}v_2) \right] \Bigg\} \\
&+ \frac{4G^2 m_1 m_2}{5c^5 r_{12}^3} \Bigg\{ n_{12}^i (n_{12}v_{12}) \left[-6\frac{Gm_1}{r_{12}} + \frac{52}{3}\frac{Gm_2}{r_{12}} + 3v_{12}^2 \right] \\
&+ v_{12}^i \left[2\frac{Gm_1}{r_{12}} - 8\frac{Gm_2}{r_{12}} - v_{12}^2 \right] \Bigg\} + O\left(\frac{1}{c^6}\right) ,
\end{aligned} \tag{87}
$$

[where $n_{12}^i = (y_1^i - y_2^i)/r_{12}$; $v_{12}^i = v_1^i - v_2^i$; and e.g. $(n_{12}v_1)$ denotes the Euclidean scalar product]. At the 1PN or $1/c^2$ level the equations were obtained before by Lorentz an Droste [20], and by Einstein, Infeld and Hoffmann [21]. The 2.5PN or $1/c^5$ term represents the radiation damping in harmonic coordinates [correct because the metric (63) we started with matches to the post-Minkowskian exterior field]. In the case of circular orbits, the equations simplify drastically:

$$
\frac{dv_{12}^i}{dt} = -\omega_{2\mathrm{PN}}^2 y_{12}^i - \frac{32G^3m^3\nu}{5c^5 r_{12}^4} v_{12}^i + O\left(\frac{1}{c^6}\right) , \tag{88}
$$

where the orbital frequency $\omega_{2\mathrm{PN}}$ of the 2PN circular motion reads

$$\omega^2_{2\mathrm{PN}} = \frac{Gm}{r^3_{12}} \left[1 + (-3+\nu)\gamma + \left(6 + \frac{41}{4}\nu + \nu^2 \right) \gamma^2 \right] \tag{89}$$

(the post-Newtonian parameter is $\gamma = Gm/c^2 r_{12}$; and $\nu = \mu/m$).

7.4 Gravitational Waveforms of Inspiralling Compact Binaries

The gravitational radiation field and associated energy flux are given by (52) and (57) in terms of time-derivatives of the radiative multipole moments, themselves related to the source multipole moments by formulas such as (56). Furthermore, at a given post-Newtonian order, the source moments admit some explicit though complicated expressions such as (68)-(69), which, when specialized to (non-spinning) point-mass circular binaries, yield e.g. (83)-(84).

Now, for insertion into the radiation field and energy flux, one must compute the *time-derivatives* of the binary moments, with appropriate order-reduction using the binary's equations of motion (87)-(89). This yields in particular the fully reduced (up to the prescribed post-Newtonian order) gravitational waveform of the binary, or more precisely the two independent "plus" and "cross" polarization states h_+ and $h_\times$. The result to 2PN order is written in the form

$$h_{+,\times} = \frac{2Gm\nu x}{c^2 R} \left\{ H^{(0)}_{+,\times} + x^{1/2} H^{(1/2)}_{+,\times} + x H^{(1)}_{+,\times} + x^{3/2} H^{(3/2)}_{+,\times} + x^2 H^{(2)}_{+,\times} \right\} , \tag{90}$$

where, for convenience, we have introduced a post-Newtonian parameter which is directly related to the orbital frequency: $x = (Gm\omega_{2\mathrm{PN}}/c^3)^{2/3}$, where $\omega_{2\mathrm{PN}}$ is given for circular orbits by (89). The various post-Newtonian coefficients in (90) depend on the cosine and sine of the "inclination" angle between the detector's direction and the normal to the orbital plane ($c_i = \cos i$ and $s_i = \sin i$), and on the masses through the ratios $\nu = \mu/m$ and $\delta m/m$, where $\delta m = m_1 - m_2$. The result for the "plus" polarization (collaboration with Iyer, Will and Wiseman [106]) is

$$H^{(0)}_+ = -(1+c_i^2)\cos 2\psi , \tag{91a}$$

$$H^{(1/2)}_+ = -\frac{s_i}{8}\frac{\delta m}{m} \left[(5+c_i^2)\cos\psi - 9(1+c_i^2)\cos 3\psi \right] , \tag{91b}$$

$$\begin{aligned} H^{(1)}_+ &= \frac{1}{6}\left[19 + 9c_i^2 - 2c_i^4 - \nu(19 - 11c_i^2 - 6c_i^4) \right] \cos 2\psi \\ &\quad -\frac{4}{3} s_i^2 (1+c_i^2)(1-3\nu)\cos 4\psi , \end{aligned} \tag{91c}$$

$$H^{(3/2)}_+ = \frac{s_i}{192}\frac{\delta m}{m} \left\{ \left[57 + 60c_i^2 - c_i^4 - 2\nu(49 - 12c_i^2 - c_i^4) \right] \cos\psi \right.$$

$$
\begin{aligned}
&-\frac{27}{2}\left[73+40c_i^2-9c_i^4-2\nu(25-8c_i^2-9c_i^4)\right]\cos 3\psi \\
&+\frac{625}{2}(1-2\nu)s_i^2(1+c_i^2)\cos 5\psi\Big\} - 2\pi(1+c_i^2)\cos 2\psi \ , \qquad (91d)
\end{aligned}
$$

$$
\begin{aligned}
H_+^{(2)} = &\frac{1}{120}\left[22+396c_i^2+145c_i^4-5c_i^6+\frac{5}{3}\nu(706-216c_i^2-251c_i^4+15c_i^6)\right. \\
&\left.-5\nu^2(98-108c_i^2+7c_i^4+5c_i^6)\right]\cos 2\psi \\
&+\frac{2}{15}s_i^2\left[59+35c_i^2-8c_i^4-\frac{5}{3}\nu(131+59c_i^2-24c_i^4)\right. \\
&\left.+5\nu^2(21-3c_i^2-8c_i^4)\right]\cos 4\psi \\
&-\frac{81}{40}(1-5\nu+5\nu^2)s_i^4(1+c_i^2)\cos 6\psi \\
&+\frac{s_i}{40}\frac{\delta m}{m}\left\{\left[11+7c_i^2+10(5+c_i^2)\ln 2\right]\sin\psi - 5\pi(5+c_i^2)\cos\psi\right. \\
&\left.-27\left[7-10\ \ln(3/2)\right](1+c_i^2)\sin 3\psi + 135\pi(1+c_i^2)\cos 3\psi\right\} . \qquad (91e)
\end{aligned}
$$

The "cross" polarization admits a similar expression (see [106]). Here, ψ denotes a particular phase variable, related to the actual binary's orbital phase ϕ and frequency $\omega \equiv \omega_{2PN}$ by

$$
\psi = \phi - \frac{2Gm\omega}{c^3}\ln\left(\frac{\omega}{\omega_0}\right) \ ; \qquad (92)
$$

ϕ is the angle, oriented in the sense of the motion, between the vector separation of the two bodies and a fixed direction in the orbital plane (since the bodies are not spinning, the orbital motion takes place in a plane). In (92), ω_0 denotes some constant frequency, for instance the orbital frequency when the signal enters the detector's frequency bandwidth; see [106] for discussion.

The previous formulas give the waveform of point-mass binaries whenever the frequency and phase of the orbital motion take the values ω and ϕ. To get the waveform as a function of time, we must replace ω and ϕ by their explicit time evolutions $\omega(t)$ and $\phi(t)$. Actually, the frequency is the time-derivative of the phase: $\omega = d\phi/dt$. The evolution of the phase is entirely determined, for circular orbits, by the energy balance equation $dE/dt = -\mathcal{L}$ relating the binding energy E of the binary in the center of mass to the emitted energy flux $\mathcal{L}$. E is computed using the equations of motion (87), and $\mathcal{L}$ follows from (57) and application of the previous formalism [changing the radiative moments to the source moments, applying (83)-(84), etc...]; the net result for the 2.5PN orbital phase [72,75,73] is

$$\phi = \phi_0 - \frac{1}{\nu}\Bigg\{\Theta^{5/8} + \left(\frac{3715}{8064} + \frac{55}{96}\nu\right)\Theta^{3/8} - \frac{3}{4}\pi\Theta^{1/4}$$
$$+ \left(\frac{9275495}{14450688} + \frac{284875}{258048}\nu + \frac{1855}{2048}\nu^2\right)\Theta^{1/8}$$
$$+ \left(-\frac{38645}{172032} - \frac{15}{2048}\nu\right)\pi \ln\Theta\Bigg\} , \qquad (93)$$

where ϕ_0 is a constant phase (determined for instance when the frequency is ω_0), and Θ the convenient dimensionless time variable

$$\Theta = \frac{c^3\nu}{5Gm}(t_c - t) , \qquad (94)$$

t_c being the instant of coalescence at which, formally, $\omega(t)$ tends to infinity (of course, the post-Newtonian method breaks down before the final coalescence). All the results are in agreement, in the limit $\nu \to 0$, with those of black-hole perturbation theory [87,99–101].

8 Conclusion

The formalism reviewed in this article permits investigating in principle all aspects of the problem of dynamics and gravitational-wave emission of a *slowly-moving* isolated system (with, say, $v/c \sim 0.3$ at most): the generation of waves, their propagation in vacuum, the back-reaction onto the system, the structure of the asymptotic field, and most importantly the relation between the far-field and the source parameters. Of course, the formalism is merely post-Newtonian and never "exact", but in applications to astrophysical objects such as inspiralling compact binaries this should be sufficient provided that the post-Newtonian approximation is carried to high order.

Furthermore, there are several places in the formalism where some results are valid formally to any order of approximation. For instance, the source multipole moments are related to the *infinite* formal post-Newtonian expansion of the pseudo-tensor [see (18) or (36)], and the post-Minkowskian iteration of the exterior field is performed to *any* non-linear order [see (43)]. In such a situation, where an infinite approximate series can be defined, there is the interesting question of its relation to a corresponding element in the exact theory. For the moment the only solid work concerns the post-Minkowskian approximation of the exterior vacuum field, which has been proved to be asymptotic [26]. Likewise it is plausible that the expressions of the source multipole moments could be valid in the case of exact solutions.

The most important part of the formalism where a general prescription for how to proceed at *any* approximate step is missing, is the post-Newtonian expansion for the field inside the isolated system. For instance, though the multipole moments are given in terms of the formal post-Newtonian expansion of the pseudo-tensor, no general algorithm for computing *explicitly* this post-Newtonian expansion is known. An interesting task would be to define such an algorithm, in a manner similar to the post-Minkowskian algorithm in Sect. 4. In the author's opinion, the post-Newtonian algorithm should be defined conjointly with the post-Minkowskian algorithm, and should rely on the matching equation (16), so as to convey into the post-Newtonian field the information about the exterior metric.

Note that even if a general method for implementing a complete approximation series is defined, this method may be unworkable in practical calculations, because not explicit enough. For instance the post-Minkowskian series (43) is defined in terms of "iterated" retarded integrals, but needs to be suplemented by some formulas, to be used in applications, for the retarded integral of a multipolar extended source. In this respect it would be desirable to develop the formulas generalizing (50)-(51) to any non-linear order. This should permit in particular the study of the general structure of tails, tails of tails, and so on.

For the moment the only application of the formalism concerns the radiation and motion of point-particle binaries. Of course it is important to keep the formalism as general as possible, and not to restrict oneself to a particular type of source, but this application to point-particles offers some interesting questions. Indeed, it seems that the post-Newtonian approximation used conjointly with a regularization *à la* Hadamard works well, and that one is getting closer and closer to an exact (numerical) solution corresponding to the dynamics and radiation of two black-holes. So, in which sense does the post-Newtonian solution (corresponding to point-masses without horizons) approach a true solution for black-holes? Does the adopted method of regularizing the self-field play a crucial role? Is it possible to define a regularization consistently with the post-Newtonian approximation to all orders?

Acknowledgments

The author is especially grateful to Bernd Schmidt for discussions and for remarks which led to improvement of this article. Stimulating discussions with Piotr Crusciel, Thibault Damour and Gerhard Schäfer are also acknowledged. The Max-Planck-Institut für Gravitationsphysik (Albert-Einstein-Institut) in Potsdam is thanked for an invitation during which the writing of the article was begun.

References

1. J.H. Taylor, L.A. Fowler and P.M. Mc Culloch (1979) Nature 277, 437
2. J.H. Taylor and J.M. Weisberg (1982) Astrophys. J. 253, 908
3. J.H. Taylor (1993) Class. Quantum Grav. 10, S167

4. K. Thorne, in Gravitational Radiation, N. Deruelle and T. Piran (eds.) (1983) North-Holland publishing company, p. 1
5. A. Ashtekar, in Gravitational Radiation, N. Deruelle and T. Piran (eds.) (1983) North-Holland publishing company, p. 421
6. J. Ehlers (1987) in the Proc. 11th Intern. Conf. on General Relativity and Gravitation, M. Mc Callum (ed.), Cambridge U. Press
7. J. Ehlers, A. Rosenblum, J.N. Goldberg and P. Havas (1976) Astrophys. J. 208, L77
8. M. Walker and C.M. Will (1980) Astrophys. J. 242, L129
9. J. Ehlers, in Classical Mechanics and Relativity: Relationship and Consistency, G. Ferrarese (ed.) (1991) Bibliopolis, Napoli
10. J. Ehlers (1997) Class. Quant. Gravity 14, A119
11. J. Ehlers (1998) The Newtonian limit of General Relativity, in Understanding Physics, Kundt-Festschrift
12. M. Lottermoser (1990) Ph. D. dissertation, Munich University, unpublished
13. A.D. Rendall (1992) Proc. R. Soc. Lond. A438, 341
14. T. Futamase (1983) Phys. Rev. D28, 2373
15. T. Futamase and B.F. Schutz (1983) Phys. Rev. D28, 2363
16. L. Bel, T. Damour, N. Deruelle, J. Ibañez and J. Martin (1981) Gen. Relativ. Gravit. 13, 963
17. T. Damour and N. Deruelle (1981) Phys. Lett. 87A, 81
18. T. Damour (1982) C. R. Acad. Sc. Paris, 294, 1355
19. T. Damour, in Gravitational Radiation, N. Deruelle and T. Piran (eds.), (North-Holland publishing Company, 1983), p. 59.
20. H.A. Lorentz and J. Droste (1917) Versl. K. Akad. Wet. Amsterdam 26, 392 and 649; in the collected papers of H.A. Lorentz, vol. 5, The Hague, Nijhoff (1937)
21. A. Einstein, L. Infeld and B. Hoffmann (1938) Ann. Math. 39, 65
22. T. Damour (1983) Phys. Rev. Lett. 51, 1019
23. L. Blanchet and T. Damour (1986) Philos. Trans. R. Soc. London A 320, 379
24. L. Blanchet (1995) Phys. Rev. D51, 2559
25. L. Blanchet (1998) Class. & Quantum Grav., 15, 1971
26. T. Damour and B. Schmidt (1990) J. Math. Phys. 31, 2441
27. D. Christodoulou and B.G. Schmidt (1979) Commun. Math. Phys. 68, 275
28. A.D. Rendall (1990) Class. & Quantum Gravity 7, 803
29. L. Blanchet (1987) Proc. R. Soc. Lond. A 409, 383
30. L. Blanchet and T. Damour (1989) Ann. Inst. H. Poincaré (Phys. Théorique) 50, 377
31. L. Blanchet (1993) Phys. Rev. D 47, 4392
32. L. Blanchet (1997) Phys. Rev. D 55, 714
33. B.R. Iyer and C.M. Will (1993) Phys. Rev. Lett. 70, 113
34. B.R. Iyer and C.M. Will (1995) Phys. Rev. D52, 6882
35. T. Damour, in Gravitation in Astrophysics, B. Carter and J.B. Hartle (eds.) (Plenum Press, New York and London, 1986).
36. T. Damour, in 300 Years of Gravitation, S.W. Hawking and W. Israel (eds.) (Cambridge Univ. Press, Cambridge, 1987).
37. K.S. Thorne, in 300 Years of Gravitation, S.W. Hawking and W. Israel (eds.), (Cambridge Univ. Press, Cambridge, 1987).

38. C.M. Will, in Proc. of the 8th Nishinomiya-Yukawa Symposium on Relativistic Cosmology, ed. M. Sasaki (Universal Acad. Press, Japan, 1994).
39. L. Blanchet (1996) in Relativistic gravitation and gravitational radiation, J.A. Marck and J.P. Lasota (eds.), Cambridge U. Press
40. J.M. Bardeen and W.H. Press (1973) J. Math. Phys. 14, 7
41. B.G. Schmidt and J.M. Stewart (1979) Proc. R. Soc. London A367, 503
42. M. Walker and C.M. Will (1979) Phys. Rev. D 19, 3495
43. J. Porrill and J.M. Stewart (1981) Proc. R. Soc. London A376, 451
44. V.A. Fock (1959) Theory of Space, Time and Gravitation, Pergamon, London
45. S. Chandrasekhar (1965) Astrophys. J. 142, 1488
46. S. Chandrasekhar and Y. Nutku (1969) Astrophys. J. 158, 55
47. S. Chandrasekhar and F.P. Esposito (1970) Astrophys. J. 160, 153
48. J. Ehlers (1977) in the Proc. of the International school of Relativistic Astrophysics, J. Ehlers (ed.), Erice, MPI:München
49. J. Ehlers (1980) Ann. N.Y. Acad. Sci. 336, 279
50. G.D. Kerlick (1980) Gen. Rel. Grav. 12, 467
51. G.D. Kerlick (1980) Gen. Rel. Grav. 12, 521
52. A. Caporali (1981) Nuovo Cimento 61B, 181
53. R. Breuer and E. Rudolph (1981) Gen. Rel. Grav. 13, 777
54. R. Breuer and E. Rudolph (1982) Gen. Rel. Grav. 14, 181
55. J.L. Anderson and T.C. DeCanio (1975) Gen. Relat. Grav. 6, 197
56. A. Papapetrou and B. Linet (1981) Gen. Relat. Grav. 13, 335
57. G. Schäfer (1985) Ann. Phys. (N.Y.) 161, 81
58. G. Schäfer (1986) Gen. Rel. Grav. 18, 255
59. W.B. Bonnor (1959) Philos. Trans. R. Soc. London A 251, 233
60. K.S. Thorne (1980) Rev. Mod. Phys. 52, 299
61. W.L. Burke and K.S. Thorne, in Relativity, M. Carmeli et al. (eds) (Plenum Press, New York, 1970) p. 208-209.
62. W.L. Burke (1971) J. Math. Phys. 12, 401
63. R.E. Kates (1980) Phys. Rev. D22, 1853
64. R.E. Kates (1980) Phys. Rev. D22, 1871
65. J.L. Anderson, R.E. Kates, L.S. Kegeles and R.G. Madonna (1982) Phys. Rev. D25, 2038
66. T. Damour and B.R. Iyer (1991) Ann. Inst. H. Poincaré (Phys. Théorique) 54, 115
67. L. Blanchet and T. Damour (1984) Phys. Lett. 104A, 82
68. L. Blanchet and T. Damour (1988) Phys. Rev. D37, 1410
69. L. Blanchet and T. Damour (1992) Phys. Rev. D46, 4304
70. L. Blanchet (1998) Class. & Quantum Grav., 15, 89
71. L. Blanchet (1998) Class. & Quantum Grav., 15, 113
72. L. Blanchet, T. Damour and B.R. Iyer (1995) Phys. Rev. D 51, 5360
73. L. Blanchet (1996) Phys. Rev. D 54, 1417
74. L. Blanchet, G. Faye and B. Ponsot (1998) Phys. Rev. D 58, 124002
75. C.M. Will and A.G. Wiseman (1996) Phys. Rev. D 54, 4813
76. R. Epstein and R.V. Wagoner (1975) Astrophys. J. 197, 717
77. L. Blanchet, B.R. Iyer and B. Joguet, in preparation.
78. T. Damour and B.R. Iyer (1991) Phys. Rev. D43, 3259
79. W.B. Campbell, J. Macek and T.A. Morgan (1977) Phys. Rev. D15, 2156
80. W.B. Bonnor and M.A. Rotenberg (1966) Proc. R. Soc. London A289, 247

81. A.J. Hunter and M.A. Rotenberg (1969) J. Phys. A 2, 34
82. W. G. Dixon (1979) in Isolated Gravitating systems in general relativity, J. Ehlers (ed.), North Holland, Amsterdam
83. J. Ehlers and E. Rudolph (1977) Gen. Rel. Grav. 8, 197
84. R. Schattner (1979) Gen. Rel. Grav. 10, 377
85. R. Schattner and M. Streubel (1981) Ann. Inst. H. Poincaré 34, 117
86. M. Streubel and R. Schattner (1981) Ann. Inst. H. Poincaré 34, 145
87. E. Poisson (1993) Phys. Rev. D 47, 1497
88. L. Blanchet and G. Schäfer (1993) Class. & Quantum Grav. 10, 2699
89. P.N. Payne (1983) Phys. Rev. D28, 1894
90. L. Blanchet (1990) thèse d'habilitation, Université P. et M. Curie (unpublished).
91. A.G. Wiseman and C.M. Will (1991) Phys. Rev. D44, R2945
92. K.S. Thorne (1992) Phys. Rev. D45, 520
93. D. Christodoulou (1991) Phys. Rev. Lett. 67, 1486
94. J. Frauendiener (1992) Class. & Quantum Gravity 9, 1639
95. C. Cutler, T.A. Apostolatos, L. Bildsten, L.S. Finn, E.E. Flanagan, D. Kennefick, D.M. Markovic, A. Ori, E. Poisson, G.J. Sussman and K.S. Thorne (1993) Phys. Rev. Lett. 70, 2984
96. C. Cutler, L.S. Finn, E. Poisson and G.J. Sussman (1993) Phys. Rev. D47, 1511
97. J. Hadamard (1932) Le problème de Cauchy et les équations aux dérivées partielles linéaires hyperboliques, Paris: Hermann
98. L. Schwartz, Théorie des distributions (Hermann, Paris, 1978).
99. M. Sasaki (1994) Prog. Theor. Phys. 92, 17
100. H. Tagoshi and M. Sasaki (1994) Prog. Theor. Phys. 92, 745
101. T. Tanaka, H. Tagoshi and M. Sasaki (1996) Prog. Theor. Phys. 96, 1087
102. T. Ohta, H. Okamura, T. Kimura and K. Hiida (1974) Progr. Theor. Phys. 51, 1220
103. T. Ohta, H. Okamura, T. Kimura and K. Hiida (1974) Progr. Theor. Phys. 51, 1598
104. S.M. Kopejkin (1985) Astron. Zh. 62, 889
105. L.P. Grishchuk and S.M. Kopejkin (1986) in Relativity in Celestial Mechanics and Astrometry, J. Kovalevsky and V.A. Brumberg (eds.), Reidel, Dordrecht
106. L. Blanchet, B.R. Iyer, C.M. Will and A.G. Wiseman (1996) Class. Quantum Grav. 13, 575

Duality and Hidden Symmetries in Gravitational Theories

Dieter Maison

Max-Planck-Institut für Physik
— Werner Heisenberg Institut —
Föhringer Ring 6
80805 Munich (Fed. Rep. Germany)

1 Introduction

Duality is presently considered the key to the Holy Grail of String Theory – it is supposed to provide links between the five known different superstring theories in ten dimensions, hoped to be just different limits of one unique eleven dimensional theory [1]. The main role of duality is to relate two different regimes – e.g. one of weak and one of strong coupling – of these theories. In most cases duality is not a very precise concept, just because the strong coupling regime is a matter of speculation. This is rather different from the duality transformations in the gravitational theories considered in this work, which have a very precise meaning – not the least because we are dealing with classical field theories (compare, however, [2] and for a modest attempt to use duality symmetry in Quantum Gravity, see [3]). The historical example of all these dualities is the duality between electric and magnetic fields in electrodynamics, which, when expressed in terms of the field strength, is just an example of the mathematical notion of Hodge duality for differential forms. Actually, the source-free Maxwell equations are not only invariant under a discrete duality, but under a continous one-parameter group of duality rotations. It is this kind of transformations, which is the subject of this article. Whereas, in general, the electromagnetic duality rotations are an 'on-shell' symmetry, i.e. a symmetry of the equations of motion and not of the action, the situation changes, if one considers time-independent solutions. In this case also the magnetic field can be derived from a (pseudo) scalar potential and the duality rotations expressed in terms of scalar potentials become a bona fide 'off-shell' symmetry of the 'dimensionally reduced' three dimensional theory. This replacement of the vector potential by a scalar one has analogues in higher dimensions playing an important role in the construction of supergravity theories through the process of dimensional reduction. A typical example is the (pseudo) scalar 'axion', obtained as the dual of a gauge field 2-form in 4 dimensions. This scalar axion combines nicely with another scalar, the dilaton to a doublet giving rise to an $SL(2)$ group of non-linear duality transformations [4]. A particular element of this group, replacing the dilaton by its inverse, lies at the heart of string duality ('S-duality')[5], where the expectation value of the dilaton plays the role of a string coupling constant.

Descending from four to three dimensions the group of continous electromagnetic duality transformations is extended from the compact group $SO(2)$ to the full non-compact Euclidean group in two dimensions. The reason for this enhancement is simply the possibility to shift the potentials by a constant without changing the field strengths.

Let us now turn to (vacuum) gravity, also restricting ourselves to stationary (time-independent) solutions and introduce the 'twist potential', playing a similar role as the magnetic potential in electrodynamics. It was Ehlers [6], who first observed that there is a discrete duality symmetry, relating the twist potential and the tt-component of the metric (the latter playing a similar role as the electric potential). Later it was recognized [7] that there is a whole $SL(2)$ group acting on this doublet of potentials transforming stationary solutions of Einstein's equations into each other and that the potentials themselves can be interpreted as coordinates of a 2-dimensional 'potential space' with the geometry of the coset space $SL(2)/SO(2)$. In fact there is a strong similarity between electric charge and mass and magnetic charge and the so-called NUT charge on the other hand.

This analogy becomes even closer in the Einstein–Maxwell theory. Combining the two scalar potentials of gravity with those of the stationary EM field, one may use them to parametrize the coset space $SU(2,1)/S(U(1,1) \times U(1))$ equipped with a natural action of the non-compact group $SU(2,1)$ [8]. This group contains 'Harrison' transformations [9] transforming mass into charge, e.g. the Schwarzschild solution into the Reissner-Nordströ m solution.

The appearance of a non-compact group of 'hidden' duality transformations like $SU(2,1)$ is a rather general phenomenon in the process of 'Kaluza–Klein' reduction of the maximal eleven dimensional supergravity theory [10]. Many different 'hidden' duality symmetries were found in various dimensions and under various truncations, all having in common the peculiar property that the related coset spaces parametrized by the scalars are non-compact *Cartan Symmetric Spaces* [11]. The underlying reason for this fact is still unknown. In the paper [12] with Breitenlohner and Gibbons we were able to determine a large class of four-dimensional theories containing abelian vector fields and scalars besides the gravitational field yielding suitable symmetric spaces upon reduction to three dimensions. Such field theories for scalar fields taking their values in a Riemannian space are familiar in Quantum Field Theory as non-linear σ-models.

In view of the situation in the Maxwell theory one may wonder, whether it is possible to lift the off-shell symmetries to corresponding on-shell symmetries in four (or even higher) dimensions. An attempt in this direction for $N = 8$ supergravity has been made by Nicolai and de-Wit [13]. But let us proceed in the opposite direction. Even more remarkable is the enhancement of 'hidden' duality symmetries, when the process of dimensional reduction is performed one step further leading to an effective 2-dimensional theory. The simplest example is provided by solutions of vacuum gravity with a two-dimensional abelian isometry group, i.e. with two commuting Killing vector fields. It was Geroch [14] who observed that each given stationary, axially symmetric solution of the vacuum Einstein equations is accompanied by an infinite family of potentials, which in turn allowed for an infinite parameter set of infinitesimal transformations acting on the initial solution. What remained unclear in Geroch's work was the precise Lie algebra structure of these infinitesimal transformations and even more so the structure of a corresponding group (the 'Geroch' group) of finite transformations. Later the problem of getting finite transformations found some seemingly different solutions ('Solution Generating Methods') – the HKX-transformations [15], the Bäcklund-Transformations of Harrison [16], Kramer and Neugebauer [17] and the Riemann–Hilbert Method of Ernst and Hauser [18]. While the first authors were mainly concerned with the actual construction of hitherto unknown solutions ('Multi-Kerr'), Ernst and Hauser made the first serious effort to give the notion of the 'Geroch' group a more precise mathematical foundation. A deeper understanding of these somewhat mysterious constructions and their interrelations is provided by the observation – made already before the advent of the 'Solution Generating Methods' – that the corresponding 2-dimensional reduced theory is *completely integrable*. This property is expressed by the existence of a 'Lax Pair' [19,20], a system of linear differential equations with a 'spectral parameter', whose compatibility is equivalent to the non-linear field equations. It is important to observe that this fact is essentially based on the non-linear σ-model structure of the reduced field equations and thus is also valid for the 2-d reductions of the large class of 4-dimensional models considered in [12]. In fact, for the analysis of the group-theoretical significance of various steps in the implementation of the infinite dimensional Geroch group this was a valuable guiding principle [10]. In [21] it was shown how the group theoretical structure of the symmetric-space non-linear σ-models provides the clue to the analysis of the Geroch group, resp. its implementation. It is possible to construct a natural extension $(G^{(\infty)}, H^{(\infty)}, \tau^{(\infty)})$ of the triple (G, H, τ) defining the symmetric space G/H (where τ is the involutive automorphism leaving H invariant [11]). Here $G^{(\infty)}$ and $H^{(\infty)}$ are infinite dimensional groups of holomorphic functions (of the spectral parameter of the Lax pair) with values in the complexification of G. The elements of the coset space $G^{(\infty)}/H^{(\infty)}$ are solutions $\mathcal{P}(t,x)$ of the system of linear differential equations mentioned before. Implementing the group $G^{(\infty)}$ on the coset space $G^{(\infty)}/H^{(\infty)}$ turns out to be equivalent to the solution of a factorization problem for group-valued analytic functions, a so-called Riemann–Hilbert problem. While the Bäcklund transformations correspond to meromorphic group elements, the HKX transformations may be understood as exponentials of certain nil-potent elements of the underlying affine Lie algebra.

Whereas for most of the algebraic structures connected with this complete integrability the signature of the two Killing vectors providing the reduction to the 2-dimensional theory is irrelevant, the analytical properties of the solutions are rather different and thus the results of the stationary, axially symmetric case cannot be directly transfered to e.g. cylindrical or plane waves. Nevertheless some attempts with interesting results have also been made for these cases [22]. Another interesting development – in particular in view of a possible quantization – is the inclusion of fermions for the supersymmetric models [23].

Although it may be very satisfactory for esthetical reasons to see all these beautiful 'hidden' duality symmetries emerge, there is also some practical 'spin-off'. The duality groups are, in a sense, large enough to act transitively on certain families of solutions. This not only allows to actually construct all solutions of such a family from the simplest member (usually some solution of the vacuum theory like Minkowski or Schwarzschild), but also allows to prove uniqueness theorems for these families. This has to do with another aspect besides the group structure – the solutions of the 3-d resp. 2-d field equations are harmonic maps [24,25]. Such maps enjoy rather strong regularity and uniqueness properties, which can be used to generalize the well known uniqueness theorems [26] for static resp. stationary, axially symetric black hole solutions of the vacuum theory resp. Einstein-Maxwell theory to the large class of theories considered here. The corresponding proofs are remarkably simple and elegant. As to be expected for harmonic maps these results are strongest when the 3-space and the target space have definite curvature, happening for stationary, axially symmetric solutions [27].

The organization of this article is as follows:

- Section 2 is devoted to a short discussion of the duality symmetry of the flat space Maxwell equations.
- In Sect. 3 the emergence of hidden duality symmetries in the process of Kaluza–Klein reduction of higher dimensional (super) gravity theories is demonstrated.
- In Sect. 4 we discuss the complete integrability of the theories obtained by reduction to 2 dimensions, in particular the implementation of the infinite dimensional Geroch group.
- In Sect. 5 the σ-model structure obtained in the previous section is applied to construct spherically symmetric solutions and to show existence and uniqueness properties of static resp. stationary, axially symmetric black hole solutions.
- In the Appendix A we collect some important group theoretical properties of symmetric space non-linear σ-models and Appendix B contains the results of a recent structural analysis [28] of the models considered in [12].

2 Electromagnetic Duality

The best known example of duality transformations are those of the Maxwell field (see [29] for a recent exposition and further references). In vacuum the Maxwell equations are

$$\partial_\mu F^{\mu\nu} = 0 \qquad \partial_\mu {}^*F^{\mu\nu} = 0 \quad \text{with} \quad {}^*F^{\mu\nu} = \frac{1}{2}\epsilon^{\mu\nu\kappa\lambda}F_{\kappa\lambda}\,, \tag{1}$$

or, when written in terms of the electric field E and the magnetic field B,

$$\begin{aligned} \partial \cdot E &= 0, \qquad & \partial \wedge E + \dot{B} &= 0, \\ \partial \cdot B &= 0, \qquad & \partial \wedge B - \dot{E} &= 0\,. \end{aligned} \tag{2}$$

These equations are not only invariant under the obvious discrete exchange $F \to {}^*F$, ${}^*F \to -F$ resp. $E \to B$, $B \to -E$, but even under the continuous rotations (conveniently written in complex $U(1)$ form)

$$F + i{}^*F \to e^{i\theta}(F + i{}^*F) \qquad \text{resp.} \qquad E + iB \to e^{i\theta}(E + iB)\,. \tag{3}$$

In contrast to the energy density $\frac{1}{2}|E + iB|^2$ and the momentum density $E \wedge B = \frac{1}{2i}(E - iB) \wedge (E + iB)$, which are invariant under the duality rotations, the Lagrangean density $E^2 - B^2$ and the toplogical density $E \cdot B$ – real and imaginary part of $(E + iB)^2$ – are not, but transform into each other. Yet, since $E \cdot B = \frac{1}{4}\partial_\mu(A_\nu {}^*F^{\mu\nu})$, the variation of the action under infinitesimal duality transformations is a surface term vanishing under suitable boundary conditions. There exists also a conserved Noether current corresponding to this variation. But since the second variation yields again the action density, the action is not invariant under finite such transformations. Thus, although the equations of motion are invariant under duality rotations, the action is not, and hence they are what is called an 'on-shell' symmetry in contrast to a true 'off-shell' symmetry, leaving the action invariant.

There is another aspect of asymmetry, connected with the introduction of a vector potential A_μ. The usual choice, valid also in the presence of electrical charge, is to interprete the equation $\partial \cdot {}^*F = 0$ as the integrability condition (Bianchi identity) for the existence of a vector potential $F_{\mu\nu} = \partial_\mu A_\nu - \partial_\nu A_\mu$. In vacuum, however, nothing prevents us to introduce a similar potential $\tilde{A}_\mu$ for the dual of the field strength, ${}^*F_{\mu\nu} = \partial_\mu \tilde{A}_\nu - \partial_\nu \tilde{A}_\mu$, interpreting $\partial \cdot F = 0$ as Bianchi identity. Combining A and $\tilde{A}$ to a 2-component potential $\underline{A}$ we may write Eq. (1) as [30]

$$\partial_\mu \underline{F}^{\mu\nu} = 0 \qquad \text{and} \qquad \underline{F}^{\mu\nu} = \Omega {}^*\underline{F}^{\mu\nu} \qquad \text{with} \qquad \Omega = \begin{pmatrix} 0 & 1 \\ -1 & 0 \end{pmatrix}. \tag{4}$$

The action can be expressed either through A or, equivalently, through $\tilde{A}$, but not in terms of $\underline{A}$.

The situation changes, if we consider stationary solutions, i.e. solutions with a time-independent field strength. Then the Maxwell equations (2) tell us that (at least locally) there are scalar potentials ϕ and ψ for the electric resp. magnetic field. Combining them into a complex potential $\Phi \equiv \phi + i\psi$ such that $E + iB = \partial\Phi$ the duality transformations can be implemented through $\Phi \to e^{i\theta}\Phi$. Using a gauge in which the vector potential A_μ is time-independent, the electric potential ϕ can be identified with A_0 (and similarly the magnetic potential ψ with $\tilde{A}_0$). The remaining Maxwell equations $\partial \cdot E = \partial \cdot B = 0$ combine into $\Delta\Phi = 0$, which can be derived from the 3-dimensional ('dimensionally reduced') action $S_{\text{red}} = \frac{1}{2}\int d^3x \; |\partial\Phi|^2$, explicitly invariant under the duality rotations. Thus the on-shell symmetry has become an off-shell one in three dimensions, provided we use the magnetic potential ψ, i.e. exchange a Bianchi identity with a field equation. In fact, the reduced action is even invariant under the full 2-dimensional Euclidean group $ISO(2)$, since we can shift the potential Φ by a constant. Considering the potentials ϕ and ψ as elements of the coset space $ISO(2)/SO(2)$ we may interprete the action for Φ as describing a non-linear σ-model. Although this may appear somewhat fancy at this point, this way of describing the reduced theory nicely fits into the structure found in the more general cases to be considered below.

One may wonder, how the 4-dimensional Lagrangean $\frac{1}{2}(E^2 - B^2)$ has turned into the 3-dimensional one $\frac{1}{2}|\partial\Phi|^2 = \frac{1}{2}(E^2 + B^2)$. One way to obtain this is to add the Bianchi identity $\partial \cdot B = 0$ for F_{ij} with a Lagrangean multiplier ψ to the dimensionally reduced action

$$\int d^3x \left(\frac{1}{2}(E^2 - B^2) + \psi\partial \cdot B\right) = \int d^3x \, \frac{1}{2}\Big((\partial\phi)^2 + (\partial\psi)^2 - (B - \partial\psi)^2\Big), \quad (5)$$

and to independently vary with respect to B and ψ. Variation with respect to B then just yields the algebraic equation $B = \partial\psi$ and thus the last term can be omitted, if this identity is used as a definition for B.

The duality symmetry is broken, as soon as an electric current j_μ is added as a source for the Maxwell field. In order to maintain the symmetry a corresponding magnetic current $\tilde{j}_\mu$ has to be added as a source for $^*F$. Yet, the latter violates the integrability condition for the existence of A_μ, as does the electrical current for $\tilde{A}_\mu$. On the other hand we need the potentials as dynamical variables for the local variational principle. However, the integrability conditions are only violated, where the currents are non-vanishing. Hence, supposing that we are dealing only with point-like electric and magnetic charges we can still maintain duality, if we cut out the positions of the charges from space. Although this destroys the simply-connectedness of the space, vector potentials nevertheless exist, if we allow for Dirac string singularities (resp. connections on non-trivial $U(1)$ bundles [31]). Denoting electric resp. magnetic charges by q resp. p, we have to transform $q + ip \to e^{i\theta}(q + ip)$ in order to maintain the duality invariance of the eqs.(1) in the presence of the corresponding point sources. Classically q and p can take any value, but quantum mechanically electric and magnetic charges of two particles have to obey the Dirac quantization condition

$$q_1 p_2 = 2\pi n\hbar \qquad n \in \mathbf{Z} \,, \tag{6}$$

again violating duality invariance. This can, however, be remedied introducing dyons carrying both types of charge and obeying the Dirac–Schwinger–Zwanziger quantization condition [32,33]

$$q_1 p_2 - q_2 p_1 = 2\pi\hbar \qquad n \in \mathbf{Z} \,, \tag{7}$$

defining a 2-dimensional lattice in the q, p-plane. There is one more important aspect of the above quantization condition related to the fact that the charges in a gauge theory act not only as sources of the gauge field, but also as coupling constants of the minimal coupling of the point particles to gauge fields, yielding the forces acting on the charges in the presence of the gauge field. The relation (6) shows that weak coupling of electric charges (i.e. small values of q in natural units) implies strong coupling in the magnetic sector and vice versa. It is this aspect of duality that carries over to the modern developments in string theory.

With the rise of non-abelian gauge groups in particle physics it was natural to attempt an extension of the abelian duality transformations to Yang–Mills theories. However, as was shown by Deser and Teitelboim [34], there is no such extension acting on the potentials or the field strength. Nevertheless, as was argued by Montonen and Olive [35], there might emerge a new form of duality in the quantized theory relating the original gauge theory to its soliton sector. Since in quantum field theory particles are represented by local fields there is no distinction at this point between quantum particles corresponding to solitons and those corresponding to the original gauge and Higgs fields. The proposal of Montonen and Olive was to find a duality that exchanges the two types of particles resp. fields. A specially favourable candidate for such a duality is the N=4 extended supersymmetric gauge theory. More recently the proposal of Montonen and Olive has been extended and made more concrete by Seiberg and Witten [36].

3 Duality in Kałuza–Klein Theories

Kałuza-Klein theories in four space-time dimensions are obtained from Einstein's theory in D dimensions assuming the existence of an isometry group generated by $D-4$ Killing vector (KV) fields. Expressing the D-dimensional theory in terms of 4-dimensional fields one obtains by this 'dimensional reduction' an effective four dimensional theory with gravity, vector fields and scalar fields. Whereas for non-abelian isometry groups this is in general an intricate procedure (except if the group acts freely, i.e. the orbits are homeomorphic to the group itself), it is more or less straightforward for abelian groups generated by commuting KV fields, which we will consider from now on.

For stationary resp. stationary, axially symmetric solutions of this four dimensional theory further dimensional reductions yield three resp. two dimensional theories. In the following sections we will study the mechanism of dimensional reduction first generally going from D to d dimensions and subsequently for the special cases d=4, 3 and 2.

3.1 Dimensional Reduction from D to d Dimensions

We shall first investigate the general dimensional reduction of the gravitational field in a D-dimensional space invariant under a $(D-d)$-dimensional abelian isometry group. Under dimensional reduction we understand a parametrization of the D-dimensional gravitational field invariant under the isometry group in terms of suitable fields defined on the space of orbits Σ_d of the isometry group. In order for this orbit space to have a reasonable manifold structure the action of the isometry group has to fulfil certain regularity conditions [37]. However, since the decomposition of the gravitational field is based on the Killing vector fields describing the action of the Lie algebra of the isometry group such global conditions play no rôle in the following discussion. They become however relevant, when one wants to reconstruct the D-dimensional manifold. Although the process of reduction could clearly be formulated in a completely geometrical, coordinate independent language it turns out to be convenient to use an adapted coordinate system

$$x^M = (x^m, \bar{x}^{\bar{m}}) \quad \text{with} \quad x^m \in \mathrm{R}^d \quad \text{and} \quad \bar{x}^{\bar{m}} \in \mathrm{R}^{D-d} \tag{8}$$

such that the Killing vector fields are $K^{\bar{m}} = \frac{\partial}{\partial \bar{x}^{\bar{m}}}$ and hence the fields considered are $\bar{x}$-independent. The coordinates x^m parametrize the d-dimensional orbit space Σ_d — the space-time manifold of the reduced theory. The gravitational field in the D-dimensional space is described by the metric tensor g_{KL} or alternatively by the D-bein fields E_K^A related to the metric by $g_{KL} = (E^{\mathrm{T}})_K{}^A \eta_{AB} E^B{}_L$, where $\eta_{AB} = (+-\ldots-)$ is the D-dimensional Minkowski metric. The invariance under local Lorentz transformations acting on $E^A{}_K$ from the left can be used to bring the D-bein into the form

$$E^A{}_K = \begin{pmatrix} \lambda\, \tilde{e}^a{}_k & 0 \\ \bar{e}^{\bar{a}}{}_{\bar{m}}\, A^{\bar{m}}{}_k & \bar{e}^{\bar{a}}{}_{\bar{k}} \end{pmatrix} , \tag{9}$$

adapted to the decomposition into $(D-d)$ and d-dimensional components. The field $\bar{e}^{\bar{a}}{}_{\bar{k}}$ is a $(D-d)$-bein for the isometry group, whereas $\tilde{e}^a{}_k$ is a d-bein for the orbit space, λ is a suitable conformal factor chosen later in order to simplify the Lagrangean. The fields $A^{\bar{m}}{}_k$ are a column of $D-d$ vector potentials. The vanishing of the field strength $F^{\bar{m}}{}_{kl} = \partial_k A^{\bar{m}}{}_l - \partial_l A^{\bar{m}}{}_k$ is the condition for the hypersurface orthogonality of the Killing vector fields $K^{\bar{m}}$. From $\bar{e}^{\bar{a}}{}_{\bar{m}}$ we can build the metric m on the isometry group as usual $m = -\bar{e}^{\mathrm{T}}\bar{\eta}\bar{e}$ with $\bar{\eta} = (-\ldots-)$ in case all the Killing vectors are space-like resp. $\bar{\eta} = (+-\ldots-)$ if one of them is time-like. The D-bein field $E^A{}_K$ transforms covariantly under diffeomorphisms and local Lorentz transformations, but the special triangular form eq.(9) is preserved only by a subgroup consisting of

- diffeomorphisms and local Lorentz transformations in d dimensions,
- local Lorentz transformations in $D-d$ dimensions depending only on x and not on $\bar{x}$,
- global $GL(D-d)$ transformations acting linearly on the coordinates $\bar{x}$,
- diffeomorphisms of the special form $\bar{x} \to \bar{x} + \bar{f}(x)$ acting as gauge transformations $A^{\bar{m}}{}_k \to A^{\bar{m}}{}_k + \partial_k \bar{f}^{\bar{m}}(x)$ on the vector potentials.

The Lagrangean for the D-dimensional gravitational theory can be expressed by the d-dimensional fields of the parametrization eq.(9)

$$\begin{aligned}\mathcal{L}^{(D,d)} &= -\frac{1}{2}ER \\ &= \tilde{e}\rho\lambda^{d-2}\Big[-\frac{1}{2}\tilde{R} - \frac{1}{8}\lambda^{-2}F^{\mathrm{T}}_{kl}mF^{kl} + \frac{1}{8}h^{kl}\Big(\mathrm{Tr}(m^{-1}\partial_k m m^{-1}\partial_l m) \\ &\quad -4\rho^{-2}\partial_k\rho\partial_l\rho - 8(d-1)\lambda^{-1}\partial_k\lambda\rho^{-1}\partial_l\rho \\ &\quad -4(d-1)(d-2)\lambda^{-2}\partial_k\lambda\partial_l\lambda\Big)\Big] \end{aligned} \tag{10}$$

where $h_{kl} = \eta_{ab}\tilde{e}^a{}_k\tilde{e}^b{}_l$ and $\rho = \det(\bar{e})$. R resp. $\tilde{R}$ are the scalar curvature of $E^A{}_K$ resp. $\tilde{e}^a{}_k$.

For $d \geq 3$ we introduce $\mu = \lambda^{-2}m$. In addition we can choose $\lambda = \rho^{-\frac{1}{d-2}}$ and eliminate the scalar factor in front of $\tilde{R}$ and hence obtain the conventional form of the gravitational Lagrangean in d dimensions

$$\begin{aligned}\mathcal{L}^{(D,d)} = \tilde{e}\Big[-\tfrac{1}{2}\tilde{R} - \tfrac{1}{8}F^{\mathrm{T}}_{kl}\mu F^{kl} + \tfrac{h^{kl}}{8}\Big(&\mathrm{Tr}(\mu^{-1}\partial_k\mu\mu^{-1}\partial_l\mu) \\ &- \tfrac{1}{D-2}\mathrm{Tr}(\mu^{-1}\partial_k\mu)\mathrm{Tr}(\mu^{-1}\partial_l\mu)\Big)\Big] .\end{aligned} \tag{11}$$

The Lagrangean $\mathcal{L}^{(D,d)}$ describes the interaction of a d-dimensional gravitational field $\tilde{e}^a{}_k$ with $D-d$ abelian vector fields $(A^{\bar{m}})_k$ and a symmetric matrix μ of scalar fields, which can be considered as taking values in the coset spaces $GL(D-d)/SO(D-d)$ resp. $GL(D-d)/SO(1, D-d-1)$ depending on the signature of the metric $\bar{\eta}$. The expressions $c_1\mathrm{Tr}(\mu^{-1}d\mu)^2 + c_2(\mathrm{Tr}(\mu^{-1}d\mu))^2$ are invariant metrics for these spaces. The action for μ describes a so-called non-linear σ-model. Non-linear σ-models are field theories over a space-time manifold Σ with coordinates x^α and metric $g_{\alpha\beta}(x)$ and fields assuming values in a target space $\bar{\Phi}$ with coordinates $\bar{\phi}^i$ and metric $\bar{\gamma}_{ij}$. The action for such a non-linear σ-model

$$S_{\bar{\Phi}} = \frac{1}{2}\int_{\Sigma} \sqrt{g}dx g^{\alpha\beta}(x)\partial_\alpha\bar{\phi}^i(x)\partial_\beta\bar{\phi}^j(x)\bar{\gamma}_{ij}(\bar{\phi}(x)) \tag{12}$$

leads to the field equations

$$D^\alpha\partial_\alpha\phi^i = \frac{1}{\sqrt{h}}\partial_\alpha(\sqrt{h}h^{\alpha\beta}\partial_\beta\phi^i) + \Gamma^i_{jk}(\phi)\partial_\alpha\phi^j\partial_\beta\phi^k h^{\alpha\beta} = 0\,, \tag{13}$$

where Γ^i_{jk} are the Christoffel symbols of the metric γ_{ij}. Solutions of eq.(13) are known as harmonic maps in the mathematical literature [24]. Many of the well-known (existence and uniqueness) results on harmonic functions could be generalized to this class of maps. Some interesting applications in gravity can be found in [25]. The target spaces considered here are very special – they are riemannian resp. pseudo-riemannian symmetric spaces G/H [11]. In Appendix A we have collected some important group theoretical aspects of these spaces. Using coset representatives M with values in the group G (compare Appendix A for details and notation) we can rewrite eqs.(12,13) in the form

$$S_{\bar G/\bar H} = \frac{1}{2}\int\limits_\Sigma \sqrt{g}dxg^{\alpha\beta}\langle \bar J_\alpha, \bar J_\beta\rangle\,, \tag{14}$$

$$D^\alpha \bar J_\alpha = 0\,. \tag{15}$$

where $\bar J_\alpha = \frac{1}{2}\bar M^{-1}\partial_\alpha \bar M$.

The vector fields $(A^{\bar m})$ transform with a vector representation of $GL(D-d)$ and are coupled non-trivially to the scalars, which play the rôle of space-time dependent dielectric constants. In order to get a positive energy density T^{00} it is essential to have $\bar\eta = (-\ldots-)$, which we have already anticipated in taking the additional dimensions greater than four all space-like.

3.2 Reduction to $d = 4$ Dimensions

The special case $D = 5$, $d = 4$ is the original Kałuza theory [38] with the Maxwell field A_κ and the scalar field $m = \rho^2$.

There is an important aspect of the case $d = 4$ (D arbitrary) connected with generalized duality transformations of the Maxwell fields ${A^{\bar m}}_\kappa$. The Lagrangean eq.(11) contains the vector fields ${A^{\bar m}}_\kappa$ only through their field strengths ${F^{\bar m}}_{\kappa\nu}$, since the scalar fields are neutral with respect to the gauge group acting on the A's. Correspondingly the field equation for the vector fields is simply

$$(\mu F^{\kappa\lambda})_{;\kappa} = 0 \tag{16}$$

without any currents on the r.h.s.. In addition the field strengths obey the Bianchi-identities

$${^*F^{\kappa\nu}}_{;\kappa} = 0 \quad \text{with} \quad {^*F^{\kappa\nu}} = \frac{1}{2\tilde e}\epsilon^{\kappa\lambda\mu\nu}F_{\mu\nu}\,. \tag{17}$$

Putting $\mu F_{\kappa\lambda} = {^*\tilde F_{\kappa\lambda}}$ we can interprete eq.(16) as the Bianchi-identity for $\tilde F_{\kappa\lambda}$, which accordingly are the field strengths of potentials $(\tilde A_{\bar m})_\kappa$ obeying the field equation

$$(\mu^{-1}\tilde{F}^{\kappa\lambda})_{;\kappa} = 0 \,. \tag{18}$$

The potentials $\tilde{A}_{\bar{m}\kappa}$ provide an alternative description of the vector fields. They transform, however, contragrediently under $GL(D-4)$ similarly as μ and μ^{-1}

$$A_\kappa \longrightarrow gA_\kappa \tag{19a}$$
$$\tilde{A}_\kappa \longrightarrow (g^{\mathrm{T}})^{-1}\tilde{A}_\kappa \tag{19b}$$
$$\mu \longrightarrow (g^{\mathrm{T}})^{-1}\mu g^{-1} \tag{19c}$$
$$\mu^{-1} \longrightarrow g\mu^{-1}g^{\mathrm{T}} \,. \tag{19d}$$

Like in the case of source free electrodynamics in flat space discussed in Sect. 2 we can render the symmetry of the field equations (16) and (18) under the discrete duality transformation $A^{\bar{m}}{}_\kappa \rightarrow \tilde{A}_{\bar{m}\kappa}$ explicit putting them together to form the $2(D-4)$ components of a vector potential

$$\underline{A}_\kappa = \begin{pmatrix} A_\kappa \\ \tilde{A}_\kappa \end{pmatrix} \tag{20}$$

with the field strength $\underline{F}_{\kappa\nu} = \partial_\kappa \underline{A}_\nu - \partial_\nu \underline{A}_\kappa$ obeying the twisted self-duality constraint

$$\underline{F}_{\kappa\nu} = \Omega\, {}^*\underline{F}_{\kappa\nu} \quad \text{with} \quad \Omega = Y\bar{M} = \begin{pmatrix} 0 & 1 \\ -1 & 0 \end{pmatrix} \begin{pmatrix} \mu & 0 \\ 0 & \mu^{-1} \end{pmatrix} \tag{21}$$

satisfying again $\Omega^2 = -1$. Due to (21) the field equation

$$(\Omega \underline{F}^{\kappa\lambda})_{;\kappa} = 0 \tag{22}$$

is obviously equivalent to the Bianchi identity

$$({}^*\underline{F}^{\kappa\lambda})_{;\kappa} = 0. \tag{23}$$

Again we may look for a Lagrangean describing the interaction of the $\tilde{A}_\kappa$'s. As described in Chap. 2 this can be achieved considering the field strength $F_{\kappa\nu}$ as independent dynamical variable besides the potential A_κ and adding the Bianchi identities with a Lagrange multiplier $\tilde{A}_\kappa$

$$\mathcal{L} = -\frac{\tilde{e}}{8} F^{\mathrm{T}}_{\kappa\lambda}\mu F^{\kappa\lambda} \rightarrow \mathcal{L}' = -\frac{\tilde{e}}{8} F^{\mathrm{T}}_{\kappa\lambda}\mu F^{\kappa\lambda} + \frac{\tilde{e}}{2}\tilde{A}^{\mathrm{T}}_\kappa {}^*F^{\kappa\lambda}{}_{;\lambda} \tag{24}$$

to obtain

$$\mathcal{L}'' = -\frac{\tilde{e}}{8}\tilde{F}^{\mathrm{T}}_{\kappa\lambda}\mu^{-1}\tilde{F}^{\kappa\lambda} \,. \tag{25}$$

As before there is no way to express the Lagrangean symmetrically in A_κ and $\tilde{A}_\kappa$ and hence the discrete duality transformation $A_\kappa \rightarrow \tilde{A}_\kappa$ is an 'on-shell' symmetry valid for the equations of motion, but not for the action.

The fact that a vector field gives rise to another vector field via dualization is clearly special to 4 dimensions. In general a p-form will lead to a $(d-p-2)$-form by dualization. Certain supergravity models [4] for instance lead to gauge potentials which are 2-forms yielding by dualization a scalar 'axion' field in 4 dimensions. In fact, it turns out to be possible to generalize the structure of the four-dimensional theory we have obtained through the Kaluza–Klein reduction in a way that allows to include all the known supergravity models with abelian gauge fields [12]. Besides the gravitational field these theories contain scalar fields $\bar{\phi}^i$ of a non-linear σ-model with a target space $\bar{\Phi}$ which is assumed to be a non-compact Riemannian symmetric space $\bar{G}/\bar{H}$. In addition there is a k-dimensional column $A_\alpha = (A^I_\alpha)$ $(I = 1, \ldots, k)$ of real vector fields with field strengths $F_{\alpha\beta} = \partial_\alpha A_\beta - \partial_\beta A_\alpha$ and their duals ${}^*F^{\alpha\beta} = \frac{1}{2\sqrt{g}}\epsilon^{\alpha\beta\gamma\delta}F_{\gamma\delta}$ $({}^{**}F = -F)$ satisfying Bianchi identities ${}^*F^{\alpha\beta}{}_{;\alpha} = 0$. The most general gauge invariant action quadratic in the field strengths is

$$S_V = \int\limits_{\Sigma} \sqrt{g}dx\Big(-\frac{1}{4}F^{\mathrm{T}}_{\alpha\beta}(\mu(\bar{\phi})F^{\alpha\beta} - \nu(\bar{\phi}){}^*F^{\alpha\beta})\Big) \tag{26}$$

where $\mu = (\mu_{IJ})$ and $\nu = (\nu_{IJ})$ are real symmetric matrices depending on the fields $\bar{\phi}^i$. This action yields field equations

$$(\mu F^{\alpha\beta} - \nu{}^*F^{\alpha\beta})_{;\alpha} = 0 \tag{27}$$

Introducing in the by now familiar way dual potentials $\tilde{A}_\alpha = (\tilde{A}_{\alpha I})$ we have obtained $2k$ vectors

$$\underline{A}_\alpha = \begin{pmatrix} A_\alpha \\ \tilde{A}_\alpha \end{pmatrix} \tag{28}$$

and field strengths $\underline{F}_{\alpha\beta} = \partial_\alpha \underline{A}_\beta - \partial_\beta \underline{A}_\alpha$ satisfying the linear relation

$$\underline{F}_{\alpha\beta} = \bar{Y}\bar{M}{}^*\underline{F}_{\alpha\beta} = \Omega{}^*\underline{F}_{\alpha\beta} \tag{29}$$

with

$$\bar{Y} = \begin{pmatrix} 0 & \eta^{\mathrm{T}} \\ -\eta & 0 \end{pmatrix}, \quad \bar{M} = \begin{pmatrix} \mu + \nu\,\mu^{-1}\,\nu & \nu\,\mu^{-1}\,\eta^{-1} \\ (\eta^{\mathrm{T}})^{-1}\,\mu^{-1}\,\nu & (\eta^{\mathrm{T}})^{-1}\,\mu^{-1}\,\eta^{-1} \end{pmatrix} \tag{30}$$

where η is some real $k \times k$ matrix and the symmetric matrix $\bar{M}$ satisfies $\bar{Y}\bar{M}\bar{Y} = -\bar{M}^{-1}$ and hence $\Omega^2 = -1$. The case of the coupled Einstein-Maxwell equations corresponds to putting $\mu = 1$, $\nu = 0$ and $\bar{G} = \bar{H} = SO(2)$. The equations

$${}^*\underline{F}^{\alpha\beta}{}_{;\alpha} = -(\bar{Y}\bar{M}\underline{F}^{\alpha\beta})_{;\alpha} = 0 \tag{31}$$

can be interpreted as field equations and Bianchi identities for either $F_{\alpha\beta}$ or ${}^*F_{\alpha\beta}$. These equations are $\bar{G}$-invariant under the transformation

$$\bar{G} \ni \bar{g}\colon\ \underline{F}_{\alpha\beta} \to \bar{\rho}(\bar{g})\underline{F}_{\alpha\beta} \tag{32}$$

provided the real representation $\bar{\rho}$ satisfies $\bar{\rho}(\bar{g}^{-1})^{\mathrm{T}} = \bar{Y}^{-1}\,\bar{\rho}(\bar{g})\,\bar{Y}$ and we can construct a matrix $\bar{M}(\bar{\phi})$ of the form (30) depending on the scalars $\bar{\phi}^i$ in such a way that

$$\bar{G} \ni \bar{g}\colon\ \bar{M}(\bar{\phi}) \to \bar{\rho}(\bar{g}^{-1})^{\mathrm{T}}\,\bar{M}(\bar{\phi})\,\bar{\rho}(\bar{g}^{-1})\,. \tag{33}$$

The contribution from S_V to the stress tensor, which takes a particularly simple form using the 'doubled' field strength $\underline{F}$

$$\begin{aligned} T^{(V)}_{\alpha\beta} &= -F^{\mathrm{T}}_{\alpha\gamma}\,\mu\,F_\beta{}^\gamma + \tfrac{1}{4}g_{\alpha\beta}\,F^{\mathrm{T}}_{\gamma\delta}\,\mu\,F^{\gamma\delta} \\ &= -\tfrac{1}{2}\underline{F}^{\mathrm{T}}_{\alpha\gamma}\,\bar{M}\,\underline{F}_\beta{}^\gamma = -\tfrac{1}{2}\underline{F}^{\mathrm{T}}_{\alpha\gamma}\,\bar{Y}\,{}^*\underline{F}_\beta{}^\gamma \end{aligned} \tag{34}$$

as well as the contribution to the scalar field equations

$$\frac{\delta S_V}{\delta\bar{\phi}^i} = -\frac{1}{4}\sqrt{g}F^{\mathrm{T}}_{\alpha\beta}\Big(\frac{\delta\mu}{\delta\bar{\phi}^i}F^{\alpha\beta} - \frac{\delta\nu}{\delta\bar{\phi}^i}{}^*F^{\alpha\beta}\Big) = -\frac{1}{8}\sqrt{g}\underline{F}^{\mathrm{T}}_{\alpha\beta}\frac{\delta\bar{M}}{\delta\bar{\phi}^i}\underline{F}^{\alpha\beta} \tag{35}$$

will then be explicitly $\bar{G}$-invariant. Since $T^{(V)}_{00}$ must be positive for a physically meaningful theory the matrix $\bar{M}$ (and hence μ) must be positive definite.

Collecting all the terms in the Lagrangean and the field equations describing the coupling of the gravitational field $g_{\alpha\beta}$, the vector field strenghts and their duals comprised in $\underline{F}_{\alpha\beta}$ and the scalars $\bar{\phi}^i$ parametrizing $\bar{M}$ we obtain

$$\begin{aligned} \mathcal{L}^{(4)} = \sqrt{g}\Big(&-\frac{1}{2}R^{(4)} + \frac{1}{8}g^{\alpha\beta}\langle\bar{M}^{-1}\partial_\alpha\bar{M}, \bar{M}^{-1}\partial_\beta M\rangle \\ &-\frac{1}{4}F^{\mathrm{T}}_{\alpha\beta}\Big(\mu(\bar{\phi})F^{\alpha\beta} - \nu(\bar{\phi}){}^*F^{\alpha\beta}\Big)\Big)\,, \end{aligned} \tag{36}$$

$$R_{\alpha\beta} = -\frac{1}{2}\underline{F}^{\mathrm{T}}_{\alpha\gamma}\,\bar{M}\,\underline{F}_\beta{}^\gamma + \frac{1}{4}\langle\bar{M}^{-1}\partial_\alpha\bar{M}, \bar{M}^{-1}\partial_\beta\bar{M}\rangle\,, \tag{37}$$

$$(\bar{M}\,\underline{F}^{\alpha\beta})_{;\alpha} = 0\,, \tag{38}$$

$$(\bar{M}^{-1}\partial_\alpha\bar{M})_{;\alpha} = -\frac{1}{2\bar{c}}\Big(\underline{F}^{\alpha\beta}\underline{F}^{\mathrm{T}}_{\alpha\beta}\bar{M}\Big)_{\mathrm{pr}} \tag{39}$$

where $(\ldots)_{\mathrm{pr}}$ denotes a projection on the Lie algebra of G (see Appendix A).

3.3 Reduction to $d = 3$ Dimensions

In this section we shall see how field configurations of the 4-dimensional theories discussed above allowing one Killing vector field give rise to a 'dimensionally reduced' 3-dimensional field theory. Since this reduction can be performed irrespective of the signature of the KV field, it can be applied to time-translations or a space-like isometry like axial rotations, but for reasons of simplicity we shall use the notations for time-translations. In case we want to explicitly refer to a space-like KV we shall indicate this with a prime on the corresponding field.

Since our 4-dimensional models contain gauge fields and scalars besides the gravitational field, we have to require that the Lie derivative of the field strengths and of the scalar fields vanish. Then we can and will use a gauge in which also the Lie derivative of the gauge potentials A^I_α vanishes. As before we choose adapted coordinates such that the isometry is just a translation (e.g. $K = \frac{\partial}{\partial t}$). All the fields $F_{\alpha\beta}$, $\bar{\phi}^i$ and A^I_α will then depend only on the remaining three coordinates x^m ($m = 1, 2, 3$) parametrizing the orbit space Σ_3 of the action of K. In these coordinates K_α has the form $K_\alpha = (\Delta k_m, \Delta)$ with $\Delta \equiv K^2$ and the metric $g_{\alpha\beta}$ can be decomposed as (compare eq.(9) for the 4-bein)

$$g_{\alpha\beta} = \begin{pmatrix} -\frac{1}{\Delta}h_{mn} + \Delta k_m k_n & \Delta k_n \\ \Delta k_m & \Delta \end{pmatrix} . \tag{40}$$

We will use the rescaled metric h_{mn} on the three dimensional orbit space Σ_3, since it leads to the standard form of the Lagrangean in 3 dimensions. In order to do this we have to require $\Delta \neq 0$. h_{mn} is positive definite if K is time-like, i.e. $\Delta > 0$ and has the signature $(+--)$ if K is space-like, i.e. $\Delta < 0$. Similarly we decompose the vector fields $A^I_\alpha = (\hat{A}^I_m + k_m A^I, A^I)$ into pieces $\hat{A}^I_m$ and A^I perpendicular and parallel to K. We can now rewrite the Lagrangean (36) in the form (apart from surface terms)

$$\begin{aligned}\tilde{\mathcal{L}} = \sqrt{h}\Big(\frac{1}{2}R^{(3)} - \frac{1}{2}h^{mn}\Big(\langle \bar{J}_m, \bar{J}_n\rangle - \frac{1}{\Delta}\partial_m A^{\mathrm{T}}\mu(\bar{\phi})\partial_n A \\ + \frac{1}{2\Delta^2}\partial_m\Delta\partial_n\Delta\Big) + \frac{\Delta^2}{8}k_{mn}k^{mn} \\ + (\hat{F}_{mn} + k_{mn}A)^{\mathrm{T}}\Big(-\frac{\Delta}{4}\mu(\bar{\phi})(\hat{F}^{mn} + k^{mn}A) + \frac{1}{2\sqrt{h}}\epsilon^{mnp}\nu(\bar{\phi})\partial_p A\Big)\Big)\end{aligned} \tag{41}$$

where $R^{(3)}$ is the scalar curvature for h_{mn}, $k_{mn} = \partial_m k_n - \partial_n k_m$, $\hat{F}_{ab} = \partial_m \hat{A}_n - \partial_n \hat{A}_m$ and all indices are raised or lowered with the metric h. The Killing vector field is hypersurface orthogonal iff $k_{mn} = 0$.

If the original field configuration was a solution of the 4-dimensional field equations eqs.(37) then h_{mn}, Δ, k_m, A^I, $\hat{A}^I_m$ and $\bar{\phi}^i$ are a solution of the three dimensional field equations derived from the action $\int_{\Sigma_3}\tilde{\mathcal{L}}\,d^3x$ and vice versa. The field equations for the 3-dimensional vector fields $\hat{A}^I_m$ and k_m

$$\Big(-\Delta\mu(\bar{\phi})(\hat{F}^{mn} + k^{mn}A) + \tfrac{1}{\sqrt{h}}\epsilon^{mnp}\nu(\bar{\phi})\partial_p A\Big)_{;m} = 0 \tag{42}$$

$$\Big(\tfrac{\Delta^2}{2}k^{mn} + A^{\mathrm{T}}\Big(-\Delta\mu(\bar{\phi})(\hat{F}^{mn} + k^{mn}A) + \tfrac{1}{\sqrt{h}}\epsilon^{mnp}\nu(\bar{\phi})\partial_p A\Big)_{;m} = 0 \tag{43}$$

can be considered as Bianchi identities for dual potentials $\tilde{A}_I$ (which are just the time components $K^\alpha \tilde{A}_{\alpha I}$ of the dual potentials $\tilde{A}_{\alpha I}$) and $\tilde{\psi}$ (the twist potential).

$$\hat{F}^{mn} + k^{mn}A = \frac{1}{\sqrt{h}}\epsilon^{mnp}\frac{1}{\Delta}\mu^{-1}(\nu\partial_p A - \eta^{-1}\partial_p\tilde{A}) \tag{44}$$

$$k^{mn} = \frac{1}{\sqrt{h}}\epsilon^{mnp}\frac{1}{\Delta^2}\omega_p \qquad \text{with} \tag{45}$$

$$\partial_m\tilde{\psi} = \omega_m - (\tilde{A}^{\mathrm{T}}(\eta^{\mathrm{T}})^{-1}\partial_m A - A^{\mathrm{T}}\eta^{-1}\partial_m\tilde{A}) \, . \tag{46}$$

Putting $\underline{A} = (A, \tilde{A})$ we can rewrite the twist vector in the explicitly $\bar{G}$-invariant form

$$\omega_m = \partial_m\tilde{\psi} + \underline{A}^{\mathrm{T}}\bar{Y}^{-1}\partial_m\underline{A} \, . \tag{47}$$

The Lagrangean of the dimensionally reduced three dimensional theory becomes

$$\begin{aligned}\mathcal{L}^{(3)} &= \sqrt{h}\Big(\tfrac{1}{2}R^{(3)} - \tfrac{1}{2}h^{mn}\Big(\tfrac{1}{2\Delta^2}(\partial_m\Delta\partial_n\Delta + \omega_m\omega_n) \\ &\qquad + \langle\bar{J}_m, \bar{J}_n\rangle - \tfrac{1}{\Delta}\partial_m\underline{A}^{\mathrm{T}}\bar{M}\partial_n\underline{A}\Big)\Big) \\ &\equiv \sqrt{h}\Big(\tfrac{1}{2}R^{(3)} - \tfrac{1}{2}h^{mn}\partial_m\phi^i\partial_n\phi^j\gamma_{ij}(\phi)\Big)\end{aligned} \tag{48}$$

where $\gamma_{ij}(\phi)$ is a metric on the 'potential space' Φ parametrized by $\phi = (\Delta, \tilde{\psi}, \underline{A}, \bar{\phi})$. Thus we have obtained a non-linear σ-model with a target space Φ coupled to (three dimensional) gravity.

For a space-like Killing vector ($\Delta < 0$) the metric on Φ is positive definite, but for a time like-Killing vector ($\Delta > 0$, stationary solutions) the metric is indefinite with $2k$ negative terms due to the fields $\underline{A}$ originating from the k vector fields in the 4-dimensional theory.

In terms of the various components the field equations are

$$R^{(3)}_{mn} = \partial_m\phi^i\partial_n\phi^j\gamma_{ij}(\phi) \, , \tag{49a}$$

$$\left(\tfrac{\omega_m}{\Delta^2}\right)_{;m} = 0 \, , \tag{49b}$$

$$\left(\tfrac{\bar{M}}{\Delta}\partial^m\underline{A} + \bar{Y}^{-1}\underline{A}\tfrac{\omega^m}{\Delta^2}\right)_{;m} = 0 \, , \tag{49c}$$

$$\left(\tfrac{\partial^m\Delta}{\Delta} - \underline{A}^{\mathrm{T}}\tfrac{\bar{M}}{\Delta}\partial^m\underline{A} + \tilde{\psi}\tfrac{\omega^m}{\Delta^2}\right)_{;m} = 0 \, , \tag{49d}$$

$$\left(\tfrac{1}{2}\bar{M}^{-1}\partial^m\bar{M} + \tfrac{1}{c}\left(\underline{A}\,\partial^m\underline{A}^{\mathrm{T}}\tfrac{\bar{M}}{\Delta} - \tfrac{1}{2}\underline{A}\tfrac{\omega^m}{\Delta^2}\underline{A}^{\mathrm{T}}\bar{Y}^{-1}\right)_{\mathrm{pr}}\right)_{;m} = 0 \, . \tag{49e}$$

If all potentials $\underline{A}$ for the vector fields vanish then $\omega_m = \partial_m\tilde{\psi}$ and $(\Delta,\tilde{\psi})$ parametrize the well known $SL(2)/SO(2)$ σ-model of pure gravity [7]. Before we proceed to the general case let us shortly recall the results about this simplest case. In order to recognize the metric $ds^2 = \Delta^{-2}(d\Delta^2 + d\tilde{\psi}^2)$ as the invariant metric of $SL(2)/SO(2)$, it is necessary to find a suitable parametrisation of the coset space in terms of the coordinates Δ and $\tilde{\psi}$. This can be done parametrizing suitably selected coset-representatives in the group itself or using some matrix representation. In the former case we may choose a basis for the Lie algebra of $SL(2)$ and use the exponential map. A canonical choice for the Lie algebra of $SL(2)$ consists of the three elements (d, e, k) with the commutation relations

$$[d, e] = e, \quad [d, k] = -k, \quad [e, k] = 2d \tag{50}$$

The choice

$$P = \mathrm{e}^{\frac{1}{2}\ln\Delta d}\mathrm{e}^{\tilde{\psi}k} \tag{51}$$

may be termed 'triangular' in view of a corresponding representation by triangular matrices

$$\hat{P} = \begin{pmatrix} \sqrt{\Delta} & 0 \\ \frac{\tilde{\psi}}{\sqrt{\Delta}} & \frac{1}{\sqrt{\Delta}} \end{pmatrix} \tag{52}$$

It is easily checked that the Killing metric of $SL(2)$ restricted to the coset space yields the desired expression in terms of Δ and $\tilde{\psi}$. As explained in Appendix A the group $SL(2)$ acts in a non-linear way on the triangular representatives. It is easy to check that the Lie algebra is realized on Δ and$\tilde{\psi}$ through the infinitesimal transformations

$$d: \quad \delta\Delta = -\Delta, \quad \delta\tilde{\psi} = -\tilde{\psi} \tag{53}$$

$$k: \quad \delta\Delta = 0\,, \quad \delta\tilde{\psi} = -1 \tag{54}$$

$$e: \quad \delta\Delta = 2\tilde{\psi}\Delta, \quad \delta\tilde{\psi} = \tilde{\psi}^2 - \Delta^2\,. \tag{55}$$

Obviously the elements k resp. d correspond to a trivial shift of $\tilde{\psi}$ resp. a simple rescaling of the KV, whereas the action of e really changes the solution. The latter is usually referred to as 'Ehlers' transformation.

Another convenient parametrization is in terms of the complex 'Ernst' potential $\mathcal{E} = \tilde{\psi} + i\Delta$ undergoing a Möbius transformation under the group action of $SL(2)$. It is easy to check that the whole group is generated by the shift $\mathcal{E} \to \mathcal{E}+c$ in combination with the discrete inversion $\mathcal{E} \to 1/\mathcal{E}$ discovered by Ehlers [6].

Turning back to the general case the metric on the space Φ and the field equations (49a-49e) are obviously invariant under

- $\bar{G}$-transformations (acting on $\bar{\phi}$ and $\underline{A}$),
- 'electromagnetic' gauge transformations (with a constant vector a)

$$\tilde{\psi} \to \tilde{\psi} + \underline{A}^{\mathrm{T}}\bar{Y}^{-1}a\,, \qquad \underline{A} \to \underline{A} + a\,, \tag{56}$$

- twist gauge transformations

$$\tilde{\psi} \to \tilde{\psi} + b\,, \tag{57}$$

- scale transformations

$$\Delta \to c^2\Delta\,, \qquad \tilde{\psi} \to c^2\tilde{\psi}\,, \qquad \underline{A} \to c\underline{A}\,. \tag{58}$$

It turns out that for many interesting theories (e.g. Einstein-Maxwell, various supergravities) it is possible to extend the Ehlers transformation to an invariance of the whole target space Φ. Commuting this generalized Ehlers transformation with (infinitesimal) electromagnetic gauge transformations yields generalized Harrison transformations [9]. The structure of the Lie algebras corrsponding to this extensions given in Appendix B. Their action on the scalar potentials is to lowest order (consult [28] for the complete transformations)

- (infinitesimal) generalized Ehlers transformations

$$\begin{aligned}
\delta\Delta &= 2\tilde{\psi}\Delta \\
\delta\tilde{\psi} &= \tilde{\psi}^2 - \Delta^2 + \tfrac{c}{2}\underline{A}^{\mathrm{T}}\bar{M}\underline{A}\,\Delta + O(\underline{A}^4) \\
\delta\bar{\phi} &= O(\underline{A}^2) \\
\delta\underline{A} &= (\tilde{\psi} - \Delta\bar{Y}\bar{M})\underline{A} + O(\underline{A}^3)\ ,
\end{aligned} \tag{59}$$

- (infinitesimal) generalized Harrison transformations

$$\begin{aligned}
\delta\Delta &= -c\underline{A}^{\mathrm{T}}\bar{Y}a\Delta \\
\delta\tilde{\psi} &= \tfrac{c}{2}\underline{A}(\Delta\bar{M} - \tilde{\psi}\bar{Y})a + O(\underline{A}^3) \\
\delta\bar{\phi} &= O(\underline{A}) \\
\delta\underline{A} &= (\tilde{\psi} - \Delta\bar{Y}\bar{M})a + O(\underline{A}^2)\ .
\end{aligned} \tag{60}$$

The Lie algebra of infinitesimal transformations gives rise to a noncompact Lie group G with maximal compact subgroup H' and depending on the sign of Δ the target space is either the Riemannian symmetric space $\Phi' = G/H'$ or the pseudo Riemannian symmetric space $\Phi = G/H$, where H is a noncompact real form of H'. The dimensions of G and H are

$$\dim G = \dim\bar{G} + \dim SL(2) + 4k\ , \quad \dim H = \dim\bar{H} + \dim SO(2) + 2k\ . \tag{61}$$

The submanifold where $\underline{A} = 0$ (i.e. $G/H \cap G/H'$) is $SL(2)/SO(2) \times \bar{G}/\bar{H}$ and the generators of the coset space transform under the two dimensional representation of $SL(2)$ in order to reproduce the scale transformations eq.(58). In Appendix A we give a complete list of all combinations of groups G, H, $\bar{G}$ and $\bar{H}$ corresponding to a time-like KV. For each of them we can start from the Lagrangean eq.(48) for the three dimensional theory and reconstruct the Lagrangean eq.(36) of the corresponding 4-dimensional theory.

In order to identify the potentials with suitable coordinates for the coset-space abstract 'triangular' generators may be exponentiated yielding [28]

$$P = \mathrm{e}^{\frac{1}{2}\ln\Delta d}\mathrm{e}^{\underline{A}^{\mathrm{T}}a + \tilde{\psi}k}\ . \tag{62}$$

Or we proceed in a similar way as we did previously for $\bar{G}$ (compare Appendix A) and choose a (possibly complex) irreducible matrix representation ρ of G and a hermitian matrix X satisfying $\rho(\tau(g)) = X^{-1}\,\rho(g^{-1})\,X$ and define $\hat{P} = \rho(P)$, $M = \hat{P}^{+}\,X\,\hat{P} = X\,\rho(\hat{M})$. Introducing the currents $J_m = \frac{1}{2}M^{-1}\partial_m M$ the Lagrangean eq.(48) can be rewritten in the form

$$\begin{aligned}
\mathcal{L}^{(3)} &= \sqrt{h}\Big(\frac{1}{2}R^{(3)} - \frac{1}{2}\langle J^m, J_m\rangle\Big) \\
&\equiv \sqrt{h}\Big(\frac{1}{2}R^{(3)} - \frac{\hat{c}}{8}h^{mn}\mathrm{Tr}(\hat{M}^{-1}\partial_m\hat{M}\,\hat{M}^{-1}\partial_n\hat{M})\Big)
\end{aligned} \tag{63}$$

with a constant $\hat{c}$ depending on the representation ρ and on the particular σ-model.

The field equations eqs.(49a-49e) become

$$R^{(3)}_{mn} = \langle J_m, J_n \rangle \tag{64}$$

$$J^m{}_{;m} = 0 \,. \tag{65}$$

but not all of the conserved currents J_m are independent, since $\tau(J) = -MJM^{-1}$. An alternative way writing the field equations related to the triangular representers P is (compare Appendix A)

$$D^\alpha \mathcal{J}_\alpha = 0 \quad \text{with} \quad D_\alpha \mathcal{J}_\beta = \mathcal{J}_{\beta;\alpha} - [\mathcal{A}_\alpha, \mathcal{J}_\beta] \tag{66}$$

In Sect. 5 we will study solutions with asymptotically Euclidean Σ_3 and σ-model fields which have an asymptotic multipole expansion

$$M(x) \sim \sum_{n=0}^{\infty} r^{-n} M_n(\vartheta, \varphi) \,. \tag{67}$$

Without loss of generality we can assume that we have the asymptotic values $\Delta_0 = 1$, $\tilde{\psi}_0 = \underline{A}_0 = 0$, $\bar{M}_0 = \mathbf{1}$ corresponding to $P_0 = \mathbf{1}$, $M_0 = \mathbf{1}$ or $\hat{M}_0 = X$. From the asymptotic expansion of M we can read off the 'gauge' charges

$$Q = \frac{1}{4\pi} \int_\infty J_m d\Sigma^m = -\frac{1}{2} M_1 \tag{68}$$

taking values in the Lie algebra of G. In fact those in the Lie algebra of H vanish, due to the relation $\tau(M) = M^{-1}$. The subgroup $H \subset G$ leaves the asymptotic value M_0 invariant and transforms the charges according to

$$H \ni h\colon\ M_1 \to h\, M_1\, h^{-1};. \tag{69}$$

The precise form of this action of H on the charges depends on the particular σ-model under consideration. There are, however, in all cases, generators of H which differ from the transformations eqs.(59-60) by suitable multiples of the transformations eqs.(56-57). These we can use to transform the (electric and magnetic) vector charges and the NUT charge into the mass and the scalar charges and vice versa.

3.4 Reduction to $d = 2$ Dimensions

Stationary, axially symmetric configurations of the 4-dimensional theory are characterized by their invariance under two commuting Killing vector fields $K = \frac{\partial}{\partial t}$ resp. $K' = \frac{\partial}{\partial \varphi}$ describing time-translations resp. axial rotations. Similar to the case of one KV we can perform a 'dimensional reduction' from 4 to 2 dimensions. Since we want to make use of the σ-model structure obtained for the 3-dimensional theory after suitable dualizations we prefer to employ a two step procedure. First we use one KV field to reduce from

4 to 3 dimensions and then the second one to do the step to 2 dimensions. Depending on whether we take the KV K or K' to perform the reduction from 4 to 3 dimensions we end up with two different σ-models corresponding to G/H resp. G/H', parametrized by the matrices M resp. M'. In the first case we obtain a pseudo-Riemannian coset space in the second a Riemannian one, whereas the the opposite holds for the remaining 2-d orbit spaces Σ_2.

In order to be able to do the step from 3 to 2 dimensions we first have to make sure that the Lie derivative of the 3-dimensional fields h_{mn} and M with respect to the second KV vanishes. For h_{mn} this is a direct consequence of the commutativity of the two KVs. For the electromagnetic potentials $\underline{A}$ we find that their Lie derivative has to be constant. This constant has to vanish for asymptotically trivial configurations with a regular rotation axis. In adapted coordinates the vanishing Lie derivative of h_{mn} and M means they are independent of t and φ. Hence the only non-trivial step in the reduction from 3 to 2 dimensions concerns the decomposition of the 3-metric h_{ab} which we parametrize in analogy to eq.(40) in the form

$$h_{mn} = \begin{pmatrix} \lambda^2 \tilde{h}_{kl} + \rho^2 b_k b_l & \rho^2 b_l \\ \rho^2 b_k & \rho^2 \end{pmatrix} . \tag{70}$$

In terms of these new fields the Lagrangian (63) becomes

$$\mathcal{L}^{(2)} = \rho\sqrt{\tilde{h}}\Big[-\frac{1}{2}\tilde{R} - \frac{\rho^2}{8\lambda^2} b_{kl} b^{kl} + \frac{c}{8}\langle M^{-1}\partial M, M^{-1}\partial M\rangle - \lambda^{-1}\partial\lambda\rho^{-1}\partial\rho\Big] . \tag{71}$$

A novel feature for $d = 2$ is that λ cannot be used to remove the factor ρ multiplying $\sqrt{\tilde{h}}$ in front of the Lagrangian. This is a consequence of the conformal invariance of the 2-dimensional theory.

A simplification occurs with the vector fields b_k which loose their dynamical degrees of freedom in 2 dimensions. From the field equation

$$(\rho^3 \lambda^{-2} b^{kl})_{;l} = 0 \tag{72}$$

it follows that the dual field strength ${}^*b = \frac{1}{2\sqrt{\tilde{h}}} \epsilon^{kl} \rho^3 \lambda^{-2} b_{kl}$ obeys the equation

$$\epsilon^{kl} \partial_l {}^*b = 0 \tag{73}$$

and hence ${}^*b$ is constant that vanishes for asymptotically Minkowskian solutions and consequently $b_{kl} = 0$, i.e. the orbits of the KVs are orthogonal to 2-surfaces Σ_2. This establishes a 'Generalized Papapetrou Theorem' [39]. Assuming the vanishing of b in the following $\mathcal{L}$ simplifies to

$$\mathcal{L}^{(2)} = \rho\sqrt{\tilde{h}}\Big[-\frac{1}{2}\tilde{R} + \frac{1}{8}\langle M^{-1}\partial M, M^{-1}\partial M\rangle - \lambda^{-1}\partial\lambda\rho^{-1}\partial\rho\Big] . \tag{74}$$

The field equations derived from $\mathcal{L}^{(2)}$ are

$$\tilde{R}_{kl} - \frac{1}{2}\tilde{h}_{kl}\tilde{R} = +\frac{1}{4}\langle M^{-1}\partial_k M, M^{-1}\partial_l M\rangle - 2\lambda^{-1}\partial_{(k}\lambda\rho^{-1}\partial_{l)}\rho$$

$$- \frac{1}{2}\tilde{h}_{kl}\Big(\frac{1}{4}\langle M^{-1}\partial M, M^{-1}\partial M\rangle - 2\lambda^{-1}\partial\lambda\rho^{-1}\partial\rho\Big) \quad (75a)$$

$$(\rho M^{-1}\partial^k M)_{;k} = 0\,, \quad (75b)$$

$$(\partial^k \rho)_{;k} = 0\,. \quad (75c)$$

We have omitted the field equation for λ, because it is a consequence of eq.(75a) as we will be argued below.

The 2-dimensional metric $\tilde{h}_{kl}$ can locally be brought to the conformally flat form $\tilde{h}_{kl} = \tilde{h}\delta_{kl}$ by a suitable choice of coordinates. Finally $\tilde{h}$ can be absorbed into the conformal factor λ leading to $\tilde{h}_{kl} = \delta_{kl}$. With this choice Eq. (75a) turns into a system of first order equations for λ, since the left hand side vanishes. Since the field equations for M and ρ are conformally invariant they are independent of λ and hence the same as in flat space. In particular ρ is a harmonic function on $\mathbf{R}^2$. Together with its conjugate harmonic function z defined by $\partial z = -{}^*\partial\rho$ (note that ${}^{**}\partial = -\partial$ in a 2 dimensional space with definite metric) it provides a canonical coordinatization of the 2-dimensional reduced manifold as long as $\partial\rho \neq 0$ (Weyl's canonical coordinates). Eq.(75a) then becomes

$$\lambda^{-1}\partial_z\lambda = \frac{\rho}{4}\langle M^{-1}\partial_\rho M, M^{-1}\partial_z M\rangle$$

$$\lambda^{-1}\partial_\rho\lambda = \frac{\rho}{8}\Big(\langle M^{-1}\partial_\rho M, M^{-1}\partial_\rho M\rangle - \langle M^{-1}\partial_z M, M^{-1}\partial_z M\rangle\Big)\,. \quad (76)$$

Using $\partial_\pm = \partial_z \pm i\partial_\rho$ with the property ${}^*\partial_\pm = \pm i\partial_\pm$ we can write these equations in the form

$$\partial_\pm \ln\lambda = \mp\frac{i\rho}{2}\langle \mathcal{J}_\pm, \mathcal{J}_\pm\rangle. \quad (77)$$

This equation allows to get λ from M (once the latter has been obtained solving eq.(75a)) through simple integration.

4 Geroch Group

In [19,20] it was recognized that the equation of motion (75c) for $M(x)$ can be obtained as the integrability condition of a system of linear differential equations involving a spectral parameter ('Lax Pair') in analogy to well-known 'Completely Integrable Systems' like KdV or Sine-Gordon [41]. Although originally formulated for the Einstein vacuum theory with $M \in SL(2)$ the same construction works for arbitrary symmetric spaces G/H. The linear system used in [20] can be written as

$$\partial_\pm U U^{-1} = \pm\frac{it}{1 \pm it} J_\pm \quad (78)$$

with $U(t, x)$ taking values in G (for real t) resp. some matrix representation of G. A second, even more useful form of this Linear Spectral Problem (LSP) is obtained for $\mathcal{P}(x,t) \equiv P(x)U(t,x)$ obeying [21] (compare Appendix A for the definition of $\mathcal{A}$ and $\mathcal{J}$)

$$\partial_{\pm}\mathcal{P}(x,t)\mathcal{P}(x,t)^{-1} = \mathcal{A}_{\pm} + \frac{1 \mp it}{1 \pm it}\mathcal{J}_{\pm} \qquad \text{(LSP)} \tag{79}$$

In order to reproduce the explicit ρ-dependence of the equation for P resp. M the 'spectral parameter' t has to depend on ρ and z. One finds that $t(\rho, z)$ has to obey the pair of differential equations

$$\partial_{\pm}t = t\frac{1 \mp it}{1 \pm it} \tag{80}$$

implying

$$\partial_{\pm}\left(z + \frac{\rho}{2}(\frac{1}{t} - t)\right) = 0 \tag{81}$$

and thus

$$z + \frac{\rho}{2}(\frac{1}{t} - t) = w \tag{82}$$

(where w is a constant of integration) with the solutions

$$t_{\pm}(w,x) = \frac{1}{\rho}\left((z-w) \pm \sqrt{(z-w)^2 + \rho^2}\right) = -\frac{1}{t_{\mp}} \, . \tag{83}$$

The pairs $(w, t_{\pm})$ define for each given $x = (z, \rho)$ with $\rho \neq 0$ a two-sheeted Riemann surface $\mathcal{R}_x$, with the branch points $w = z \pm i\rho$. The replacement $t \to -\frac{1}{t}$ exchanges the two sheets. We can choose the branch cut along the line segment $(z = \mathrm{Re}\, w,\ \rho \leq |\,\mathrm{Im}\, w|)$. For $z < \mathrm{Re}\, w$ resp. $z > \mathrm{Re}\, w$ the value of t_+ lies inside resp. outside and t_- lies outside resp. inside of the unit circle. For $\rho = 0$ the Riemann surface degenerates and splits into two disconnected planes $w = z$ and the function $t(w)$ becomes singular. One finds in particular

$$t_+ \underset{\rho \to 0}{\longrightarrow} \begin{cases} 0 \\ \infty \end{cases} \qquad t_- \underset{\rho \to 0}{\longrightarrow} \begin{cases} \infty & \text{for} \quad z < \mathrm{Re}\, w \, , \\ 0 & \text{for} \quad z > \mathrm{Re}\, w \end{cases} \tag{84}$$

Since t depends on x it is important to recognize that $\partial\mathcal{P}$ is to be interpreted as differentiation for fixed w, i.e.

$$\partial_{\pm}\mathcal{P}(t,x) = \partial_{\pm}\mathcal{P}(t,x)\Big|_t + \partial_{\pm}t(w,x)\frac{\partial\mathcal{P}(t,x)}{\partial t} \tag{85}$$

This implies that to any given value of w there will be two solutions $\mathcal{P}_{\pm}(w,x)$ according to the choice $t_{\pm}(w)$. Under suitable conditions on $P(x)$ (see below) $\mathcal{P}_{\pm}$ will however define a single valued function $\mathcal{P}(t,x)$ on some domain on $\mathcal{R}_x$. We still have to supplement the equation for $\mathcal{P}$ with a suitable normalization condition. Since for $t = 0$ (resp. $w = \infty$) the r.h.s. of eq.(79) becomes $\partial_{\pm}PP^{-1}$ we can impose $\mathcal{P}(0,x) = P(x)$. More precisely we shall try to find solutions $\mathcal{P}(t,x)$ having a Taylor series expansion around $t = 0$ starting with $P(x)$.

For real t eq.(79) is invariant under the transformation

$$\tau^{(\infty)}\colon \mathcal{P}(t,x) \to \tau(\mathcal{P}(-\frac{1}{t},x)) . \tag{86}$$

This transformation $\tau^{(\infty)}$ plays an important group-theoretical role. We may consider the r.h.s. of eq.(79) as an element of the infinite-dimensional loop algebra with respect to the Lie algebra of G [40]. Applied to elements of this Lie algebra $\tau^{(\infty)}$ is a natural extension of the involutive automorphism τ of G. Assuming for the moment there is a corresponding infinite-dimensional group $G^{(\infty)}$ (the 'Geroch group'), the function $\mathcal{P}(t,x)$ considered as an element of that group is again 'triangular' in the sense that it has a Taylor series expansion at $t=0$ (i.e. only positive powers of t) and the t-independent term is triangular.

A central object for completely integrable systems is the 'Scattering' or 'Monodromy' matrix of the LSP [41]. As a kind of non-linear Fourier transform it encodes the properties of the solution of the non-linear equations of motion in its dependence on the spectral parameter of the LSP. Usually this object is obtained through some limiting procedure from special ('scattering') solutions of the equations of motion with specified asymptotic behaviour. Although it is well known that there are problems with this definition of the Monodromy matrix for non-linear σ-models in flat space, it was recently shown [2] that the x-dependence of the spectral parameter t allows to overcome this problem. We will however avoid this problem altogether by a purely 'algebraic' definition. Guided by eq.(A.214) we can use the automorphism $\tau^{(\infty)}$ to construct

$$\mathcal{M}(t,x) \equiv \tau^{(\infty)}(\mathcal{P}^{-1}(t,x))\mathcal{P}(t,x) . \tag{87}$$

Clearly, in order for this definition to make sense, we have to assume for the moment that the two factors have an overlapping domain of existence as functions of t for given x. This is highly non-trivial, since $\mathcal{P}(t,x)$ is assumed to be analytic at $t=0$ and thus $\tau^{(\infty)}\mathcal{P}(t,x) = \tau\mathcal{P}(-1/t,x)$ is analytic at $t=\infty$. Since $\mathcal{P}(t,x)$ and $\tau(\mathcal{P}(-\frac{1}{t},x))$ are solutions of the same linear differential equation we find, as a consequence, that $\mathcal{M}(t,x)$ satisfies $\partial_\pm\mathcal{M}(t,x) = 0$ and thus is independent of x, but may and in fact does depend on w. The corresponding G-valued function $\mathcal{M}(w)$ is our definition of the Monodromy matrix of the LSP (compare [42] for a similar construction for the KdV equation).

The inverse transformation ('Inverse Scattering Transform') requires the factorization of $\mathcal{M}$ considered as a function of $w(t,x)$ in the form (87). In order to give this factorization a more precise meaning, we have to determine the analytical properties of $\mathcal{P}(t,x)$. To be specific, let us assume that $P(x)$ is regular outside some compact region containing possible singularities or sources. Regularity is here to be understood with respect to the manifold structure of G as a Lie group or a suitable matrix representation thereof. Thus let us assume that $P(x)$ is asymptotically regular in the sense that

- $P(x)$ is analytic in a simply connected domain $\mathcal{X}$ in the closure $\bar{\Sigma}_2$ of Σ_2 whose complement $\bar{\mathcal{X}} = \bar{\Sigma}_2 \setminus \mathcal{X}$ is closed and contained in a semi-disk

$$\mathcal{D}_R = \{(z, \rho) \colon \rho \geq 0,\, r \equiv \sqrt{z^2 + \rho^2} < R\} \tag{88}$$

  of radius R;
- the configuration is asymptotically flat and sufficiently regular at infinity, i.e.

$$P(x) = \mathbf{1} + O\left(\frac{1}{r}\right), \qquad \partial P(x) = O\left(\frac{1}{r^2}\right) . \tag{89}$$

For a solution to have a physical interpretation one should add further requirements to those put down in the preceding conditions as e.g. positivity of the mass or boundary conditions at the horizon for black holes. Note however that the regularity on the rotation axis is guaranteed by the regularity of $P(x)$.

Given some asymptotically regular solution $P(x)$ we integrate (79) and obtain the corresponding functions $\mathcal{P}_\pm(w, x)$. We choose the w-dependent constants of integration in $\mathcal{P}_\pm(w, x)$ such that in the limit $\rho \to 0$, $z \to -\infty$ (i.e. $t_+ \to 0$)

$$\mathcal{P}_+(w, x) \underset{\substack{\rho\to 0 \\ z\to-\infty}}{\longrightarrow} \mathbf{1} . \tag{90}$$

Due to the asymptotic behaviour (89) of $P(x)$ we can integrate the differential equation (79) along a large circle and find

$$\mathcal{P}_+(w, x) \underset{r\to\infty}{\longrightarrow} \mathbf{1} \tag{91}$$

in the whole asymptotic region. In order to characterize the domain of analyticity of $\mathcal{P}(t, x)$ we need some definitions. Let $\mathcal{X}$ be a domain of the type described in the preceding definition and $\mathcal{W}$ the domain in the complex w plane given by $\mathcal{W} = \{w \colon (\operatorname{Re} w, |\operatorname{Im} w|) \in \mathcal{X}\}$. For each point x in the (z, ρ) half plane we define the domains $\bar{\mathcal{T}}^x_\pm$, $\mathcal{T}^x_\pm$ and $\mathcal{T}^x$

$$\bar{\mathcal{T}}^x_\pm = \{t \colon t = t_\mp(w, x)\, ,\ w \in \bar{\mathcal{W}}\}\, , \qquad \mathcal{T}^x_\pm = \mathbf{C} \setminus \bar{\mathcal{T}}^x_\pm\, , \qquad \mathcal{T}^x = \mathcal{T}^x_+ \cap \mathcal{T}^x_- . \tag{92}$$

where $\bar{\mathcal{W}}$ is the complement of $\mathcal{W}$. The transformation $t \to -\frac{1}{t}$ will clearly map $\mathcal{T}^x_\pm$ onto $\mathcal{T}^x_\mp$ and will therefore leave their intersection $\mathcal{T}^x$ invariant. Furthermore let $\mathcal{D}_\pm$ and $\mathcal{D}$ be the domains

$$\mathcal{D}_\pm = \{(t, x) \colon \quad t \in \mathcal{T}^x_\pm, x \in \mathcal{X}\}\, , \qquad \mathcal{D} = \mathcal{D}_+ \cap \mathcal{D}_- . \tag{93}$$

Starting from the asymptotic value $\mathcal{P}_+(w, x) \underset{r\to\infty}{\longrightarrow} \mathbf{1}$ we can, for a fixed value of w, determine $\mathcal{P}_+(w, x)$ by integration of (79) along a suitable path in the (z, ρ) half plane. The r.h.s. of (79) is analytic (in x and w) for $x \in \mathcal{X}$ except for the branch cut starting at $x = (\operatorname{Re} w, |\operatorname{Im} w|)$. The resulting $\mathcal{P}_+(w, x)$ will therefore be an analytic function of w and x as long as we can find a path of integration avoiding these singularities. This is possible for all $x \in \mathcal{X}$ except those on the branch cut. In order to determine $\mathcal{P}_-$ we have to reach the second sheet, i.e. w must lie in the domain $\mathcal{W}$. For the moment we need, as before, the additional assumption that $\operatorname{Im} w \neq 0$ but it is easily seen that the results remain true for $\operatorname{Im} w = 0$. Except for the branch cut the function $\mathcal{P}_-(w,x)$ will be analytic if $(w,x) \in \mathcal{W} \times \mathcal{X}$. We can finally analyze, for $w \in \mathcal{W}$, the behaviour near the branch point for $w \in \mathcal{W}$ and find

$$\mathcal{P}_\pm(w,x) = \mathcal{P}_1(w,x) + t_\pm(w,x)\mathcal{P}_2(w,x) \tag{94}$$

with $\mathcal{P}_1$ and $\mathcal{P}_2$ analytic in a neighborhood of the branch point.

Using the analyticity of $\mathcal{P}_\pm(w,x)$ and the behaviour (94) near the branch point we deduce that $\mathcal{P}(t,x)$ is an analytic function in the domain $\mathcal{D}_+$ and that $\mathcal{P}(-\frac{1}{t},x)$ is analytic in $\mathcal{D}_-$. $\mathcal{M}(t,x)$ will therefore be analytic in $\mathcal{D}$, but the domain of analyticity is in fact much larger. Since $\mathcal{M}_\pm(w)$ is x-independent it cannot have branch points at $w = z \pm \rho$, i.e. $\mathcal{M}_+(w) = \mathcal{M}_-(w) = \mathcal{M}(w)$. $\mathcal{M}(w)$ is analytic in $\mathcal{W}$ and $\mathcal{M}(t,x)$ is therefore analytic in the domain

$$\tilde{\mathcal{D}} = \{(t,x): \quad t \in \mathcal{T}^x\} \supseteq \mathcal{D}\,. \tag{95}$$

Let us now turn to the factorization problem for $\mathcal{M}$. For that reason it is important to analyze the position of the sets $\bar{\mathcal{T}}^x_\pm$. As long as $z < -R$, the domain $\bar{\mathcal{T}}^x_+$ lies entirely outside the unit circle and to the left of the imaginary axis whereas $\bar{\mathcal{T}}^x_-$ lies inside the unit circle and to the right of the imaginary axis. For these x all singularities of $\mathcal{M}(t,x)$ in $\bar{\mathcal{T}}^x_+$ should be due to $\mathcal{P}(t,x)$ and those in $\bar{\mathcal{T}}^x_-$ due to $\mathcal{P}(-\frac{1}{t},x)$. If we vary x, these domains and the singularities of $\mathcal{P}(t,x)$ will move in the complex t plane and will eventually cross the unit circle but the two domains $\bar{\mathcal{T}}^x_\pm$ will never intersect as long as $z + i\rho \in \mathcal{W}$ or equivalently $x \in \mathcal{X}$. As soon as $z > R$, $\bar{\mathcal{T}}^x_+$ lies entirely inside the unit circle (but still to the left of the imaginary axis) and $\bar{\mathcal{T}}^x_-$ lies outside the unit circle.

We can, therefore, choose a family of contours C^x in the t plane which have, for all $x \in \mathcal{X}$, the following properties:

- C^x is invariant under the mapping $t \to -\frac{1}{t}$ and will thus pass through the two fixed points $t = \pm i$ of this map.
- The domain $\mathcal{T}^x$ contains a neighborhood of C^x and the domains $\bar{\mathcal{T}}^x_+$ resp. $\bar{\mathcal{T}}^x_-$ lie in the exterior resp. interior of C^x.
- The contours C^x depend continously on x.

There remains a lot of ambiguity in the choice of these contours, and we could e.g. choose the unit circle for $z < -R$. What we really need is the existence of a contour which separates the domains $\bar{\mathcal{T}}^x_\pm$. No such contour can exist for $x \in \bar{\mathcal{X}}$ because, in this case, both $\bar{\mathcal{T}}^x_+$ and $\bar{\mathcal{T}}^x_-$ will contain the points $t = \pm i$ and this non-existence of a suitable contour will lead to singularities of $\mathcal{P}(t,x)$.

In view of these properties of the domains of analyticity the factorization problem maybe considered as a group- or matrix-valued Riemann–Hilbert problem for the group-valued function $\mathcal{M}(t,x)$ of t for fixed x given in a neighbourhood of the curve C^x with factors analytic in domains $\mathcal{T}_+^x$ resp. $\mathcal{T}_-^x$ with the geometrical properties described above. A simple argument shows [21] that a solution $\mathcal{P}(t,x)$ of the factorization problem automatically solves a LSP of type (79) with suitable currents $\mathcal{A}(x)$ and $\mathcal{J}(x)$ and thus yields a solution of the eq.(75a) for $M(x)$.

In general the solution of such a Riemann–Hilbert problem is difficult. A useful strategy [41,18] is to reduce it to the solution of a system of Fredholm equations. In the present case there is however a very helpful connection between $\mathcal{M}(w)$ and $M(z=w,0)$, i.e. $M(x)$ on the axis $\rho=0$, derived from the relations (84).

$$\mathcal{M}(w) = \begin{cases} M(w,0) & \text{for} \quad w < -R\,, \\ M^{-1}(w,0) & \text{for} \quad R < w\,. \end{cases} \tag{96}$$

We see from (96) that although $\mathcal{M}(w)$ is x-independent, it contains enough information to determine the behaviour of $M(x)$, and thus $P(x)$ on the axis ($|z|>R$, $\rho=0$). Together with the assumed analyticity properties this is in principle sufficient to determine $P(x)$ everywhere in $\mathcal{X}$. As shown by Ernst and Hauser [18] the factorisation problem can be explicitly solved along a suitable piece of the $\rho=0$ axis employing eq.(96). From this they are able to conclude its solvability in the whole domain $\mathcal{X}$.

Next we turn to the definition of the 'Geroch group' acting on the solutions of (75a-75c). Given a solution of the linear equation LSP (79) and a function $g(w)$ with values in G resp. the complexification of G for complex w we can obviously get another one by the transformation

$$\mathcal{P}(t,x) \to \mathcal{P}(t,x)g(w)^{-1} \tag{97}$$

where $g(w)$ is considered as a function $g(t,x) \equiv g(w(t,x))$. The only problem with this transformation is that in general it destroys the 'triangularity' of $\mathcal{P}(t,x)$, since in view of (83) w considered as a function of t is

$$w = z + \frac{\rho}{2}(\frac{1}{t} - t)\,, \tag{98}$$

which is singular at $t=0$. In order to compensate for that one can try to act from the left with some group element $h(t,x)$ depending on g and $\mathcal{P}$ removing the negative powers of t and bring $P(x) = \mathcal{P}(0,x)$ to triangular form. For the x-independence of $\mathcal{M}$ it is necessary to choose h invariant under $\tau^{(\infty)}$. This is quite analogous to the standard action of G on the symmetric space G/H parametrized with the triangular representatives P. In [21] it was shown that the resulting transformations

$$\mathcal{P}(t,x) \to \mathcal{P}_g(t,x) \equiv h(t,x)\mathcal{P}(t,x)g(w)^{-1} \tag{99}$$

correspond exactly to the group proposed by Geroch [14]. The latter emerges in the attempt to act on the potentials parametrizing the matrix M with the group $GL(2)$ mixing the two Killing vector fields K and K' linearly. According to Geroch this requires the introduction of an infinite set of further potentials – represented here with the 'generating functions' $U(t,x)$. The action of the Geroch group on these functions was derived in [21] using the so-called Kramer–Neugebauer mapping [43]. Since there is no natural extension of this mapping to the general class of theories considered in this context we prefer to postulate the transformations through (99). This requires the determination of $h(t,x)$, which can however be reduced to the solution of the factorization problem for $\mathcal{M}(w)$: The point is that the transformation of $\mathcal{M}(w)$ corresponding to (99) requires only the knowledge of $g(w)$, because due to $\tau^{(\infty)}h = h$ we get

$$\mathcal{M}_g(w) = \tau g(w)\mathcal{M}(w)g(w)^{-1} . \tag{100}$$

Hence we can construct $\mathcal{P}_g$ making a detour via $\mathcal{M}_g$ expressed by the diagram

$$\mathcal{P} \to \mathcal{M} \to \mathcal{M}_g \to \mathcal{P}_g . \tag{101}$$

Once we have succeeded to factorize $\mathcal{M}_g(w)$ as

$$\mathcal{M}_g(t,x)) = \tau^{(\infty)}\left(\mathcal{P}_g(t,x)^{-1}\right)\mathcal{P}_g(t,x) \tag{102}$$

we just *define*

$$h \equiv \mathcal{P}_g \, g \, \mathcal{P}^{-1} . \tag{103}$$

It is straightforward to show that this h is invariant under $\tau^{(\infty)}$:

$$\tau^{(\infty)}h\, h^{-1} = \tau^{(\infty)}\mathcal{P}_g \tau^{(\infty)} g \tau^{(\infty)}\mathcal{P}^{-1}\, \mathcal{P} g^{-1}\mathcal{P}_g^{-1} = \tau^{(\infty)}\mathcal{P}_g \, \mathcal{M}_g \, \mathcal{P}_g^{-1} = \mathbf{1} . \tag{104}$$

Since we are only interested in $\mathcal{P}(t,x)$ corresponding to asymptotically regular solutions $P(x)$ we have to impose the condition that $g(w)$ is holomorphic in a neighbourhood of $w = \infty$ (taking values in G for real w) and $g(\infty) = \mathbf{1}$. In order to give this set the structure of a group we have to form equivalence classes of $g(w)$'s related through analytic continuation or equivalently take the maximal analytic continuation.

As first demonstrated by Ernst and Hauser [18] the Geroch group acts transitively on the class of stationary, axially symmetric solutions of the vacuum Einstein eqs. ($G = SL(2)$) which are regular in the neighbourhood of some given point on the axis. This implies in particular that all asymptoically regular solutions can be obtained from Minkowski space. Since the latter is represented by $\mathcal{M}(w) = \mathcal{P}(t,x) = \mathbf{1}$ this requires the factorisation of $\mathcal{M}(w)$ in the form $\mathcal{M}(w) = g(w)^{\mathrm{T}} g(w)$ with $g(w)$ analytic at $w = \infty$. This can easily be done explicitely in terms of triangular matrices in the case of $SL(2)$, but in general no such explicit solution exists. In fact, for pseudo-riemannian symmetric spaces such a factorisation in 'triangular' group elements can only exist in a neighbourhood of the unit element. Fortunately this is sufficient in our case in view of the normalisation condition $g(\infty) = \mathbf{1}$. Thus the transitivity of the Geroch group remains valid for the general class of models under consideration.

It is sometimes convenient to express the new solution $\mathcal{P}_g(t,x)$ as

$$G^{(\infty)} \ni g(s): \quad \mathcal{P}(t,x) \to \mathcal{P}_g(t,x) = Z^g_+(t,x)\,\mathcal{P}(t,x) \tag{105}$$

and use a slightly different Riemann–Hilbert problem in order to determine directly $Z^g_+(t,x)$ [44]. Clearly $Z^g_+(t,x)$ is analytic in the domain $\mathcal{D}_+$ (for a suitably chosen $\mathcal{X}$) as are $\mathcal{P}(t,x)$ and $\mathcal{P}_g(t,x)$. We find that we have to construct

$$\mathcal{G}(t,x) \;\; = \mathcal{P}(t,x)\, g^{-1}(w(t,x))\, \mathcal{P}^{-1}(t,x) \tag{106}$$

$$\mathcal{Z}^g(t,x) = (\tau^{(\infty)}\mathcal{G}(t,x))^{-1}\, \mathcal{G}(t,x) \tag{107}$$

and decompose $\mathcal{Z}^g(t,x)$ in the form

$$\mathcal{Z}^g(t,x) = Z^g_-(t,x)\, Z^g_+(t,x) \tag{108}$$

where $Z^g_\pm(t,x)$ is analytic in $\mathcal{D}_\pm$ and $Z^g_+(0,x) = Z^g_-(\infty,x)$ is 'triangular'.

If the function $\mathcal{Z}^g(t,x)$ in the Riemann–Hilbert problem is meromorphic (e.g. with $2N$ poles) then the transformation (105) corresponds to an N-fold Bäcklund transformation [41] adding N poles to the solution $\mathcal{P}(t,x)$ of the linear system. For a general $\mathcal{P}(t,x)$ it is, unfortunately, practically impossible to find an element $g(w) \in G^{(\infty)}$ such that the corresponding $\mathcal{Z}^g(t,x)$ is meromorphic. One can, however, investigate what are the conditions on a meromorphic $Z^g_+(t,x)$ in order that $\mathcal{P}_g(t,x)$ determined by (105) is again a solution of the LSP. This method yields algebraic equations for the residues of $Z^g_+(t,x)$. This is, in fact, the procedure of Belinskiĭ and Zakharov [19] employed in the special case $G = SL(2)$. For this special case Cosgrove [45] has shown the equivalence of this method with the Bäcklund transformations introduced independently by Harrison resp. Kramer and Neugebauer [16,17].

For the full description of the 4-dimensional solution we need also the conformal factor λ related to $P(x)$ resp. $M(x)$ through (76). The question arises how λ changes with the transformations $P \to P_g$. In [46] Julia showed that λ transforms under infinitesimal transformations like a Lie algebra cocycle defining the central extension of the affine Lie algebra. In [21] this action was 'exponentiated' and it was demonstrated that there is a central extension $G^{(\infty)}_{\mathrm{ce}}$ of $G^{(\infty)}_t$ defined through a group 2-cocycle Ω [47] acting on the pairs $(\mathcal{P},\lambda)$ considered again as 'triangular' elements of this extended group $G^{(\infty)}_{\mathrm{ce}}$. The multiplication law for this central extension of $G^{(\infty)}$ using the group 2-cocycle Ω is given by

$$(a, e^{\alpha}) \circ (b, e^{\beta}) = \left(ab, e^{\alpha+\beta+\Omega(a,b)}\right). \tag{109}$$

There is a corresponding extension of the transformation law (99) to

$$(\mathcal{P},\lambda^{-1}) \to \Big(h(\mathcal{P},g),1\Big)\circ(\mathcal{P},\lambda^{-1})\circ(g,e^{\gamma})^{-1} \qquad \text{for} \quad (g,e^{\gamma}) \in G^{(\infty)}_{\mathrm{ce}}\,. \tag{110}$$

Similar to the construction of $\mathcal{M}$ from $\mathcal{P}$ we can define

$$(\mathcal{M},\mu) \equiv \tau^{(\infty)}\Big((\mathcal{P},\lambda^{-1})^{-1}\Big) \circ (\mathcal{P},\lambda^{-1}) = \Big(\mathcal{M},\lambda^{-2}e^{\Omega(\tau^{(\infty)}\mathcal{P}^{-1},\mathcal{P})}\Big) \quad (111)$$

where the last expression is obtained using (110). Above we concluded that $\mathcal{M}$ was x-independent due to the invariance of $\partial\mathcal{P}\mathcal{P}^{-1}$ under $\tau^{(\infty)}$. The natural extension of this invariance to $\mathcal{G}_{\rm ce}^{(\infty)}$ the Lie algebra of $G_{ce}^{(\infty)}$ is

$$\partial(\mathcal{P},\lambda^{-1}) \circ (\mathcal{P},\lambda^{-1})^{-1} = (\partial\mathcal{P}\mathcal{P}^{-1},0) \quad (112)$$

since $\tau^{(\infty)}\Omega = -\Omega$.

As a consequence of (112) we find $\partial\mu = 0$, i.e. not only $\mathcal{M}$ but the whole pair $(\mathcal{M},\mu)$ is x-independent. This remarkable fact provides us with a nice explicit formula for λ as a function of $\mathcal{P}$

$$\lambda^2 = c\, e^{\Omega(\tau^{(\infty)}\mathcal{P}^{-1},\mathcal{P})} \quad (113)$$

(where the constant c can be determined from the asymptotic behaviour for $|x| \to \infty$). Making use of the properties of the 2-cocycle Ω it is possible to show that this expression for λ indeed satisfies (76).

Next we shall demonstrate the factorization of $\mathcal{M}(w)$ in a simple example yielding the Schwarzschild solution. Let us take $G = SL(2)$ and put

$$\mathcal{M}(w) = \begin{pmatrix} \frac{w-c}{w+c} & 0 \\ 0 & \frac{w+c}{w-c} \end{pmatrix} . \quad (114)$$

Factorizing

$$\begin{aligned} w-c &= -\tfrac{\rho}{2t_1}(t-t_1)(\tfrac{1}{t}+t_1) \quad \text{with} \quad t_1 = \tfrac{1}{\rho}\Big((z-c)+\sqrt{(z-c)^2+\rho^2}\Big) \\ w+c &= -\tfrac{\rho}{2t_2}(t-t_2)(\tfrac{1}{t}+t_2) \quad \text{with} \quad t_2 = \tfrac{1}{\rho}\Big((z+c)+\sqrt{(z+c)^2+\rho^2}\Big) \end{aligned} \quad (115)$$

allows to separate the poles of $\mathcal{M}(t,x)$ in the complex t-plane and to obtain

$$\mathcal{P}(t,x) = \begin{pmatrix} \sqrt{\frac{t_2}{t_1}\frac{t-t_1}{t-t_2}} & 0 \\ 0 & \sqrt{\frac{t_1}{t_2}\frac{t-t_2}{t-t_1}} \end{pmatrix} \quad (116)$$

The corresponding solution of the vacuum field equation is

$$\Delta = \frac{t_1}{t_2}, \qquad \tilde{\psi} = 0, \qquad \lambda = \frac{(1+t_1t_2)^2}{(1+t_1^2)(1+t_2^2)} \quad (117)$$

which is nothing but the Schwarzschild solution of mass $m = c$ in Weyl's canonical coordinates. These expressions simplify to

$$\Delta = \frac{u-c}{u+c}, \qquad \lambda = \frac{u^2-c^2}{u^2-c^2v^2} \quad (118)$$

using prolate spherical coordinates (u, v)

$$z = uv\ , \qquad \rho = \sqrt{(u^2-c^2)(1-v^2)}\ , \qquad \begin{array}{c} c \leq u < \infty \\ -1 \leq v \leq 1 \end{array} \tag{119}$$

Hence the Schwarzschild solution can be regarded as some kind of soliton obtained through a Bäcklund transformation from the trivial solution $\mathcal{M}(w) = 1$ using the element

$$g(w) = \begin{pmatrix} \sqrt{\frac{w-m}{w+m}} & 0 \\ 0 & \sqrt{\frac{w+m}{w-m}} \end{pmatrix} \tag{120}$$

of the Geroch group. The Kerr solution with mass m and angular momentum $J = am$ is obtained replacing $\mathcal{M}(w)$ of eq.(114) by

$$\mathcal{M}(w) = \frac{1}{w^2-c^2}\begin{pmatrix} (w+m)^2+a^2 & a \\ a & (w-m)^2+a^2 \end{pmatrix}\ , \tag{121}$$

where $c^2 = m^2 - a^2$. Performing again the now rather more complicated factorisation of $\mathcal{M}(w)$ one obtains ($\zeta = \frac{c+a-m}{c-a+m}$)

$$\begin{aligned}
\lambda^2 &= \frac{\frac{t_2-\zeta^2 t_1}{t_2-t_1}\frac{t_1-\zeta^2 t_2}{t_2-t_1} + \left(\zeta\frac{t_1t_2-1}{t_1t_2+1}\right)^2}{\left(\frac{1-\zeta^2}{2}\right)^2\left[\left(\frac{t_2+t_1}{t_2-t_1}\right)^2 - \left(\frac{t_1t_2-1}{t_1t_2+1}\right)^2\right]} = \frac{u^2-m^2+(av)^2}{u^2-(cv)^2}\ , \\
\Delta &= \frac{\frac{t_2-\zeta^2 t_1}{t_2-t_1}\frac{t_1-\zeta^2 t_2}{t_2-t_1} + \left(\zeta\frac{t_1t_2-1}{t_1t_2+1}\right)^2}{\left(\frac{t_2-\zeta^2 t_1}{t_2-t_1}\right)^2 + \left(\zeta\frac{t_1t_2-1}{t_1t_2+1}\right)^2} = \frac{u^2-m^2+(av)^2}{(u+m)^2+(av)^2}\ , \\
\tilde{\psi} &= -\frac{(1+\zeta)^2\zeta\frac{t_1t_2-1}{t_1t_2+1}}{\left(\frac{t_2-\zeta^2 t_1}{t_2-t_1}\right)^2 + \left(\zeta\frac{t_1t_2-1}{t_1t_2+1}\right)^2} = -\frac{2Jv}{(u+m)^2+(av)^2}\ ,
\end{aligned} \tag{122}$$

By repeated application of such group transformations one can obtain 'Multi-Kerr' solutions first constructed by Kramer and Neugebauer [17] with the help of Bäcklund transformations. Unfortunately none of these seem to be free of unphysical singularities [48].

In addition to the transformations of the Geroch group we can define a non-linear realization of the conformal group of the complex plane (resp. the Riemann sphere) on the solutions. For any holomorphic function $f(w)$ leaving the point at infinity fixed we may put $\mathcal{M}_f(w) \equiv \mathcal{M}(f^{-1}(w))$ [42]. Again there corresponds a transformation $\mathcal{P} \to \mathcal{P}_f$ obtained through factorization of $\mathcal{M}_f$. The infinitesimal form (Virasoro algebra) of these transformations on the solutions of the Ernst equation was introduced in [49]. More recently Julia and Nicolai [50] discussed an extension of the group $\mathcal{G}_{\rm ce}^{(\infty)}$ to the semi-direct product with this conformal group acting on a triple $(Y(t,x), \mathcal{P}(t,x), \lambda(x))$, where $Y(t,x)$ is the generating function for a set of potentials generalizing the function $1/w(t,x)$.

Finally let us remark that the 'Inverse Scattering Method' has also been successfully applied to solutions with two space-like Killing vectors, describing gravitational plane waves [22].

5 Stationary Black Holes

The remaining part of this article is devoted to applications of the 3-d σ-model structure found in the preceding sections. We will concentrate on stationary space-times which are asymptotically flat (or possibly asymptotically NUT). In addition we require that the matter fields tend asymptotically to the 'classical vacuum' configuration, i.e. the scalar fields tend to constants and the vector field strengths to zero. Furthermore we shall assume that the space-times considered fulfil the standard causality requirements (compare [51], p. 323 for a precise formulation).

A space-time manifold is said to be stationary if it admits a Killing vector field K which is time-like near infinity and strictly stationary if K is everywhere time-like. Scalar and vector fields living on a stationary space-time are called stationary, if their Lie derivative with respect to K vanishes.

A stationary space-time is said to be static, if it is invariant under time-reflections. This implies that the Killing vector field K is hypersurface orthogonal, i.e. the twist vector vanishes.

In order to define staticity for the scalar and vector fields it is necessary to assign to each field a time reflection parity, such that the action (36) is time reflection invariant. This assignment is in general not unique. Consider, e.g., the Einstein-Maxwell system: the electromagnetic field $F_{\alpha\beta}$ is usually considered as temporal vector field strength $F^{(+)}$, i.e. the magnetic field $B_\alpha = {}^*F_{\alpha\beta}K^\beta$ vanishes for a static solution. Due to the duality invariance of the field equations $F_{\alpha\beta}$ could as well be considered as temporal pseudo vector field strength $F^{(-)}$ with the consequence that a solution would be called static if the electric field $E_\alpha = F_{\alpha\beta}K^\beta$ vanishes. Actually we prefer to define staticity in terms of the scalar potentials $\underline{A}$, which we split into temporal scalars $\underline{A}^{(+)}$ and pseudoscalars $\underline{A}^{(-)}$. Analogously we proceed with the scalars of $\bar{G}/\bar{H}$. In view of the σ-model structure we will however make the additional assumption, that the truncation of G/H to temporal scalars is a consistent truncation of G/H as defined in Appendix A leading to a truncated σ-model G_{st}/H_{st}.

Let us now turn to the study of stationary solutions of the class of gravitational models introduced in the previous chapter. According to a result of Lichnerowicz [52] there are no non-trivial globally regular stationary solutions of the vacuum Einstein equations – a result that can be generalized to the class of theories considered here [12]. From the view-point of harmonic maps this is not an unexpected phenomenon [24]. This means that interesting stationary solutions have to have singularities. In case the singularities are hidden behind event horizons we are dealing with black holes, otherwise the singularities are called naked. From the physical view-point our interest clearly centers around the first alternative – stationary black holes. In the following we shall use the σ-model structure to extend in a simple way a number of classical existence and uniqueness results for these solutions.

The uniqueness theorems on static black holes are usually expressed as 'No-Hair Theorems' [26] stating that these black holes are characterized uniquely by their 'gauge' charges – comprising besides the mass only electric resp. magnetic charges, but no 'scalar' charges or charges referring to global invariances. These 'gauge' charges can be read off from the asymptotic behaviour of the corresponding gauge fields and are related to Gauss-type conservation laws. Hence we have to discuss the asymptotic conditions for stationary black holes in terms of the reduced 3-dimensional formulation. For the Einstein-Maxwell theory this question has been thoroughly investigated by Simon [53]. As mentioned by Simon his results can be extended to a larger class of theories including those considered here. Accordingly we require the following asymptotic behaviour at space-like infinity (holding in a suitable coordinate system) of the fields defined on Σ_3:

$$h_{ab} = \delta_{ab} + O(\frac{1}{r}) \tag{123}$$

$$M = M_0 + \frac{1}{r}M_1 + O(\frac{1}{r^2}) \tag{124}$$

where M_0, corresponds to the 'vacuum' solution. The behaviour of the individual fields (after a suitable choice of gauge) parametrizing M is

$$\Delta = 1 - \frac{2m}{r} + O(\frac{1}{r^2}) \tag{125}$$

$$\tilde{\psi} = \frac{n}{r} + O(\frac{1}{r^2}) \tag{126}$$

$$\underline{A} = \frac{Q_A}{r} + O(\frac{1}{r^2}) \tag{127}$$

$$\bar{M} = \mathbf{1} - \frac{2\bar{Q}}{r} + O(\frac{1}{r^2}) \tag{128}$$

where m is the total mass, n is the NUT-charge, Q_A a vector of 'electric' and 'magnetic' charges and $\bar{Q}$ a hermitian matrix of scalar charges. In order to have an asymptotically Minkowskian geometry the NUT-charge has to vanish. Thus solutions with non-vanishing NUT-charge are unphysical, nevertheless they appear naturally in our 3-dimensional formalism (their nontrivial topology becomes only visible if one reconstructs the 4-dimensional space-time!).

Next we turn to the event horizon. According to Hawking's strong rigidity theorem [51] the event horizon of a stationary black hole is a Killing horizon, i.e. its null generator L coincides with a Killing vector field. Although there has been some criticism concerning the strong technical assumptions going into its proof [54], we shall assume the validity of this theorem here. Now, either L coincides with the KV K for time-translations or there must exist a second independent one K', which is then necessarily space-like near infinity. As argued by Hawking it generates axial rotations and correspondingly such horizons are called rotating. Since its orbits are closed, causality requires that the axial Killing vector K' is space-like on and outside the horizon, apart from the rotation axis where it vanishes.

For non-rotating horizons the exterior domain is either static or not. Hawking [51] showed for the vacuum case that strict stationarity of the exterior domain implies staticity. The generalization of this result to electromagnetism has been achieved only recently by Sudarsky and Wald [55]. In contrast to our treatment of stationary solutions based on the 3-d orbit space these authors prefer a 'Hamiltonian' treatment using fields restricted to a 3-d (maximal) hyper-surface. There is no doubt that the 'Staticity Theorem' holds also for the class of models considered here and it should be possible to extend the proof of [55].

Regardless of whether the horizon is a Killing horizon or not it follows that on the horizon the energy-momentum tensor satisfies [39]

$$T_{\alpha\beta}L^\alpha L^\beta = 0\ . \tag{129}$$

This implies that

- The scalar fields are constant along the null generators

$$L^\alpha \partial_\alpha \bar{\phi}^i = 0 \tag{130}$$

- The electric and magnetic fields satisfy

$$L^\gamma F_{\gamma[\alpha} L_{\beta]} = 0 \tag{131}$$

$$L^{\gamma\,*}F_{\gamma[\alpha} L_{\beta]} = 0\ . \tag{132}$$

If L coincides on the horizon with a KV then the surface gravity κ defined by

$$\kappa = \sqrt{\frac{1}{2} L_{\alpha;\beta} L^{\beta;\alpha}} \tag{133}$$

is constant on each connected component of the horizon. Similarly the vector potentials $\underline{A}$ are constant there; note however that the scalar fields $\bar{\phi}$ have to be only finite but not necessarily constant on the horizon.

Let us now consider black holes which are the strictly stationary in their exterior domain, which includes in particular the static ones and switch to the 3-dimensional formulation. The horizon occurs at the boundary of Σ_3 where Δ vanishes. The surface gravity of the horizon is given by

$$\kappa^2 = \frac{1}{4} h^{ab} \partial_a \Delta \partial_b \Delta \tag{134}$$

which vanishes only for a degenerate horizon. Moreover, being orthogonal to the horizon, K satisfies there the hypersurface orthogonality condition $K_{[\alpha;\beta} K_{\gamma]} = 0$. This implies that ω_m vanishes and $\tilde{\psi}$ is constant on each connected component of the horizon. In addition the geometry determined by the physical 3-metric $\Delta^{-1} h_{mn}$ should remain regular as one approaches the boundary $\Delta \downarrow 0$. If $\kappa \neq 0$, i.e. for a non-degenerate horizon, this implies that Δ vanishes as the geodesic distance (in the scaled 3-metric h_{mn}) from the boundary and that (in suitable coordinates) the same holds true for the induced 2-metric on the surfaces Δ = const. In terms of the physical 3-metric the boundary is a totally geodesic 2-surface [57] and each connected component has the topology of a 2-sphere [58,59].

The boundary conditions in the *stationary axisymmetric* case are more complicated because using the adapted coordinates ρ and z introduced in Sect. 3.4 we must consider the axis as well as the horizon. They are essentially the same as those discussed by Carter [39] in the Einstein-Maxwell case. The 2-dimensional space is the open half-plane $\Sigma_2 = \{(z, \rho)\colon\ \rho > 0\}$. Both the horizon and (the exterior part of) the rotation axis are represented by the the z-axis ($\rho = 0$). Each connected component of the horizon corresponds to a segment of the z-axis with $\Delta' < 0$ whereas $\Delta' = 0$ on the rotation axis.

The boundary conditions at infinity ($r = \sqrt{z^2 + \rho^2} \to \infty$) are now

$$\lambda = 1 + O(\frac{1}{r}) \tag{135}$$

$$M = M_0 + \frac{1}{r} M_1 + O(\frac{1}{r^2})\ . \tag{136}$$

The corresponding behaviour of M' is more complicated due to the non-trivial asymptotic behaviour of the axial Killing vector field and involves the total angular momentum J. For vanishing NUT-charge one gets

$$\Delta' = \rho^2 \left(1 + \frac{2m}{r} + O\left(\frac{1}{r^2}\right)\right) \tag{137}$$

$$\tilde{\psi}' = \frac{2J(3\rho^2 z + 2z^3)}{r^3} + O\left(\frac{1}{r}\right) \tag{138}$$

$$\underline{A}' = \frac{Q_A z}{r} + O\left(\frac{1}{r}\right) \tag{139}$$

$$\bar{M}' = \bar{M}\ . \tag{140}$$

The conditions at the horizon $\mathcal{H}$ are simply that M' has to attain a finite limit with $\Delta' < 0$ there (distinguishing the horizon from the rotation axis where Δ' vanishes.

At the rotation axis $\Delta' = \rho^2 O(1)$ and $\bar{M}'$ tends to a finite limit quadratically with ρ. The behaviour of $\tilde{\psi}'$ and $\underline{A}'$ can be determined using eqs.(75a-75c):

$$\underline{A}' = \pm Q_A + a(z)\rho^2 + O(\rho^4) \tag{141}$$

$$\tilde{\psi}' = \pm 4J \mp {Q_A}^{\mathrm{T}} \bar{Y}^{-1} a(z)\rho^2 + O(\rho^4) \tag{142}$$

with some function $a(z)$ and the signs refering to the upper and lower part of the axis.

5.1 Spherically Symmetric Solutions

A very important class of solutions of the theories considered in the previous section are the stationary, spherically symmetric ones. According to the Birkhoff Theorem spherically symmetric solutions of the vacuum Einstein resp. the EM equations have a further Killing vector and thus are automatically static if the latter is time-like. But as soon as scalar fields are included this theorem doesn't hold anymore. Usually one assumes that the orbits of the isometry group $SO(3)$ in four dimensions are 2-spheres, implying a vanishing twist potential $\tilde{\psi}$ [51]. Since we will however use the 3-dimensional reduced theory, it would be unnecessarily restrictive to make this assumption. Hence we make this assumption on the orbits only for the 3-dimensional theory manifold Σ_3. Using polar coordinates with some arbitrary radial coordinate τ the metric can then be parametrized as

$$ds^2 = h_{mn} dx^m dx^n = N^2 d\tau^2 + f(\tau)^2 (d\vartheta^2 + sin^2\vartheta d\varphi^2). \tag{143}$$

Substitution of this ansatz into the action (48) resp. (63) and integrating over the angles yields an effective 1-dimensional Lagrangean

$$\mathcal{L}^{(1)} = N\left(\frac{f'^2}{N^2} + 1 - \frac{f^2}{2N^2}\gamma_{ij}(\phi)\phi^{i'}\phi^{j'}\right) \tag{144}$$

where primes denote derivatives with respect to τ. Varying with respect to N yields

$$1 - \frac{(f')^2}{N^2} + \frac{f^2}{2N^2}\gamma_{ij}\phi^{i'}\phi^{j'} = 0 \tag{145}$$

whereas the variation of ϕ gives

$$\frac{N}{f^2}(\frac{f^2}{N}\phi^{i'})' + \Gamma^i_{jk}(\phi)\phi^{j'}\phi^{k'} = 0 \tag{146}$$

With the gauge choice $N = f^2$ this is just the equation for a geodesic of the target space Φ with τ as the affine parameter. Putting $\gamma_{ij}(\hat{\phi})\frac{d\phi^i}{d\tau}\frac{d\phi^j}{d\tau} = 2v^2$

one can integrate eq.(145) for f^2 to obtain

$$f^2 = v^2 \sinh(v\tau)^{-2} \quad \text{for} \quad v^2 > 0 \tag{147}$$
$$f^2 = -v^2 \sin(v\tau)^{-2} \quad \text{for} \quad v^2 < 0 \tag{148}$$
$$f^2 = \tau^{-2} \quad \text{for} \quad v^2 = 0 \,. \tag{149}$$

Since for asymptotically flat solutions Δ turns to 1 at infinity the function f^2 tends to r^2, where r is the geometrical radius of the 2-spheres. Hence infinity corresponds to $\tau = 0$. Thus each of the geodesics corresponding to a solution with the required asymptotic behaviour has to pass through $\phi_0 = (\Delta(0), \tilde{\psi}(0), \underline{A}(0), \bar{\phi}(0)) = (1, 0, \underline{0}, \mathbf{1})$ for $\tau = 0$ and is uniquely determined by its tangent vector $\phi'(0)$ there. Using the fact that we are studying geodesics of a coset space represented by group elements M the geodesics obviously correspond to 1-parameter subgroups. In terms of M the geodesic equation is simply

$$\frac{d}{d\tau}\left(M^{-1}\frac{d}{d\tau}M\right) = 0 \tag{150}$$

From the asymptotic behaviour we find $M^{-1}\frac{d}{d\tau}M = -2Q$ where Q is the matrix of global charges lying in the Lie algebra part of G/H. Thus the geodesics have the form

$$M = M_0 e^{-2\tau Q} \,. \tag{151}$$

Parametrizing the matrix Q of charges one can in principle get explicit expressions for general spherically symmetric black holes, although in practice this may turn out quite cumbersome. In the following we shall demonstrate this on a not completely trivial example, the case of the original Kaluza–Klein theory yielding the coset space $SL(3)/SO(2,1)$.

Using the methods described in the Appendix one finds that $\hat{M}$ has the parametrization

$$\hat{M} = \mu^{-\frac{1}{3}} \begin{pmatrix} \Delta - A\mu A + \Delta^{-1}\hat{\psi}^2 & -A\mu + \Delta^{-1}\hat{\psi}\tilde{A} & \Delta^{-1}\hat{\psi} \\ -\mu A + \Delta^{-1}\hat{\psi}\tilde{A} & -\mu + \Delta^{-1}\tilde{A}\tilde{A} & \Delta^{-1}\tilde{A} \\ \Delta^{-1}\hat{\psi} & \Delta^{-1}\tilde{A} & \Delta^{-1} \end{pmatrix} \tag{152}$$

with $\hat{\psi} = \tilde{\psi} + \frac{1}{2}A\tilde{A}$. Δ plays the rôle of a gravitational potential, A resp. $\tilde{A}$ are the electric resp. magnetic potential, μ is the scalar field and $\tilde{\psi}$ the twist-potential.

In this case $\hat{Q}$ is an element of $sl(3)$ satisfying

$$\hat{Q} = -\tau\hat{Q} \equiv D^{-1}\hat{Q}^{\mathrm{T}}D \quad \text{with} \quad D = \begin{pmatrix} 1 & & \\ & -1 & \\ & & 1 \end{pmatrix} . \tag{153}$$

This suggests the ansatz

$$\hat{Q} = \frac{1}{2}\begin{pmatrix} -2m - s & -q & n \\ q & 2s & -p \\ n & p & 2m - s \end{pmatrix} \tag{154}$$

where m is the mass, s the scalar charge, q the electric and p the magnetic charge and n the NUT-parameter. Since we are only interested in asymptotically Minkowskian solutions we put $n = 0$.

For black holes Δ has to vanish at the horizon approached for $\tau \to -\infty$ and the other fields have to stay finite there [12]. This is only possible if the characteristic equation of $\hat{Q}$ has the form [56]

$$\hat{Q}^3 - \frac{1}{2}(\mathrm{Tr}\hat{Q}^2)\hat{Q} = 0 \tag{155}$$

where

$$\mathrm{Tr}\hat{Q}^2 = \frac{1}{2}(4m^2 + 3s^2 - q^2 - p^2) = 2v^2 \,. \tag{156}$$

From (155) we get the additional relation

$$\det \hat{Q} = \frac{1}{8}\Big(-2s(4m^2 - s^2) + (2m - s)q^2 - (2m + s)p^2\Big) = 0 \,. \tag{157}$$

Solving for q^2 and p^2 we find

$$q^2 = \frac{(2m + s)\Big((2m + s)^2 - 4v^2\Big)}{4m} \,, \qquad p^2 = \frac{(2m - s)\Big((2m - s)^2 - 4v^2\Big)}{4m} \,. \tag{158}$$

In order to simplify the discussion we shall study only some special cases in more detail. For $v^2 > 0$ we put $p = 0$, i.e. we restrict ourselves to strictly static solutions lying in $GL(2)/SO(1,1)$.

Exponentiating the matrix

$$\hat{Q} = \frac{1}{2}\begin{pmatrix} -2m - s & -q & 0 \\ q & 2s & 0 \\ 0 & 0 & 2m - s \end{pmatrix} \tag{159}$$

with the additional relations $2v = 2m - s$ and $q^2 = 2s(2m + s)$ derived from (156,157) we get

$$\hat{M} = M_0 e^{-2\tau\hat{Q}} = M_0\Big(1 - \frac{\sinh 2v\tau}{v}\hat{Q} + \frac{(\cosh 2v\tau - 1)}{v^2}\hat{Q}^2\Big) \tag{160}$$

and hence

$$\mu^{-\frac{1}{3}}\Delta^{-1} = \exp -2v\tau \tag{161}$$

$$\mu^{\frac{2}{3}} = \frac{1}{2v}(2m + s - 2s\exp 2v\tau) \tag{162}$$

$$\mu^{\frac{2}{3}} A = \frac{q}{2v}(1 - \exp 2v\tau) \,. \tag{163}$$

We can rewrite this in the form

$$\mu^{\frac{2}{3}} = \cosh^2\xi - \sinh^2\xi \exp 2v\tau \tag{164}$$

$$\mu^{\frac{2}{3}} A = \cosh\xi \sinh\xi (1 - \exp 2v\tau) \tag{165}$$

$$\text{with} \quad \cosh^2\xi = \frac{(1 + \frac{s}{2m})}{(1 - \frac{s}{2m})} \,. \tag{166}$$

This way of writing the solution displays it as a $GL(2)$-transformed version of the Schwarzschild solution given by

$$\Delta = \exp 2v\tau \ , \qquad \mu = 1 \ , \qquad A = 0 \tag{167}$$

with the transformation matrix

$$\begin{pmatrix} \cosh\xi & \sinh\xi & 0 \\ \sinh\xi & \cosh\xi & 0 \\ 0 & 0 & 1 \end{pmatrix} \tag{168}$$

in $SO(1,1)$, the stability group of the point M_0. As we will show in the following this possibility to obtain the most general static black hole from the Schwarzschild solution is in fact a rather general phenomenon.

For $v^2 = 0$ we are not allowed to put $p = 0$, otherwise we would violate the regularity of the solution on the horizon. On the other hand the exponentiation of $\hat{Q}$ becomes particularly simple, since $\hat{Q}^3 = 0$:

$$\hat{M} = M_0(1 - 2\tau\hat{Q} + 2\tau^2\hat{Q}^2) \tag{169}$$

resulting in

$$\Delta = \left[\frac{1 - \frac{s^2}{4m^2}}{\left(1 - (1 - \frac{s^2}{4m^2})\sigma\right)^4 - \frac{s^2}{4m^2}}\right]^{\frac{1}{2}} , \tag{170}$$

$$\mu = \frac{|q|}{|p|}\left[\frac{\left(1 - (1 - \frac{s^2}{4m^2})\sigma\right)^2 - \frac{s}{2m}}{\left(1 - (1 - \frac{s^2}{4m^2})\sigma\right)^2 + \frac{s}{2m}}\right]^{\frac{3}{2}} , \tag{171}$$

$$A = -\frac{q\sigma}{m}\frac{(1 - \frac{s}{2m})\left(1 - (1 - \frac{s}{2m})\sigma\right)}{\left(1 - (1 - \frac{s^2}{4m^2})\sigma\right)^2 - \frac{s}{2m}} , \tag{172}$$

$$\tilde{A} = -\frac{p\sigma}{m}\frac{(1 + \frac{s}{2m})\left(1 - (1 + \frac{s}{2m})\sigma\right)}{\left(1 - (1 - \frac{s^2}{4m^2})\sigma\right)^2 + \frac{s}{2m}} , \tag{173}$$

$$\tilde{\psi} = \frac{sqp\sigma^2}{4m^3}\frac{\left(1 - (1 - \frac{s^2}{4m^2})\sigma\right)^2 - 1}{\left(1 - (1 - \frac{s^2}{4m^2})\sigma\right)^4 - \frac{s^2}{4m^2}} \tag{174}$$

where $|q| = \sqrt{2}m(1+\frac{s}{2m})^{\frac{3}{2}}$, $|p| = \sqrt{2}m(1-\frac{s}{2m})^{\frac{3}{2}}$ and $\sigma = m\tau$. The behaviour at the horizon $\tau = -\infty$ is

$$\Delta = \frac{2}{|qp|\tau^2} + O\left(\frac{1}{\tau^3}\right) , \tag{175}$$

$$\mu = \frac{|q|}{|p|} + O\left(\frac{1}{\tau}\right) , \tag{176}$$

$$A = \frac{\sqrt{2}q}{|q|}(1 + \frac{s}{2m})^{-\frac{1}{2}} + O\left(\frac{1}{\tau}\right) , \tag{177}$$

$$\tilde{A} = \frac{\sqrt{2}p}{|p|}(1 - \frac{s}{2m})^{-\frac{1}{2}} + O\left(\frac{1}{\tau}\right) , \tag{178}$$

$$\tilde{\psi} = \frac{sqp}{2m|qp|}(1 - \frac{s^2}{4m^2})^{-\frac{1}{2}} + O\left(\frac{1}{\tau}\right) \tag{179}$$

characteristic for a degenerate event horizon. Since $v^2 = 0$ the three dimensional space Σ_3 is flat for this type of solution. The solution has no strictly static limit ($p = 0$), because we have to require $pq \neq 0$.

All the solutions with $v^2 < 0$ have naked singularities [56], hence there are no black hole solutions in that case. This remains so also for $D > 5$, where the behaviour of the geodesics is quite analogous. For light-like geodesics ($v^2 = 0$) the condition $p \cdot q \neq 0$ is replaced by more general conditions to be derived in the following.

After this instructive example let us return to the general case for which we obtain the following equations for the geodesics describing spherically symmetric black holes

$$\tfrac{d}{d\tau}\left(\Delta^{-1}\bar{M}\tfrac{d\underline{A}}{d\tau}\right) = 0 \tag{180}$$

$$\tfrac{d}{d\tau}\left(\Delta^{-1}\tfrac{d\Delta}{d\tau} - \Delta^{-1}\underline{A}^{\mathrm{T}}\bar{M}\tfrac{d\underline{A}}{d\tau}\right) = 0 \tag{181}$$

$$\tfrac{d}{d\tau}\left(\tfrac{1}{2}\bar{M}^{-1}\tfrac{d\bar{M}}{d\tau} + \Delta^{-1}\tfrac{1}{\bar{c}}(\underline{A}\tfrac{d\underline{A}^{\mathrm{T}}}{d\tau}\bar{M})_{\mathrm{pr}}\right) = 0 . \tag{182}$$

From the asymptotic behaviour of the geodesics we learn how to fix the arbitrary constants obtained by integrating these equations.

$$\Delta^{-1}\bar{M}\tfrac{d\underline{A}}{d\tau} = -Q_A \tag{183}$$

$$\Delta^{-1}\tfrac{d\Delta}{d\tau} - \Delta^{-1}\underline{A}^{\mathrm{T}}\bar{M}\tfrac{d\underline{A}}{d\tau} = 2m \tag{184}$$

$$\tfrac{1}{2}\bar{M}^{-1}\tfrac{d\bar{M}}{d\tau} + \Delta^{-1}\tfrac{1}{\bar{c}}(\underline{A}\tfrac{d\underline{A}^{\mathrm{T}}}{d\tau}\bar{M})_{\mathrm{pr}} = \bar{Q} . \tag{185}$$

From eq.(150) we get

$$\frac{1}{2}(\Delta^{-1}\frac{d\Delta}{d\tau})^2 + \frac{1}{4}\langle\bar{M}^{-1}\frac{d\bar{M}}{d\tau}, \bar{M}^{-1}\frac{d\bar{M}}{d\tau}\rangle - \Delta^{-1}\frac{d\underline{A}^{\mathrm{T}}}{d\tau}\bar{M}\frac{d\underline{A}}{d\tau} = 2v^2 . \tag{186}$$

Putting $\tau = 0$ in this equation we obtain the quadratic relation

$$m^2 + \frac{1}{2}\langle\bar{Q}, \bar{Q}\rangle - \frac{1}{2}Q_A^{\mathrm{T}}Q_A = v^2 . \tag{187}$$

Recently this 'Quadratic Mass Formula' has been generalized by Heusler [60] to arbitrary strictly stationary single black holes of the EM theory and a more general model involving a dilaton and an axion. A very simple proof for all the models considered here is given recently in [28].

Although it seems quite impossible to extract the potentials in closed form from (151) in the general case it is possible to study the behaviour of the geodesics qualitatively. We are particularly interested in the asymptotic behaviour when $\tau \to -\infty$, where we approach the horizon. We have to recall the boundary conditions on the horizon

$$\Delta(-\infty) = 0 \, , \quad \bar{M}(-\infty) = \bar{M}_{\mathcal{H}} < \infty \, , \quad \underline{A}(-\infty) = \underline{A}_{\mathcal{H}} < \infty \, . \tag{188}$$

These boundary conditions imply in particular that $\frac{d\bar{M}}{d\tau}$ must vanish in the limit $\tau \to -\infty$ and we obtain the following relations between the charges and boundary values at the horizon:

$$2m - \underline{A}_{\mathcal{H}}^{\mathrm{T}} Q_A = \lim_{\tau \to -\infty} \frac{d \ln \Delta}{d\tau} = 2m_{\mathcal{H}} \, , \qquad \bar{Q} = -\frac{1}{\bar{c}} (\underline{A}_{\mathcal{H}} Q_A^{\mathrm{T}})_{\mathrm{pr}} \, . \tag{189}$$

There are two essentially different situations:

- $m_{\mathcal{H}} > 0$. This is only possible for $v^2 > 0$, because the negative contribution in eq.(186) vanishes for $\Delta \to 0$. Obviously Δ behaves like

$$\Delta \sim e^{2 m_{\mathcal{H}} \tau} \quad \text{for} \quad \tau \to -\infty \, . \tag{190}$$

  Considering $\underline{A}$ and $\bar{M}$ as functions of Δ one finds $\underline{A} = \underline{A}_{\mathcal{H}} + O(\Delta)$ and $\bar{M} = \bar{M}_{\mathcal{H}} + O(\Delta)$. The quantity $4\pi m_{\mathcal{H}}$ is the surface gravity κ times the area of the horizon.
- $m_{\mathcal{H}} = 0$ describes solutions with a degenerate horizon. Eqs.(180) together with the boundary conditions eq.(188) show that this is only possible for $v^2 = 0$. The asymptotic behaviour is, in this case,

$$\Delta \sim \frac{2}{Q_A^{\mathrm{T}} \bar{M}_{\mathcal{H}}^{-1} Q_A} \tau^{-2} \quad \text{for} \quad \tau \to -\infty \tag{191}$$

  with $\underline{A} = \underline{A}_{\mathcal{H}} + O(\sqrt{\Delta})$ and $\bar{M} = \bar{M}_{\mathcal{H}} + O(\Delta)$. In addition to the relations eq.(189) there is a series of constraints on $\bar{M}$. The coefficients C_k of the expansion $\bar{M} = \bar{M}_{\mathcal{H}} \exp \sum_{k=1}^{\infty} C_k \Delta^k$ must satisfy the equations

$$k(k+\frac{1}{2})(Q_A^{\mathrm{T}} \bar{M}_{\mathcal{H}}^{-1} Q_A) C_k - \frac{1}{\bar{c}} (\hat{C}_k \bar{M}_{\mathcal{H}}^{-1} Q_A Q_A^{\mathrm{T}})_{\mathrm{pr}} = r.h.s. \quad k = 1, 2, \ldots \tag{192}$$

  where the r.h.s. depends only on C_l with $l < k$ and vanishes for $k = 1$. This is an inhomogeneous linear system for the C_k. For k large enough there will be no solutions of the homogeneous system and we can solve the equations recursively. For small k (and in particular for $k = 1$) we must, however, look for solutions of the homogeneous system. A very simple example of this situation is provided by the extremal solution of the σ-model with $G/H = SL(3)/SO(2,1)$ and $\bar{G}/\bar{H} = GL(1)$ discussed above.

We may summarize the results of this section stating that stationary spherically symmetric black hole solutions are characterized by a time-like or light-like geodesic segment in G/H starting at the point M_0 (corresponding to $r = \infty$) and running to the boundary $\Delta = 0$. Due to extra constraints on the scalar charges $\bar{Q}$ in general not every time-like geodesic leads to a black hole solution.

5.2 Uniqueness Theorems for Static Black Holes

In the following we consider static (as defined above), non-degenerate single black hole solutions of eqs.(49a-49e). The term single means that the horizon of the black hole is connected. The claim is that these solutions can be completely determined in the present setting – being just the spherically symmetric ones already described above. For pure gravity (no vectors, no scalars) according to a classical result of Israel [57] a static non-degenerate single non-degenerate black hole is necessarily spherically symmetric and thus given by the Schwarzschild solution. Israel's argument was later improved and simplified by Robinson [61] and Bunting and Masood-ul-Alam [62] showed that the connectedness of the horizon need not be required, but follows. This can be interpreted as the impossibilty to bring several black holes into static equilibrium. Israel was also able to generalize his theorem to the Einstein-Maxwell theory [63]. In this case his proof consists of two steps. First he shows that the electric potential $\underline{A}$ is a function of the gravitational potential Δ and then he proceeds similarly to the case of pure gravity. Note however, that there are static superpositions of degenerate charged black holes, the Papetrou-Majumdar solutions [64]. Their uniqueness is undoubted, but not yet completely established [65]. There exists a simplified version of Israel's proof due to Bunting [66]. Making use of the fact that the matrix of conserved gauge invariant currents $J_m = M^{-1}\partial_m M$ of the corresponding $SO(2,1)/SO(1,1)$ σ-model has a component in the Lie algebra of $H = SO(1,1)$, namely $J^H_m = \Delta^{-1}[A\partial_m\Delta + (1-\Delta-A^2)\partial_m A]$, with vanishing 'charge' $\int_{r=\infty} J^H_m d\Sigma^m = 0$ he is able to derive that the current J^H_m has to vanish identically, which allows him to express A as a function of Δ. Employing an adaption of the argument of Robinson permits him to reach the the desired conclusion. It is an interesting problem to generalize Bunting's method to the type of theories including scalars considered here. Another even simpler possibility could be to generalize Israel's original method using the so-called Bach tensor directly making use of eqs.(64). We shall, however, proceed somewhat differently [12] using an argument put forward in [67] (compare also [68]). The idea is to transform the vector (electrical) charges to zero by a suitable Harrison transformation and then to show that the vector fields vanish identically if they carry no charges. In a second step one concludes that, once the vector fields vanish, also the scalar fields have to be trivial, i.e. constant, which may be considered as a kind of scalar 'No Hair

Theorem'. This way one displays any charged single static (non-degenerate) black hole as a Harrison transform of the Schwarzschild solution.

The first step of this strategy is to prove the *'No Charges - no Vectors Theorem'* [12]:
For a static single black hole solution of eqs.(49a-49e) with vanshing vector charges $Q_A = \frac{1}{4\pi}\int_{r=\infty}\partial_m \underline{A} d\Sigma^m = 0$ the potentials $\underline{A}$ vanish identically.
Proof: From eq.(49a-49e) we find

$$(\Delta^{-1}\underline{A}^{\mathrm{T}}\bar{M}\partial^m\underline{A})_{;m} = \Delta^{-1}\partial^m\underline{A}^{\mathrm{T}}\bar{M}\partial_m\underline{A} \tag{193}$$

and hence through integration

$$\int_{\mathcal{H}}\Delta^{-1}\underline{A}^{\mathrm{T}}\bar{M}\partial_m\underline{A}d\Sigma^m = \int_{\Sigma_3}\Delta^{-1}\partial^m\underline{A}^{\mathrm{T}}\bar{M}\partial_m\underline{A}d\Sigma \tag{194}$$

(The surface term at infinity vanishes, since the integrand is $O(r^{-3})$.)
On the other hand

$$\begin{aligned}\int_{\mathcal{H}}\Delta^{-1}\underline{A}^{\mathrm{T}}\bar{M}\partial_m\underline{A}d\Sigma^m &= \underline{A}_{\mathcal{H}}^{\mathrm{T}}\int_{\mathcal{H}}\Delta^{-1}\bar{M}\partial_m\underline{A}d\Sigma^m \\ &= -\underline{A}_{\mathcal{H}}^{\mathrm{T}}\int_{r=\infty}\Delta^{-1}\bar{M}\partial_m\underline{A}d\Sigma^m = 0\end{aligned} \tag{195}$$

due to the boundary conditions on the horizon $\mathcal{H}$ and the vanishing of Q_A. Since $\bar{M}$ is a positive matrix we get $\partial_a\underline{A} = 0$ as claimed and thus $\underline{A} = 0$ (being normalized to 0 at infinity).

The second step is to prove the *'Scalar No Hair Theorem'*[12]:

If for a static single black hole solution of eqs.(49a-49e) the vector potentials $\underline{A}$ vanish, then the scalar fields are constant.
Proof: For vanishing vector potentials $\underline{A}$ the scalar field equation in (49a-49e) reduces to

$$(\bar{M}^{-1}\partial^m\bar{M})_{;m} = 0 \tag{196}$$

Integrating we get

$$\bar{Q} = \frac{1}{8\pi}\int_{r=\infty}\bar{M}^{-1}\partial_m\bar{M}d\Sigma^m = 0 \tag{197}$$

since the contribution at the horizon vanishes because $d\Sigma^m$ vanishes like Δ (compare 40).

On the other hand we can multiply eq.(196) by $\bar{M}$ and obtain

$$(\partial^m\bar{M})_{;m} = \partial^m\bar{M}\bar{M}^{-1}\partial_m\bar{M} \tag{198}$$

and again by integration

$$\int_{\Sigma_3}\partial^m\bar{M}\bar{M}^{-1}\partial_m\bar{M}d\Sigma = \int_{r=\infty}\partial_m\bar{M}d\Sigma^m = 0 \tag{199}$$

From the positivity of $\bar{M}$ we deduce $\partial_m \bar{M} = 0$ and therefore $\bar{M} = \bar{M}_0 = \mathbf{1}$.
Remark: As can be seen from the proof the assumption of staticity can be replaced by the weaker assumption of strict stationarity in this theorem.
Remark: For non-vanishing vector charges Q_A we obtain from eq.(49a-49e)

$$\left(\frac{1}{2}\bar{M}^{-1}\partial^m\bar{M} + \Delta^{-1}\frac{1}{c}\left(\underline{A}\,\partial^m\underline{A}^{\mathrm{T}}\bar{M}\right)_{\mathrm{pr}}\right)_{;m} = 0 \tag{200}$$

yielding a relation between the Q_A's and the scalar charges $\bar{Q}$

$$\bar{Q} = -\frac{1}{c}(\underline{A}_{\mathcal{H}} Q_A^{\mathrm{T}})_{\mathrm{pr}} \,. \tag{201}$$

Combining the two theorems with Israel's theorem for pure gravity we obtain the *'Generalized Israel Theorem'*:
Any static single black hole solution of eqs.(49a-49e) with a non-degenerate horizon is a member of the 'Schwarzschild family', i.e. can be obtained from the Schwarzschild solution through a Harrison transformation and hence is necessarily spherically symmetric.

As already mentioned the strategy to prove this theorem is to transform an arbitrary single static black hole solution of eqs.(49a-49e) via a suitable Harrison transformation into one with vanishing vector charge Q_A and thus vanishing $\underline{A}$ and constant $\bar{M}$. In [12] it was claimed that a suitable Harrison transformation doing the job exists without giving a detailed proof. This will gap will be hopefully fixed in the final version of [28].

5.3 Stationary, Axially Symmetric Black Holes

The prototype of a stationary, axisymmetric black hole is the Kerr solution, depending on two parameters, mass and angular momentum. As proved by Robinson [69], this is, *in vacuum*, the only such single black hole solution. Also for the Einstein-Maxwell theory the questions of *existence* and *uniqueness* have a simple, positive answer in the form of the Kerr-Newman solution, depending on four parameters – mass, angular momentum, electric and magnetic charge. The uniqueness of the Kerr-Newman solution has been proven by Mazur [27] and Bunting [66]. From our experience with the static solutions we expect the uniqueness problem for our class of more general theories to be more delicate. In fact, in the static case the charges $\bar{Q}$ for the scalars cannot be given freely, but are determined by the vector charges Q_A. On the other hand we were able to show that all the static single black holes are members of the 'Schwarzschild family', i.e. could be generated from the Schwarzschild solution via a suitable Harrison transformation acting on M. It is tempting to speculate that this situation prevails in the case of rotating black holes, if we replace Schwarzschild by Kerr.

In order to prove the uniqueness of the solutions with given $c = m_H$ ('irreducible mass' or size of the horizon), angular momentum J and charges Q_A

we shall generalize the the uniqueness theorem of Mazur [27]. As already observed by Mazur himself [70] his theorem can in fact be rather easily extended to theories of the type considered here, the essential ingredient being the 3-dimensional σ-model structure for axisymmetric solutions. In contrast to the uniqueness proofs for static black holes using the G/H σ-model the Mazur theorem uses the G/H' σ-model based on the matrix M' instead of M. This is necessary because only G/H' has positive definite metric as required in the proof of the theorem. Moreover this formulation avoids the singularities of the matrix M at the ergosurface.

As discussed in Sect. 3.4 we can choose canonical coordinates ρ and z such that the horizon meets the rotation axis at the two points $z = \pm c$. The horizon itself is then given by $\mathcal{H} = \{\rho = 0, |z| \leq c\}$. The exterior region of the black hole is the union of the two pieces of the rotation axis $\{\rho = 0, |z| > c\}$ with $\Sigma_2 = \{(z, \rho) : \rho > 0\}$.

The uniqueness of single stationary rotating black holes is based on the 'Generalized Mazur Theorem':
Two solutions $\hat{M}'_i = (P'_i)^+ P'_i$ $(i = 1, 2)$ of the equation $(\rho(\hat{M}'_i)^{-1}\partial^m \hat{M}'_i)_{;m} = 0$ taking their values in the Riemannian symmetric space G/H' and obeying the same boundary conditions for single black holes (i.e. equal values for c, J and Q_A) are identical.

Since the proof of this theorem is astonishingly simple in the present formalism we shall present it here. Putting $J_i = \frac{1}{2}(\hat{M}'_i)^{-1}\partial \hat{M}'_i$ and using $(\rho J^m)_{;m} = 0$ and $J^+_i = \hat{M}'_i J_i (\hat{M}'_i)^{-1}$ we find

$$\frac{1}{4}\left(\rho \mathrm{Tr} \partial^m ((\hat{M}'_1)^{-1}\hat{M}'_2)\right)_{;m} = \frac{1}{2}\left(\rho \mathrm{Tr}\left((\hat{M}'_1)^{-1}\hat{M}'_2(J^m_2 - J^m_1)\right)\right)_{;m}$$
$$= \rho \mathrm{Tr}\left((\hat{M}'_1)^{-1}(J^m_2 - J^m_1)^+ \hat{M}'_2(J_{2m} - J_{1m})\right) = \rho \mathrm{Tr}(\mathcal{J}^+_{12m}\mathcal{J}^m_{12}) \quad (202)$$

with $\mathcal{J}_{12} \equiv P'_2(J_2 - J_1)(P'_1)^{-1}$. Integrating this identity over Σ_2 and using $\mathrm{Tr} J_i = 0$ we obtain

$$\int_{\Sigma_2} \mathrm{Tr}(\mathcal{J}^+_{12m}\mathcal{J}^m_{12})\rho \, d\rho dz$$
$$= \int_{\partial\Sigma_2} \frac{1}{2}\mathrm{Tr}\left((\hat{M}'_1)^{-1}\hat{M}'_2(J^m_2 - J^m_1)\right)\rho \, d\Sigma_m$$
$$= \int_{\partial\Sigma_2} \frac{1}{2}\mathrm{Tr}\left(\left((P'_2(P'_1)^{-1})^+ - P'_1(P'_2)^{-1}\right)\mathcal{J}^m_{12}\right)\rho \, d\Sigma_m \, . \quad (203)$$

In order to evaluate the boundary term we use the boundary conditions discussed in the beginning of this section. There is no contribution from the horizon due to the explicit factor ρ. At infinity one finds $P'_2(P'_1)^{-1} = \mathbf{1} + O(\frac{1}{r})$ and the normal component of $\mathcal{J}_{12}$ is $O(\frac{1}{r^2})$; hence the boundary term at infinity vanishes for any two solutions. In order to analyze the contribution from the axis we need the detailed form of the boundary conditions (compare eq.(141). For two solutions with the same angular momentum J and vector charges Q_A it follows that both $P_2'(P_1')^{-1}$ and $\mathcal{J}_{12}$ stay finite at the axis. Hence the boundary term at the axis vanishes as well. This implies $\mathcal{J}_{12} = 0$, i.e., $\partial((\hat{M}_1')^{-1}\hat{M}_2') = 0$ leading to $\hat{M}_1' = \hat{M}_2'$ since $(\hat{M}_1')^{-1}\hat{M}_2' \to \mathbf{1}$ for $r \to \infty$.

Although the parameters m, J, Q_Aq and $\bar{Q}$ determine the black hole uniquely they cannot be chosen arbitrarily as the discussion of the spherically symmetric black holes has already shown. In fact, it appears that the scalar charges $\bar{Q}$ are completely fixed through the regularity requirements once the other parameters have been chosen.

Similarly to the case of static black holes a possible strategy for a uniqueness proof could be to transform the vector charges to zero by a Harrison transformation. The problem with that is that the Harrison transformation acts on $\hat{M}$, based on the reduction according to the stationary KV K. This formulation involving Δ becomes singular at the ergo-surface lying in the interior of the domain Σ_2. The problem now is that we do not know, if the transformed solution still fulfils all the regularity requirements for a black hole ($\Delta' < 0$ and boundary conditions for $\rho = 0$). In principle this could be checked, since the action on $\hat{M}'$ can be obtained using the Geroch group. If this step is successfully done, we can proceed as in the static case and conclude that the transformed solution solves the vacuum problem and thus is the Kerr solution. Reversing the argument shows that it is possible (under the provisions made above) to obtain the most general single black hole solution from the Kerr solution through a suitable Harrison transformation, establishing a *'Generalized Robinson Theorem'*.

6 Acknowledgments

I am deeply indebted to P. Breitenlohner for years of collaboration. Many as yet unpublished results mentioned in this article emerged from this collaboration. Further I would like to thank H.Nicolai for many fruitful discussions.

7 Non-linear σ-Models and Symmetric Spaces

In this appendix we shall study group theoretical aspects of non-linear σ-models whith a target space Φ which is a non-compact Riemannian or pseudo-Riemannian symmetric space G/H.

7.1 Non-compact Riemannian Symmetric Spaces

Let G be a non-compact real form of some compact Lie group. There is an involutive automorphism τ: $G \to G$, $\tau^2 = 1$ such that

$$H = \Big\{ h \in G\colon\ \tau(h) = h \Big\} \tag{A.204}$$

is the maximal compact subgroup of G and the coset space G/H is a non-compact Riemannian symmetric space [11]. In order to parametrize the coset space G/H we can choose a group element P as representative for each coset, i.e. introduce fields $P(x)$

$$P\colon \Sigma \ni x \to P(x) \in G\,. \tag{A.205}$$

Customarily, G/H denotes the space of left cosets gH. For purely historical reasons we use the space of right cosets which is sometimes denoted by $H\backslash G$. As these two spaces are isomorphic we prefer the notation G/H. The choice eq.(A.205) of representatives is obviously not unique and the freedom to choose representatives leads to a gauge invariance with gauge group H, in addition to the group action of G on G/H

$$P(x) \to h(x)P(x)g^{-1}\,, \quad h(x) \in H\,, \quad g \in G\,. \tag{A.206}$$

We can eliminate the gauge group H by choosing one standard representative for each coset. We will obtain a particularly simple parametrization of G/H if we choose a solvable subgroup (represented by matrices 'triangular' in the sense of the Iwasawa decomposition [11]) which intersects each coset once. Given such a gauge choice (or any other one) the action (A.206) of a $g \in G$ will lead to non-standard representatives and must therefore be accompagnied by an induced gauge transformation $h(P(x), g) \in H$ (depending in general non-linearly on P and g) in order to maintain the gauge choice

$$P(x) \to h(P(x), g)P(x)g^{-1}\,, \quad g \in G\,. \tag{A.207}$$

In order to construct an invariant metric on G/H we consider the 1-form $dP(x)P(x)^{-1}$ with value in the Lie algebra $\mathcal{G}$ of G and its decomposition

$$dP\,P^{-1} = \mathcal{A} + \mathcal{J} = (\mathcal{A}_a + \mathcal{J}_a)dx^a\,, \qquad \begin{aligned} \tau(\mathcal{A}) &= \mathcal{A} \\ \tau(\mathcal{J}) &= -\mathcal{J}\,. \end{aligned} \tag{A.208}$$

The transformation laws (induced by eq.(A.206)) for $\mathcal{A}$ and $\mathcal{J}$ are

$$\mathcal{A} = \tfrac{1}{2}(dP\,P^{-1} + \tau(dP\,P^{-1})) \to h\,\mathcal{A}\,h^{-1} + dh\,h^{-1} \tag{A.209}$$

$$\mathcal{J} = \tfrac{1}{2}(dP\,P^{-1} - \tau(dP\,P^{-1})) \to h\,\mathcal{J}\,h^{-1}\,, \tag{A.210}$$

i.e. $\mathcal{A}$ can be interpreted as connection for H whereas $\mathcal{J}$ transforms H-covariantly and both are G-invariant. Given any invariant scalar product $\langle\cdot,\cdot\rangle$ on $\mathcal{G}$ we can define an invariant metric γ on G/H by

$$d\phi^i d\phi^j \gamma_{ij}(\phi) \equiv \langle\mathcal{J},\mathcal{J}\rangle\,. \tag{A.211}$$

If G is simple any such $\langle\cdot,\cdot\rangle$ is a (positive) multiple of the Killing metric and for each faithful representation $\rho\colon \mathcal{G} \ni \mathcal{J} \to \hat{\mathcal{J}} \equiv \rho(\mathcal{J})$ there is a (positive) constant $\hat{c}$ such that

$$\langle\mathcal{J},\mathcal{J}\rangle = \hat{c}\,\mathrm{Tr}(\hat{\mathcal{J}}\hat{\mathcal{J}})\,. \tag{A.212}$$

In case G consists of several simple factors one gets corresponding constants $\hat{c}$. The situation remains essentially unchanged if G contains *one* abelian factor but becomes more complicated in the presence of several such factors ($U(1)$ or $SO(1,1) \sim \mathbf{R}$).

Note that the scalar product $\langle\cdot,\cdot\rangle$ on $\mathcal{G}$ is indefinite but the metric γ on G/H is (positive) definite.

Given an r-dimensional faithful representation ρ of $\mathcal{G}$ we can project any $r \times r$ matrix $\hat{T}$ onto $\mathcal{G}$, $\hat{T} \to (\hat{T})_{\mathrm{pr}}$, such that for every $\mathcal{J} \in \mathcal{G}$

$$\langle(\hat{T})_{\mathrm{pr}}, \mathcal{J}\rangle = \hat{c}\,\mathrm{Tr}(\hat{T}\hat{\mathcal{J}}) \tag{A.213}$$

and thus $(\hat{\mathcal{J}})_{\mathrm{pr}} = \mathcal{J}$. This projection satisfies $(\hat{g}\hat{T}\hat{g}^{-1})_{\mathrm{pr}} = g(\hat{T})_{\mathrm{pr}}g^{-1}$ for all $g \in G$ [28].

The automorphism τ provides us with a canonical embedding of G/H in G [71]

$$P \to M = \tau(P^{-1})P\,, \quad \tau(M) = M^{-1} \tag{A.214}$$

where M is H-invariant and transforms covariantly under G

$$M \to \tau(g)\,M\,g^{-1}\,, \quad g \in G\,. \tag{A.215}$$

The corresponding current $J = \frac{1}{2}M^{-1}dM$ is related to $\mathcal{J}$ by

$$\mathcal{J} = DP\,P^{-1} = \frac{1}{2}P\,M^{-1}\,dM\,P^{-1} \equiv P\,J\,P^{-1} \tag{A.216}$$

where $DP = dP - \mathcal{A}P$ is the H-covariant derivative of P.

The line element eq.(A.211) on G/H can be reexpressed in terms of M

$$d\phi^i d\phi^j \gamma_{ij}(\phi) = \frac{1}{4}\langle M^{-1}\,dM, M^{-1}\,dM\rangle \tag{A.217}$$

Using this expression or the one of eq.(A.211) for the metric in the σ-model action eq.(12) leads to field equations in the form of conservation equations for the currents

$$D^\alpha \mathcal{J}_\alpha = 0 \quad \text{or} \quad J^\alpha_{;\alpha} = 0 \quad \text{or} \quad (M^{-1}\,\partial^\alpha M)_{;\alpha} = 0 \tag{A.218}$$

where $D_\alpha \mathcal{J}_\beta = \mathcal{J}_{\beta;\alpha} - [\mathcal{A}_\alpha, \mathcal{J}_\beta]$. There are $\dim G$ currents J and conservation equations, but clearly only $\dim G/H$ of them are independent. In fact, the currents J obey the identity $\tau(J) = -MJM^{-1}$ due to $\tau(M) = M^{-1}$.

We will choose the basis for a representation ρ such that $\tau(\hat{g}) = (\hat{g}^+)^{-1}$ for all $g \in G$ and thus $\tau(\hat{\mathcal{J}}) = -\hat{\mathcal{J}}^+$ for all $\mathcal{J} \in \mathcal{G}$. In such a basis the matrix $\hat{M} \equiv \rho(M) = \hat{P}^+\hat{P}$ is hermitian and positive definite.

Note that M and P are very close analogues of the metric and the moving frame (tetrad) in general relativity. We see that the 'metric' M is sufficient to formulate the σ-model and this remains true if we include vector fields. Nevertheless the 'moving frame' P with a 'triangular' gauge choice yields a very convenient and simple parametrization of G/H. The situation changes if we add fermion fields to some of our models (e.g. $N = 1$ supergravity) because these fermion fields transform with some representation of H and not of G. The 'moving frame' P is, therefore, necessary in order to describe these fermions.

7.2 Pseudo-Riemannian Symmetric Spaces

The situation becomes slightly more complicated if we consider a pseudo-Riemannian symmetric space G/H [72]. In this case we have a non-compact Lie group G and an involutive automorphism τ which defines the subgroup H of G as before. In addition there is a different involutive automorphism τ' commuting with τ ($\tau\tau' = \tau'\tau = \tau''$) which determines the maximal compact subgroup H' of G

$$H' = \left\{ h \in G \colon \tau'(h) = h \right\} . \tag{A.219}$$

In contrast to the case before the parametrisation with 'triangular' P's is now no more possible globally due to the fact that some compact generators are remaining in G/H. Thus this parametrisation will be singular for certain points (e.g. there may be regular points of G/H with $\Delta = 0$). The basis for a representation ρ will be chosen such that $\tau(\hat{g}) = \hat{X}^{-1}(\hat{g}^{+})^{-1}\hat{X}$ with a hermitian matrix $\hat{X}$. If we define $\hat{M} \equiv \hat{X}\rho(M) = \hat{P}^{+}\hat{X}\hat{P}$ the matrix $\hat{M}$ will again be hermitian. Due to the non-compactness of H the matrix $\hat{X}$ and thus $\hat{M}$ and the metric γ will, however, not be positive definite.

7.3 Consistent Truncations

Given a non-linear σ-model with target space Φ and metric γ we may be interested to study a truncated σ-model with target space $\tilde{\Phi} \subset \Phi$ and the induced metric $\tilde{\gamma} = \gamma|_{\tilde{\Phi}}$.

We say that a non-linear σ-model with target space $\tilde{\Phi}$ is a consistent truncation of another σ-model with target space Φ if $\tilde{\Phi} \subset \Phi$ and if every solution of the field equations for the $\tilde{\Phi}$ σ-model is a solution of the field equations for the Φ σ-model as well.

This is the case iff $\tilde{\Phi}$ is a totally geodesic subspace of Φ.

If Φ is a Riemannian symmetric space G/H then every totally geodesic subspace of Φ is again a Riemannian symmetric space $\tilde{G}/\tilde{H}$ [11]. The automorphism τ maps the subgroup $\tilde{G} \subset G$ onto itself, $\tilde{\tau} = \tau|_{\tilde{G}}$ and therefore $H \cap \tilde{G} = \tilde{H} \subset H$. Similar statements hold true if Φ is a pseudo-Riemannian symmetric space.

8 Structure of the Lie Algebra

We shall now describe in some detail the structure of the Lie algebras of the groups G corresponding to the theories discussed in this work [28]. Due to

Table 1. List of all symmetric spaces obtained by dimensional reduction from four to three dimensions of theories with scalars and vectors (reproduced from Table 2 in [12]).

#	G/H	$\bar{G}/\bar{H}$	$\dim \bar{G}/\bar{H}$	k
1	$SL(n+2)/SO(n,2)$	$GL(n)/SO(n)$	$\frac{n(n+1)}{2}$	n
2	$\frac{SU(p+1,q+1)}{S(U(p,1)\times U(1,q)}$	$U(p,q)/(U(p)\times U(q))$	$2pq$	$p+q$
3	$\frac{SO(p+2,q+2)}{SO(p,2)\times SO(2,q)}$	$\frac{SO(p,q)}{SO(p)\times SO(q)}\times\frac{SO(2,1)}{SO(2)}$	$pq+2$	$p+q$
4	$SO^*(2n+4)/U(n,2)$	$\frac{SO^*(2n)}{U(n)}\times\frac{SU(2)}{SU(2)}$	$n(n-1)$	$2n$
5	$Sp(2n+2;\mathbf{R})/U(n,1)$	$Sp(2n;\mathbf{R})/U(n)$	$n(n+1)$	n
6	$\frac{G_{2(+2)}}{SU(1,1)\times SU(1,1)}$	$SU(1,1)/U(1)$	2	2
7	$\frac{F_{4(+4)}}{Sp(6;\mathbf{R})\times SU(1,1)}$	$Sp(6;\mathbf{R})/U(3)$	12	7
8	$E_{6(+6)}/Sp(8;\mathbf{R})$	$SL(6)/SO(6)$	20	10
9	$\frac{E_{6(-2)}}{SU(3,3)\times SU(1,1)}$	$\frac{SU(3,3)}{S(U(3)\times U(3))}$	18	10
10	$\frac{E_{6(-14)}}{SO^*(10)\times SO(2)}$	$SU(5,1)/U(5)$	10	10
11	$E_{7(+7)}/SU(4,4)$	$\frac{SO(6,6)}{SO(6)\times SO(6)}$	36	16
12	$\frac{E_{7(-5)}}{SO^*(12)\times SO(2,1)}$	$SO^*(12)/U(6)$	30	16
13	$\frac{E_{7(-25)}}{E_{6(-14)}\times SO(2)}$	$\frac{SO(10,2)}{SO(10)\times SO(2)}$	20	16
14	$E_{7(+8)}/SO^*(16)$	$E_{7(+7)}/SU(8)$	70	28
15	$\frac{E_{8(-24)}}{E_{7(-25)}\times SU(1,1)}$	$\frac{E_{7(-25)}}{E_{6(-78)}\times SO(2)}$	54	28

the possibilities to use a space-like resp. a time-like Killing vector for the reduction from 4 to 3 dimensions the coset spaces G/H are characterized by two commuting involutive automorphisms τ and τ': All elements of G that are invariant under τ form the subgroup H, all elements invariant under τ' form the subgroup H', and all elements invariant under $\tau\tau'$ form the subgroup $\bar{G}\otimes SL(2)$. The restriction of τ or τ' to $\bar{G}$ is the original automorphism $\bar{\tau}$; the restriction to $SL(2)$ is the automorphism defining the maximal compact subgroup $SO(2)$. We may assume that the representation $\bar{\rho}$ of $\bar{G}$ is faithful, i.e., that all scalars of the 4-dimensional $\bar{G}/\bar{H}$ σ-model couple to vector fields.

The group G will then be simple, and Table 1 reproduction of Table 2 in [12]) lists all possible cases. We choose a basis t_i for the Lie algebra $sl(2)$, where $t_+ = e$, $t_0 = d$ and $t_- = k$ satisfy the commutation relations

$$[d,e] = e\ , \quad [d,k] = -k\ , \quad [e,k] = 2d \tag{B.220}$$

and

$$\tau(e) = -k\ , \quad \tau(d) = -d\ , \quad \tau(k) = -e\ . \tag{B.221}$$

In addition we have the generators s_i of $\bar{g}$ and additional generators h_i, a_i, $i = 1, \ldots, 2k$ for Harrison resp. gauge transformations, which must form doublets under $SL(2)$ in order to reproduce the scale transformations eq.(58).

$$\begin{aligned} [d,h_i] &= \tfrac{1}{2}h_i\ , \quad [e,h_i] = 0\ , \quad [k,h_i] = -a_i\ , \\ [d,a_i] &= -\tfrac{1}{2}a_i\ , [e,a_i] = -h_i\ , [k,a_i] = 0\ , \end{aligned} \tag{B.222}$$

and with a $2k$-dimensional real matrix representation $\bar{\rho}$ of $\bar{g}$

$$\bar{\rho} : s_i \mapsto \bar{\rho}(s_i) = R_i\ , \quad [s_i, h\cdot\alpha] = h\cdot R_i\alpha\ , \quad [s_i, a\cdot\alpha] = a\cdot R_i\alpha\ . \tag{B.223}$$

The remaining commutators can be determined from the Jacobi identities $J(x,y,z) \equiv [[X,Y],Z] + [[Y,Z],X] + [[Z,X],Y] = 0$:

$$\begin{aligned} J(h\cdot\alpha, h\cdot\beta, d) &\Rightarrow [h\cdot\alpha, h\cdot\beta] = \alpha\cdot y\beta\, e\ , \\ J(h\cdot\alpha, h\cdot\beta, k) &\Rightarrow [h\cdot\alpha, a\cdot\beta] = \alpha\cdot y\beta\, d + \alpha\cdot x^l\beta\, s_l\ , \\ J(h\cdot\alpha, a\cdot\beta, k) &\Rightarrow [a\cdot\alpha, a\cdot\beta] = -\alpha\cdot y\beta\, k\ , \end{aligned} \tag{B.224}$$

with numerical matrices $y^T = -y$ and $x^{lT} = x^l$ obeying

$$\begin{aligned} & yR_i + R_i^T y = 0\ , \quad x^l R_i + R_i^T x^l = f_{ij}{}^l x^j\ , \\ & \tfrac{1}{2}\alpha\,(\beta\cdot y\gamma) + R_l\alpha\,(\beta\cdot x^l\gamma) - (\alpha\leftrightarrow\beta) = \gamma\,(\alpha\cdot y\beta)\ . \end{aligned} \tag{B.225}$$

The trace of this 'Completeness relation' yields

$$x^l R_l - (x^l R_l)^T + (2k+1)y = 0\ . \tag{B.226}$$

The explicit form of the matrices appearing in these commutation relations can be found in [28] for the various coset spaces of Table 1.

References

1. Schwarz, J.H.: hep-th/9607201
2. Korotkin, D. and Samtleben, H.: *Class. Quantum Gravity***14** (1997) L151. Nicolai, H. and Samtleben, H.: *Nucl. Phys.* **B 533** (1998) 210.
3. Hollmann, H.: *Phys. Lett.* **B 388** (1996) 702, Breitenlohner, P., Hollmann, H. and Maison, D.: gr-qc/9804030
4. Cremmer, E., Ferrara, S. and Scherk, J.: *Phys. Lett.* **B 74** (1978) 341,

5. Font,A., Ibañez, L., Lüst, D. and Quevedo, F.: *Phys. Lett.* **B 249** (1990) 35.
6. Ehlers, J.:*Konstruktion und Charakterisierungen von Lösungen der Einstein'schen Gravitationsgleichungen*, Dissertation, Hamburg, 1957.
7. Neugebauer, G. and Kramer, D.: *Ann. d. Physik* **24** (1969) 62,
   Geroch, R.: *J. Math. Phys.* **12** (1971) 918.
8. Kinnersley, W.: *J. Math. Phys.* **14** (1973) 651.
9. Harrison, B.K.: *J. Math. Phys.* **9** (1968) 1744.
10. Julia B. in:*Superspace and Supergravity*, ed. Hawking, S. and Rocek, M., Cambridge Univ. Press (1981).
11. Helgason, S.: *Differential Geometry and Symmetric Spaces*, Academic, New York, 1962.
    Kobayashi, S. and Nomizu, K.; *Foundations of Differential Geometry*, Interscience, New York, 1969.
12. Breitenlohner, P., Gibbons, G. and Maison, D.: *Commun. Math. Phys.* **120** (1988) 295.
13. deWit, B. and Nicolai, H.: *Nucl. Phys.* **B 274** (1986) 363.
14. Geroch, R.: *J. Math. Phys.* **13** (1972) 394.
15. Hoenselaers, C., Kinnersley, W. and Xanthopoulos, B.: *J. Math. Phys.* **20** (1979) 2530.
16. Harrison, B.K.: *Phys. Rev. Lett.* **41** (1978) 1197.
17. Kramer, D. and Neugebauer, G.: *Phys. Lett.* **A 75** (1980) 259.
18. Hauser, I. and Ernst, F.J.: *J. Math. Phys.* **22** (1981) 1051.
19. Belinskiĭ, V.A. and Zakharov, V.E.: *JETP* **75** (1978) 1955 and *JETP* **77** (1979)3.

20. Maison, D.: *Phys. Rev. Lett.* **41** (1978) 521.
21. Breitenlohner, P. and Maison, D.: *Ann. Inst. Henri Poincaré* **46** (1987) 215.
22. Li, W., Hauser, I. and Ernst, F.J.: *J. Math. Phys.* **32** (1991) 723
    Hauser, I. and Ernst, F.J.: gr-qc/9303021
23. Nicolai, H.: *Phys. Lett.* **B 194** (1987) 402.
24. Eells, J. Jr. and Sampson, J.H.: *Am. J. Math.* **86** (1964) 109,
    Eells, J. and Lemaire, L.: *Bull. London Math. Soc.* **10** (1978) 1.
25. Misner, C.W.: *Phys. Rev.* **D 18** (1978) 4510.
26. Heusler, M.: *Black hole uniqueness theorems.* Cambridge Univ. Press, Cambridge, 1996.
27. Mazur, P.O.:*J. Phys.* **A 15** (1982) 3173.
28. Breitenlohner, P. and Maison, D.: *On Nonlinear σ-Models arising in (Super-) Gravity.*
    gr-qc/9806002, to appear in Commun. Math. Phys.
29. Olive, D.: *Nucl. Phys. Proc. Suppl.***45A** (1996) 88;
    *Nucl. Phys. Proc. Suppl.***46** (1996) 1.
30. Cremmer, E. and Julia, B.: *Nucl. Phys.* **B 159** (1979) 141.
31. Wu, T.T. and Yang, C.N.: *Phys. Rev.* **D 12** (1975) 3845.
32. Schwinger, J.: *Science***165** (1969) 757.
33. Zwanziger, D.: *Phys. Rev. 176* (1968) 1489.
34. Deser, S. and Teitelboim, C.: *Phys. Rev.* **D 13** (1976) 1592.
35. Montonen, and Olive,D.: *Phys. Lett.* **B 72** (1977) 117.
36. Seiberg, N. and Witten, E.: *Nucl. Phys.* **B 426** (1994) 19.
37. Chruściel, P.T.: *Class. Quantum Grav. 10* (1993) 2091.

38. Kałuza, T.: *Sitzungsber. Preuss. Akad. Wiss.* (1921) 966.
39. Carter, B.: in *Black Holes*, B. De Witt and C. De Witt eds., Gordon and Breach, New York, 1973.
40. Kac, V.G.:*Infinite Dimensional Lie algebras*, Birkhäuser, Boston, 1984.
41. Novikov, S., Manakov, S.V., Pitaevskii, L.P. and Zakharov, V.E.: *Theory of Solitons*, Consultants Bureau, New York, 1984. Faddeev, L.D. and Takhtajan, L.A.: *Hamiltonian Methods in the Theory of Solitons*, Springer-Verlag, Berlin, 1987.
42. Maison,D.: in *Nonlinear evolution equations: integrability and spectral methods*, eds. Degasperis, Fordy and Lakshamanan, Manchester University Press, 1990.
43. Kramer, D. and G. Neugebauer, G.: *Commun. Math. Phys.* **10** (1968) 132.
44. Ueno, K. and Nakamura, Y.:*Phys. Lett.* **B 117** (1982) 208.
45. Cosgrove, C.M.: *J. Math. Phys.* **21** (1980) 2417.
46. Julia, B.: in *Proceedings of the Johns Hopkins Workshop on Particle Theory*, Baltimore, 1981.
47. Cartan, H. and Eilenberg, S.:*Homological Algebra*, Princeton Univ. Press, 1956.
48. Bičak, J. and Hoenselaers, C.: *Phys. Rev.* **D 31** (1985) 2476.
49. Hou, B.Y. and Li, W.: *Lett. Math. Phys.*.**13** (1987) 1.
50. Julia, B. and Nicolai H.: *Nucl. Phys.* **B 482** (1996) 431.
51. Hawking, S.W. and Ellis, G.F.R.: *The large scale structure of space-time*. Cambridge University Press, 1973.
52. Lichnerowicz, A.: *Théories Relativiste de la Gravitation et de l'Électromagnétisme*, Masson, Paris, 1955.
53. Simon, W.: *J. Math. Phys.* **25** (1984) 1038.
54. Chruściel, P.T.: *Commun. Math. Phys.* **189** (1997) 1.
55. Sudarsky, D. and Wald, R.M.: *Phys. Rev.* **D 46** (1992) 1453.
56. Dobiasch, P. and Maison, D.: *Gen. Rel. Grav.* **14** (1981) 231.
57. Israel, W.: *Phys. Rev. 164* (1967) 1776.
58. Hawking, S.W.: *Commun. Math. Phys.* **25** (1972) 152.
59. Chruściel, P.T. and Wald, R.M.: *Class. Quantum Grav. 11* (1994) L147.
60. Heusler, M.: *Phys. Rev.* **D 56** (1997) 961.
61. Robinson, D.C.: *Gen. Rel. Grav.* **8** (1977) 695.
62. Bunting, G.L. and Masood-Ul-Alaam, A.K.M.: *Gen. Rel. Grav.* **19** (1987) 147.
63. Israel, W.: *Commun. Math. Phys.* **8** (1968) 245.
64. Papapetrou, A.:*Proc. Roy. Irish Acad.* **A51** (1947) 191; S.D. Majumdar, S.D.: *Phys. Rev. 72* (1947) 390.
65. Chruściel, P.T. and Nadirashvili, N.S.: *Class. Quant. Grav.* **12** L17 (1995) 12.
66. Bunting, G: *Proof of the uniqueness conjecture for black holes*, PHD thesis, Dept. of Math., University of New England, Armidale N.S.W., 1983.
67. Maison, D: in *Non-linear Equations in Classical and Quantum Field Theory*, N. Sanchez ed., Springer, Berlin, 1985.
68. Simon, W.: *Gen. Rel. Grav.* **17** (1985) 761.
69. Robinson, D.C. : *Phys. Rev. Lett.* **34** (1975) 905.
70. Mazur, P.: *Phys. Lett.* **A 100** (1984) 341.
71. Eichenherr, H. and Forger, M.: *Nucl. Phys.* **B 155** (1979) 382.
72. Gilmore, R.: *Lie Groups, Lie Algebras and their Applications.* Wiley, New York, 1974.

Time-Independent Gravitational Fields

Robert Beig[1] and Bernd Schmidt[2]

[1] Institut für Theoretische Physik, Universität Wien, Boltzmanngasse 5, A–1090 Wien, Österreich
[2] Max-Planck-Institut für Gravitationsphysik
Albert-Einstein-Institut
Mühlenweg 1, D-14476 Golm b. Potsdam, Deutschland

1 Introduction

In this article we want to describe what is known about time independent spacetimes from a global point of view. The physical situations we want to treat are isolated bodies at rest or in uniform rotation in an otherwise empty universe. In such cases one expects the gravitational field to have no "independent degrees of freedom". Very loosely speaking, the spacetime geometry should be uniquely determined by the matter content of the model under consideration. In a similar way, for a given matter model (such as that of a perfect fluid), there should be a one-to-one correspondence between Newtonian solutions and general relativistic ones.

The plan of this paper is as follows. In Sect. 2.1 we collect the information afforded on one hand by the Killing equation obeyed by the vector field generating the stationary isometry and, on the other hand, that by the Einstein field equations. Throughout this paper we assume this stationary isometry to be everywhere timelike. Thus ergoregions are excluded. We try as much as possible to write the resulting equations in terms of objects intrinsic to the space (henceforth simply called "the quotient space") obtained by quotienting spacetime by the action of the stationary isometry. Much of this is standard. But since none of the references known to us meets our specific purposes, we give a self-contained treatment starting from scratch. Since all the models we treat are axially symmetric, we add in Sect. 2.2 a second, axial Killing vector to the formalism of Sect. 2.1.

In Sect. 2.3 we first introduce new dependent variables for the vacuum gravitational field, namely a conformally rescaled metric on the quotient space and two potentials, using which the field equations have an interpretation in terms of harmonic maps from the quotient space into the Poincaré half plane. These potentials, originally due to Hansen, are used in our treatment of asymptotics in Sects. 3.1,2. We then formulate the boundary conditions at spatial infinity appropriate for isolated systems and prove two basic theorems due to Lichnerowicz on stationary solutions obeying these conditions. These theorems are manifestations of the above-mentioned principle concerning the lack of gravitational degrees of freedom. The first result, the "staticity theorem", basically states that the gravitational field is static when the matter is non-rotating. The second one, the "vacuum theorem", states that spacetime is Minkowski when no matter is present. In Sect. 2.4 we restore c, the velocity of light, in the field equations and show that these tend to the Newtonian ones as $c \to \infty$.

In Chap. 3 we we study solutions only "near infinity". (Note that by the Lichnerowicz vacuum theorem, such solutions can not be extended to all of $\mathbf{R}^4$ exept for flat spacetime.) In the Newtonian case such solutions are known to have a convergent expansion in negative powers of the radius where the coefficients are given by multipole moments. The relativistic situation is slightly at variance with our statement at the beginning concerning the Newton–Einstein correspondence: namely, there are now, corresponding to the presence of two potentials rather than one, two infinite sequences of multipole moments, the "mass moments" which have a Newtonian analogue and the "angular momentum moments" which do not. One may now study the two potentials and the rescaled quotient space metric in increasing powers of $1/r$, where r is the radius corresponding to a specific coordinate gauge on the quotient space which has to be readjusted at each order in $1/r$.

The results one finds are sufficient for the existence of a chart in the one-point "compactification" of the quotient space (i.e. the union of the quotient space and the point-at-infinity), in terms of which yet another conformal rescaling of the 3-metric, together with a corresponding rescaling of the two potentials, admit regular extensions to the compactified space. As summarized in Sect. 3.2 one is then able to find field equations for these "unphysical" variables which are regular at the point-at-infinity and in addition can be turned into an elliptic system. From this it follows that the unphysical quantities are in fact analytic near infinity and this, in turn, implies convergence for a suitable $1/r$-expansion (r being a "physical" radius) for the original physical variables. Furthermore the structure of the unphysical equations yields the result that the (physical) spacetime metric is uniquely characterized by the two sets of multipole moments.

It is remarkable that stationary vacuum solutions satisfying rather weak fall-off conditions at spatial infinity, by the very nature of the field equations, have to have a convergent multipole expansion. We believe that the topic of far-field behaviour of time-independent gravitational fields is by now reasonably well understood. The main open problem is to characterize an a priori given sequence of multipole moments for which the expansion converges.

In Chap. 4 we review global rotating solutions. In Sect. 4.1 we outline a result due to Lindblom which shows that stationary rotating spacetimes with a one-component fluid source with phenomenological heat conduction and viscosity have to be axisymmetric. In Sect. 4.2 we describe a theorem of Heilig which proves the existence of axisymmetric, rigidly rotating perfect-fluid spacetimes with polytropic equation of state, provided the parameters are sufficiently close to ones for a nonrotating Newtonian solution. In Sect. 4.3 we present the solution of Neugebauer and Meinel representing a rigidly rotating infinitely thin disk of dust. In the final chapter we treat global nonrotating solutions.

In Sect. 5.1 we outline the essentials of a relativistic theory of static elastic bodies. The remaining sections are devoted to spherical symmetry. It has long been conjectured that nonrotating perfect fluids are spherical whence Schwarzschild in their exterior region. In Sect. 5.2 we discuss the present status of this conjecture. A proof exists when the allowed equations of state are limited by a certain inequality. While this inequality covers many cases of physical interest, the Newtonian situation suggests that the conjecture is probably true without this restriction. In Sect. 5.3 we review spherically symmetric perfect fluid solutions. The final Sect. 5.4 gives a short description of self-gravitating Vlasov matter in the sperically symmetric case.

In the subject of time-independent gravitational field of isolated bodies there are some topics we do not cover. We do not address the question of the conjectured non-existence of solutions with more than one body. (Müller zum Hagen [58] has some results on this in the static case.) Furthermore we limit ourselves to "standard matter" sources. Thus Black Holes are excluded. (For this see the article of Maison in this volume.) We also could not cover the interesting case of soliton-like solutions for "non-linear matter sources", starting with the discovery of the Bartnik–McKinnon solutions of the Einstein–Yang Mills system (see Bizon [11].)

Acknowledgements

Part of this work was carried out at the Institute for Theoretical Physics at Santa Barbara, where the authors took part in the program on "Classical and Quantum Physics of Strong Gravitational Fields". We thank the ITP for its support and kind hospitality. R. Beig was in part supported by "Fonds zur Förderung der wissenschaftlichen Forschung in Österreich", project P12626-PHY.

2 Field Equations

2.1 Generalities

Let $(M, g_{\mu\nu})$ be a 4-dimensional smooth connected manifold with Lorentz metric $g_{\mu\nu}$ of signature $(- + ++)$. We assume M to be chronological, i.e. to admit no closed timelike curves. Let ξ^μ be an everywhere timelike Killing vector field with complete orbits. Thus we do not allow points where ξ^μ turns null, i.e. we exclude horizons and ergospheres. It follows (see [26]) that the quotient of M by the isometry group generated by ξ^μ is a Hausdorff manifold N and that M is a principal $\mathbf{R}^1$-bundle over N. Furthermore this bundle is trivial, i.e. M is diffeomorphic to $\mathbf{R}^1 \times N$. The fact that this diffeomorphism is non-natural (whereas of course the projection π mapping M onto N is) plays a role in the formalism we shall now develop.

Let us introduce the differential geometric machinery necessary for writing the stationary Einstein equations in a way naturally adapted to ξ^μ. As far as possible, we will be interested in quantities and equations intrinsic to N ("dimensional reduction"). For a similar treatment see the Appendix of [23]. We define the fields V and $\omega_{\lambda\nu\lambda} = \omega_{[\mu\nu\lambda]}$ by

$$V := \xi_\mu \xi^\mu \Rightarrow V < 0 \tag{2.1}$$

$$\omega_{\mu\nu\lambda} := 3\xi_{[\mu}\nabla_\nu \xi_{\lambda]}. \tag{2.2}$$

The 3-form $\omega_{\mu\nu\lambda}$ vanishes if and only if ξ^μ is hypersurface orthogonal – in which case $(M, g_{\mu\nu})$ is called static. More important than $\omega_{\mu\nu\lambda}$ will be the 2-form $\sigma_{\mu\nu}$, given by

$$\sigma_{\mu\nu} := \omega_{\mu\nu\lambda}\xi^\lambda. \tag{2.3}$$

Given ξ^μ, the fields $\sigma_{\mu\nu}$ and $\omega_{\mu\nu\lambda}$ carry the same information, since

$$\omega_{\mu\nu\lambda} = 3V^{-1}\xi_{[\mu}\sigma_{\nu\lambda]}. \tag{2.4}$$

Equation (2.4) is obtained by expanding the identity $\xi_{[\mu}\omega_{\nu\lambda\rho]} = 0$, which follows from (2.2), and contracting with ξ^μ. In a similar way we obtain the relations

$$\omega_{\mu\nu\lambda}\omega^{\mu\nu\lambda} = 3V^{-1}\sigma_{\mu\nu}\sigma^{\mu\nu} \tag{2.5}$$

$$\omega_{\mu\nu\lambda}\sigma^{\nu\lambda} = \frac{1}{3}\omega_{\rho\nu\lambda}\omega^{\rho\nu\lambda}\xi_\mu. \tag{2.6}$$

We now invoke the Killing equation for ξ^μ, i.e.

$$\mathcal{L}_\xi g_{\mu\nu} = \nabla_\mu \xi_\nu + \nabla_\nu \xi_\mu = 0. \tag{2.7}$$

Expanding $\omega_{\mu\nu\lambda}$ in terms of ξ_μ, we easily see that

$$\nabla_\mu \xi_\nu = V^{-1}[\sigma_{\mu\nu} + (\nabla_{[\mu}V)\xi_{\nu]}], \tag{2.8}$$

or, equivalently,

$$\sigma_{\mu\nu} = V^2 \nabla_{[\mu}(V^{-1}\xi_{\nu]}). \tag{2.9}$$

In the static case we have $\sigma_{\mu\nu} = 0$, whence there exist global cross sections given by $t =$ const, where $\xi_\mu = V\nabla_\mu t$.

Equation (2.9) implies that

$$\nabla_{[\mu}(V^{-2}\sigma_{\nu\lambda]}) = 0. \tag{2.10}$$

Clearly we have $\mathcal{L}_\xi \tau = 0$, where τ is the 3-form given by $\tau_{\mu\nu\lambda} = V^{-2}\omega_{\mu\nu\lambda}$. By (2.10) and the identity $\mathcal{L}_\xi = \xi \rfloor d\tau + d(\xi \rfloor \tau)$, this implies $\xi \rfloor d\tau = 0$. Since

$d\tau$ is a 4-form and ξ^μ is nowhere zero, we infer in 4 dimensions that $d\tau$ is zero, i.e.

$$\nabla_{[\mu}(V^{-2}\omega_{\nu\lambda\rho]}) = 0. \tag{2.11}$$

Equations (2.10,11) are integrability conditions for the Killing equations (2.7) which are "purely geometric" in that they do not involve the Ricci (whence: energy-momentum) tensor. Now recall the relation

$$\nabla_\mu\nabla_\nu\xi_\lambda = -R_{\nu\lambda\mu}{}^\rho\xi_\rho, \tag{2.12}$$

which follows from (2.7) and its corollary

$$g^{\nu\rho}\nabla_\nu\nabla_\rho\xi_\mu = -R_\mu{}^\nu\xi_\nu. \tag{2.13}$$

From (2.2), (2.7) and (2.13) we find that

$$\nabla^\mu\omega_{\mu\nu\lambda} = 2\xi_{[\nu}R_{\lambda]\mu}\xi^\mu, \tag{2.14}$$

which, using (2.8), implies

$$\nabla^\mu(V^{-1}\sigma_{\mu\nu}) = 2V^{-1}\xi_{[\nu}R_{\lambda]\mu}\xi^\mu\xi^\lambda - V^{-3}\sigma_{\mu\lambda}\sigma^{\mu\lambda}\xi_\nu, \tag{2.15}$$

where we have also used (2.6,7). Interpreting $G_{\mu\nu} = R_{\mu\nu} - \frac{1}{2}g_{\mu\nu}R$ as the energy-momentum tensor of matter, the r.h. side of (2.14) is zero iff the matter current, for an observer at rest relative to ξ^μ, is zero. In that case, and provided that M is simply connected, there exists a scalar field ω, called twist potential, such that

$$\omega_{\mu\nu\lambda} = \frac{1}{2}\varepsilon_{\mu\nu\lambda}{}^\rho\nabla_\rho\omega, \tag{2.16}$$

and then (2.11) implies

$$\nabla^\mu(V^{-2}\nabla_\mu\omega) = 0. \tag{2.17}$$

Note that, by virtue of $\xi_{[\mu}\omega_{\nu\lambda\rho]} = 0$, ω satisfies $\mathcal{L}_\xi\omega = 0$.

Next, using the definition (2.1) and eq. (2.8), it is straightforward to show that

$$\begin{aligned}\nabla_\mu\nabla_\nu V = &-2R_{\mu\lambda\nu\rho}\xi^\lambda\xi^\rho + 2V^{-2}[\sigma_{\mu\lambda}\sigma_\nu{}^\lambda - (\nabla^\lambda V)\xi_{(\mu}\sigma_{\nu)\lambda} \\ &+ \frac{1}{4}V\nabla_\mu V\nabla_\nu V + \frac{1}{4}\xi_\mu\xi_\nu(\nabla V)^2]\end{aligned} \tag{2.18}$$

and whence

$$\nabla^\mu\nabla_\mu V = -2R_{\mu\nu}\xi^\mu\xi^\nu + V^{-1}(\nabla V)^2 + 2V^{-2}\sigma_{\mu\nu}\sigma^{\mu\nu}. \tag{2.19}$$

Now recall (see e.g. the Appendix of [23]) that there is a $1-1$ correspondence between tensor fields on M with vanishing Lie derivative with respect to ξ^μ and such that all their contractions with ξ^μ and ξ_μ are zero – and ones of the same type on N. In the case of covariant tensor fields on N, this correspondence is the same as pull-back under π. Examples of such tensor fields on M are the scalar field V, the symmetric tensor field

$$h_{\mu\nu} := g_{\mu\nu} - V^{-1}\xi_\mu\xi_\nu \tag{2.20}$$

and the 2-form $\sigma_{\mu\nu} = \omega_{\mu\nu\lambda}\xi^\lambda$. Note that $\sigma_{\mu\nu}$ can also be written as

$$\sigma_{\mu\nu} = V h_\mu{}^{\mu'} h_\nu{}^{\nu'} \nabla_{\mu'}\xi_{\nu'}. \tag{2.21}$$

The tensor $h_{\mu\nu}$ is, of course, the natural Riemannian metric on N. The covariant derivative D_μ associated with $h_{\mu\nu}$ acting, say, on a covector X_μ living on N, is given by

$$D_\mu X_\nu = h_\mu{}^{\mu'} h_\nu{}^{\nu'} \nabla_{\mu'} X_{\nu'}. \tag{2.22}$$

Denoting by $\mathbb{R}_{\mu\nu\lambda\sigma}$ the curvature associated with D_μ, we find, using (2.9), that

$$\mathbb{R}_{\mu\nu\lambda\sigma} = h_\mu{}^{\mu'} h_\nu{}^{\nu'} h_\lambda{}^{\lambda'} h_\sigma{}^{\sigma'} R_{\mu'\nu'\lambda'\sigma'} + 2V^{-3}\sigma_{\mu\nu}\sigma_{\lambda\rho} - V^{-3}(\sigma_{\lambda[\mu}\sigma_{\nu]\rho} - \sigma_{\rho[\mu}\sigma_{\nu]\lambda}). \tag{2.23}$$

Since N is 3-dimensional, there holds

$$\sigma_{\mu[\nu}\sigma_{\lambda\rho]} = 0, \tag{2.24}$$

so that

$$\mathbb{R}_{\mu\nu\lambda\rho} = h_\mu{}^{\mu'} h_\nu{}^{\nu'} h_\lambda{}^{\lambda'} h_\rho{}^{\rho'} R_{\mu'\nu'\lambda'\rho'} + 3V^{-3}\sigma_{\mu\nu}\sigma_{\lambda\rho}. \tag{2.25}$$

Thus

$$\mathbb{R}_{\mu\nu} = h_\mu{}^{\mu'} h_\nu{}^{\nu'} R_{\mu'\nu'} - V^{-1} R_{\mu\nu'\lambda\rho'}\xi^{\nu'}\xi^{\rho'} + 3V^{-3}\sigma_{\mu\lambda}\sigma_\nu{}^\lambda. \tag{2.26}$$

Using (2.18), (2.26) finally leads to

$$\mathbb{R}_{\mu\nu} = h_\mu{}^{\mu'} h_\nu{}^{\nu'} R_{\mu'\nu'} + \frac{1}{2}V^{-1}D_\mu D_\nu V + 2V^{-3}\sigma_{\mu\lambda}\sigma_\nu{}^\lambda - \frac{1}{4}V^{-2}(D_\mu V)(D_\nu V). \tag{2.27}$$

From (2.19) we deduce that

$$D^2V := h^{\mu\nu}D_\mu D_\nu V = -2R_{\mu\nu}\xi^\mu\xi^\nu + \frac{1}{2}V^{-1}(DV)^2 + 2V^{-2}\sigma_{\mu\nu}\sigma^{\mu\nu}. \tag{2.28}$$

We now make the following observation: when $\tau_{\mu\ldots\lambda}$ is an arbitrary tensor on N, there holds

$$h_\nu{}^{\nu'} \ldots h_\lambda{}^{\lambda'} \nabla^\mu \tau_{\mu\nu'\ldots\lambda'} = (-V)^{-1/2} D^\mu[(-V)^{1/2}\tau_{\mu\nu\ldots\lambda}]. \tag{2.29}$$

Applying (2.29) to (2.15) it follows that

$$(-V)^{-1/2}D^\mu[(-V)^{1/2}\sigma_{\mu\nu}] = h_\nu{}^{\nu'} R_{\nu'\mu}\xi^\mu. \tag{2.30}$$

Finally, projecting (2.10) down to N, it follows that

$$D_{[\mu}(V^{-2}\sigma_{\nu\rho]}) = 0. \tag{2.31}$$

Given the spacetime $(M, g_{\mu\nu})$ with the Killing vector ξ^μ, under the conditions stated at the beginning of this section, there are coordinates (t, x^i) on M, such that the canonical projection π takes the form $\pi : (t, x^i) \mapsto (x^i)$, with x^i local coordinates on N and such that the Killing vector ξ^μ takes the form $\xi = \partial/\partial t$. In terms of such coordinates tensor fields on N, say $\tau_{i\ldots j}(x)$, can be viewed as the tensor fields

$$\tau_{\mu\ldots\nu}(t, x) = \delta_\mu{}^i \ldots \delta_\nu{}^j \tau_{i\ldots j}(x). \tag{2.32}$$

Since $\xi_\mu \xi^\mu = V$, there holds

$$\xi_\mu dx^\mu = V(dt + \varphi_i dx^i), \tag{2.33}$$

for some 1-form φ_i. Note that, in the tangent space at each point $(t, x^i) \in M$, the $g_{\mu\nu}$-orthogonal complement of ξ_μ is spanned by $\varphi_i \partial/\partial t + \partial/\partial x^i$ and the orthogonal complement of ξ_μ in the cotangent space is spanned by dx^i. From the definition $h_{\mu\nu} = g_{\mu\nu} - V^{-1}\xi_\mu\xi_\nu$ it follows that

$$g_{\mu\nu} dx^\mu dx^\nu = V(dt + \varphi_i dx^i)^2 + h_{ij} dx^i dx^j, \tag{2.34}$$

where V, φ_i, h_{ij} on the r.h. side of (2.34) are all independent of t. It is now straightforward to check that

$$\sigma_{\mu\nu} dx^\mu dx^\nu = 3(\xi_{[\mu} \nabla_\nu \xi_{\lambda]}) dx^\mu dx^\nu = V^2 \partial_{[i}\varphi_{j]} dx^i dx^j. \tag{2.35}$$

Thus $\sigma_{\mu\nu}$, viewed as a tensor on N, is given by

$$\sigma_{ij} = V^2 \partial_{[i}\varphi_{j]}. \tag{2.36}$$

In the static case t can be chosen so that $\varphi_i = 0$.

Conversely, let us start from the 3-manifold $(N, h_{ij}, V, \sigma_{ij})$ with Riemannian metric h_{ij}, a negative scalar field V and the 2-form σ_{ij}, subject to

$$D_{[i}(V^{-2}\sigma_{jk]}) = 0, \tag{2.37}$$

which corresponds to (2.31). Suppose, moreover, that N has trivial second cohomology. Then there exists a covector φ_i on N with

$$\sigma_{ij} = V^2 D_{[i}\varphi_{j]}. \tag{2.38}$$

Define $M = \{t \in \mathbf{R}\} \times N$ and define on N the Lorentz metric $g_{\mu\nu}$ by (2.34) and ξ^μ by $\xi = \partial/\partial t$. Then one checks that $\xi_\mu \xi^\mu = V$, that, under the projection $\pi : M \to N$, $h_{\mu\nu}$ is the pull-back of h_{ij} and that $\sigma_{\mu\nu}$ is the pull-back of $\sigma_{ij} = V^2 D_{[i}\varphi_{j]}$. The fact that the product structure of M as $M = \mathbf{R}^1 \times N$ is not natural is reflected in the above construction by the fact that φ_i, solving (2.38), is given only up to $\varphi_i \mapsto \bar{\varphi}_i = \varphi_i + D_i F$, with F a scalar field on N. Under this change $g_{\mu\nu}$ given (2.34) remains unchanged only when we set $t \mapsto \bar{t} = t - F$.

Given the fields (h_{ij}, V, σ_{ij}) on N, we can define the fields r, r_i, r_{ij} by the following equations:

$$D^2 V = -2r + \frac{1}{2} V^{-1} (DV)^2 + 2V^{-2} \sigma_{ij} \sigma^{ij} \tag{2.39}$$

$$D^i[(-V)^{-1/2} \sigma_{ij}] = (-V)^{1/2} r_j \tag{2.40}$$

$$\mathbb{R}_{ij} = r_{ij} + \frac{1}{2} V^{-1} D_i D_j V + 2V^{-3} \sigma_{ik} \sigma_j{}^k - \frac{1}{4} V^{-2} (D_i V)(D_j V). \tag{2.41}$$

It then follows from our previous considerations that the spacetime $(M, g_{\mu\nu})$ satisfies

$$R_{\mu\nu} dx^\mu dx^\nu = r(dt + \varphi_\ell dx^\ell)^2 + 2r_i dx^i (dt + \varphi_\ell dx^\ell) + r_{ij} dx^i dx^j. \tag{2.42}$$

In particular, iff r, r_i, r_{ij} are all zero, $(M, g_{\mu\nu})$ is a vacuum spacetime. In this case we refer to (2.39,40,41) as 'the vacuum equations'.

For later use we record another form of the field equations

$$G_{\mu\nu} = \kappa T_{\mu\nu}, \tag{2.43}$$

where

$$T_{\mu\nu} dx^\mu dx^\nu = \tau(dt + \varphi_i dx^i)^2 + 2\tau_i (dt + \varphi_j dx^j) dx^i + \tau_{ij} dx^i dx^j, \tag{2.44}$$

and where we set

$$g_{\mu\nu} dx^\mu dx^\nu = -e^{2U} (dt + \varphi_i dx^i)^2 + e^{-2U} \bar{h}_{ij} dx^i dx^j, \tag{2.45}$$

given by

$$\bar{D}^2 U = \frac{\kappa}{2} (e^{-4U} \tau + \bar{\tau}_\ell{}^\ell) - e^{4U} \bar{\omega}_{ij} \bar{\omega}^{ij} \tag{2.46}$$

$$\bar{D}^i \bar{\omega}_{ij} = \kappa e^{-4U} \tau_j \tag{2.47}$$

$$\bar{\mathbb{R}}_{ij} = 2(D_i U)(D_j U) - 2e^{4U} \bar{\omega}_{ik} \bar{\omega}_j{}^k + \bar{h}_{ij} e^{4U} \bar{\omega}_{k\ell} \bar{\omega}^{k\ell} + \kappa(\tau_{ij} - \bar{h}_{ij} \bar{\tau}_\ell{}^\ell). \tag{2.48}$$

Here

$$\bar{\omega}_{ij} = \omega_{ij} = \partial_{[i} \varphi_{j]} \tag{2.49}$$

and indices are raised with $\bar{h}^{ij}$.

2.2 Axial Symmetry

We now assume the existence of a second, spacelike Killing vector η^μ on $(M, g_{\mu\nu})$. There is the following identity

$$4\nabla^\mu(\eta_{[\rho}\omega_{\mu\nu\lambda]}) = -\mathcal{L}_\eta \omega_{\nu\lambda\rho} + 6\xi^\mu R_{\mu[\nu}\xi_\lambda \eta_{\rho]}, \tag{2.50}$$

where $\omega_{\mu\nu\lambda}$ is given by (2.2) and we have used (2.14). Suppose, in addition, that ξ and η commute. Then the first term on the right in (2.50) vanishes so that

$$4\nabla^\mu(\eta_{[\rho}\omega_{\mu\nu\lambda]}) = 6\xi^\mu R_{\mu[\nu}\xi_\lambda \eta_{\rho]}. \tag{2.51}$$

In an analogous manner

$$4\nabla^\mu(\xi_{[\rho}\omega'_{\mu\nu\lambda]}) = -6\eta^\mu R_{\mu[\nu}\xi_\lambda \eta_{\rho]}, \tag{2.52}$$

where $\omega'_{\mu\nu\lambda}$ is given in terms of η in the same way as $\omega_{\mu\nu\lambda}$ is given in terms of ξ. The r.h. sides of (2.51,52) are zero (at points where ξ and η are linearly independent) iff the timelike 2-plane spanned by ξ and η is invariant under $R_\mu{}^\nu$. These conditions will be satisfied when the energy momentum tensor is that of a rotating perfect fluid. We now assume that η^μ has an axis, i.e. vanishes on a timelike 2-surface which is tangent to ξ^μ. Then, and when the r.h. sides of (2.51,52) are zero, it follows that

$$\eta_{[\rho}\omega_{\mu\nu\lambda]} = \xi_{[\rho}\omega'_{\mu\nu\lambda]} = 0. \tag{2.53}$$

The relations (2.53), in turn, are nothing but the conditions for the 2-plane elements orthogonal to ξ and η to be integrable ("surface transitivity of ξ and η"). The above result is due to Kundt and Trümper [38].

For the purposes of Sect. 4.2 we need to transcribe the relations satisfied by η^μ on the quotient manifold N. Writing the 1-form $\eta_\mu = g_{\mu\nu}\eta^\nu$ as

$$\eta_\mu dx^\mu = \eta(dt + \varphi_i dx^i) + \eta_i dx^i, \tag{2.54}$$

so that

$$\eta^\mu \frac{\partial}{\partial x^\mu} = (V^{-1}\eta - \varphi_i \eta^i)\frac{\partial}{\partial t} + \eta^i \frac{\partial}{\partial x^i}, \tag{2.55}$$

the Killing equations

$$\eta^\lambda \partial_\lambda g_{\mu\nu} + 2g_{\lambda(\mu}\partial_{\nu)}\eta^\lambda = 0 \tag{2.56}$$

are equivalent to

$$\eta^i D_i V = 0 \tag{2.57}$$

$$2\omega_{ij}\eta^j = D_i(V^{-1}\eta) \tag{2.58}$$

$$\mathcal{L}_\eta h_{ij} = 0, \tag{2.59}$$

where $\omega_{ij} := D_{[i}\varphi_{j]}$. The surface transitivity conditions (2.53) get translated into

$$\eta_{[i}D_j\eta_{k]} = 0 \tag{2.60}$$
$$\eta_{[i}\omega_{jk]} = 0. \tag{2.61}$$

In particular, η^i is a hypersurface-orthogonal Killing vector on (N, h_{ij}). Suppose, now, that the energy momentum tensor is that of a rigidly rotating perfect fluid, i.e.

$$T_{\mu\nu} = (\rho + p)u_\mu u_\nu + p g_{\mu\nu}, \tag{2.62}$$

with

$$u_\mu = f(\xi_\mu + \Omega\eta_\mu), \qquad \Omega = \text{const}, \tag{2.63}$$

and f is chosen so that u^μ is future-pointing and $u_\mu u^\mu = -1$. With this specialization the quantities τ, τ_i, τ_{ij} entering in the field equation (2.46,47,48) become

$$\tau = f^2(\rho + p)(-e^{2U} + \Omega\eta)^2 - pe^{2U} \tag{2.64}$$
$$\tau_i = f^2(-e^{2U} + \Omega\eta)(\rho + p)\Omega\eta_i \tag{2.65}$$
$$\tau_{ij} = pe^{-2U}\bar{h}_{ij} + f^2\Omega^2(\rho + p)\eta_i\eta_j, \tag{2.66}$$

where $\eta_i = h_{ij}\eta^j$. The normalization factor f is given by

$$f = [e^{-2U}(-e^{2U} + \Omega\eta)^2 - \Omega^2\eta_\ell\eta^\ell]^{-1/2}. \tag{2.67}$$

The field equations have to be supplemented by the Killing relations (2.57,58,59). Note that these imply that ρ and p are invariant under η^i (in addition of course to being invariant under $\partial/\partial t$). Under these circumstances the contracted Bianchi identities, which imply that

$$\nabla_\mu T^{\mu\nu} = 0, \tag{2.68}$$

boil down to the relation

$$(\rho + p)f^{-1}D_i f = D_i p, \tag{2.69}$$

the remaining condition, namely $\bar{D}^j(e^{-4U}\tau_j) = 0$, being identically satisfied.

2.3 Asymptotic Flatness: Lichnerowicz Theorems

Before stating the conditions for stationary spacetimes to be asymptotically flat, we elaborate somewhat more on the vacuum field equations. First recall from (2.17) that, when M (or equivalently: N) is simply connected, there exists a field ω on N such that

$$\sigma_{ij} = \frac{1}{2}(-V)^{1/2}\varepsilon_{ij}{}^k D_k\omega, \tag{2.70}$$

where we have used (2.16) and

$$\varepsilon_{ijk}dx^i dx^j dx^k = (-V)^{1/2}\xi^\mu \varepsilon_{\mu\nu\lambda\sigma}dx^\nu dx^\lambda dx^\sigma. \tag{2.71}$$

(Of course, the existence of ω could have also been inferred from (2.40) for $r_i = 0$.) We now rewrite the vacuum equations in terms of the conformally rescaled metric $\bar{h}_{ij}$ (see (2.45)), given by

$$\bar{h}_{ij} = (-V)h_{ij}. \tag{2.72}$$

Then (2.39,40,41), together with (2.70) lead to

$$\bar{D}^2 V = V^{-1}(\bar{D}V)^2 - V^{-1}(\bar{D}\omega)^2 \tag{2.73}$$

$$\bar{D}^2 \omega = 2V^{-1}(\bar{D}\omega)(\bar{D}V) \tag{2.74}$$

and

$$\bar{\mathbb{R}}_{ij} = \frac{1}{2}V^{-2}[(D_iV)(D_jV) + (D_i\omega)(D_j\omega)], \tag{2.75}$$

or

$$\bar{G}_{ij} = \frac{1}{2}V^{-2}\left\{(D_iV)(D_jV) + (D_i\omega)(D_j\omega) - \frac{1}{2}\bar{h}_{ij}[(\bar{D}V)^2 + (\bar{D}\omega)^2]\right\}. \tag{2.76}$$

We can now give an interesting geometric interpretation of the vacuum equations (2.73,74,76). Namely, let $\mathcal{P}$ be the Poincaré half-plane with metric q_{AB} given by

$$q_{AB}dz^A dz^B = V^{-2}(dV^2 + d\omega^2) \quad (V > 0, -\infty < \omega < \infty). \tag{2.77}$$

Viewing $(z^1(x), z^2(x)) = (V(x), \omega(x))$ as a map from $(N, \bar{h}_{ij})$ to $(\mathcal{P}, q_{AB})$, one easily checks that (2.73,74) are exactly the conditions in order for this map to be harmonic, in other words

$$\bar{D}^2 z^A + \Gamma^A_{BC} z^B{}_{,j} z^C{}_{,j} \bar{h}^{ij} = 0, \tag{2.78}$$

where Γ^A_{BC} denotes the Christoffel symbols of q_{AB}, composed with $z^C(x)$. The metric $\bar{h}_{ij}(x)$, of course, is not given, but has to satisfy (2.51). The r.h. side of (2.51), in turn, is nothing but the energy momentum tensor of the harmonic map. $(\mathcal{P}, q_{AB})$ can also be viewed as a spacelike hyperboloid in (2+1)-dimensional Minkowski space. Namely, define fields

$$\Phi_M = \frac{V^2 + \omega^2 - 1}{-4V}, \qquad \Phi_S = -\frac{\omega}{2V}, \qquad \Phi_K = -\frac{V^2 + \omega^2 + 1}{4V}. \tag{2.79}$$

Then

$$-\Phi_K^2 + \Phi_M^2 + \Phi_S^2 = -\frac{1}{4}. \tag{2.80}$$

Viewing (Φ_K, Φ_M, Φ_S) as coordinates on $\mathbf{R}^3$ with Lorentz metric $4(-d\Phi_K^2 + d\Phi_M^2 + d\Phi_S^2)$, the induced metric under the map (2.79) is nothing but q_{AB}.

The fields Φ_M, Φ_S are the potentials first introduced by Hansen [25] which we shall use in Sects. 3.1 and 3.2.

As with any harmonic map, we can associate a conserved current on N with any Killing vector on the target space $\mathcal{P}$. Since $\mathcal{P}$ has $SO(2,1)$ as isometry group, there are three independent such Killing vectors, namely

$$\overset{1}{\eta}{}^A \frac{\partial}{\partial z^A} = \frac{\partial}{\partial \omega} \tag{2.81}$$

$$\overset{2}{\eta}{}^A \frac{\partial}{\partial z^A} = V\frac{\partial}{\partial V} + \omega\frac{\partial}{\partial \omega} \tag{2.82}$$

$$\overset{3}{\eta}{}^A \frac{\partial}{\partial z^A} = \omega V\frac{\partial}{\partial V} + \frac{1}{2}(\omega^2 - V^2)\frac{\partial}{\partial \omega}. \tag{2.83}$$

We note in passing that the $SO(2,1)$ isometry of $\mathcal{P}$ is closely related to the "Ehlers transformation" discussed in the article by Maison in this volume. The conserved current j_i associated with any Killing vector η^A on $\mathcal{P}$ is given by

$$j_i = z^A{}_{,i}\eta^B q_{AB}. \tag{2.84}$$

Hence

$$\overset{1}{j}_i = V^{-2}D_i\omega \tag{2.85}$$

$$\overset{2}{j}_i = V^{-1}D_iV + V^{-2}\omega D_i\omega \tag{2.86}$$

$$\overset{3}{j}_i = V^{-1}\omega D_iV + (2V)^{-2}(\omega^2 - V^2)D_i\omega \tag{2.87}$$

are all divergence-free on $(N, \bar{h}_{ij})$. By (2.70) and (2.38), $\overset{1}{j}_i$ is also equal to

$$\overset{1}{j}_i = \bar{\varepsilon}_i{}^{jk}D_j\varphi_k, \tag{2.88}$$

and so the "charge" associated with $\overset{1}{j}_i$ is always zero. In the asymptotically flat we shall turn to later, (2.87) will be identically zero.

The quantity (2.86) has the following spacetime interpretation (compare [23]). Let Σ be a 2-surface in M which projects down to a smooth 2-surface on N. Then there exist local coordinates $(x^\mu) = (t, x^i)$ such that Σ is given by

$$x^\mu(y^A) = (0, x^i(y^A)), \qquad A = 1,2. \tag{2.89}$$

Now integrate the quantity $\varepsilon_{\mu\nu\rho\sigma}\nabla^\rho\xi^\sigma$ over Σ. After some computation one finds

$$\varepsilon_{\mu\nu\rho\sigma}(\nabla^\rho\xi^\sigma)\frac{\partial x^\mu}{\partial y^1}\frac{\partial x^\nu}{\partial y^2} = (-V)^{-1/2}(\partial_iV - 2\sigma_{ij}\varphi^j)\varepsilon^i{}_{k\ell}\frac{\partial x^k}{\partial y^1}\frac{\partial x^\ell}{\partial y^2} \tag{2.90}$$

where, as before, $\sigma_{ij} = V^2\partial_{[i}\varphi_{j]}$. Now, using (2.70),

$$-2(-V)^{-1/2}\sigma_{ij}\varphi^j + (-V)^{-3/2}\omega D_i\omega = D^j(\omega\varepsilon_{ij}{}^k\varphi_k). \tag{2.91}$$

Thus, when Σ is closed, the integral I of the expression (2.91) is given by $(\bar{h}_{ij} = (-V)h_{ij})$

$$I = \int_{\Sigma} (-V)^{-1}(D_i V + V^{-1}\omega D_i \omega) d\bar{S}^i. \tag{2.92}$$

The fact that this integral in vacuum only depends on the homology class of Σ arises, in the spacetime picture, from the fact that

$$\nabla^\mu \nabla_{[\mu}\xi_{\nu]} = 0, \text{ when } R_{\mu\nu} = 0. \tag{2.93}$$

The quantity

$$M = \frac{1}{8\pi} I \tag{2.94}$$

is called the Komar mass of $(M, g_{\mu\nu})$. For the Schwarzschild solution it coincides with the Schwarzschild mass when the "outward" orientation is chosen for $d\bar{S}^i$.

We now come to the

Boundary Conditions

Recall that we require $(M, g_{\mu\nu})$ to be connected, simply connected and chronological. Let, in addition, M contain a compact subset K and let $M \setminus K$ be an "asymptotically flat end". (The results of this subsection will remain to be true if $M \setminus K$ consists of finitely many asymptotic ends.) This means that $M \setminus K$ should be diffeomorphic to M_R $(R > 0)$ with

$$M_R = \{(x^0, x^i) \in \mathbf{R}^1 \times (\mathbf{R}^3 \setminus B(R))\} \tag{2.95}$$

with $B(R)$ a closed ball of radius R. In terms of this diffeomorphism, the metric $g_{\mu\nu}$ in $M \setminus K$ has to satisfy that there exists a constant $C > 0$ such that (see [3])

$$|g_{\mu\nu}| + |g^{\mu\nu}| + r^\alpha |g_{\mu\nu} - \eta_{\mu\nu}| + r^{1+\alpha}|\partial_\sigma g_{\mu\nu}| + r^{2+\alpha}|\partial_\sigma \partial_\rho g_{\mu\nu}| \leq C \tag{2.96}$$

$$g_{00} \leq -C^{-1}, \qquad g^{00} \leq -C^{-1} \tag{2.97}$$

$$\forall\, X^i \in \mathbf{R}^3 \quad g_{ij}X^i X^j \geq C^{-1} \sum (X^i)^2. \tag{2.98}$$

We assume $\alpha > 1/2$. Furthermore we require $R_{\mu\nu}$ to be zero in $M \setminus K$. (This latter condition could be considerably relaxed.) It now follows that the level set $x^0 = 0$ is a spacelike submanifold of $M \setminus K$ which has a finite ADM-momentum p^μ (see [3]). If p^μ is a timelike vector (which it will be for 'reasonable' matter except in the vacuum case), it now follows from the timelike character of ξ^μ that it has to be an asymptotic time translation, i.e.

$$|\xi^\mu - A^\mu| + r|\partial_\sigma \xi^\mu| + r^2 |\partial_\rho \partial_\sigma \xi^\mu| \leq C r^{-\alpha}, \tag{2.99}$$

where the constants A^μ satisfy

$$A^\mu A^\nu \eta_{\mu\nu} < 0 \tag{2.100}$$

(see [3]). Furthermore it follows from [3], that in $M \setminus K$, or a subset thereof diffeomorphic to $M_{R'}$ for sufficiently large $R' > R$, there are coordinates (t, y^i) in terms of which $g_{\mu\nu}$ is again asymptotically flat with the same $\alpha > 1/2$ and so that ξ^μ is of the form $\xi^\mu\, \partial/\partial x^\mu = \partial/\partial t$. Hence, in the coordinates (t, y^i), which we now call (t, x^i), the metric

$$g_{\mu\nu} dx^\mu dx^\nu = V(dt + \varphi_i dx^i)^2 + h_{ij} dx^i dx^j \tag{2.101}$$

satisfies

$$|V+1| + r|\partial_i V| + r^2|\partial_i \partial_j V| \leq Cr^{-\alpha} \tag{2.102}$$
$$|\varphi_i| + r|\partial_j \varphi_i| + r^2|\partial_k \partial_j \varphi_i| \leq Cr^{-\alpha} \tag{2.103}$$
$$|h_{ij} - \delta_{ij}| + r|\partial_k h_{ij}| + r^2|\partial_k \partial_\ell h_{ij}| \leq Cr^{-\alpha} \tag{2.104}$$

in $M \setminus C$. It follows that

$$r|\sigma_{ij}| + r^2|\partial_k \sigma_{ij}| \leq Cr^{-\alpha}. \tag{2.105}$$

We remark that the time coordinate t, which is at first only defined in the open subset $M \setminus K$ of M, can be (Kobayashi–Nomizu [36]) extended to a smooth global cross section of $\pi : M \to N$.

We now state and prove two uniqueness theorems due to Lichnerowicz [41], which are basic for the theory of stationary solutions.

Staticity theorem: Let $(M, g_{\mu\nu}, \xi^\lambda)$ be asymptotically flat with $\alpha > 1/2$, ξ^μ be an asymptotic time translation. If the matter is non-rotating relative to ξ^μ, i.e. r_i in (2.40) is zero, then the spacetime is static.

Proof: From (2.40) we have

$$D^i[(-V)^{-1/2}\sigma_{ij}] = 0. \tag{2.106}$$

Contract (2.106) with φ^j, using $\sigma_{ij} = V^2 D_{[i}\varphi_{j]}$. It follows that

$$D^i[(-V)^{-1/2}\sigma_{ij}\varphi^j] = (-V)^{-5/2}\sigma_{ij}\sigma^{ij}. \tag{2.107}$$

Now integrate (2.107) over N. Since the term in brackets on the left is $O(r^{-2-2\alpha})$, the boundary term at infinity gives zero. Consequently $\sigma_{ij} = 0 \Rightarrow \omega_{\mu\nu\lambda} = 0$.

Remark: Since $\sigma_{ij} = 0$, the field φ_i is of the form $\varphi_i = D_i F$, where $F = O(r^{1-\alpha})$. In the coordinates $\bar{t} = t - F$, $g_{\mu\nu}$ takes the form

$$g_{\mu\nu} dx^\mu dx^\nu = V dt^2 + h_{ij} dx^i dx^j. \tag{2.108}$$

Vacuum theorem: Let $(M, g_{\mu\nu}, \xi^\lambda)$ satisfy the conditions in the staticity theorem and let $(M, g_{\mu\nu})$ in addition be vacuum. Then $(M, g_{\mu\nu})$ is the Minkowski space.

Proof: Firstly, by the staticity theorem, we have that $\sigma_{ij} = 0$. Using this in (2.39) for $r = 0$ we have with $v := (-V)^{1/2}$ that

$$D^2 v = 0. \tag{2.109}$$

By the maximum principle, or multiplying (2.109) by μ, integrating by parts and using $\mu - 1 = O(r^{-\alpha})$, $\partial_i \mu = O(r^{-1-\alpha})$, we infer that $\mu \equiv 1$. Now (2.41) implies $\mathbb{R}_{ij} = 0 \Rightarrow \mathbb{R}_{ijk\ell} = 0$ since $\dim N = 3$. Since N is simply connected, it follows that (N, h_{ij}) is flat $\mathbf{R}^3$. Thus

$$g_{\mu\nu} dx^\mu dx^\nu = -dt^2 + \delta_{ij} dx^i dx^j. \tag{2.110}$$

2.4 Newtonian Limit

Ehlers showed (unpublished, see [49]) that one can write the field equation containing a parameter $\lambda = c^{-2}$ such that the equation remain meaningful for $\lambda = 0$ and then they are equivalent to the Newtonian equations. The variables for which this is true in the time dependent case have to be chosen in a quite sophisticated way. The stationary case can be treated in a direct and simple way as follows.

We write the metric as

$$g_{\mu\nu} dx^\mu dx^\nu = -e^{-\frac{2U}{c^2}} (cdt + \varphi_i dx^i)^2 + e^{\frac{2U}{c^2}} \bar{h}_{ik} dx^i dx^k \tag{2.111}$$

where we inserted "c" by dimensional analysis. The field equations decomposed in section 2.1

$$R_{\mu\nu} = \frac{8\pi G}{c^4}(T_{\mu\nu} - \frac{1}{2} T g_{\mu\nu}) \tag{2.112}$$

for the energy momentum tensor

$$T_{\mu\nu} = c^2 \tau (cdt + \varphi_i dx^i)^2 + 2c\tau_i (cdt + \varphi_j dx^j) dx^i + \tau_{ij} dx^i dx^j \tag{2.113}$$

become

$$\bar{D}^2 U = 4\pi G (e^{-\frac{4U}{c^2}} \tau + c^{-2} \bar{\tau}_l^l) - e^{\frac{4U}{c^2}} \bar{\omega}_{ij} \bar{\omega}^{ij} \tag{2.114}$$

$$\bar{D}^i \bar{\omega}_{ij} = 8\pi G c^{-3} e^{-\frac{4U}{c^2}} \tau_j \tag{2.115}$$

$$\bar{R}_{ij} = 2c^{-4} D_i U D_j U - 2e^{\frac{4U}{c^2}} \bar{\omega}_{ik} \bar{\omega}_j{}^k + \bar{h}_{ij} e^{\frac{4U}{c^2}} \bar{\omega}_{kl} \bar{\omega}^{kl} + 8\pi G c^{-4} (\tau_{ij} - \bar{h}_{ij} \bar{\tau}_l^l) \tag{2.116}$$

Considered as equations for $U, \bar{h}_{ij}, \phi_i, \tau, \tau_i, \tau_{ij}$, (2.114–116) have a limit for $c \to \infty$.

In the static case the limit is

$$\bar{D}^2 U = 4\pi G \tau \tag{2.117}$$

$$\bar{R}_{ik} = 0 \tag{2.118}$$

Hence we obtain immediately that the metric $\bar{h}_{ik}$ of the quotient is flat and therefore (2.117) is the Poisson equation of Newton's theory. The connection has also a limit and the only non vanishing Christoffel symbol is $\Gamma^i_{tt} = \bar{D}^i U$. The equation of motion $\nabla_\nu T^{\mu\nu} = 0$ becomes in the limit the Newtonian equilibrium condition $\bar{D}^j \tau_{ij} = -\tau D_j U$.

Now to the stationary case: Because the right hand side of (2.115) vanishes in the Newtonian limit, the Lichnerowicz theorem implies that $\omega_{ik} = 0$ which in turn implies by (2.116) that $\bar{R}_{ik} = 0$ whence the metric on the quotient is again flat.

For the metric written in the form (1) the connection has no limit for $c \to \infty$. If we use however as a consequence of the field equations that $\omega_{ij} = 0$, the connection has a limit and the equations of motion become the Newtonian equilibrium conditions, i.e.

$$\bar{D}^i \tau_i = 0 \quad , \quad \bar{D}^j \tau_{ij} = -\tau D_j U \; . \tag{2.119}$$

2.5 Existence Issues and the Newtonian Limit

The fact that the equations can be written to contain $\lambda = c^{-2}$ in such a way that they are analytic in λ and are the Newtonian equations for $\lambda = 0$, suggests to use this structure for existence theory. In this section we will make some remarks about the static case. In section 4.2 an existence theorem for a rigidly rotating body by Heilig will be discussed which exploits the fact that the equations have a nice Newtonian limit.

To obtain partial differential equations for which there is an existence theory we write (2.104) and (2.105) in the static case in harmonic coordinates on N, defined by $\bar{D}^2 x^i = 0$, for the unknowns U and Z^{ij} defined by $\bar{h}^{ij} = \delta^{ij} + \lambda^2 Z^{ij}$ and obtain:

$$\Delta U := \delta^{ij}\partial_i\partial_j U = 4\pi G\tau + A(\lambda, \tau, \tau_{ij}, Z^{ij}) \tag{2.120}$$

$$\begin{aligned} \Delta Z^{ij} \quad &= -4\partial_k U \partial_l U \delta^{ik}\delta^{lj} - 16\pi G(\tau^{ij} - \tau\delta^{ij}) \\ &\quad + \lambda^2 B^{ij}(\lambda, \tau, \tau_{kl}, Z^{kl}, \partial_m Z^{kl}, \partial_m\partial_n Z^{kl}) \end{aligned} \tag{2.121}$$

here we used the well known expression for the Ricci tensor in harmonic coordinates

$$\bar{R}^{ij} = -\frac{1}{2}\bar{h}^{kl}\partial_k\partial_l \bar{h}^{ij} + H^{ij}(\partial\bar{h}, \partial\bar{h}) \tag{2.122}$$

where H^{ij} is quadratic in the first derivatives of $\bar{h}^{ij}$. As usual we call (2.120), (2.121) the reduced field equations. These form a quasilinear elliptic system with the property that for given small sources τ, τ_{ij} of compact support and small λ there exist unique solutions U, Z^{ij} which tend to 0 at infinity.

In particular we can choose for τ, τ_{ij} a Newtonian solution and determine then for small λ a relativistic solution of the reduced field equations which have a Newtonian limit. Is is to be expected that the solution will be analytic in λ. Then the Taylor expansion in λ can be considered as a converging post Newtonian expansion.

A solution of the reduced field equations is only solution of the field equation if it satisfies the harmonicity condition or equivalently if $\nabla_\mu T^{\mu\nu} = 0$ holds. To solve the reduced equations and the equation of motion is a much harder problem. It makes only sense once a matter model is chosen.

In the static case only matter models of elasticity lead to new interesting problems because, as we will see in Sect. 5.2 , static fluids are spherically symmetric and can be investigated by ordinary differential equations (see Sect. 5.3). Some remarks on static, small self gravitating bodies can be found in Sect. 5.1.

For a stationary rigidly rotating fluid Heilig has given an existence theorem by perturbing away from a Newtonian solution. We will describe this result in Sect. 4.2.

3 Far Fields

3.1 Far-Field Expansions

While, as we have seen, little is known so far about globally regular, asymptotically flat solutions to the stationary field equations with reasonable matter sources, there is an almost complete understanding of the behaviour of general asymptotically flat solutions near spatial infinity, which we now describe.

Here the quotient manifold N is of the form

$$N = \mathbf{R}^3 \setminus B(R).$$

On N there are given the fields (h_{ij}, V, φ_i) satisfying (2.39,40,41). The whole discussion is "local-at-infinity". In particular one has to allow for R in $B(R)$ to be made suitably large, as one proceeds. We will do so tacitly without changing the letter "R". The Einstein equations are given by

$$\bar{D}^2 V = V^{-1}(\bar{D}V)^2 - V^{-1}(\bar{D}\omega)^2 \tag{3.1}$$

$$\bar{D}^2 \omega = 2V^{-1}(\bar{D}\omega)(\bar{D}V) \tag{3.2}$$

$$\bar{G}_{ij} = \frac{1}{2}V^{-2}\{(D_i V)(D_j V) + (D_i\omega)(D_j\omega) - \frac{1}{2}\bar{h}_{ij}[(\bar{D}V)^2 + (\bar{D}\omega)^2]\} \tag{3.3}$$

By the asymptotic conditions (2.102–104), ω tends to a constant at infinity. Subtracting this from ω, and calling the result again ω, we find that

$$|\omega| + r|\partial_i\omega| + r^2|\partial_i\partial_j\omega| \le Cr^{-\alpha}. \tag{3.4}$$

In short, we have that

$$V = -1 + O(r^{-\alpha}), \qquad \omega = O(r^{-\alpha}), \qquad \bar{h}_{ij} - \delta_{ij} = O(r^{-\alpha}), \qquad 1 > \alpha > 1/2 \tag{3.5}$$

and that these relations may be differentiated twice. The condition $\alpha > 1/2$ could be relaxed (see Kennefick and Ó Murchadha [34]). It now follows that

$$\Delta V = O(r^{-2-2\alpha}) \tag{3.6}$$

$$\Delta\omega = O(r^{-2-2\alpha}). \tag{3.7}$$

Since the r.h. sides of (3.6,7) decay stronger than $O(r^{-3})$, it follows from standard results in potential theory [33,73] that there exist constants M, S such that

$$V = -1 + \frac{2M}{r} + O(r^{-1-\alpha}) \tag{3.8}$$

$$\omega = \frac{2S}{r} + O(r^{-1-\alpha}). \tag{3.9}$$

But the existence of φ_i in (2.36) implies that S has to be zero.

Equation (3.3), which involves second derivatives of the metric $\bar{h}_{ij}$, yields

$$\Delta(k_{ij} - \frac{1}{2}\delta_{ij}k) - 2\partial_{(i}\Gamma_{j)} + \delta_{ij}\partial_\ell\Gamma_\ell = O(r^{-2-2\alpha}), \tag{3.10}$$

where $k_{ij} = \bar{h}_{ij} - \delta_{ij}$, $k = k_{ii}$ and

$$\Gamma_i = \partial_j k_{ij} - \frac{1}{2}\partial_i k. \tag{3.11}$$

This equation can be viewed in two ways, both of which recur in the higher-order steps leading to the theorem below. Firstly, in the gauge where $\Gamma_i = 0$, i.e. the linearized harmonic gauge for $\bar{h}_{ij}$, it is an elliptic equation, namely essentially the componentwise Laplace equation, for the leading-order part of k_{ij}. Secondly, (3.10) can be rewritten as

$$\varepsilon_{i\ell m}\varepsilon_{jnp}\partial_\ell\partial_n k_{mp} = O(r^{-2-2\alpha}), \tag{3.12}$$

which expresses the fact that the linearized Riemann tensor of $\bar{h}_{ij}$ decays faster than $O(r^{-3})$. Note that (3.12) makes essential use of the three-dimensionality of space. It follows [73] that there exists $g_i = O(r^{1-\alpha})$ such that

$$k_{ij} = \partial_i g_j + \partial_j g_i + O(r^{-2\alpha}). \tag{3.13}$$

Thus the leading-order contribution to the metric $\bar{h}_{ij}$ is "pure gauge".

To next order in $1/r$ one finds that there is a gauge, namely

$$\Gamma_i = O(r^{-3-\alpha}), \tag{3.14}$$

for which

$$V = -1 + \frac{2M}{r} - \frac{2M_i x^i}{r^3} + \frac{2M^2}{r^2} + O(r^{-1-2\alpha}) \tag{3.15}$$

$$\omega = \frac{2S_i x^i}{r^3} + O(r^{-1-2\alpha}) \tag{3.16}$$

and for which $\bar{h}_{ij}$ can be brought into the form

$$k_{ij} = -\frac{M^2(\delta_{ij}r^2 - x_i x_j)}{r^4} + O(r^{-1-2\alpha}). \qquad (3.17)$$

In the above M, M_i, S_i are constants. All indices are lowered and raised with δ_{ij}. When $M \neq 0$ one can, by a rigid translation, arrange for $M_i = 0$. In that case the metric $g_{\mu\nu}$ obtained from (3.15–17) coincides, to order $1/r^2$, with that of the Kerr spacetime with $|S| = -Ma$, M being the mass and a being the Kerr parameter.

In order to extend the above result to higher orders in $1/r$, it is convenient to replace (V, ω) by some other choice of scalar potentials. One choice, due to Hansen [25], is to set (see (2.79))

$$\phi_M = -\frac{V^2 + \omega^2 - 1}{4V} \qquad (3.18)$$

$$\phi_S = -\frac{\omega}{2V} \qquad (3.19)$$

$$\phi_K = -\frac{V^2 + \omega^2 + 1}{4V} \qquad (3.20)$$

It then turns out that $(\phi_\alpha) = (\phi_M, \phi_S, \phi_K)$ all satisfy

$$\bar{D}^2 \phi_\alpha = 2\bar{\mathbb{R}}\phi_\alpha. \qquad (3.21)$$

Then one has [73] the following

Theorem: There exists a gauge, namely that where $\Gamma_i = O(r^{-m-1-\alpha})$, for which there are constants $A\ldots, B\ldots, \ldots, G\ldots$ such that

$$\phi_M = \sum_{\ell=0}^{m-1} \frac{E_{i_1\ldots i_\ell} x^{i_1}\ldots x^{i_\ell}}{\ell! r^{2\ell+1}} + O^\infty(r^{-m+1-2\alpha}) \qquad (3.22)$$

$$\phi_S = \sum_{\ell=0}^{m-1} \frac{F_{i_1\ldots i_\ell} x^{i_1}\ldots x^{i_\ell}}{\ell! r^{2\ell+1}} + O^\infty(r^{-m+1-2\alpha}) \qquad (3.23)$$

$$\phi_K = \frac{1}{2} + \sum_{\ell=0}^{m-1} \frac{G_{i_1\ldots i_{\ell-1}} x^{i_1}\ldots x^{i_{\ell-1}}}{\ell! r^{2\ell}} + O^\infty(r^{-m+1-2\alpha}) \qquad (3.24)$$

Note that $E = M$,

$$\bar{h}_{ij} = \delta_{ij} + \sum_{\ell=2}^{m} \Bigg(\frac{x_i x_j A_{i_1\ldots i_{\ell-2}} x^{i_1}\ldots x^{i_{\ell-2}}}{r^{2\ell}} + \frac{\delta_{ij} B_{i_1\ldots i_{\ell-2}} x^{i_1}\ldots x^{i_{\ell-2}}}{r^{2\ell-2}}$$
$$+ \frac{x_{(i} C_{j) i_1\ldots i_{\ell-3}} x^{i_1}\ldots x^{i_{\ell-3}}}{r^{2\ell-2}} + \frac{D_{ij i_1\ldots i_{\ell-4}} x^{i_1}\ldots x^{i_{\ell-4}}}{r^{2\ell-4}}$$
$$+ O^\infty(r^{-m+1-2\alpha}) \Bigg). \qquad (3.25)$$

All constants are symmetric in their $i_1 \dots$ indices. The constants D are also symmetric in i and j. The constants $C \dots$ appear only for $m \geq 3$, the constants $D \dots$ only for $m \geq 4$. The symbol $O^\infty(r^k)$ means that the quantity in question is of $O(r^k)$, its derivative is $O(r^{k-1})$, a.s.o. Furthermore all constants are determined by the tracefree parts of $E \dots, F \dots$ in a way which does not depend on the solution at hand. The tracefree parts of $E \dots$ are the analogues of the Newtonian multipole moments. The constants $F \dots$ play an analogous role for the "angular-momentum aspect", which does not have a Newtonian counterpart. The three-metric $\bar{h}_{ij}$, for reasons explained after (3.14), has no independent degrees of freedom.

This theorem shows, in essence, that any stationary, asymptotically flat solution to the Einstein vacuum equations is uniquely determined by the "moments" $E \dots, F \dots$. However no statement concerning convergence of series like the ones appearing in (3.22–25) can be made. In order to do that it is necessary to use "conformal compactification" of three-space N.

3.2 Conformal Treatment of Infinity, Multipole Moments

Before turning to the situation G.R., it is instructive to recall the Newtonian situation. Suppose we are given a Newtonian potential near infinity, i.e. a function ϕ with

$$\Delta\phi = 0 \qquad \text{on } \mathbf{R}^3 \setminus B(R). \tag{3.26}$$

Extending ϕ smoothly to all of $\mathbf{R}^3$, we thus have that

$$\Delta\phi = 4\pi\rho \qquad \text{with } \rho \in C_0^\infty(\mathbf{R}^3) \tag{3.27}$$

and $\phi \to 0$ at infinity. Thus ϕ is of the form

$$\phi(x) = -\int_{\mathbf{R}^3} \frac{\rho(x')}{|x - x'|} dx'. \tag{3.28}$$

In $\mathbf{R}^3 \setminus B(R)$ this can (see e.g. [33]) be expanded in a standard fashion in powers of $1/r$. One obtains an expansion of the form

$$\phi = \sum_{\ell=0}^{\infty} \frac{E_{i_1 \dots i_\ell} x^{i_1} \dots x^{i_\ell}}{\ell! r^{2\ell+1}}, \tag{3.29}$$

with $E_{i_1 \dots i_\ell}$ totally symmetric and tracefree. One shows [33,73] that this series converges absolutely and uniformly in $\mathbf{R}^3 \setminus B(R)$ for sufficiently large R.

As a warm-up for G.R. it is useful to rephrase the Newtonian situation using "conformal compactification". First observe that there is a positive smooth function Ω on $N = \mathbf{R}^3 \setminus B(R)$ with the following properties. The metric

$$\widetilde{h}_{ij} = \Omega^2 \delta_{ij} \tag{3.30}$$

extends to a smooth metric on the one-point compactification

$$\widetilde{N} = N \cup \{r = \infty\} = N \cup \{\Lambda\}, \tag{3.31}$$

where

$$\Omega|_\Lambda = 0, \qquad \widetilde{D}_i \Omega|_\Lambda = 0 \tag{3.32}$$

and

$$\widetilde{D}_i \widetilde{D}_j \Omega - 2\widetilde{h}_{ij} = 0. \tag{3.33}$$

To prove this, take $\Omega = 1/r^2$ and introduce

$$\widetilde{x}^i = \frac{x^i}{r^2} \tag{3.34}$$

as coordinates on $\widetilde{N}$. One also sees that $\widetilde{h}_{ij}$ is again the standard flat metric in the coordinates $\widetilde{x}^i$. (This would also follow from (3.33) and the standard formula for the behaviour of R_{ij} under conformal rescalings.) As for the potential, rewrite (3.26) as

$$\left(D^2 - \frac{\mathbb{R}}{6}\right)\phi = 0, \tag{3.35}$$

and observe that the operator in (3.35) obeys

$$\left(\widetilde{D}^2 - \frac{\widetilde{\mathbb{R}}}{6}\right)\widetilde{\phi} = \Omega^{-5/2}\left(D^2 - \frac{\mathbb{R}}{6}\right)\phi, \tag{3.36}$$

when $\widetilde{h}_{ij} = \Omega^2 h_{ij}$ and $\widetilde{\phi} = \Omega^{-1/2}\phi$ for arbitrary $\Omega > 0$. Thus we again have

$$\left(\widetilde{D}^2 - \frac{\widetilde{\mathbb{R}}}{6}\right)\widetilde{\phi} = \widetilde{D}^2\widetilde{\phi} = 0, \tag{3.37}$$

at first only on N.

In the case of G.R. we were unable to prove convergence of the multipole series, but only an asymptotic estimate like

$$\phi = \sum_{\ell=0}^{m-1} \frac{E_{i_1 \dots i_\ell} x^{i_1} \dots x^{i_\ell}}{\ell! r^{2\ell+1}} + O^\infty(r^{-m+1-2\alpha}). \tag{3.38}$$

But, from (3.38) for $m = 4$, it follows immediately that $\widetilde{\phi}$ extends to a C^3-function on $\widetilde{N}$. Thus, by continuity

$$\widetilde{D}^2\widetilde{\phi} = 0 \qquad \text{on } \widetilde{N}. \tag{3.39}$$

But it is a standard fact that solutions to the Laplace equation and, more generally, for non-linear elliptic systems with analytic coefficients [55], are themselves analytic. Thus $\widetilde{\phi}$ has a convergent Taylor expansion at the point Λ. But this is nothing but (3.29) in inverted coordinates. Furthermore the multipole moments $E_{i_1 \dots i_\ell}$ can now be viewed as the Taylor coefficients of $\widetilde{\phi}$ at Λ. It follows from (3.39) that they have to be tracefree, and it is trivial that they determine $\widetilde{\phi}$ uniquely.

Suppose Ω is just required to satisfy (3.32,33). Then, given h_{ij}, there is in $(\widetilde{h}_{ij}, \Omega)$ the following 3-parameter gauge freedom

$$\Omega' = \omega \Omega, \tag{3.40}$$

$$\widetilde{h}'_{ij} = \omega^2 \widetilde{h}_{ij}, \tag{3.41}$$

where

$$\omega = (1 - b^i \widetilde{D}_i \Omega + b^i b_i \Omega)^{-1}, \tag{3.42}$$

with $\widetilde{D}_i b^j = 0$, which, in the compactified picture, corresponds to the freedom of choosing an origin in the "physical" space $\mathbf{R}^3$, w.r.t. to which the inversion $\widetilde{x}^i = x^i / r^2$ can be made. Therefore the Taylor coefficients of $\widetilde{U}$ at Λ behave under (3.40,41) in a way which precisely corresponds to their dependence on the choice of origin.

In G.R. it is impossible to require a conformal compactification for which (3.33) holds everywhere. We call a 3-metric $\bar{h}_{ij}$ on a manifold $N \cong \mathbf{R}^3 \setminus B(R)$ conformally C^k, when there exists a C^k-function $\Omega > 0$ on N such that $\widetilde{h}_{ij} = \Omega^2 \bar{h}_{ij}$ extends to a C^k-metric on $\widetilde{N} = N \cup \{\Lambda\}$ and

$$\Omega|_\Lambda = 0, \qquad \widetilde{D}_i \Omega|_\Lambda = 0, \tag{3.43}$$

$$\left.(\widetilde{D}_i \widetilde{D}_j \Omega - 2\widetilde{h}_{ij})\right|_\Lambda = 0. \tag{3.44}$$

A scalar potential ϕ is called conformally C^k, when $\widetilde{\phi} = \Omega^{-1/2} \phi$ extends to a C^k-function on $\widetilde{N}$. Given (3.43,44) there is now a much larger gauge freedom involved in constructing the unphysical from the physical quantities, namely

$$\Omega' = \omega \Omega, \qquad \widetilde{h}'_{ij} = \omega^2 \widetilde{h}_{ij}, \qquad \widetilde{\phi}' = \omega^{-1/2} \phi \tag{3.45}$$

where ω satisfies $\omega|_\Lambda = 1$. Now consider, following Geroch [22], this recursively defined set of tensor fields on $\widetilde{N}$:

$$P_0 = \widetilde{\phi} \tag{3.46}$$

$$P_i = D_i \widetilde{\phi} \tag{3.47}$$

$$P_{ij} = TS \left[\widetilde{D}_i D_j \widetilde{\phi} - \frac{1}{2} \widetilde{\mathbb{R}}_{ij} \widetilde{\phi} \right] \tag{3.48}$$

$$P_{i_1 \dots i_{m+1}} = TS \left[\widetilde{D}_{i_{m+1}} P_{i_1 \dots i_m} - \frac{s(2s-1)}{2} \widetilde{\mathbb{R}}_{i_1 i_2} P_{i_3 \dots i_{m+1}} \right], \tag{3.49}$$

where TS denotes the operation of taking the symmetric, trace-free part. It turns out that the tensors

$$E_{i_1 \dots i_m} = P_{i_1 \dots i_m}|_\Lambda \tag{3.50}$$

behave under (3.45) in exactly the same way as the Newtonian moments under the restricted gauge freedom (3.40–42) with $b_i = \widetilde{D}_i \omega|_\Lambda$. Thus the Ricci terms in (3.46–49) cancel out unwanted dependencies from higher-than-first derivatives of ω at Λ.

Now return to the expansions (3.22–25) for some fixed $m \geq 1$. Performing, again, an inversion $\widetilde{x}^i = x^i/r^2$ and setting, in these coordinates,

$$\widetilde{\phi}_M = \Omega^{-1/2} \phi_M, \qquad \widetilde{\phi}_S = \Omega^{-1/2} \phi_S, \tag{3.51}$$

$$\widetilde{h}_{ij} = \Omega^2 \bar{h}_{ij} \tag{3.52}$$

with $\Omega = 1/r^2$ we find that $(\widetilde{\phi}_M, \widetilde{\phi}_S, \widetilde{h}_{ij})$ are all C^m. Furthermore Ω is C^∞. Thus we have obtained a C^m conformal compactification. Our proof would be complete if we could find an elliptic system satisfied by $(\widetilde{h}_{ij}, \widetilde{\phi}_M, \widetilde{\phi}_S)$ or quantities derived from them. Doing this is not completely trivial. We explain the essentials in the static case where $\phi_S = 0$. Thus

$$\bar{D}^2 \phi_M = 2 \bar{\mathbb{R}} \phi_M \tag{3.53}$$

$$\bar{\mathbb{R}}_{ij} = \frac{2}{1 + 4\phi_M^2} (D_i \phi_M)(D_j \phi_M). \tag{3.54}$$

Let us assume that $M \neq 0$. Define, instead of $1/r^2$ as above, a conformal factor also called Ω by

$$\Omega = \frac{[(-V)^{1/2} - 1]^2}{(-V)^{1/2}}. \tag{3.55}$$

It is not hard to see from (3.22–25) that this yields a C^m-compactification $(\widetilde{\phi}_M, \widetilde{h}_{ij})$ where, however, we have for convenience replaced (3.44) by

$$\left(\widetilde{D}_i \widetilde{D}_j \Omega - \frac{2}{M^2} \widetilde{h}_{ij} \right) \Big|_\Lambda = 0. \tag{3.56}$$

It is useful to employ, as the scalar variable in the unphysical picture neither $\widetilde{\phi}_M$ nor Ω, but the quantity σ defined by

$$\sigma := \left[\frac{(-V)^{1/2} - 1}{(-V)^{1/2} + 1} \right]^2. \tag{3.57}$$

After some labor we find from (3.53,54) that

$$\widetilde{\mathbb{R}} = 0 \tag{3.58}$$

and

$$-\sigma(1-\sigma)\widetilde{\mathbb{R}}_{ij} = \widetilde{D}_i\widetilde{D}_j\sigma - \frac{1}{3}\widetilde{h}_{ij}\widetilde{D}^2\sigma. \tag{3.59}$$

The scalar σ satisfies

$$\sigma|_\Lambda = 0, \qquad \widetilde{D}_i\sigma|_\Lambda = 0, \qquad \widetilde{D}^2\sigma|_\Lambda = \frac{3}{2M^2}. \tag{3.60}$$

Taking a "curl" of (3.59) we obtain

$$(1-\sigma)\widetilde{D}_{[i}\widetilde{\mathbb{R}}_{j]k} = 2(\widetilde{D}_{[i}\sigma)\widetilde{R}_{j]k} - \widetilde{h}_{k[i}\widetilde{R}_{j]\ell}\widetilde{D}^\ell\sigma. \tag{3.61}$$

If we take $\widetilde{D}^i$ of the quantity $\widetilde{D}_{[i}\widetilde{\mathbb{R}}_{j]k}$ and use the Ricci and Bianchi identities we find the relation

$$\widetilde{D}^2\widetilde{\mathbb{R}}_{jk} = \frac{1}{2}\widetilde{D}_j\widetilde{D}_k\widetilde{\mathbb{R}} + 2\widetilde{D}^i\widetilde{D}_{[i}\widetilde{\mathbb{R}}_{j]k} + 3(\widetilde{\mathbb{R}}_{ji}\widetilde{\mathbb{R}}^i{}_k - \frac{1}{2}\widetilde{\mathbb{R}}\widetilde{\mathbb{R}}_{jk}) - \frac{1}{2}\widetilde{h}_{jk}(\widetilde{\mathbb{R}}_{i\ell}\widetilde{\mathbb{R}}^{i\ell} - \frac{1}{2}\widetilde{\mathbb{R}}^2). \tag{3.62}$$

Using that $\widetilde{\mathbb{R}}$ is zero and (3.62), writing $\widetilde{\mathbb{R}}_{ij} = \tau_{ij}$, and using (3.59) to eliminate second derivatives of σ, we obtain an equation of the form

$$\widetilde{D}^2\tau_{ij} = \text{non-linear terms}, \tag{3.63}$$

where these non-linear terms depend at most on τ_{ij}, σ and their first derivatives and on $\widetilde{D}^2\sigma$. We call $\widetilde{D}^2\sigma = \rho$. From the divergence of (3.59) we infer that

$$\rho\sigma = \frac{3}{2}(\widetilde{D}\sigma)^2, \tag{3.64}$$

and from this after some work that

$$\widetilde{D}^2\rho = 3\sigma(1-\sigma)^2\widetilde{\mathbb{R}}_{ij}\widetilde{\mathbb{R}}^{ij} + 3\widetilde{\mathbb{R}}_{ij}(\widetilde{D}^i\sigma)(\widetilde{D}^j\sigma). \tag{3.65}$$

Now (3.63) can be completed as follows:

$$\widetilde{\mathbb{R}}_{ij} = \tau_{ij} \tag{3.66}$$

$$\widetilde{D}^2\tau_{ij} = \text{non-linear terms} \tag{3.67}$$

$$\widetilde{D}^2\sigma = \rho \tag{3.68}$$

$$\widetilde{D}^2\rho = 3\sigma(1-\sigma)^2\widetilde{\mathbb{R}}_{ij}\widetilde{\mathbb{R}}^{ij} + 3\widetilde{\mathbb{R}}_{ij}(\widetilde{D}^i\sigma)(\widetilde{D}^j\sigma). \tag{3.69}$$

Going over to harmonic coordinates, the "non-elliptic" terms in the expression of $\widetilde{\mathbb{R}}_{ij}$ in (3.66) in terms of the metric go away, and the whole set of (3.66-69) becomes an elliptic system. Note that the point of the whole manœuvre was that the original eq. (3.59), when written in terms of $\widetilde{h}_{ij}$ is singular since $\sigma|_\Lambda = 0$. The miracle was that, in the transition from (3.59) to (3.61) a factor σ is obtained on both sides of (3.61) which can be cancelled since σ is nonzero outside Λ by (3.57).

Thus, taking m sufficiently large and appealing to the theorem of Morrey [55], we have the

Theorem: When $M \neq 0$, there is a chart in a neighbourhood of Λ for which $(\sigma, \widetilde{h}_{ij})$ are analytic. Consequently, from (3.57), Ω is also analytic, and so is $\widetilde{\phi}_M = (1 - \sigma)^{-3/2}$.

An analogous result can be proved for a suitable set $(\widetilde{h}_{ij}, \Omega, \widetilde{\phi}_M, \widetilde{\phi}_S)$ in the stationary case [6], see also [39]. The equations one obtains imply in particular that the "physical" quantities $(\bar{h}_{ij}, \phi_M, \phi_S)$ have an analytic chart in a neighbourhood of each point of N and thus entail the "classic" result of Müller zum Hagen on the analyticity of stationary vacuum solutions [57].

By smoothness of $(\widetilde{h}_{ij}, \widetilde{\phi}_M, \widetilde{\phi}_S)$ we can define multipole moments for each of $\widetilde{\phi}_M, \widetilde{\phi}_S$, following (3.46–50). One can show [73] that they coincide with the quantities $E\ldots$ and $F\ldots$ in the expansions (3.22–25). (These, in turn, coincide with the ones in Thorne [74], as shown in [24]). One can now prove [6], that these moments determine the stationary solution uniquely up to isometries. We give a more careful formulation of this result only in the static case.

Theorem: Let there be two static solutions with the same $\widetilde{h}_{ij}|_\Lambda$, the same $M \neq 0$ and the same set of (mass-centered) multipole moments. Then the corresponding physical solutions $(\bar{h}_{ij}, \phi_M)$ are isometric.

The proof is a not-too-difficult inductive argument based on (3.61,62), (3.68,69) and (3.59,60).

There remains the question to what degree the multipole moments of stationary solutions can be prescribed. It is fairly easy to see, e.g. from the asymptotic analysis of Sect. 3.1, that the multipole moments are "algebraically independent", i.e. for a given finite number of them, there always exists a spacetime having those moments which solves the stationary field equations to arbitrary order in $1/r$. It is not known what conditions on moments for high order have to be imposed in order for the multipole expansion to converge. In particular, convergence is not even known when only finitely many moments are non-zero.

There are of course solution-generating techniques to in principle write down the general stationary axisymmetric spacetime. To date the only result on existence of stationary asymptotically flat solutions without any further symmetry is that of Reula [66].

We note, in passing that the above equations lend themselves to an easy proof of a result which is often used in black-hole uniqueness theorems (see [31]). Namely an asymptotically flat, static vacuum solution with $M \neq 0$, which is spatially conformally flat, has to be isometric to the Schwarzschild metric near Λ. To see this, use that now the Cotton tensor of $\widetilde{h}_{ij}$ is zero. Thus, since $\widetilde{\mathbb{R}} = 0$, the left-hand side of (3.61) vanishes. Contracting the r.h. side of (3.61) with $(\widetilde{D}^i\sigma)\widetilde{\mathbb{R}}^{jk}$ we find that

$$2(\widetilde{\mathbb{R}}_{ij}\widetilde{\mathbb{R}}^{ij})(\widetilde{D}\sigma)^2 = (\widetilde{\mathbb{R}}_{ij}\widetilde{D}^j\sigma)(\widetilde{\mathbb{R}}^i{}_\ell\widetilde{D}^\ell\sigma). \tag{3.70}$$

But, by Cauchy–Schwarz, the right-hand side of (3.70) is bounded above by

$$(\widetilde{\mathbb{R}}_{ij}\widetilde{\mathbb{R}}^{ij})(\widetilde{D}\sigma)^2,$$

which has hence to be zero. Since σ can not have critical points near Λ except at Λ itself, it follows that

$$\widetilde{\mathbb{R}}_{ij} = 0, \tag{3.71}$$

whence, from (3.59), $\widetilde{D}^2\sigma = 3/2M^2$ and thus, in a chart $\widetilde{x}^i$ for which $\widetilde{h}_{ij} = \delta_{ij}$ we have $\sigma = |\widetilde{x}|^2/4M^2$, from which it easily follows that $(\bar{h}_{ij}, \phi_M)$ corresponds to Schwarzschild with mass M.

4 Global Rotating Solutions

4.1 Lindblom's Theorem

Lindblom showed in his thesis [45] that stationary asymptotically flat dissipative fluid configuration are axisymmetric. In this section we want to outline and discuss this theorem.

There are three ingrediences of the proof:

(i) The local fluid field equations imply that the fluid flow is proportional to a Killing vector t^μ provided the divergence of the entropy current vanishes.
(ii) The Killing field t^μ has an extension into the vacuum field of the solution.
(iii) If the manifold of orbits of the stationary Killing vector ξ^μ is R^3 and asymptotically flat, then ξ^μ is linearly independent of t^μ. The two Killing fields commute and there is a linear combination of the two Kiliing fields which has fixed points near which it act like a rotation.

(i) Theorem: Let $g_{\mu\nu}$, $T_{\mu\nu}$ be a stationary local solutions of the Einstein field equations for a one–component fluid with phenomenological heat conduction and viscosity laws and vanishing of the divergence of the entropy current. Then the fluid flow is proportional to a Killing vector.

Proof: The energy momentum tensor for a fluid with shear and bulk viscosity is [54] (θ and $\sigma_{\mu\nu}$ are the expansion and shear of the fluid; q^μ is the heat flow [20])

$$T^{\mu\nu} = \rho u^\mu u^\nu + (p - \zeta\theta)h^{\mu\nu} - 2\eta\sigma^{\mu\nu} + q^\mu u^\nu + q^\nu u^\mu \tag{4.1}$$

with

$$h^{\mu\nu} = g^{\mu\nu} + u^\mu u^\nu \ , \quad q_\mu u^\mu = 0, \quad \sigma_{\mu\nu}u^\mu = 0 \ . \tag{4.2}$$

This implies (a dot denotes the covariant derivative in the direction of the fluid flow u^μ)

$$0 = -(\nabla_\mu T^{\mu\nu})u_\nu = \dot{\rho} + (\rho + p)\theta - \zeta\theta^2 - 2\eta\sigma^{\mu\nu}\sigma_{\mu\nu} + \nabla_\mu q^\mu + q_\mu \dot{u}^\mu \ . \tag{4.3}$$

Introducing n, the conserved rest-mass density, and the specific volume $v = \frac{1}{n}$ and the specific internal energy $u = \frac{\rho}{n}$ we can rewrite this, using $\nabla_\mu(nu^\mu) = 0$, as

$$n(\dot{u} + p\dot{v}) - \zeta\theta^2 - 2\eta\sigma^{\mu\nu}\sigma_{\mu\nu} + \nabla_\mu q^\mu + q_\mu \dot{u}^\mu = 0 \; . \tag{4.4}$$

For a one-component fluid we have an equation of state $u = u(p, v)$ and consequently there exist scalar functions $T(p, v)$ and $s(p, v)$ with the interpretation of temperature and specific entropy such that

$$du + pdv = Tds \; . \tag{4.5}$$

Hence $n(\dot{u} + p\dot{v}) = nT\dot{s}$ can be used to rewrite (4.4) as

$$nT\dot{s} - \zeta\theta^2 - 2\eta\sigma^{\mu\nu}\sigma_{\mu\nu} + \nabla_\mu q^\mu + q_\mu \dot{u}^\mu = 0 \tag{4.6}$$

or

$$n\dot{s} + T^{-1}\nabla_\mu q^\mu = T^{-1}(\zeta\theta^2 + 2\eta\sigma^{\mu\nu}\sigma_{\mu\nu} - q_\mu \dot{u}^\mu) = 0 \; . \tag{4.7}$$

Using again $\nabla_\mu(nu^\mu) = 0$ we obtain

$$\nabla_\mu(nsu^\mu + T^{-1}q^\mu) = T^{-1}[\zeta\theta^2 + 2\eta\sigma^{\mu\nu}\sigma_{\mu\nu} - q_\mu(\dot{u}^\mu + T^{-1}\nabla^\mu T)] = 0 \; . \tag{4.8}$$

Inserting the phenomenological law of heat conduction

$$q_\mu = -\kappa h^\nu{}_\mu (T_{,\nu} + T\dot{u}_\nu) \tag{4.9}$$

we obtain

$$\nabla_\mu(nsu^\mu + T^{-1}q^\mu) = T^{-1}(\zeta\theta^2 + 2\eta\sigma^{\mu\nu}\sigma_{\mu\nu} + \kappa T^{-1} q_\mu q^\mu) = 0 \; . \tag{4.10}$$

The left-hand side of this equation is the conserved entropy current $\nabla_\mu s^\mu$ which vanishes according to our assumptions. Hence the positivity of λ, ζ and κ implies $\theta = \sigma^{\mu\nu} = q^\mu = 0$ and $\dot{u}_\mu = -T^{-1}T_{,\mu}$.

Assume $T \neq 0$ and consider $\xi^\mu = T^{-1}u^\mu$, the candidate for the Killing vector. We have

$$\nabla_{(\mu}\xi_{\nu)} = -T^{-2}\nabla_{(\mu}T u_{\nu)} + T^{-1}\nabla_{(\mu}u_{\nu)} \; . \tag{4.11}$$

The vanishing of $\theta = \sigma_{\mu\nu} = q^\mu = 0$ implies $\nabla_{(\mu}u_{\nu)} = -\dot{u}_{(\mu}u_{\nu)} = T^{-1}(\nabla_{(\mu}T)u_{\nu)}$, hence $\nabla_{(\mu}\xi_{\nu)} = 0$.

Now we come to the most complicated part, the extension of the Killing vector proportional to the fluid flow from the fluid into the surrounding vacuum region.

(ii) Conjecture: Let $g_{\mu\nu}, T_{\mu\nu}$ be a strictly stationary, asymptotically flat perfect fluid solution where the matter is a ball of finite extent and the fluid flow is proportional to a Killing vector t^μ. Then t^μ has a unique extension into the vacuum region, provided certain differentiability conditions are satisfied at the boundary.

In Lindblom's original treatment this conjecture was shown to be true under the assuption that the outside metric is analytic up to and including the boundary Σ of the fluid. Then one can propagate the Killing vector into a neighbourhood of the boundary using the Cauchy Kowalevskaja theorem because a Killing vector satisfies a wave type equation. Finally a theorem by Nomizu [61] can be used to obtain a global Killing vector field.

One might, however argue, that analyticity up to and including Σ is too strong an assumption. On physical grounds one would like to treat also non-analytic equations of state. In this case it is unlikely that the metric is analytic in the boundary.

Finally we show that the new Killing vector t^μ is actually different from the stationary Killing vector ξ^μ.

(iii) Theorem: Under the assumption of the above conjecture we have:

(1) The Killing vectors ξ^μ and t^μ are linearly independent.

(2) Both Killing vectors commute.

(3) There exists a linear combination $\eta^\mu = t^\mu + a\xi^\mu$ which has fixed points and acts like a rotation with closed orbits.

Proof: (1) Suppose t^μ would be linearly dependent of ξ^μ. Then there would be a timelike Killing vector, namely t^μ, which is asymptotically a translation and relative to which the matter does not rotate. Hence, by the Licherowicz staticity theorem, spacetime would have to be static.

(2) As T and u^μ are invariant objects we have $\mathcal{L}_\xi T = 0$ and $\mathcal{L}_\xi u^\nu = 0$ which imply immediately $[\xi, t] = 0$ on the support of the matter. To show that this is also true outside the matter one can use the analyticity of the outer metric up to and including the boundary or one can use a theorem by Beig and Chrusciel [4] classifying all possible group action on asymptotically flat spacetimes.

(3) As ξ^μ commutes with t^μ there is a Killing vector $\hat{t}^i$ on the manifold of orbits of ξ^μ. The corresponding group acts in the 2-surface of constant pressure, in particular in the boundary, $p = 0$. As this is topologically S^2, there must be a point where $\hat{t}^i$ vanishes. A Killing vector on a Riemannian space with a fixed point acts always as a rotation with closed circular orbits. At a point q in spacetime projecting on the fixed point of $\hat{t}^i$, t^μ must be proportional to ξ^μ and therefore a linear combination $\eta^\mu = t^\mu + a\xi^\mu$ with constant coefficients vanishing at q exist such that $\eta^\mu(q) = 0$. We have a fixed point and because also the timelike direction of ξ^μ is fixed, η^μ acts like a rotation and has therefore closed orbits.

We see that we can obtain the existence of the axis working only on the body, provided we know that both Killing vectors are independent. Lindblom [43] obtains the axis and commutativity of the Killing vectors from the asymptotic symmetry group. The key property that the two Killing vectors are linearly independent is only implied by a global argument and uses asymptotic flatness.

4.2 Existence of Stationary Rotating Axi-symmetric Fluid Bodies

Following work by Liapunoff and Poincaré, Lichtenstein [42] demonstrated at the beginning of this century the existence of rotating fluid bodies in Newtonian theory. An account of this almost forgotten work can be found in [71]. Using implicit function theorem techniques – as we would say today – he shows the existence of solutions near known solutions or approximately known solutions: starting with a static fluid ball, he obtains a slowly rotating fluid ball; starting from a self gravitating 2-body point particle solution, he obtains a solution for two small fluid bodies orbiting their center of mass on a circle. Furtheremore, there is a number of exact solutions in Newtonian theory: the Maclaurin spheroids, the Jacobi and the Dedekind ellipsoids and the Riemann ellipsoids [18].

In Einstein's theory we do not know any stationary exact solution describing an extended rotating body. Spacetimes describing such solutions can be characterized as follows: Besides a timelike Killing vector ξ^μ there is a further symmetry, the axial symmetry generated by η^μ, whose orbits are circles (Remember that we showed in Sect. 4.1 that such an extra symmetry exists on physical grounds) The body is spatially compact and the spacetime with topology R^4 is assumed to be asymptotically flat. We assume that there is an axis where η^μ vanishes. Then we can use a result of Carter [16] which states that under these circumstances the two Killing vectors commute. Such spacetimes are called "stationary axisymmetric", the orbits of the axial Killing vector are circles.

We showed in Sect 2.2 that for stationary axisymmetric perfect fluids with an axis and a fluid flow vector contained in the two-surface spanned by the two Killing vectors, the two-surface elements orthogonal to the two-dimensional group orbit are surface forming (the group action is orthogonally transitive). The same holds in the vacuum region. The property of orthogonal transitivity is equivalent to the existence of a discrete isotropy group [70].

To introduce a global coordinate system let us assume that outside the 2-dimensional axis the spacetime is the product of the orbits of the isometry group and the orthogonal 2-surface which we assume to have topology R^2.

Using coordinate adapted to the Killing vectors the metric can be written as

$$ds^2 = g_{AB}(x^c)dx^A dx^B + g_{00}(x^C)dt^2 + 2g_{0\phi}(x^C)dt d\phi + g_{\phi\phi}(x^C)d\phi^2. \quad (4.12)$$

Locally we can always introduce coordinates $(x^A) = (r, z)$ such that g_{AB} is conformal to the flat metric in standard coordinates and can therefore write the metric as

$$ds^2 = e^{2k-2U}(dr^2 + dz^2) + e^{-2U}W^2 d\phi^2 - e^{2U}(dt + Ad\phi)^2. \quad (4.13)$$

There is the freedom in (r, z) of an arbitrary conformal transformation which is given by the real part of analytic function.

The function W^2 is the volume element of the group orbits. As a consequence of the field equations in vacuum one can locally achieve $W = r'$ such that

$$ds^2 = e^{2k'-2U}(dr'^2 + dz'^2) + e^{-2U}r'^2 d\phi^2 - e^{2U}(dt + Ad\phi)^2. \qquad (4.14)$$

These coordinates are called Weyl's canonical coordinates. Matters can be arranged so that $r' = 0$ is the axis. Then the coordinates are fixed up to a translation in z'.

It is tempting to try to extend the Weyl coordinates from the outside of the body to the interior such that the two-surface orthogonal to the group orbit is covered by one (r', z') system with $r' = 0$ describing the axis and $W \neq r'$ in the interior. However, Müller zum Hagen has demonstrated [56] that this is impossible in the case of static spherically symmetric solutions. (r' becomes negative inside the body and the axis is reached for $\rho' \to \infty$.) There is no reason to assume that this would be different in the stationary case.

Numerical codes work successfully with a global (r, z) coordinate system such that $r = 0$ is the axis but it is not assumed that one has Weyl's canonical coordinate in vacuum.

For perfect fluids whose velocity is proportional to a constant linear combination of the two Killing vectors, the case of rigid rotation, $\nabla_\nu T^{\mu\nu} = 0$ becomes particularly simple. (See equation (2.69).)

$$0 = \nabla_\nu T_\mu{}^\nu = (\rho + p)\frac{1}{2}(\ln f^{-2})_{,\mu} + p_{,\mu}. \qquad (4.15)$$

where $f^\mu = f^2(\xi^\mu + \Omega\eta^\mu)$ is the four velocity of the fluid. This shows that, provided an equation of state $\rho(p)$ is given, the matter variables p and ρ can be expressed as functions of the quantity f which is determined by the geometry. This property of rigidly rotating fluids is essential for all the numerical schemes as well as for all the attempts to prove existence.

Various authors have developed codes to calculate numerically stationary, axisymmetric rigidly rotating fluid solutions [12]. Today this can be done with very high presicion by different numerical techniques. These numerical solutions are also the basis for investigations of oscillations of rotating stars.

Schaudt and Pfister [68] try to obtain an existence theorem working in the above coordinates adapted to the symmetry. This approach is attractive because the field equations become semilinear elliptic. One has, however, to control the singularities in the equations on the axis. This is possible and two Dirichlet problems have been solved, which give existence of outside, asymptotically flat solutions and existence of inner parts of bodies, provided appropriate boundary values are given [67]. Up to now this was only possible for the "outer" and the "inner" problem separately and work is in progress which tries to combine the inner and outer solution.

Let us now turn to the discussion of the only existence theorem for rotating fluids in Einstein's theory. It is remarkable that the first existence theorem for rotating fluids, proved by Heilig in 1995 [30], uses Lichtenstein's technique and does not adapt the coordinate to the axial Killing vector to avoid difficulties at the axis.

Let us formulate one particular case of the theorem proved by Heilig [30].

Theorem: Let $\rho(p) = Cp^\gamma$ be a polytropic equation of state with $1 < \gamma < 6/5$. The central density ρ_0 determines a unique Newtonian static fluid ball solution of finite extent. Then there exist a positive constant Ω_0 such that for all Ω with $0 < \Omega < \Omega_0$ a stationary axisymmetric rigidly rotating solution with angular velocity Ω of the Einstein field equations for a perfect fluid exists. The solution is geodesically complete, asymptotically flat with finite mass and angular momentum. The matter is of finite extent and has the same equation of state and central density as the Newtonian solution.

The theorem holds also for more general equations of state. It is not clear whether the case of positive boundary density may be treated by this method.

Heilig uses the observation of Jürgen Ehlers [21] that it is possible to write the field equations as an elliptic system with a parameter $\lambda = c^{-2}$ – interpreted as the velocity of light – such that the equations for , $\lambda \to 0$ give the Newtonian equations and the limit is regular. This can be achieved by a particular choice of unknowns for which the field equations are formulated.

We will describe the structure of Heiligs proof using the equations formulated in Sect. 2.4 because these are much simpler. We want, however, to stress that we expect that Heilig's result could be proved more easily using these equations, but this is not certain before all the functional analysis has been done properly.

Let us first adapt the field equations to a rigidly rotating fluid. We write the axial Killing vector as in (2.54)

$$\eta_\mu dx^\mu = \eta(cdt + \phi_i dx^i) + \eta_i dx^i \ . \tag{4.16}$$

The Killing equation in spacetime is equivalent to the equations (2.57–59) on the quotient N. For a rigidly rotating perfect fluid with fluid flow vector u^μ we have

$$u_\mu = f(\xi_\mu + \Omega\eta_\mu), \quad \Omega = \text{const} \ , \quad u_\mu u^\mu = -c^2, \tag{4.17}$$

where

$$f^{-2} = e^{-\frac{2U}{c^2}}(-e^{\frac{2U}{c^2}} + c^{-1}\Omega\eta)^2 - c^{-2}\Omega^2\eta_l\eta^l \ . \tag{4.18}$$

To obtain the field equation we replace in (2.64–66) η by $c^{-1}\eta$ and p by $c^{-2}p$ to obtain from (2.46–48) using $U \to c^{-2}U$

$$\begin{aligned}
\bar{D}^2U &= 4\pi G[f^2(-e^{\frac{2U}{c^2}} + c^{-1}\Omega\eta)^2(\rho + c^{-2}p) + 2c^{-2}pe^{-\frac{2U}{c^2}}] \\
&\quad + c^{-2}e^{\frac{4U}{c^2}} f^2\Omega^2(\rho + c^{-2}p)\eta_i\eta_j\bar{h}^{ij}]e^{\frac{4U}{c^2}} - e^{\frac{4U}{c^2}}\bar{\omega}_{ij}\bar{\omega}^{ij} \qquad (4.19) \\
\bar{D}^i\bar{\omega}_{ij} &= 8\pi Gc^{-3}e^{-\frac{4U}{c^2}} f^2(-e^{\frac{2U}{c^2}} + c^{-1}\Omega\eta)(\rho + c^{-2}p)\Omega\eta_j \qquad (4.20) \\
\bar{R}_{ij} &= 2c^{-4}D_iUD_jU - 2e^{\frac{4U}{c^2}}\bar{\omega}_{ik}\bar{\omega}_j{}^k + \bar{h}_{ij}e^{\frac{4U}{c^2}}\bar{\omega}_{kl}\bar{\omega}^{kl}
\end{aligned}$$

$$+ 8\pi G c^{-4}[-2pe^{-\frac{2U}{c^2}}\bar{h}_{ij} + f^2\Omega^2(\rho + c^{-2}p)\eta_i\eta_j - \bar{h}_{ij}f^2\Omega^2(\rho + c^{-2}p)\eta_l\eta_m\bar{h}^{lm}] \tag{4.21}$$

The above field equations have to be supplemented by the Killing equations (2.57–2.59). The equations of motion are

$$\nabla_\mu T^{\mu\nu} = 0 \Longleftrightarrow (c^2\rho + p)f^{-1}D_i f = D_i p \tag{4.22}$$

For $c \to \infty$ we have from (4.18) that $f^2 = 1$ which implies by (4.20) that $\bar{D}^i\omega_{ij} = 0$. The staticity theorem now gives that $\bar{\omega}_{ij} = 0$. Using (2.58) this implies $D_i(e^{-\frac{2U}{c^2}}\eta) = 0$. The vanishing of η on the axis implies $\eta = 0$. Using all this the field equations reduce to

$$\bar{R}_{ij} = 0\ , \quad \bar{D}^2 U = 4\pi G\rho \tag{4.23}$$

Therefore the metric on N is flat. Using $lim_{c\to\infty}(c^2 D_i f) = -D_i(U - \frac{1}{2}\Omega^2\eta_l\eta^l)$ the equation of motion become the Newtonian equation

$$-\rho D_i(U - \frac{1}{2}\Omega^2\eta_l\eta^l) = D_i p \tag{4.24}$$

equation ($\eta_l\eta^l = x^2 + y^2$ in Cartesian coordinates).

As discussed in Sect. 2.5 for the static case, the field equations become again a quasilinear elliptic system for U, Z^{ij}, φ_i in harmonic coordinates ($\nabla_\mu\nabla^\mu t = 0 \Longleftrightarrow \bar{D}^i\varphi_i = 0$, $\nabla_\mu\nabla^\mu x^i = 0 \Longleftrightarrow \bar{D}^2 x^i = 0$). Namely, the condition that the time function is harmonic turns the left-hand side of Equ.(4.20) into an elliptic operator acting on φ_i. Harmonicity of x^i has the same effect on the left-hand side of Equ.(4.21). Theorem 4.1 of Heilig [30] can be adapted to show that for small λ, Ω a solution of the reduced field equation exists near the Newtonian solution. Such a solution satisfies only the harmonicity condition if the equation of motion holds. So, this has to be solved simultaneously. This is possible because given a equation of state (4.22) can be integrated such that the matter quantities can be expressed in terms of the geometrical quantity f. Therefore the following iteration procedure is well defined: begin with a Newtonian solution U^0, p^0; choose some λ, Ω and use ρ^0, p^0, U^0 as a source in the field equations in harmonic coordinates to obtain $U^1, Z^{1ij}, {\varphi^1}_i$. Calculate f from $U^1, Z^{1ij}, {\varphi^1}_i, \lambda, \Omega$ and determine p^1 from the equation of motion. Then one solves again the field equation with the new source and so on. Heilig has shown that for sufficiently small λ and Ω such an iteration converges in his variables. It should also converge in the variables used here.

It is remarkable that we have used $\eta_i dx^i = xdy - ydx$ as a given field. At the end one has to check that the solution is axisymmetric and satisfies the harmonicity condition.

Note that only for $\lambda = c^{-2}$ with some fixed value of the velocity of light in some units the above field equations are Einstein's equations. It is however possible to reinterpret solutions with any λ as solutions of Einstein's equation expressed in different units [30]. With this interpretation the theorem above demonstates the existence of slowly – the theorem does not control the range of ω – rotating fluid configurations.

4.3 The Neugebauer–Meinel Disk

The only known global solution describing a rotating object in Einstein's theory, is the relativistic analogue of the rigidly rotating Maclaurin disk in Newton's theory [10].

An axisymmetric surface density distribution (in cylindrical coordinates (r, ϕ, z))

$$\sigma(r) = \sigma_0 \sqrt{1 - \frac{r^2}{r_0^2}} , \quad 0 < r < a , \tag{4.25}$$

generates a gravitational potential $\Phi(r, z)$, which is determined by the Poisson integral from σ. At the disk the potential is

$$\Phi(r, 0) = \frac{1}{2} \Omega^2 r^2 + \text{const} , \quad 0 < \mathrm{r} < \mathrm{a} , \quad \Omega^2 = \frac{\pi^2 \mathrm{G} \sigma_0^2}{\mathrm{r}_0^2} . \tag{4.26}$$

Outside the disk the potential can be expressed, for example, in terms of integrals over Bessel functions.

The centifugal force acting on rigidly rotating particles balances the gravitational force in the disk. Therefore, we can interpret the density distribution as formed by self gravitating, rigidly rotating dust. The two parameters σ_0 and r_0 determine such disks uniquely. The total mass of the disk is $M = \frac{2}{3}\pi\sigma_0 r_0^2$.

Neugebauer and Meinel found the relativistic analog of these disks [59].

There is a well known formalism available in General Relativity to describe matter surface distributions [31]. In the particular case of a reflection symmetric disk, we have to find solutions of the stationary vacuum field equations, defined outside the disk such that the difference of the normal derivatives of the metric at of the disk have a certain algebraic structure [31].

The general stationary axisymmetric metric can be parametrized as

$$ds^2 = e^{-2U} \left[e^{2k}(dr^2 + dz^2) + r^2 d\phi^2 \right] - e^{2U}(dt + a d\phi)^2 . \tag{4.27}$$

The metric coefficients U, k and a depend only on r, z; the vector fields $\xi^\mu \partial/\partial x^\mu = \partial/\partial t$ and $\eta^\mu \partial/\partial x^\mu = \partial/\partial\phi$ are Killing vector fields. We assume that the orbits of the axial Killing vector are circles; $r = 0$ is the axis.

Let us assume that the disk is located at $z = 0$, $0 \le r < r_0$. A rigidly rotating flow forming the disk is described by a vector field (which is defined at the disk)

$$u^\mu = e^{-V}(\xi^\mu + \Omega\eta^\mu) , \quad u^\mu u_\mu = -1 , \tag{4.28}$$

where Ω is constant. The definition of a dust disk implies that the metric is continuous across the disk and that $\tau^{\mu\nu} = \sigma u^\mu u^\nu$, where σ is the surface density, satisfies $\tau^{\mu\nu}{}_{;\nu} = 0$ with respect to the Levi Civita connection of the metric induced on the disk. As $v^\mu = \xi^\mu + \Omega\eta^\mu$ is a Killing vector, it holds $v^\mu{}_{;\mu} = 0$, $\sigma_{,\mu}v^\mu = 0$, $V_{,\mu}v^\mu = 0$ and we obtain

$$\tau^{\mu\nu}{}_{;\nu} = \left(\sigma e^{-2V} v^\mu v^\nu\right)_{;\nu} = \sigma e^{-2V} v^\mu{}_{;\nu} v^\nu . \tag{4.29}$$

Finally $e^{2V} = g_{\mu\nu}v^\mu v^\nu$ implies $e^{2V} 2V_{,\gamma} = 2g_{\mu\nu}v^\mu v^\nu{}_{;\gamma} = 2v^\nu v_{\gamma;\nu}$ and we see that V must be constant on a disk formed of rigidly rotating dust, $V = V_0$.

It is natural to introduce comoving coordinates

$$t' = t;\ \ \phi' = \phi - \Omega t\ ,\ \ \xi^{\mu'} = \xi^\mu + \Omega\eta^\mu\ ,\ \ \eta^{\mu'} = \eta^\mu\ ,\ u^{\mu'} = e^{-V}\delta^{\mu'}_{t'} . \tag{4.30}$$

The vacuum field equations can be expressed in terms of the following quantities:

$$e^{2U'} = -\xi_{\mu'}\xi^{\mu'} = e^{2V}\ ,\quad a' = -e^{-2U'}\eta_{\mu'}\xi^{\mu'}\ ,\ \ U'(r,\phi,z=0) = V_0 = const \tag{4.31}$$

and $b'(r,z)$ determined by

$$a'_{,r} = re^{-4U'}b'_{,z}\ ,\quad a'_{,z} = -re^{-4U'}b'_{,r}\ . \tag{4.32}$$

Using the Ernst potential $f' = e^{2U'} + ib'$ the key field equation is the semi-linear elliptic Ernst equation [37]

$$Re(f')(f'_{,rr} + f'_{,zz} + \frac{1}{r}f'_{,r}) = f'^2_{,r} + f'^2_{,z}\ . \tag{4.33}$$

For a solution of the Ernst equation the integrability condition of (4.32) is satisfied and one can solve for a'. The remaining metric coefficient k' follows from the equations

$$k'_{,r} = r\left[U'^2_{,r} - U'^2_{,z} + \frac{1}{4}e^{-4U'}(b'^2_{,r} - b'^2_{,z})\right]\ ,\ \ k'_{,z} = 2r\left[U'_{,r}U'_{,z} + \frac{1}{4}e^{-4U'}(b'_{,r}b'_{,z})\right] \tag{4.34}$$

whose integrability condition is again satisfied for solutions of the Ernst equations.

We can perform an integral in the $(r-z)$ -plane around the disk of the integrability condition of (4.32), namely

$$(r^{-1}e^{4U'}a'_{,r})_{,r} + (r^{-1}e^{4U'}a'_{,z})_{,z} = 0\ , \tag{4.35}$$

which can be replaced by a surface integral. As we assume that the tangential derivatives of the metric are continuous at the disk, we obtain at the disk

$$a'_{,z}|_{z=0^+} = a'_{,z}|_{z=0^-}\ . \tag{4.36}$$

On the other hand reflection symmetry at the disk implies

$$a'_{,z}|_{z=0^+} = -a'_{,z}|_{z=0^-} \tag{4.37}$$

on the disk, hence,

$$a'_{,z}|_{z=0^+} = a'_{,z}|_{z=0^-} = 0 \ , \tag{4.38}$$

which by (4.32) implies $b' = const$ on the disk.

Now it is easy to calculate the second fundamental form $k_{cd} = \frac{1}{2}e^U g_{cd,z}$ of the disk $z = 0$ ($c, d, \ldots = (t, r, \phi)$), because we have at the disk that $a'_{,z} = k'_{,z} = 0$, as a consequence of (4.38),(4.35) and (4.34). We find

$$k_{rr} = -2U'_{,z} g_{rr} \tag{4.39}$$

$$k_{\phi'\phi'} = -2U'_{,z}(a'^2 e^{2U'} + e^{-2U'} r^2) \tag{4.40}$$

$$k_{t't'} = 2U'_{,z} g_{t't'} \tag{4.41}$$

$$k_{t'\phi'} = 2U'_{,z} g_{t'\phi'} . \tag{4.42}$$

Now we can check the condition for a disk of dust [31], namely

$${}^+k_{cd} - {}^-k_{cd} = 2{}^+k_{cd} = -8\pi(\tau_{cd} - \frac{1}{2}g_{cd}\tau^e_e) = -8\pi(\sigma u_c u_d + \frac{1}{2}\sigma g_{cd}) \ , \tag{4.43}$$

which, in the primed coordinates (because of $u^{\mu'} = \delta^{\mu'}_{t'}$), reads

$$k_{c'd'} = -8\pi\sigma(g_{c't'}g_{d't'} + g_{c'd'}) \ . \tag{4.44}$$

Because of the form of the metric (4.27) in primed coordinates and by (4.39–42), this is satisfied if we define the surface density by

$$\sigma = \frac{1}{2\pi} U'_{,z} \ . \tag{4.45}$$

Thus we have shown that a rigidly rotating disk of dust is determined by a solution of the Ernst equation which satisfied at the disk $U' = const$ and $b' = const$. Outside the disk the solutions of the Ernst equation must be regular. For a well-posed elliptic problem we need furthermore asymptotic flatness at infinity and regularity conditions at the axis.

In Newton's theory there is a 2-parameter family of disks (4.25), (4.26). If we use the 2-parameter group of similarity transformations – or dimensionless quantities – we can assume $r_0 = 1$ and $\sigma_0 = 1$ and we have just one disk.

Because of the appearence of the velocity of light there is only a 1-parameter group of similarity transformations in Einstein's theory. Hence, after we put $r_0 = 1$, we expect a 1-parameter family of disk solutions.

The investigations of Neugebauer and Meinel suggest that

$$\mu = 2\omega^2 r_0^2 e^{-2V_0} \tag{4.46}$$

is an appropriate parameter.

Neugebauer and Meinel prove by the so called inverse scattering method of soliton physics (compare the contribution of Maison in this volume) that the boundary value problem for f' has a unique global solution provided V_0 and ω are such that $\mu < \mu_{crit} = 4.629\ldots$.

The solution f' can be expressed in terms of hyperelliptic theta functions [60]. The remaining metric coefficients a' and k' are determined by integration from the equations (4.32) and (4.34).

If we put the velocity of light, c, in the appropriate places we obtain the MacLaurin disk as a Newtonian limit.

Further properties of these disks are discussed in [60].

Many global stationary solutions with disk sources may be constructed from known stationary vacuum solutions by "cutting out" a region containing singularities an making appropriate identifications. This method was first used by Bicak and Ledvinka [9] to produce physically plausible sources for the Kerr metric with arbitrary values of the parameters a, M. These disks are made of two streams of particles circulating in opposite directions with differential velocities. They are extending to infinity but have finite mass. See Sect. 6 of the article by Bicak in this volume, where this procedure is related back to the "method of images" in Newtonian galactic dynamics. In the static case these methods yield an infinite number of such static disk solutions. Solutions corresponding to stationary counterrotating dust disks of finite extent have been constructed by Klein and Richter [35].

5 Global Non-rotating Solutions

5.1 Elastic Static Bodies

No doubt, Einstein's theory should allow for the description of static, solid bodies. It is useful to make the following distinction:

(i) small bodies, whose shape is not dominated by gravitational forces, like a piece of sugar or an iron ball. If we ignore gravity, the structure of the body is determined by the laws of quantum mechanics. This is true in a Galilei invariant formulation as well as in a special relativistic one. Linear and non-linear elasticity theory describes the deformation of such a configuration under external forces.

Suppose we now want to add the gravitational field. This is straightforward for linear elasticity in Newtonian theory; we just have to insert the gravitational field calculated from the Poisson integral as an external force into the equations of elasticity.

To pass from special relativity to Einstein's theory is more complicated. Now the deformed configuration has to satisfy Einstein's field equations, and the elasticity equations are a consequence of the latter!

(ii) bodies like stars whose shape is dominated by gravitational forces. There a relaxed state does not really exist and one has to modify the desciption of elasticity. This holds in Newtonian theory as well as in Einstein' theory.

Elastostatics can be described in Einstein's theory as follows [17]. The collection of particles which form the body is described by the three-dimensional "body manifold" B. The essential dynamical variable is a map $\Phi : M^4 \to B$. such that $\Phi^{-1}(y^i)$ is the world line of the particle in spacetime labeled by y^i in B. In the static case the world lines of the particles are the integral curves of the Killing vector, and we can consider Φ as a 1-1 map $N \to B$. We assume that we have given on B a Riemannian metric $\bar{\kappa}_{ij}$. Its physical interpretation may be different: for small bodies it describes a relaxed state; for big bodies which go never into a relaxed state, it could be an "isotropic state of minimal energy".

We need now information about the energy momentum tensor of the material. Let

$$T^{\mu\nu} = \rho u^\mu u^\nu + p^{\mu\nu} \ , \quad p^{\mu\nu} u_\nu = 0 \tag{5.1}$$

be such that the stress tensor $p^{\mu\nu}$ has only spatial components and can be considered as a tensor on N. We can now define

$$e_{ij} := \frac{1}{2}(h_{ij} - \Phi_* \bar{\kappa}_{ij}) \tag{5.2}$$

the "Lagrangian strain tensor". In the Hookian approximation of elastcity one assumes that one has given on the body B a tensor field $\bar{K}^{ijk\ell}$ such that after moving this object with Φ into the space N' one can define

$$\rho = \rho_0 + \frac{1}{2} K^{ijk\ell} e_{ij} e_{k\ell} \tag{5.3}$$

$$p^{ij} = -K^{ijk\ell} e_{k\ell} \tag{5.4}$$

as the energy and stresses of the body B in 3-space with the metric h_{ij}.

With this energy momentum tensor we consider Einstein's field equations as differential equations for the spacetime metric and the map Φ. No general existence theorem is available for this problem. The only case treated so far is the spherically symmetic one [62].

To get some feeling for these equation let us consider some further idealisation. For small deformations we can linearize $e_{ij} := \frac{1}{2}(h_{ij} - \Phi_* \bar{\kappa}_{ij})$ as follows: Suppose that Ψ_ϵ is a 1-parameter family of diffeomorphism $N' \to N'$ such that $\epsilon = 0$ is the identity and Φ_0 is some diffeomorphism $N' \to B$. Now we assume that $\Psi_\epsilon \Phi^{-1}{}_0$ defines our deformed body and calculate the stress tensor e_{ij} to first order in ϵ If we define $\kappa^0_{ij} = \Phi_{0*} \bar{\kappa}_{ij}$ we obtain

$$e_{ij} = \frac{1}{2}(h_{ij} - \kappa^0_{ij} + \mathcal{L}_\chi \kappa^0_{ij}) = \frac{1}{2}(h_{ij} - \kappa^0_{ij} + D^0_{(i} \chi^k \kappa^0_{j)k}) \tag{5.5}$$

Here the vector field χ^i is defined by the linearization of Ψ_ϵ on N' . We see that $p^{ij}{}_{;j} = 0$ leads to second order differential equations for χ^a.

Consider first the case of special relativity which coincides with Galilei invariant classical mechanics in the static situation. Then we have $h_{\alpha\beta} = \eta_{\alpha\beta}$

and $h_{ij} = \delta_{ij}$. We choose Φ_0 to be the identity may which discribes the relaxed body in spacetime. With $\bar{\kappa}_{ij} = \delta_{ij}$ we obtain

$$e_{ij} = \frac{1}{2}\chi_{(i,j)} \tag{5.6}$$

This gives implies the equations of classical, linearized elastostatics [51].

$$0 = p^{ij}{}_{,j} = -K^{ijk\ell}\chi_{k,\ell j} \tag{5.7}$$

With the appropriate symmetry and positivity conditions on $K^{ijk\ell}$ the equations are elliptic and solutions exist for various boundary conditions.

Next we want to calculate the deformation of a small elastic body by its own gravitational field. The relaxed state is determined by solid state physics as above. To switch on gravity we assume that we have families $g_{\mu\nu} = \eta_{\mu\nu} + Gg^1_{\mu\nu} + G^2 g^2_{\mu\nu} \dots$ and $T_{\mu\nu} = T^0_{\mu\nu} + GT^1_{\mu\nu} + G^2 T^2_{\mu\nu} \dots$ satisfying the field equations.

At order G^0 we obtain the trivial solution if there are no forces at the body , the density ρ^0_0 is constant and the stresses vanish,i.e. $\chi^0_a = 0$. The field equation in order G are obtained from the equations in section 2.4 with an energy momentum tensor $T^0_{\mu\nu}$ which has only a term ρ^0_0 because the stresses vanish. We obtain U^1 as a solution of the Poisson equation with the source source ρ^0_0. The metric $\bar{h}^1_{ij}$ remains flat in this order. The expansion of the equation of motion in G gives to first order

$$p^{1ij}{}_{,j} = -K^{ijk\ell}\chi^1_{k,\ell j} = -\rho^0_0 U^1{}_{,a}\ , \qquad \Delta U^1 = 4\pi G\rho^0_0 \tag{5.8}$$

Hence we obtain classical elastostatics with the force deforming the body being the gravitational force.

One might try to obtain an existence theorem for small self gravitating elastic bodies in Einstein's theory by an implicit function theorem argument similarly as in the case of a rigidly rotating body (section 4.2).

5.2 Are Perfect Fluids $O(3)$-Symmetric?

It is intuitively "obvious" that a static, in particular non-rotating, ball of perfect fluid, due to the absence of shear stresses should have spherical symmetry, and in particular the gravitational field in its exterior should be the one described by the Schwarzschild spacetime. This result, in its most general form, is still open in G.R. (The Newtonian case was settled in Lichtenstein [42] and Carleman [15], see also Lindblom [44].) Rather, one has today a theorem which is essentially a uniqueness result in the spirit of black hole uniqueness theorems. An earlier result due to Künzle and Savage [40] states that, near a spherical solution, there is no aspherical one with the same equation of state and the same mass.

The following result, due to Beig and Simon [8], is a refinement of previous work by Masood-ul-Alam [52], see also the review of Lindblom [47].

Theorem: Let us have a static, asymptotically flat, spherically symmetric solution to the Einstein equations with a perfect fluid and barotropic equation of state $\rho = \rho(p)$. (This solution is called reference spherical solution.) Let there be given another static, asymptotically flat solution with the same equation of state and the same value $V|_{\partial S}$ of the Killing vector norm on the surface ∂S of the star. Let further $\rho(p)$ satisfy the differential inequality $I \leq 0$, specified later. Then these two spacetimes are isometric, in particular the second one is also $O(3)$-symmetric.

The condition stipulating the existence of a spherical reference solution was disposed of by Lindblom and Masood-ul-Alam [48]. The condition on the matter, besides the one stating that $\rho \geq 0$, $p \geq 0$ and $d\rho/dp \geq 0$, is that

$$I := \frac{1}{5}\kappa^2 + 2\kappa + (\rho + p)\frac{d\kappa}{dp} \leq 0, \tag{5.9}$$

where $\kappa := \frac{\rho+p}{\rho+3p}\frac{d\rho}{dp}$. One can check that it is for example satisfied for the equation of state of a relativistic ideal Fermi gas at zero temperature, but only up to densities of order $10^{15}\mathrm{gcm}^{-3}$, which is roughly the critical density where gravitational instability sets in. It is known from numerical results [69] that beyond that limit the uniqueness statement of the above theorem will fail. One believes however, that sphericity will still hold.

We will here confine ourselves to an outline of the proof to the case of the special equation of state given by [13]

$$\rho(p) = \frac{1}{6}\,\rho^{6/5}(\rho_0^{1/5} - \rho^{1/5})^{-1} \qquad (\rho_0 = \mathrm{const} > 0) \tag{5.10}$$

which is a relativistic generalization of the equation for a polytrope of index 5 in the Newtonian theory. The expression in (5.10) satisfies $I \equiv 0$. The reference spherical solution in this case is known explicitly [7]. It has the curious property that it is asymptotically flat, but the fluid extends to spatial infinity.

Introducing the variable $v = (-V)^{1/2}$, the static field equations for a perfect fluid with energy momentum tensor

$$T_{\mu\nu} = (\rho + p)u_\mu u_\nu + p g_{\mu\nu} \tag{5.11}$$

with $u_\mu = v^{-1}\xi_\mu$ read

$$D^2 v = 4\pi v(\rho + 3p) \tag{5.12}$$

$$\mathbb{R}_{ij} = v^{-1}D_i D_j v + 4\pi(\rho - p)h_{ij} \tag{5.13}$$

The asymptotic conditions (2.102,104) imply that $v \to 1$ at infinity. ¿From the maximum principle for elliptic equations it follows that $0 < v < 1$ in N. Since the surface of the star is at infinity for the Buchdahl solution, the $v|_{\partial S}$, which is always equal to one in that case, has to be replaced by the total mass M (see Sect. 4.1).

Applying the contracted Bianchi identity to (5.12,13), there follows

$$D_i p = -v^{-1}(\rho + p) D_i v. \tag{5.14}$$

Thus p and ρ are both functions of v and

$$\frac{dp}{dv} = -v^{-1}(\rho + p). \tag{5.15}$$

Define the Cotton tensor of h_{ij} B_{ijk} by

$$B_{ijk} = 2D_{[k}\left(\mathbb{R}_{j]i} - \frac{1}{2}h_{j]i}\mathbb{R}\right). \tag{5.16}$$

With the definition

$$W = (D_i v)(D^i v) \tag{5.17}$$

the equations (5.12,13) now imply (see Lindblom [46]) that

$$\begin{aligned} D^2 W = & \frac{1}{4} v^4 W^{-1} B_{ijk} B^{ijk} + v^{-1}(D^i v)(D_i W) + 8\pi v (D^i v)(D_i \rho) \\ & + \frac{3}{4} W^{-1}(D^i W)(D_i W) - 8\pi W(\rho + p) + 16\pi^2 v^2 (\rho + 3p)^2 \\ & - 4\pi v(\rho + 3p) W^{-1}(D^i v)(D_i W). \end{aligned} \tag{5.18}$$

In the spherically symmetric case $W = W_0$ has to be of the form $W_0 = W_0(v)$. The ODE resulting in that case from (5.18), has, for the equation of state (5.10), an explicit solution given by

$$W_0 = (1 - v^2)^4 \left[\frac{1}{16M^2} - \frac{\pi\rho_0}{3}\left(\frac{1-v}{1+v}\right)^2\right]. \tag{5.19}$$

We assume that $\alpha = \frac{16\pi}{3}\rho_0 M^2 > 1$. The function W_0 is defined for $v \in [0, 1]$. It is positive for $v \in (v_c, 1)$, with $v_c = (\sqrt{\alpha} - 1)/(\sqrt{\alpha} + 1)$ and $W_0(v_c) = 0$, $W_0(1) = 0$. Thus W_0 satisfies the correct boundary condition at the central value v_c of v and at infinity.

We now define, for the given solution (v, h_{ij}), the scalar function

$$\widetilde{W} - \widetilde{W}_0 = \left(\frac{1 - v^2}{2}\right)^{-4} (W - W_0) \tag{5.20}$$

and the conformally rescaled metric

$$\widetilde{h}_{ij} = v^{-2}\left(\frac{1 - v^2}{2}\right)^4 h_{ij}. \tag{5.21}$$

(The constant M occurring in W_0 is taken to be the mass of the given solution.) In the asymptotically flat, vacuum case discussed in Sect. 3.2 one finds that the metric $\widetilde{h}_{ij}$ extends smoothly to the manifold $\widetilde{N} = N \cup \{\Lambda\}$, with Λ the point at infinity. This is also true for the Buchdahl solution, and we assume it to be true for the given, a priori non-spherical one. After some calculations we find that

$$\widetilde{D}^2(\widetilde{W} - \widetilde{W}_0) = \frac{1}{4}\widetilde{W}^4 \widetilde{B}_{ijk}\widetilde{B}^{ijk} + \frac{3}{4}\widetilde{W}^{-1}\widetilde{D}^i(\widetilde{W} - \widetilde{W}_0)\widetilde{D}_i(\widetilde{W} - \widetilde{W}_0). \quad (5.22)$$

Since $\widetilde{W}, \widetilde{W}_0$ also extend smoothly to $\widetilde{N}$, the function $\widetilde{W} - \widetilde{W}_0$ satisfies the elliptic equation with nonnegative right-hand side on the compact manifold $\widetilde{N}$. After integrating (5.22) over $\widetilde{N}$ (or by the maximum principle) it follows that $\widetilde{B}_{ijk}$ is zero and

$$\widetilde{W} = \widetilde{W}_0(v). \quad (5.23)$$

It then follows from [8], that the given model is isometric to the Buchdahl solution with the same ρ_0 and the same M.

5.3 Spherically Symmetric, Static Perfect Fluid Solutions

The metric for a static spherically symmetric spacetime can be written

$$ds^2 = -c^2 e^{\nu(r)} dt^2 + e^{\lambda(r)} dr^2 + r^2(d\theta^2 + \sin^2\theta \, d\phi^2) \,. \quad (5.24)$$

For a derivation see [28,70]. Here c is a constant which plays the role of the speed of light. In Appendix B of [28] it is also demonstrated that the r^2 in front of the sphere metric is no loss of generality for a static perfect fluid with positive mass density and pressure. Hence it is impossible to have two centers or two infinities. The field equations for a perfect fluid are

$$8\pi G c^{-2} \rho r^2 = e^{-\lambda}(r\lambda' - 1) + 1 \quad (5.25)$$

$$8\pi G c^{-4} p r^2 = e^{-\lambda}(r\nu' + 1) - 1 \quad (5.26)$$

$$8\pi G c^{-4} p = \frac{1}{2} e^{-\lambda}\left(\nu'' + \frac{1}{2}\nu'^2 + r^{-1}(\nu' - \lambda') - \frac{1}{2}\nu'\lambda'\right) \quad (5.27)$$

A prime denotes a derivative with respect to r. We have written $-c^2\rho$ for the timelike eigenvector of the energy–momentum tensor to make the comparison with the Newtonian equations easier. The field equation imply 'energy–momentum conservation', which is a single equation for a static perfect fluid

$$2p' = -\nu'(p + c^2\rho) \,. \quad (5.28)$$

The first exact solution of these equation was alredy found in 1918 by Karl Schwarzschild, the solution with constant density [37]. We have three independent ordinary differential equations for for four functions. Hence, one function can be specified freely. The most physical case is to prescribe an equation of state $\rho = \rho(p)$. Equation (5.25) can easily be integrated:

$$e^{-\lambda} = 1 - \frac{8\pi G}{c^2}\frac{1}{r}\int r^2 \rho(r) dr + \text{const} . \tag{5.29}$$

As we only are interested in solutions with a regular center of spherical symmetry we define λ as follows

$$e^{-\lambda} = 1 - \frac{8\pi G}{c^2}\frac{1}{r}\int_0^r r^2 \rho(r) dr . \tag{5.30}$$

The usual definition of the 'mass up to r', namely

$$m(r) = 4\pi \int_0^r r^2 \rho(s) ds \tag{5.31}$$

gives

$$e^{-\lambda} = 1 - \frac{2G}{c^2}\frac{m(r)}{r} . \tag{5.32}$$

It is also useful to introduce the following quantity which is related to the 'mean density up to r'

$$w(r) = r^{-3} m(r) . \tag{5.33}$$

Then (5.32) becomes

$$e^{-\lambda} = 1 - \frac{2G}{c^2} r^2 w . \tag{5.34}$$

Various forms of the equations (5.25-28) will be used. Equations (5.25), (5.26) and (5.28) contain all the information. If we eliminate ν' then (5.26) and (5.28) imply the Tolman–Oppenheimer–Volkoff equation [75]

$$p' = -Gr\left(1 - \frac{2G}{c^2} r^2 w\right)^{-1} \left(\frac{4\pi p}{c^2} + w\right)\left(\frac{p}{c^2} + \rho\right) \tag{5.35}$$

If an equation of state is given we can integrate (5.28)

$$\nu(r) = -\int_{p_0}^{p(r)} \frac{2dp}{p + c^2 \rho(p)} + \text{constant} \tag{5.36}$$

In this formula p_0 denotes the central pressure. If we add the definition of w then (5.35) and (5.33) form an integro-differential system. Differentiating (5.33) we obtain

$$w' = \frac{1}{r}(4\pi\rho - 3w) \tag{5.37}$$

In [65] the following theorem is proved.

Theorem: Let an equation of state $\rho(p)$ be given such that ρ is defined for $p \geq 0$, non-negative and continuous for $p \geq 0$, C^∞ for $p > 0$ and suppose that $d\rho/dp > 0$ for $p > 0$.

Then there exists for any value of the central density ρ_0 a unique inextensible, static, spherically symmetric solution of Einstein's field equation with a perfect fluid source and equation of state $\rho(p)$. The matter either has finite extent, in which case a unique Schwarzschild solution is joined on as an exterior field, or the matter occupies the whole space, with ρ tending to 0 as r tends to infinity.

There are two parts of the proof. The equations (5.35) and (5.37) form a system of ordinary differential equations for $p(r), w(r)$. However, the system is singular at $r =$ and the first step is to demonstrate that for each value of the central density there is a unique solution such that the spacetime is regular at the center. This is shown in [65] or in [50]. This solution defines a neighborhood of a regular center and can be extended as long as $(1 - \frac{2G}{c^2} r^2 w)$ remains positive. This can be seen as follows.

Introduce the variables first used by Buchdahl [13]

$$y^2 = 1 - \frac{2G}{c^2} r^2 w \ , \quad \zeta = e^{\nu/2} \ , \quad x = r^2 \tag{5.38}$$

Rewriting the equations in these variables and eliminating p in 5.26) and (5.28) gives an equation which is linear in ζ and w,

$$(1 - \frac{2G}{c^2} x w) \zeta_{,xx} - \frac{G}{c^2} \zeta_{,x} (w + x w_{,x})_{,x} - \frac{G}{2c^2} w_{,x} \zeta = 0 \tag{5.39}$$

or

$$(y \zeta_{,x})_{,x} - \frac{G}{2c^2} \frac{w_{,x} \zeta}{y} = 0 \tag{5.40}$$

Let $0 \leq x < x_0$ be an intervall such that $y^2 = (1 - \frac{2G}{c^2} x w > 0$ and $p > 0$. As the density does not increase outwards we have $w_{,x} \leq 0$. Therefore

$$(y \zeta_{,x})_{,x} \leq 0 \tag{5.41}$$

The equation (5.26) can be rewritten as

$$y \zeta_{,x} = \frac{\zeta}{y} \frac{G}{2c^2} (w + \frac{4\pi}{c^2} p) \ . \tag{5.42}$$

¿From (5.41) and (5.42) we obtain the inequality

$$y \geq \frac{w + 4\pi p / c^2}{w_0 + 4\pi p_0 / c^2} \ . \tag{5.43}$$

Hence, we see that y cannot vanish before p.

Suppose $p(x_b) = 0$. Then we call the corresponding r_b the radius of the star. The Schwarzschild solution is given in the form $e^{-\lambda} = e^\nu = 1 - A/r$ for some constant A. Hence, we determine a unique exterior field by the condition $A = (\frac{2G}{c^2})m(r_b)$. In this way the matter solution and the outside solution are joined only in a C^0-fashion because the boundary density may be non-zero. If we introduce Gauss coordinates relative to the hypersurface $p = 0$ the metric is C^1. It is obvious that this metric cannot be extended because the area of the group orbits $r =$ constant grows from 0 to infinity.

Let us now consider the second possibility that $p(x) > 0$ for all x. Because $p(x)$ is monotonically decreasing for $x \to \infty$, $\lim_{x\to\infty} p(x) = p_\infty$ exists. This implies that p' tends to 0 for $x \to \infty$. Since $y \leq 1$, (5.35) then implies $p_\infty = 0$ and hence, using the equation of state, that $\rho \to 0$ as $x \to \infty$. As before the spacetime is not extensible.

This completes our outline of the proof. It shows in particular that for for $\rho(p)$ with $\rho(0) = \rho_b > 0$ the radius of the star has to be finite.

There are various exact global solutions known. (For a useful list of such solutions including a discussion of their physical acceptability has been given by Delgaty and Lake [19].) For the 1-parameter family of equations of state given by Equ. (5.10) the whole 1-parameter family of solutions is known. A 2-parameter family of equations of state of interest for the issue of section 3.1 is investigated by Simon [72]; all the corresponding exact solutions are given.

There are some conditions on the equation of state known, which allow to decide whether the radius of the star is finite or infinite in the case of vanishing boundary density ρ_b. In [65] it is shown that the radius of the star is finite if $\int_o^{p_0} dp/\rho(p)^2$ is finite. Conversely, $\int_o^{p_0} dp/(\rho(p)c^{-2}p < \infty$ implies that the matter distribution is infinitely extended. Both conditions depend only on the behaviour of the equation of state near the boundary $p = 0$. Makino [50] gives conditions for a finite radius in cases which are not covered by the above. He shows in particular, that for polytropic equations of state, $p = const.\rho^\gamma$ with $4/3 < \gamma < 2$ the radius is finite.

For finite distributions "Buchdahl's inequality" holds [13].

Theorem: For finite distributions with non-negative density and a monotonic equations of state there holds

$$1 - \frac{2G}{c^2}\frac{M}{r_b} > \frac{1}{9} \, . \tag{5.44}$$

Proof: To obtain the inequality one compares the solution with a solution of constant density ρ, an interior Schwarzschild solution. Equ. (5.40) implies for this solution (written with an overbar) that

$$(\bar{y}\bar{\zeta}_{,x})_{,x} = 0 \implies \bar{y}\bar{\zeta}_{,x} = a = \text{constant} \tag{5.45}$$

We normalize ζ by the condition that at the boundary we have $\zeta_b = y_b$. Then we find a if we rewrite (5.26) in the new varables

$$\frac{8\pi G}{c^2}p = 4y^2\frac{\zeta_{,x}}{\zeta} - \frac{2G}{c^2}w \tag{5.46}$$

and evaluate it at the boundary as $a = \frac{2G}{c^2}\bar{w}$.

Then (5.45) can be integrated with the result

$$\bar{\zeta}(x) = \frac{1}{2}\left(1 + 2\bar{\zeta}(0) + \sqrt{1 - \frac{2G}{c^2}x\bar{w}}\right) \tag{5.47}$$

Now (5.41) implies

$$y\zeta_{,x} > (y\zeta_{,x})_b = \bar{y}\bar{\zeta}_{,x} \tag{5.48}$$

As $\bar{y} > y$ we obtain

$$\zeta_{,x} \geq \bar{\zeta}_{,x} = \frac{1}{2}\left(1 + 2\bar{\zeta}(0) + \sqrt{1 - \frac{2G}{c^2}x\bar{w}}\right) \tag{5.49}$$

As $\bar{\zeta}$ is positive we obtain at the boundary

$$y_b \geq -\frac{1}{2}y_b + \frac{1}{2} \tag{5.50}$$

which is (5.44).

Buchdahl's inequality show that one can pack only a certain mass into a given fixed radius. The physical reason is that the pressure is also a source of the gravitational field. In Newton's theory there are constant density balls with a fixed radius for arbitrary density. In Einstein's theory the central pressure diverges if the density approaches some maximum value.

In [1] an analogue of Buchdahls inequality is derived for distributions in which the the density is only assumed to be positive. There holds $1 - \frac{2G}{c^2}\frac{M}{r_b} > 0$.

Another important topic are bounds on the total mass of the system. Suppose we know the equation of state only for $\rho < \rho_0$. Then we can estimate the mass and radius of a core in which the density is greater ρ_0 as follows: Clearly, $m(r_0) > \frac{4\pi}{3}\rho_0(r_0)^3$; because of $y > 0$ we have also $m(r_0) < \frac{c^2}{2G}r_0$. Hence the possible cores occupy a compact part of the $m(r_0)$-r_0- plane. Taking intial values from this part one can numerically integrate outwards using the known equation of state, until the pressure vanishes. This was done in [27] for $\rho_0 = 5.1 \times 10^{14} g/cm^3$ and with a certain realistic equation of state for smaller densities. All configurations had a total mass smaller then $5M_\odot$. It is quite remarkable the the knowledge of the equation of state for a finite density range allows to show such a bound on the total mass, assuming nothing but the monotonicity of the equation of state in the unknown density range. This is not possible in Newton's theory.

In the special case of bodies with a sharp edge, i.e $\rho_b > 0$, we can combine the Buchdahl inequality (5.44) with the estimate $M = m(r_b) \geq 4\pi\rho_b r_b^3$ to obtain the mass bound

$$M \leq \left(\frac{2}{3}\right)^3 \left(\frac{3c^6}{4\pi G^3 \rho_b}\right)^{1/2} . \tag{5.51}$$

Let us finally compare with Newton's theory. In (5.35) it is almost obvious that for $c \to \infty$ one obtains the Newtonian equation for the pressure. The relativistic corrections show how "the pressure enters in the active and passive gravitational mass". The first factor describes an effect of the geometry. Static fluid ball are the simplest examples of families of relativistic solutions with a Newtonian limit [21].

5.4 Spherically Symmetric, Static Einstein–Vlasov Solutions

In recent years existence and further properties of solutions of Einstein's field equations for a collisionless gas have been shown [64]. The Vlasov–Einstein system determines the spacetime metric and the distribution function $f(x^\mu, p^\mu)$ describing the particles.

$$\begin{aligned} p^\mu \partial_{x^\mu} f - \Gamma^\mu_{\nu\sigma} p^\nu p^\sigma \partial_{p^\mu} f &= 0 \\ T^{\mu\nu} &= \int p^\mu p^\nu |g|^{1/2} \frac{d^4 p}{m} \\ G^{\mu\nu} &= 8\pi T^{\mu\nu} . \end{aligned} \tag{5.52}$$

In the static spherically symmetric case and for the metric (5.24), these equations reduce to (r= $|x^i|$, v^i are the spatial frame components of p^α)

$$\frac{v^i}{\sqrt{1+v^2}} \partial_{x^i} f - \sqrt{1+v^2} \nu' \frac{x^i}{r} \partial_{v^i} f = 0 \tag{5.53}$$

$$8\pi G c^{-2} \rho r^2 = e^{-\lambda}(r\lambda' - 1) + 1 \tag{5.54}$$

$$8\pi G c^{-4} p r^2 = e^{-\lambda}(r\nu' + 1) - 1 \tag{5.55}$$

where

$$\rho(x) = \rho(r) = \int_{R^3} f(x^i, v^i) \sqrt{1+v^2}\, dv , \tag{5.56}$$

$$p(x) = p(r) = \int_{R^3} f(x^i, v^i) \left(\frac{x^i v_i}{r} \right) \frac{dv}{\sqrt{1+v^2}} . \tag{5.57}$$

The distribution function is assumed to be spherically symmetric.

Rein and Rendall [64] show the existence of asymptotically flat solutions, regular at the center, with finite total mass and finite extension of the matter and isotropic pressure. It is also possible to construct solutions with anisotropic pressure; Furthermore shells of finite extent of matter around a regular center or a black hole can be constructed [63].

References

1. Baumgarte, T.W., Rendall, A.D. (1993): Class. Quantum Grav. **10**, 327
2. Beig, R. (1981): Acta Phys. Austr. **53**, 249
3. Beig, R., Chrusciel, P.T. (1996): J. Math. Phys. **37**, 1939
4. Beig, R., Chrusciel, P.T. (1997): Commun. Math. Phys. **188**, 585
5. Beig, R., Simon, W. (1980): Commun. Math. Phys. **78**, 75
6. Beig, R., Simon, W. (1981): Proc. R. Soc. Lond. **A376**, 333
7. Beig, R., Simon, W. (1991): Lett. Math. Phys. **21**, 245
8. Beig, R., Simon, W. (1992): Commun. Math. Phys. **144**, 373
9. Bicak, J. Ledvinka, T. (1993): Phys. Rev. Lett. **71**, 1669
10. Binney, J., Tremaine, S. (1987): Galactic Dynamics (Princeton Univ. Press, Princeton)
11. Bizon, P. (1994): Acta Phys. Polon. **B25**, 877
12. Bonazzola, S., Gourgoulhon, E., Salgado, M., Marck, J. A. (1993): Astron.Astrophys. **278**, 421
13. Buchdahl, H.A. (1959): Phys. Rev. **116**, 1027
14. Buchdahl, H.A. (1964), Astrophys. J. **140**, 1512
15. Carleman, T. (1919): Math.Z. **3**, 1
16. Carter, B. (1970): Commun. Math. Phys. **17**, 233
17. Carter, B., Quintana, H. (1972): Proc. R. Soc. Lond., **A331**, 57
18. Chandrasekhar, S. (1969): Ellipsoidal Figures of Equilibrium (Yale Univ. Press, New Haven)
19. Delgaty, M.S.R., Lake, K. (1998): gr-qc/9809013
20. Ehlers, J. (1961): Abh. Math.-Naturw.Kl.Akad.Wiss. Mainz **11**, 763; English version: Gen.Rel.Grav. **25**, 1225 (1993)
21. Ehlers, J. (1997): Class.Quantum.Grav. **14** A119
22. Geroch, R. (1970): J. Math. Phys. **11**, 2580
23. Geroch, R. (1972): J. Math. Phys. **13**, 394
24. Gürsel, Y. (1983): Gen. Rel. Grav. **20**, 1540
25. Hansen, R.O. (1974): J. Math. Phys. **15**, 1
26. Harris, S. (1992): Class. Quantum Grav. **9**, 1823
27. Hartle, J. B. (1978): Phys. Rep. **46**, 202
28. Hawking, S. W., Ellis, G. F. R. (1973): The large scale structure of space-time (Cambridge University Press, Cambridge)
29. Heilig, U. (1993): Ph.D. thesis, Universität Tübingen
30. Heilig, U. (1995): Commun. Math. Phys. **166**, 457
31. Israel, W. (1966): Nuovo Cim. **44**, 1
32. Israel, W. (1967): Phys. Rev. **164**, 1776
33. Kellogg, O.D. (1926): Foundations of Potential Theory (Springer, Berlin)
34. Kennefick, D., Murchadha, N.Ó (1995): Class. Quantum Grav. **12**, 149
35. Klein, C., Richter, O. (1999): Phys. Rev. Lett. **83**, 2884
36. Kobayashi, S., Nomizu, K. (1969): Foundations of Differential Geometry, Volume I (Interscience, New York)
37. Kramer, D., Stephani, H., MacCallum, M., Herlt, E. (1980): Exact Solutions of Einstein's Field Equations (VEB Deutscher Verlag der Wissenschaften, Berlin)
38. Kundt, W., Trümper, M. (1966): Zs. f. Physik **192**, 419
39. Kundu, P. (1981): J. Math. Phys. **22**, 2006
40. Künzle, H.P., Savage, J.R. (1980): Gen.Rel.Grav. **12**, 155

41. Lichnerowicz, A. (1955): Théories relativistes de la gravitation et de l'électromagnetisme (Masson, Paris)
42. Lichtenstein, L. (1933): Gleichgewichtsfiguren rotierender Flüssigkeiten (Springer, Berlin)
43. Lindblom, L. (1976): Ap. J. **208**, 873
44. Lindblom, L. (1977): J. Math. Phys. **18**, 2352
45. Lindblom, L. (1978), Ph.D. dissertation, Univ. of Maryland
46. Lindblom, L. (1980): J. Math. Phys. **21**, 1455
47. Lindblom, L. (1992): Phil. Trans. R. Soc. Lond. **A340**, 353
48. Lindblom, L., Masood-ul-Alam, A.K.M. (1994): Commun. Math. Phys. **162**, 123
49. Lottermoser, M. (1992): Ann.Inst.Henri Poincaré **57**, 279
50. Makino, T. (1998): J. Math. Kyoto Univ. **38**, 55
51. Marsden, J.E., Hughes, T.J.R. (1983): Mathematical Foundations of Elasticity (Dover, New York)
52. Masood-ul-Alam, A.K.M. (1988): Class. Quantum Grav. **5**, 409
53. Meyers, N. (1963): J. Math. Mech. **12**, 247
54. Misner, C. W., Thorne, K. S., Wheeler, J. A. (1973): Gravitation (W. H. Freeman, New York)
55. Morrey, C.B. (1958): Am. J. Math. **80**, 198
56. Müller zum Hagen, H. (1969): Proc. Camb. Phil. Soc. **66** 155
57. Müller zum Hagen, H. (1970): Proc. Camb. Phil. Soc. **68**, 199
58. Müller zum Hagen, H. (1974): Proc. Camb. Phil. Soc. **75**, 249
59. Neugebauer, G., Meinel, R. (1993): Astrophys. J. **414**, L97
60. Neugebauer, G. Kleinwächter, A., Meinel, R. (1996): Helv. Phys. Acta **69**, 472
61. Nomizu, K. (1960), Ann. Math. **72**, 105
62. Park, J. (1998): preprint AEI-084, gr-qc 9810010
63. Rein, G. (1994): Proc. Camb. Phil. Soc. **115**, 559
64. Rein, G., Rendall, A.D. (1993): Ann. Inst. H. Poincaré (Physique Théorique) **59**, 383
65. Rendall, A.D., Schmidt, B.G. (1991): Class. Quantum Grav. **8**, 985
66. Reula, O. (1989): Commun. Math. Phys. **122**, 615
67. Schaudt, U. M. (1998): Commun. Math. Phys. **190**, 509
68. Schaudt, U.M., Pfister, N. (1996): Phys. Rev. Lett. **77**, 3284
69. Schmid, W. (1972): unpublished manuscript (Institut für Theoretische Physik der Universität Wien)
70. Schmidt, B. G. (1967) Zs. f. Naturf. **22a**, 1351
71. Schmidt, B.G. (1999) in: On Einstein's Path, Essays in Honor of Engelbert Schücking, A. Harvey, Ed. (Springer, New York)
72. Simon, W. (1994): Class. Quantum Grav. **24**, 97
73. Simon, W., Beig, R. (1983): J. Math. Phys. **24**, 1163
74. Thorne, K. (1980): Rev. Mod. Phys. **52**, 299
75. Wald, R. (1984): General Relativity (Univ. of Chicago Press, Chicago)

Gravitational Lensing from a Geometric Viewpoint

Volker Perlick

TU Berlin, Institute of Theoretical Physics, Sekr. PN 7-1
10623 Berlin, Germany. email: vper0433@@w421zrz.physik.tu-berlin.de

Abstract. The theory of gravitational lensing is discussed in a Lorentzian manifold setting. To that end we fix a point p (observer at a particular instant) and a timelike curve γ (worldline of a light source) in a 4-dimensional Lorentzian manifold (spacetime) and we investigate how many past-pointing lightlike geodesics (light rays) go from p to γ. If there is more than one such geodesic, then we are in a gravitational lensing situation. Among other things, we study the geometry of light cones and we use the theory of conjugate points and cut points to find necessary and sufficient criteria for gravitational lensing; we discuss a Morse theory, based on a general relativistic version of Fermat's principle, to characterize the number of images for gravitational lensing situations in globally hyperbolic spacetimes; and we discuss gravitational lensing in asymptotically simple and empty spacetimes, giving an elementary proof for an odd number theorem in this situation.

1 Introduction

According to general relativity the path of a light ray is influenced by the gravitational field of massive objects. The verification of this effect during a total Sun eclipse in the year 1919 made Einstein's theory famous all over the world. It was soon realized by Eddington [12] and Chwolson [10] that, in principle, this deflection of light by massive objects might lead to the effect that an observer sees two or more distinct images of one and the same light source. Also, Chwolson [10] mentioned the possibility that, in cases of axial symmetry, the light source might appear as a ring around the deflecting mass. Those effects are usually summarized under the name *gravitational lensing.* For many decades it was not clear if gravitational lensing is, indeed, realized in Nature. It was not before 1979 that a promising candidate for gravitational lensing was found. In this year Walsh, Carlswell and Weyman [81] published their results on the double quasar 0957 +561 and suggested that in this case we see two images of one and the same quasar, produced by the gravitational field of an intervening galaxy. Since then, a great number of further gravitational lens candidates have been found, including multiple quasars, radio rings and luminous arcs. This has led to the effect that gravitational lensing is one of the most rapidly developing field in astronomy, in particular from an observational but also from a theoretical point of view. There is a comprehensive monograph on the subject by Schneider, Ehlers and Falco [70] and there is a great number of review articles, including a regularly updated electronic review by Wambsganss [82] from which literature, in particular on the present status of observations, can be traced.

In this article we want to approach the theory of gravitational lensing from the viewpoint of Lorentzian geometry. This is somewhat unusual insofar as the majority of theoretical work on gravitational lensing is done in a quasi-Newtonian approximation formalism which was developed, in essence, by Sjur Refsdal in the 1960s and which is discussed in full detail, e. g., in Schneider, Ehlers and Falco [70]. In the standard version of this approximation formalism one restricts to a purely spatial description, as opposed to a spacetime description, and light rays are represented by straight lines in Euclidean 3-space, with the only exception that they may have a sharp bend when passing through a particular plane which is known as the "deflector plane". (There is also a variant with several deflector planes.) This formalism has proven very powerful for calculating particular models. On the other hand, one should keep in mind that it is only an approximation. Thus, it is perfectly fine if it is used for quantitative calculations where the approximative assumptions are satisfied, but it is not the complete story as far as qualitative aspects of the theory are concerned. Gravitational lensing, by its very nature, is a general relativistic effect and it can be understood only on the basis of a 4-dimensional spacetime description, i. e., in terms of Lorentzian geometry.

Therefore, the following strategy seems to be appropriate for studying the theory of gravitational lensing. In the beginning one should concentrate on getting an understanding of gravitational lensing in terms of spacetime diagrams and becoming familiar with the 4-dimensional geometry involved. The present article tries to serve this purpose. After that, one should study the passage to the quasi-Newtonian formalism which involves several approximative assumptions. Some of these assumptions are easily understood, such as that the gravitational field should be weak and that the deflection angles should be small. However, in addition one needs some assumptions whose interpretation is less obvious. So the passage to the quasi-Newtonian approximation is, in fact, a rather subtle issue. These problems are carefully discussed by Seitz, Schneider and Ehlers [73], related material can also be found in Schneider, Ehlers and Falco [70] and in Sasaki [68]. In the third step one is then ready to study how the quasi-Newtonian formalism is operating. This is what is done in the majority of the theoretical literature on gravitational lensing and what is reviewed in full detail, e. g., in Schneider, Ehlers and Falco [70]. For mathematical aspects of gravitational lensing in the quasi-Newtonian approximation formalism, in particular for the theory of caustics, we also refer to a forthcoming book by Petters, Levine and Wambsganss [65].

For our plan to study gravitational lensing in a Lorentzian geometry setting we assume that light propagation can be described in terms of rays and we restrict to light rays in vacuo, i. e., we exclude the case that the light rays are influenced by a medium, e. g., in terms of diffraction, on their way from the light source to the observer. Moreover, we shall restrict to the case that both the observer and the light source may be considered as pointlike. We are then naturally led to studying past-pointing lightlike geodesics (light rays) from a point (observer at a particular instant) to a timelike curve (worldline of the light source) in a Lorentzian manifold (spacetime). If there are two or more such geodesics, then we are in a gravitational lensing situation. In essence, our analysis will be kinematical throughout, although Einstein's field equation will be mentioned occasionally.

The article is organized as follows. In Sect. 2 we recall some basic notions from Lorentzian geometry and fix some conventions as to terminology and notation. These conventions will be essential for understanding the following, so the reader is kindly requested to read this section carefully. In Sect. 3 we study gravitational lensing situations in arbitrary spacetimes. In Sect. 4 we specialize to the case of globally hyperbolic spacetimes and in Sect. 5 we further specialize to asymptotically simple and empty spacetimes.

Many mathematical results will be simple corollaries of standard theorems from Lorentzian geometry which can be found in the books by Hawking and Ellis [29], Wald [80], O'Neill [50], or Beem, Ehrlich and Easley [4]. For theorems proven in one of those books the proof is not repeated here unless in cases where this seemed instructive. Also, the material on Morse theory, i. e., Sects. 3.5 and 4.2, turned out to be so technical that for proves the reader must be refered to the quoted original papers. In all other cases proves are given in (hopefully) sufficient detail.

2 Some Basic Notions of Spacetime Geometry

According to general relativity, a spacetime is a 4-dimensional Lorentzian manifold. For convenience we shall consider only Lorentzian manifolds that are *time-orientable*, i. e., we assume that it is possible to distinguish between future and past in a globally consistent way. More precisely, we use the following definition.

Definition 1. A *spacetime* is a triple $(\mathcal{M}, g, \mathcal{T}^+)$ where
(a) $\mathcal{M}$ is a connected 4-dimensional real C^∞ manifold whose topology satisfies the axiom of second countability and the Hausdorff axiom and is, thus, paracompact;
(b) g is a Lorentzian metric on $\mathcal{M}$, i. e., a symmetric covariant second rank C^∞ tensor field which has signature $(+,+,+,-)$ at each point;
(c) $\mathcal{T}^+$ is a *time orientation* for $(\mathcal{M}, g)$, i. e., the set of timelike tangent vectors $\{ X \in T\mathcal{M} \,|\, g(X,X) < 0 \}$ consists of exactly two connected components and $\mathcal{T}^+$ is one of those components.
For a spacetime $(\mathcal{M}, g, \mathcal{T}^+)$, we denote the *Levi-Civita connection* of g by ∇ and we denote the *Riemannian curvature tensor* of ∇ by R.

A linear subspace N of the tangent space $T_p\mathcal{M}$ is called (i) *spacelike* if g is positive definite on N, (ii) *lightlike* if g is positive semi-definite but not positive definite on N, and (iii) *timelike* otherwise. The property of being spacelike, lightlike or timelike is assigned to a vector in $T_p\mathcal{M}$ if the linear space generated by this vector has the respective property, to a differentiable curve if its tangent vector has the respective property everywhere and to a submanifold of $\mathcal{M}$ if its tangent space has the respective property everywhere. Moreover, a differentiable curve is called *causal* if its tangent vector is either timelike or lightlike at each point.

Here and in the following, by a *curve* in $\mathcal{M}$ we mean a map from a real interval I into $\mathcal{M}$. (A "real interval" is a connected subset of $\mathbb{R}$ that contains more than one point. I may be open, half-open or closed.) Quite generally, we shall use a lower case greek letter for a curve, e. g., $\gamma : I \longrightarrow \mathcal{M}$, and we shall use the corresponding boldface letter for the image set of this curve, e. g., $\boldsymbol{\gamma} = \{\gamma(s) \,|\, s \in I\}$.

By a *geodesic* we always mean what is more fully called an "affinely parametrized geodesic", i. e., a C^∞ curve $\lambda : I \longrightarrow \mathcal{M}$ such that $\nabla_{\lambda'}\lambda' = 0$. This leaves the freedom of changing the parameter affinely, $I \longrightarrow \tilde{I}$, $s \longmapsto as + b$ with $a, b \in \mathbb{R}$, $a \neq 0$. However, when counting geodesics we shall tacitly identify two geodesics if one is a reparametrization of the other. That is to say, in a sentence such as "There are two geodesics λ_1 and λ_2 ..." it goes without saying that λ_2 is not just a reparametrization of λ_1. Please note that, according to this rule, a periodic geodesic gives rise to infinitely many geodesics between any two points on this geodesic.

Our study will be concentrating upon lightlike geodesics, which are to be interpreted as light rays. It is well-known and easily verified that under a conformal transformation $g \longmapsto e^{2f} g$ of the metric g, with an arbitrary C^∞ function $f : \mathcal{M} \longrightarrow \mathbb{R}$, the lightlike geodesics undergo a reparametrization but are unchanged otherwise. Since we are interested only in the paths of lightlike geodesics and not in their particular parametrizations, we could therefore allow for arbitrary conformal transformations of the metric, i. e., we could prescribe a conformal equivalence class rather than a particular metric. However, we shall not do so because we want to occasionally discuss additional assumptions on the spacetime which are not conformally invariant, such as, e. g., conditions on the Ricci tensor.

The totality of all geodesics issuing from a point p give rise to the *exponential map*

$$\exp_p : \mathcal{W}_p \longrightarrow \mathcal{M} \,. \tag{1}$$

This map is defined on a subset $\mathcal{W}_p$ of the tangent space $T_p\mathcal{M}$ by setting $\exp_p(X) = \lambda(1)$ where $\lambda : [0,1] \longrightarrow \mathcal{M}$ is the geodesic with $\lambda'(0) = X$. It is well known that the maximal domain $\mathcal{W}_p$ on which this map is well-defined is an open subset of $T_p\mathcal{M}$ that contains the origin. In general, $\mathcal{W}_p$ is not all of $T_p\mathcal{M}$, thereby reflecting the fact that a geodesic may arrive at the "boundary" of $\mathcal{M}$ (using the word "boundary" in a colloquial manner) before its affine parameter has reached the value 1. In general, the map $\exp_p$ need not be injective on its maximal domain $\mathcal{W}_p$, thereby indicating that the geodesics isssuing from p may reconverge and, eventually, intersect each other. However, it is well known that there is an open neighborhood $\mathcal{W}_p^o$ of the origin in $T_p\mathcal{M}$ such that the restriction of $\exp_p$ to $\mathcal{W}_p^o$ is a diffeomorphism onto its image. Thus, sufficiently short pieces of geodesics issuing from p do not intersect. A neighborhood of p that is contained in the image of $\mathcal{W}_p^o$ under $\exp_p$ is called a *normal neighborhood.*

This notion can be used to assign the property of being timelike or causal to continuous curves which need not be differentiable. A curve $\gamma : I \longrightarrow \mathcal{M}$ is called *timelike* (or *causal*, respectively) if it is continuous and if each $s \in I$ has a neighborhood $\tilde{I}$ in I such that for any two parameter values s_1 and s_2 in $\tilde{I}$ there is a timelike (or causal, respectively) geodesic from $\gamma(s_1)$ to $\gamma(s_2)$ which is completely contained in a normal neighborhood of $\gamma(s)$. If this geodesic is future-pointing whenever $s_1 < s_2$, γ is called *future-pointing*; otherwise γ is called *past-pointing.*

Moreover, we shall frequently use the following standard definition.

Definition 2. For a point p in a spacetime $(\mathcal{M}, g, \mathcal{T}^+)$, we define the *chronological future* $\mathcal{I}^+(p)$ (and the *chronological past* $\mathcal{I}^-(p)$, respectively) of p as the set of all points $q \in \mathcal{M}$ that can be reached from p along a future-pointing (or past-pointing, respectively) timelike curve.

$\mathcal{I}^+(p)$ and $\mathcal{I}^-(p)$ are obviously open subsets of $\mathcal{M}$ for every $p \in \mathcal{M}$. However, as long as the causal structure of spacetime has not been restricted this is more or less the only statement that can be made about these two sets. The following causality notions will be of relevance for us.

Definition 3. (a) A spacetime $(\mathcal{M}, g, \mathcal{T}^+)$ is called *causal* at a point $p \in \mathcal{M}$ if there is no closed causal curve through p.
(b) A spacetime $(\mathcal{M}, g, \mathcal{T}^+)$ is called *future-distinguishing* (or *past-distinguishing*, respectively) at p if every neighborhood $\mathcal{U}$ of p contains a neighborhood $\mathcal{V}$ of p such that a future-pointing (or past-pointing, respectively) causal curve from p that has left $\mathcal{U}$ cannot reenter $\mathcal{V}$. An equivalent condition is that the equation $\mathcal{I}^+(p) = \mathcal{I}^+(q)$ (or the equation $\mathcal{I}^-(p) = \mathcal{I}^-(q)$, respectively) implies the equation $p = q$.
(c) A spacetime $(\mathcal{M}, g, \mathcal{T}^+)$ is called *strongly causal* at p if every neighborhood $\mathcal{U}$ of p contains a neighborhood $\mathcal{V}$ of p such that no causal curve intersects $\mathcal{V}$ more than once.

It is easy to check that the strong causality condition implies both the future-distinguishing and the past-distinguishing condition and that either distinguishing condition implies the causality condition. For a rather detailed discussion of these well known notions we refer to Hawking and Ellis [29], to O'Neill [50] and to Beem, Ehrlich and Easley [4]. In particular, illustrative examples of spacetimes satisfying some causality assumptions but violating others are given in Figs. 37 and 38 of Hawking and Ellis [29].

3 Gravitational Lensing in Arbitrary Spacetimes

In this section we discuss the geometry of gravitational lensing situations in arbitrary spacetimes $(\mathcal{M}, g, \mathcal{T}^+)$. To that end we fix a point $p \in \mathcal{M}$ and a timelike C^∞ curve $\gamma : I \longrightarrow \mathcal{M}$. We interpret p as an event where an observation takes place (i. e. "an astronomer here and now in his observatory") and we interpret γ as the worldline of a light source, e. g., a distant quasar. This interpretation is, of course, based on the assumption that the spatial extension of the light source and of the observer can be neglected, i. e., that they can be considered as pointlike. Our assumption of γ being timelike means that the light source moves at a subluminal velocity. There is no need to specify the parametrization of γ (e. g., to proper time parametrization $g(\gamma', \gamma') = -1$) since we are primarily interested in the set γ and not in a particular parametrization. – The question we want to discuss is the following (see Figs. 1 and 2).

> How many past-pointing lightlike geodesics are there that start at the point p and terminate on the worldline γ?

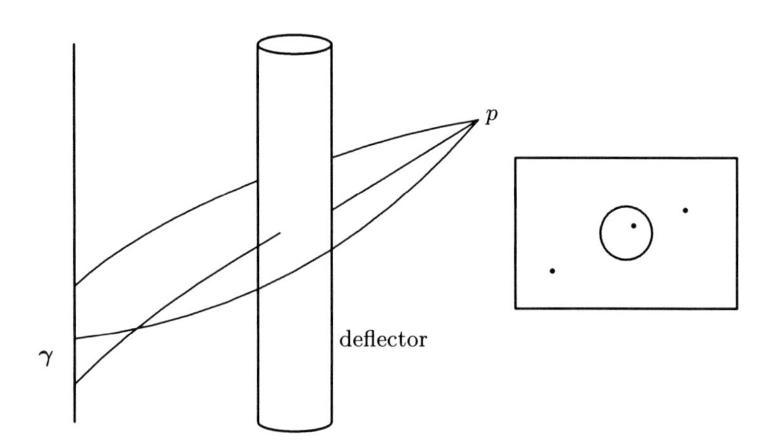

Fig. 1. This is a typical spacetime diagram of a multiple imaging situation. In correspondence with the three past-pointing lightlike geodesics from p to γ, an observer at p would see three images of a light source with worldline γ, as indicated in the insert. Actually, one of the three images is hidden behind the deflector if the latter is non-transparent.

According to the rules of general relativity, every lightlike geodesic can be interpreted as a light ray traveling under the influence of gravity alone. (As outlined in the Introduction, we shall not be concerned with light rays influenced by a medium throughout this text.) Thus, each past-pointing lightlike geodesic from p to γ gives rise to an image of the light source γ at the celestial sphere of the observer p. To justify this interpretation one may view each lightlike geodesic as representing a thin bundle of almost parallel light rays which is focused by the observer's eye lense onto his or her retina. – The following cases are to be distinguished.

Case A: *There is no past-pointing lightlike geodesic from p to γ.* Then the observer at p does not see any image of the light source γ. Situations of this kind are far from being unusual. They may occur even for an inextendible worldline γ in Minkowski space, viz., if γ asymptotically approaches the past light cone of p. Please note that, in general, the non-existence of a past-pointing lightlike geodesic from p to γ does not imply the non-existence of a past-pointing causal curve from p to γ. In other words, even if p cannot receive a freely traveling light ray from γ it is very well possible that p can be causally influenced by γ. The Gödel cosmos provides an interesting example of this kind, see, e. g. Hawking and Ellis [29], Sect. 5.7. In this spacetime any two points can be joined by a past-pointing causal curve; however, the lightlike geodesics issuing from some point are restricted to a cylindrical region.

Case B: *There is exactly one past-pointing lightlike geodesic from p to γ.* Then the observer at p sees exactly one image of the light source γ. This is the situation naively taken for granted in pre-relativistic astronomy.

Case C: *There are at least two but not more than denumerably many past-pointing lightlike geodesics from p to γ.* Then the observer at p sees finitely or infinitely many distinct images of γ at his or her celestial sphere. In view of Einstein's field equation one may think of a heavy mass ("deflector"), occupying a worldtube between γ and p, whose gravitational field causes a bending of light rays. Figure 1 shows a typical three-image-configuration. Please note that, generically, different lightlike geodesics from p to γ intersect the worldline γ at different points. In other words, the various images seen at p show the light source at different ages. Astronomers use the term *time delay* for this phenomenon.

Case D: *There are more than denumerably many past-pointing lightlike geodesics from p to γ.* E. g., there may be a continuous one-parameter family of lightlike geodesics from p to γ such that the light source γ appears at the celestial sphere of the observer p as an arc or, in situations of axially symmetry, as a ring, see Fig. 2. Such rings are often called *Einstein rings* although Chwolson [10] and not Einstein was the first to mention this phenomenon. We shall prove later in Proposition 12 that all members of a continuous one-parameter family of light rays from p to γ necessarily meet γ at the same point. In other words, all parts of such an extended image show the light source at the same age.

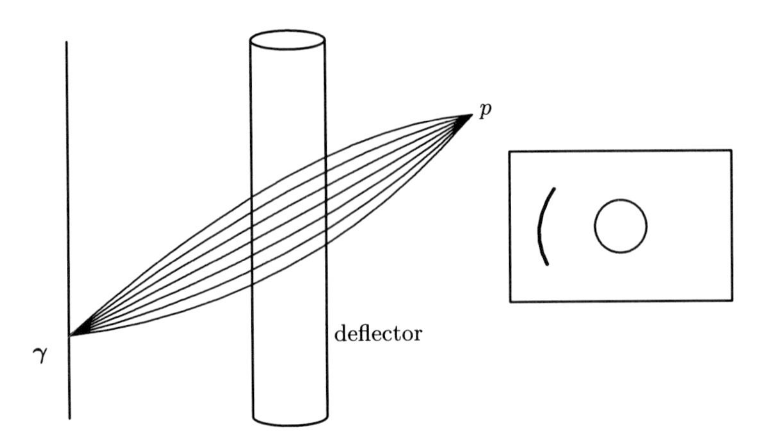

Fig. 2. In the situation depicted in this spacetime diagram there is a one-parameter family of past-pointing lightlike geodesics from p to γ. Correspondingly, an observer at p would see an extended image, such as an arc, of a light source with worldline γ. The insert indicates what is seen by the observer at p.

Whenever Case C or Case D occurs astronomers speak of *multiple imaging* by the *gravitational lens effect.* In Proposition 10 below we shall prove that Case D is "exceptional" in the sense that, under a small perturbation of the point p (keeping the worldline γ fixed), Case D always disappears. In this sense, Case C is the "generic" multiple imaging case. However, this does not mean that Case D is of no interest from a physical point of view. First, a systematic investigation of Case D situations in a spacetime will give interesting information on Case C situations. Second, the "non-genericity" of Case D situations is based on our idealization to view the light source and the observer as spatially non-extended, i. e., as pointlike. For extended light sources it is not true that Case D situations can be destroyed by an arbitrarily small perturbation. That is the reason why, actually, astronomers do observe arcs and rings produced by gravitational lensing.

It is now our goal to characterize spacetime geometries that lead to multiple imaging, with special emphasis on the difference between Case C and Case D situations. This will be done in the next three subsections. We begin with a review of the notions of conjugate points and of cut points in Sect. 3.1. These notions will be instrumental to study the local and global geometry of light cones in Sect. 3.2 and, thereupon, to give criteria for multiple imaging in Sect. 3.3. Sections 3.4 and 3.5 are devoted to a version of Fermat's principle on arbitrary spacetimes that is very useful in studying gravitational lensing situations.

3.1 Conjugate Points and Cut Points

The notion of conjugate points is used to characterize the situation that neighboring geodesics undergo a (partial) focusing effect. We are interested in lightlike geodesics in a spacetime $(\mathcal{M}, g, \mathcal{T}^+)$ that pass through a particular point p. (In applications to gravitational lensing this will be the point where the observation takes place and we shall be interested in lightlike geodesics that issue from p into the past.) In other words, we are interested in C^∞ curves $\lambda : I \longrightarrow \mathcal{M}$ that satisfy the conditions

$$\nabla_{\lambda'}\lambda' = 0\,, \tag{2}$$

$$g(\lambda', \lambda') = 0\,, \tag{3}$$

$$\lambda(s_o) = p\,, \tag{4}$$

where s_o denotes a fixed parameter value $s_o \in I$.

To investigate the behavior of geodesics which are close to λ we consider a one-parameter variation of λ, i. e., a C^∞ map $\eta :]-\varepsilon_o, \varepsilon_o[\, \times I \longrightarrow \mathcal{M}$ with $\eta(0, \cdot) = \lambda$ where ε_o is some positive real number. We assume that not only λ but also all the varied curves $\eta(\varepsilon, \cdot)$, for $0 < |\varepsilon| < |\varepsilon_o|$, satisfy the three conditions (2), (3), (4). By differentiation with respect to the variational parameter ε this implies that the variational vector field $J : I \longrightarrow T\mathcal{M}$, which is defined by $J(s) = \eta(\cdot, s)'(0)$, satisfies the three conditions

$$\nabla_{\lambda'}\nabla_{\lambda'}J - R(\lambda', J, \lambda') = 0\,, \tag{5}$$

$$g(\lambda', \nabla_{\lambda'}J) = 0\,, \tag{6}$$

$$J(s_o) = 0\,, \tag{7}$$

see Fig. 3. For any vector field J along λ that satisfies these three conditions, one may think of the "arrow-head" of J as tracing a neighboring lightlike geodesic through p in linear approximation. (5) is called the *equation of geodesic deviation* or the *Jacobi equation* and any solution J of this equation is called a *Jacobi field* along λ.

The equations (5), (6) and (7) are obviously satisfied by any multiple of the tangent field, $J(s) = f(s)\,\lambda'(s)$, with $f(s_o) = 0$. Such a solution of (5), (6) and (7) is called *trivial* since it represents an infinitesimally neighboring geodesic which is just a reparametrization of λ. We are now ready to define the notion of conjugate points. For $s_1 \in I \setminus \{s_o\}$, one says that the point $q = \lambda(s_1)$ is *conjugate* to $p = \lambda(s_o)$ along λ if there is a non-trivial solution J of (5), (6) and (7) such that $J(s_1)$ is parallel to $\lambda'(s_1)$. It is easy to check that the conjugacy of q to p along λ is independent of which affine parametrization has been chosen for λ. Also, it is known that in a compact section of a lightlike geodesic there are at most finitely many points conjugate to a given point p, see, e. g., Beem, Ehrlich and Easley [4], Theorem 10.77. (The same result

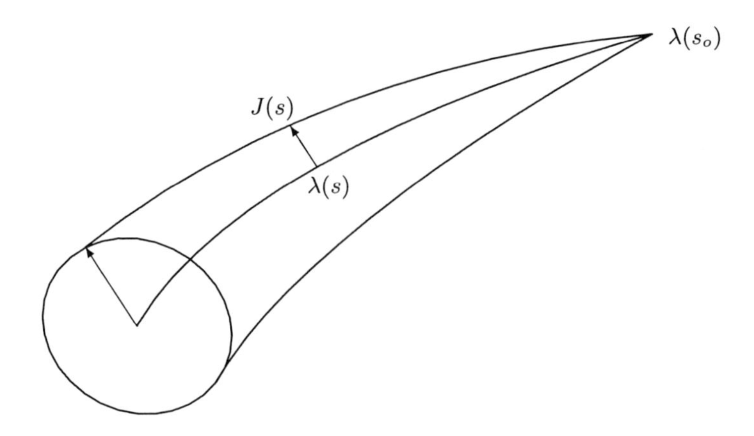

Fig. 3. The solutions J of (5), (6) and (7) form an "infinitesimal bundle of light rays" around λ. The timelike dimension is suppressed in the picture.

is true for timelike geodesics as well, but not for spacelike ones. An example where a whole interval is conjugate to a point along a spacelike geodesic was contrived by Helfer [30]. The reader is cautioned against a proof that conjugate points are always isolated, along any geodesic in a semi-Riemannian manifold of any signature, suggested by O'Neill [50], Exercise 8, p. 299. This is one of the very few mistakes in this otherwise excellent text-book; a basis of Jacobi fields with the desired properties need not exist.)

It follows directly from the definitions that a conjugate point indicates a partial focusing effect in the following sense. If, for a lightlike geodesic λ, the point $\lambda(s_1)$ is conjugate to the point $\lambda(s_o)$, then a one-parameter family of lightlike geodesics issuing from $\lambda(s_o)$ is being refocused into the point $\lambda(s_1)$ to within linear approximation.

In addition, conjugate points indicate that a geodesic loses its extremizing property, according to the following well-known proposition.

Proposition 1. *Let $\lambda : I \longrightarrow \mathcal{M}$ be a lightlike geodesic in a spacetime $(\mathcal{M}, g, \mathcal{T}^+)$ and consider two different parameter values s_o and s_1 in I. If, for some $s \in]s_o, s_1[$, the point $\lambda(s)$ is conjugate to $\lambda(s_o)$ along λ, then there is a C^∞ variation of $\lambda|_{[s_o,s_1]}$ such that all varied curves are timelike curves from $\lambda(s_o)$ to $\lambda(s_1)$. Conversely, the existence of such a variation implies that there must be a point $\lambda(s)$ conjugate to $\lambda(s_o)$ in the half-open interval $s \in]s_o, s_1]$.*

For a proof we refer to Hawking and Ellis [29], Proposition 4.5.11 and Proposition 4.5.12. It might be helpful to consult O'Neill [50], Chap. 10, Proposition 48, in addition.

Proposition 1 says that an observer moving at subluminal velocity may catch up with a light ray λ after the latter has passed through a conjugate point and that, for observers staying close to λ, this is impossible otherwise. Here the restriction to observers staying close to λ is essential. Observers "taking a short cut" may very well catch up with a lightlike geodesic even if the latter is free of conjugate points. For investigating the extremizing property of a lightlike geodesic from a global point of view, not just with respect to neighboring curves, the notion of conjugate points is not appropriate. Instead, one has to consider the notion of "cut points" which was introduced in Riemannian geometry (i. e., for positive definite metrics) by Poincaré [66] for a special situation and by Whitehead [85] in generality. For lightlike geodesics in a Lorentzian manifold, this notion can be introduced in the following way, cf. Beem, Ehrlich and Easley [4].

For any two points p and q in $\mathcal{M}$, we denote by $d(p,q)$ the Lorentzian pseudo-distance between p and q, i. e.

$$d(p,q) = \operatorname{supr}_{\beta} \int_0^1 \sqrt{\left|g\big(\beta'(s),\beta'(s)\big)\right|}\; ds\,, \tag{8}$$

where the supremum is to be taken over all causal curves $\beta : [0,1] \longrightarrow \mathcal{M}$ with $\beta(0) = p$ and $\beta(1) = q$. Owing to the well-known fact (cf. Beem, Ehrlich and Easley [4], p. 75) that any causal curve is differentiable almost everywhere, it is not necessary to restrict to differentiable causal curves to make sure that the integral in (8) does exist. Whenever p and q are causally related, such that the supremum is to be taken over a non-empty set, the existence of the supremum is guaranteed but it may be infinite.

Now let $\lambda : I \longrightarrow \mathcal{M}$ be a past-pointing lightlike geodesic. For s_o and s_1 in I with $s_o < s_1$, $\lambda(s_1)$ is called the *past cut point* of $\lambda(s_o)$ along λ if $d\big(\lambda(s_o),\lambda(s)\big) = 0$ for $s \in\,]s_o, s_1]$ and $d\big(\lambda(s_o),\lambda(s)\big) > 0$ for all $s \in I$ with $s > s_1$. Thus, the past cut point occurs where λ loses its extremizing property among causal curves with respect to the pseudo-distance. In this situation, for $s > s_1$ the point $\lambda(s)$ can be reached from $\lambda(s_o)$ along a past-pointing causal curve which is not a lightlike geodesic. By a well-known theorem (see, e. g., Hawking and Ellis [29], Proposition 4.5.10) this implies that $\lambda(s)$ can be reached from $\lambda(s_o)$ along a past-pointing timelike curve. Thus, beyond the past cut point λ intersects some past-pointing timelike curve which started together with λ at $\lambda(s_o)$.

It is not difficult to check that the past cut point of p along λ is independent of which past-pointing affine parametrization has been chosen for λ. Also, it follows directly from the definition that the past cut point is unique if it exists. The non-existence of the past cut point may have two quite different reasons. Either $d\big(\lambda(s_o),\lambda(s)\big) = 0$ for all $s \in I$ with $s > s_o$, i. e., the extremizing property is always preserved; or $d\big(\lambda(s_o),\lambda(s)\big) > 0$ for all $s \in I$ with $s > s_o$, i. e., the extremizing property never holds. Clearly, the latter case is possible only if the past distinguishing condition of Definition 3 (b) is violated. In other words, the past-distinguishing property guarantees that a sufficiently short past-pointing lightlike geodesic is extremizing.

There is, of course, a completely analogous definition of the *future cut point* along a lightlike geodesic. However, we shall concentrate upon the past cut point because this will be the relevant notion in view of gravitational lensing situations.

In Riemannian geometry the notions of cut points and conjugate points are related by the easily remembered rule: "The cut point comes first", see, e. g., Klingenberg [35], Proposition 2.1.7. The proof of this result is based on the well-known fact that a sufficiently short Riemannian geodesic always extremizes the Riemannian distance. We have just seen that for lightlike geodesics in Lorentzian manifolds the analogous fact need not be true unless the distinguishing property is satisfied. Therefore, the rule "The cut point comes first" can be proven for distinguishing spacetimes only. The precise statement reads as follows.

Proposition 2. *Let $\lambda : I \longrightarrow \mathcal{M}$ be a past-pointing lightlike geodesic in a spacetime that satisfies the past-distinguishing property at the point $p = \lambda(s_o)$. Assume that, for some parameter value $s_1 > s_o$, the point $\lambda(s_1)$ is conjugate to $\lambda(s_o)$ along λ. Then the past cut point $\lambda(s)$ of $\lambda(s_o)$ along λ exists and it is $s_o < s \leq s_1$.*

Proof. Since the past-distinguishing property is satisfied at p, the equation $d\big(p, \lambda(s)\big) = 0$ holds for $s \in [\, s_o \,, s_o + \varepsilon\, [$ whenever ε is a sufficiently small positive number. By Proposition 1, $d\big(p, \lambda(s)\big) > 0$ for $s \in \,]\, s_1 \,, s_1 + \delta\, [$ for arbitrarily small positive δ. Thus, the past cut point must lie in the parameter interval $]s_o, s_1]$. □

We end this subsection with a proposition saying that, under the past-distinguishing assumption, any intersection of two past-pointing lightlike geodesics starting from a point p is indicated by the occurence of a past cut point on each of those geodesics, please cf. Beem, Ehrlich and Easley [4], Lemma 9.13.

Proposition 3. *Let $(\mathcal{M}, g, \mathcal{T}^+)$ be a spacetime that satisfies the past-distinguishing condition at a point $p \in \mathcal{M}$. Assume that there is a point $q \in \mathcal{M}$ that can be reached from p along two past-pointing lightlike geodesics. Then the past cut point of p exists on each of those geodesics (and it comes on or before q).*

Proof. We may parametrize the two past-pointing lightlike geodesics such that $\lambda_1(0) = \lambda_2(0) = p$ and $\lambda_1(1) = \lambda_2(1) = q$. Then $\lambda_1'(1)$ and $\lambda_2'(1)$ are linearly independent since otherwise λ_2 would be a reparametrization of λ_1. This follows from the uniqueness theorem for the geodesic equation and from the fact that, owing to our past-distinguishing condition, a closed lightlike geodesic through p cannot exist. Then, for small positive ε, the curve $\lambda_1\big|_{[0,1]}$ joined to the curve $\lambda_2\big|_{[1,1+\varepsilon]}$ gives a causal curve which is not an (unbroken) lightlike geodesic. By a well-known theorem (see Hawking and Ellis [29],

Proposition 4.5.10) this implies that the point $\lambda_2(1+\varepsilon)$ can be reached from $p = \lambda_1(0)$ by a timelike curve, thus $d\big(p\,,\lambda_2(1+\varepsilon)\big) > 0$. On the other hand, the past-distinguishing condition at p guarantees that $d\big(p\,,\lambda_2(s_o+\delta)\big) = 0$ for small positive δ. Hence, the past cut point of p along λ_2 comes on or before $q = \lambda_2(1)$. □

Moreover, one might ask if the past cut point itself can be reached from p along a second past-pointing lightlike geodesic. A theorem to that effect can be proven only under the assumption of global hyperbolicity and will be postponed until Sect. 4, see Proposition 14 below. In the Riemannian case, an analogous result holds on *complete* Riemannian manifolds and is the content of the celebrated *Poincaré Theorem*, proven by Poincaré [66] for a special case and by Whitehead [85] in its generality. It is this property that gave rise to the name "cut point".

3.2 The Geometry of Light Cones

In a spacetime, the lightlike geodesics issuing from a point p into the past make up the socalled *past light cone* of p. In general, the past light cone need not be an immersed (let alone embedded) submanifold of $\mathcal{M}$, i. e., even its local structure may be very complicated. The failure of past light cones to be submanifolds is crucial for gravitational lensing. In this subsection we use the notions of conjugate points and cut points to investigate whether the past light cone is an immersed or embedded submanifold.

Please recall that the totality of all geodesics issuing from a point p are given in terms of the exponential map (1). For our study of gravitational lensing we are interested in lightlike geodesics issuing from p into the past. Therefore, we restrict the exponential map to the 3-dimensional submanifold

$$\mathcal{C}_p^- = \big\{ X \in \mathcal{W}_p \setminus \{0\} \,\big|\, X \text{ is lightlike and past-pointing} \big\} \tag{9}$$

of $T_p\mathcal{M}$. In (9) $\mathcal{W}_p \subseteq T_p\mathcal{M}$ denotes the maximal domain of $\exp_p$. For the sake of convenience, we introduce the abbreviation

$$e_p^- = \exp_p\big|_{\mathcal{C}_p^-} : \mathcal{C}_p^- \longrightarrow \mathcal{M} \tag{10}$$

for the restriction of the exponential map to $\mathcal{C}_p^-$. The image of this map e_p^- is the *past light cone* of p, i. e., the set of all events $q \in \mathcal{M}$ that can be reached from p along a past-pointing lightlike geodesic. This set determines what is visible for an observer at p. In particular, it determines whether p observes a gravitational lensing situation.

e_p^- is a C^∞ map from a 3-dimensional manifold into a 4-dimensional manifold. Thus, at any $X \in \mathcal{C}_p^-$ the rank of the differential $T_X e_p^-$ cannot be bigger than 3. If the rank is equal to 3, e_p^- is an immersion at X, i. e., the past light cone with vertex p is an immersed submanifold near the point $e_p^-(X)$. This is, of course, necessarily the case for vectors $X \in \mathcal{C}_p^- \cap \mathcal{W}_p^o$, where

$\mathcal{W}_p^o$ denotes the domain on which the exponential map is a diffeomorphism. This reflects the well-known fact that the past light cone with vertex p is a 3-dimensional manifold if we restrict to sufficiently short lightlike geodesics issuing fom p. For $X \in \mathcal{C}_p^- \setminus \mathcal{W}_p^o$, however, the number $m = 3 - \text{rank}\big(T_X e_p^-\big)$ may be bigger than 0. Comparison with the preceding subsection shows that this is the case if and only if the point $e_p^-(X)$ is conjugate to p along the geodesic generated by X. The number m is called the *multiplicity* of this conjugate point. Since, obviously, $(T_X e_p^-)(X) \neq 0$, the multiplicity m may be either 1 or 2. It is important to realize that, in any case, the image of the differential $T_X e_p^-$ is a lightlike subspace. For a proof it suffices to realize that (2), (6) and (7) imply $g(\lambda', J) = 0$, so the tangent vector of λ and all "connecting vectors" with infinitesimally neighboring lightlike geodesics starting from the same point span a lightlike subspace. (Here we make use of the well-known fact that a timelike vector cannot be orthogonal to a lightlike vector, i. e., that the equations $g(\lambda', \lambda') = 0$ and $g(\lambda', J) = 0$ imply the equation $g(J, J) \geq 0$.) In particular, this proves the well-known fact that the light cone is a 3-dimensional *lightlike* submanifold at each point where $T_X e_p^-$ has maximal rank.

The union of all points which are conjugate to p, along any past-pointing lightlike geodesic issuing from p, is called the *past lightlike conjugate locus* of p or the *caustic* of the past light cone of p. In other words, the caustic is the set of all points where the past light cone fails to be an immersed submanifold of $\mathcal{M}$. At caustic points, the light cone typically forms edges or vertices whose geometry might be arbitrarily complicated. If one restricts to caustics which are *stable* against perturbations in a certain sense, then a local classification of caustics is possible with the help of Arnold's singularity theory of Lagrangian or Legendrian maps, see Arnold, Gusein-Zade and Varchenko [3] or Arnold [2]. This formalism has been applied to *wavefronts* in general relativity, a notion which includes light cones as special cases, by Friedrich and Stewart [18], by Hasse, Kriele and Perlick [28] and, in a particularly elegant way, by Low [43]. (In [28] the proof of Theorem 4.4 is incorrect. A corrected version is going to appear.) In the case of globally hyperbolic spacetimes the formalism of Low even allows to tackle the problem of *globally* classifying the caustics of light cones, although this has not been carried through until now. For the sake of comparison the reader should also consult Petters' work [62] [64] [65] on caustics in the quasi-Newtonian approximation formalism of gravitational lensing. Unfortunately, the subject of classifying stable caustics is so technical that we cannot go into this matter here for lack of space.

It is important to realize that a light cone may fail to be an embedded submanifold of $\mathcal{M}$ even if its caustic is empty. At the end of this subsection we shall illustrate this claim by an example where a light cone develops transverse self-intersections without ever failing to be an immersed submanifold of $\mathcal{M}$, see Fig. 4 below. The relevant notion for finding out whether a light cone is an embedded submanifold is the notion of cut points, and not the notion of conjugate points. To work this out, we have to take a closer look at the chronological past $\mathcal{I}^-(p)$ of a point p, please recall Definition 2.

In Minkowski space, the lightlike geodesics issuing from p into the past make up the boundary $\partial\mathcal{I}^-(p)$ of $\mathcal{I}^-(p)$. In spacetimes with a complicated causal structure, however, those lightlike geodesics may penetrate into the open set $\mathcal{I}^-(p)$, i. e., the past light cone of p may have a non-void intersection with $\mathcal{I}^-(p)$. In spacetimes with drastic causality violations (such as, e. g., the Gödel cosmos, see Hawking and Ellis [29], Sect. 5.7) $\mathcal{I}^-(p)$ may even be all of $\mathcal{M}$ such that $\partial\mathcal{I}^-(p)$ is empty and the past light cone of p is completely contained in $\mathcal{I}^-(p)$. Quite generally, $\partial\mathcal{I}^-(p)$ can be characterized in the following way.

Proposition 4. *For any point p in a spacetime, the set $\partial\mathcal{I}^-(p)$ is either empty or a 3-dimensional achronal closed embedded C^{1-} submanifold of $\mathcal{M}$. (A subset of a spacetime is* achronal *if it is impossible to connect any two of its points by a timelike curve. A C^{1-} manifold is a topological manifold whose transition maps satisfy a Lipschitz condition.)*

For a proof we refer to Hawking and Ellis [29], Proposition 6.3.1. With the Lorentzian distance function d defined by (8), we get the following result for a past-pointing lightlike geodesic λ with $\lambda(s_o) = p$. A point $\lambda(s)$ with $s > s_o$ is in the boundary of $\mathcal{I}^-(p)$ if $d\big(p\,,\lambda(s)\big) = 0$ and it is in the open set $\mathcal{I}^-(p)$ if $d\big(p\,,\lambda(s) > 0$. Thus, the past cut point of p along a lightlike geodesic can be characterized as the point where this geodesic leaves the boundary of $\mathcal{I}^-(p)$ and penetrates into the open set $\mathcal{I}^-(p)$. The set of all past cut points of p along lightlike geodesics through p is called the *past lightlike cut locus* of p. We shall now prove that in past-distinguishing spacetimes the failure of a past light cone to be an embedded submanifold is indicated by a non-empty past lightlike cut locus.

Proposition 5. *Assume that the spacetime $(\mathcal{M}, g, \mathcal{T}^+)$ satisfies the past-distinguishing property at a point p. If the past lightlike cut locus of p is empty, then the map e_p^- defined in (10) is a C^∞ embedding, i. e., the past light cone of p is an embedded C^∞ submanifold of $\mathcal{M}$.*

Proof. If the past lightlike cut locus is empty, Proposition 2 implies that no past-pointing lightlike geodesic starting at p can have a point conjugate to p. Hence, the map e_p^- is a C^∞ immersion. Together with the past-distinguishing condition, the same assumption implies that such a geodesic must stay on $\partial\mathcal{I}^-(p)$ forever, i. e., the past light cone of p must be completely contained in $\partial\mathcal{I}^-(p)$. But then Proposition 4 guarantees that the past light cone has no self-intersection and no almost self-intersection. Hence, e_p^- must even be an embedding. □

In Sect. 4 below we shall prove that the converse of this proposition is true in globally hyperbolic spacetimes, see Proposition 15. – We now illustrate the properties of conjugate loci and cut loci with three examples.

Example 1:
Figure 4 shows the past light cone of a point p in a spacetime with a non-transparent deflector. To have a concrete example, the reader may consider the spacetime metric

$$g = -dt^2 + dz^2 + dr^2 + k^2\, r^2\, d\varphi^2 \,, \tag{11}$$

with some constant $0 < k < 1$, on $\mathcal{M} = \mathbb{R}^2 \times \big(\mathbb{R}^2 \setminus \{0\}\big)$. Here (t, z) denote Cartesian coordinates on $\mathbb{R}^2$ and (r, φ) denote polar coordinates on $\mathbb{R}^2 \setminus \{0\}$. This can be interpreted as the spacetime around a static non-transparent string, see Vilenkin [79], Hiscock [32] and Gott [24]. (Vilenkin in his pioneering paper discussed this metric in connection with the linearized Einstein field equation; it was then realized independently by Hiscock and Gott that Vilenkin's results remain true even if the full Einstein equation is used.) One should think of the string as being situated at the z-axis. Since the latter is not part of the spacetime, it is indeed justified to speak of a *non-transparent* string. It is easy to see that the metric (11) induces on each plane $t =$ const., $z =$ const. the geometry of a cone; i. e., this metric has a "conic singularity" along the z-axis.

It is an instructive exercise to verify that in this string spacetime each past light cone qualitatively looks like the one depicted in Fig. 4. (Clearly, in Fig. 4 one spatial dimension is suppressed. This missing dimension corresponds to the z direction in the string example. The fat vertical line in Fig. 4 actually indicates that a two-dimensional world sheet has been excised from spacetime, viz., the total history of the z-axis in the string example.) The caustic of this light cone is empty, i. e., there are no points conjugate to p along any past-pointing lightlike geodesic from p. The past lightlike cut locus of p, however, is not empty, thereby illustrating our earlier claim that a light cone may fail to be an embedded submanifold without failing to be an immersed submanifold. Moreover, Fig. 4 nicely exemplifies our general result that each past-pointing lightlike geodesic from p enters $\mathcal{I}^-(p)$ exactly when passing through the cut locus.

Example 2:
We now modify Example 1 by switching to a transparent deflector. In the case of the string metric (11) this can be done by changing the metric in the neighborhood of the z-axis in such a way that there is no longer a singularity, i. e., by "rounding off the tip of the cone" which represents each plane $t =$ const., $z =$ const. The region around the z-axis in which the metric has been changed can then be interpreted as the interior region of a transparent string. The resulting light cone looks like the one depicted in Fig. 5. The fact that now there are new lightlike geodesics (in comparison to Fig. 4) that pass through the interior region of the deflector gives rise to the formation of conjugate points, i. e., the light cone is no longer everywhere an immersed submanifold of $\mathcal{M}$. More precisely, the light cone develops two cuspidal edges that meet in a so-called *swallow-tail* at the point denoted by q in the figure.

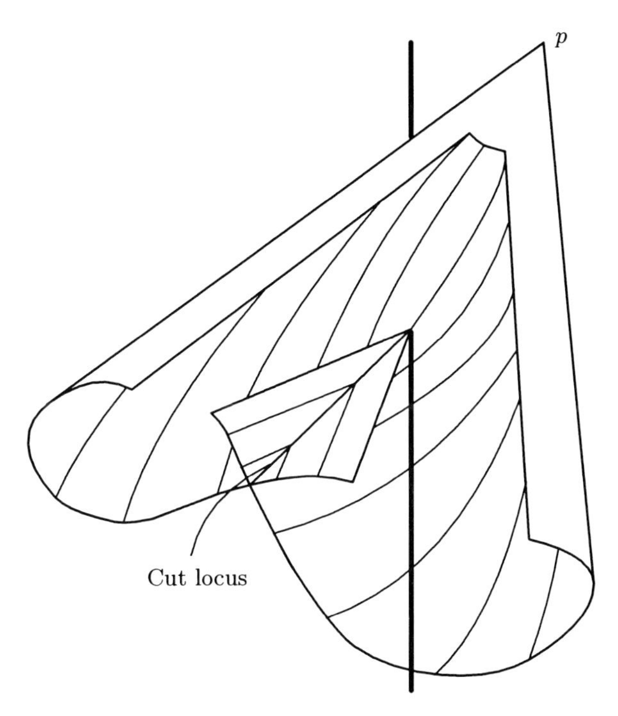

Fig. 4. In a spacetime with non-transparent deflector, e. g., a non-transparent string, light cones may develop self-intersections without failing to be immersed submanifolds. It is geometrically evident that the situation depicted here gives rise to double imaging for an observer at p.

These two cuspidal edges (including the point q) make up the caustic of the light cone. The point q also belongs to the cut locus which looks quite similar to the one in Fig. 4. A special role is played by the past-pointing lightlike geodesic λ starting from p that passes through the point q. (In Fig. 4 this geodesic was blocked by the deflector, so it did not reach a point analogous to q.) q is conjugate to p and, at the same time, the past cut point of p along λ. All neighboring geodesics emanating from p pass through the cut point first (where they enter into $\mathcal{I}^-(p)$) and reach their first conjugate point afterwards (where they smoothly slip over the cuspidal edge). Again we emphasize that one spatial dimension is suppressed in Fig. 5.

Example 3:
By excising a neighborhood of the point q from the spacetime in Fig. 5 we are led back to a light cone of the kind considered in Fig. 4. It is more interesting to leave a neighborhood of q untouched and instead to remove a worldline from spacetime that intersects the lightlike geodesic $\boldsymbol{\lambda}$ between p and q. (This may then be reinterpreted as the worldline of a non-transparent

deflector, although some adjustments are necessary to reconcile this new interpretation with Einstein's field equation.) The interesting new feature of this modified example is that the geodesic $\boldsymbol{\lambda}$ is blocked before it reaches the point q (contrary to Fig. 5) but that the point q is still part of the spacetime (contrary to Fig. 4). It is obvious that in this new situation the past light cone of p united with $\{p\}$ is no longer a closed subset of $\mathcal{M}$ since q is in the closure of the light cone but does not belong to it. By the same token, the past lightlike conjugate locus and the past lightlike cut locus of p are no longer closed subsets of $\mathcal{M}$. In Sect. 4 we shall see that this is possible only in spacetimes that are not globally hyperbolic.

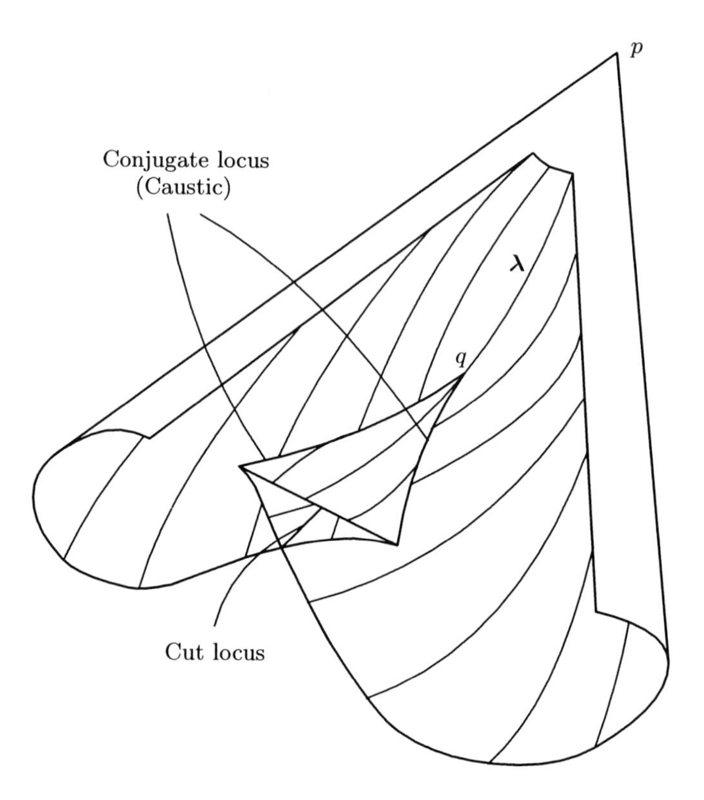

Fig. 5. In a spacetime with a transparent deflector, e. g., a transparent string, light cones typically fail to be immersed submanifolds. It is geometrically evident that the situation depicted here gives rise to triple imaging for an observer at p, cf. Fig. 1.

These three examples are quite instructive but they are, of course, not universal in view of gravitational lensing situations. Completely new features may occur if the missing spatial dimension is taken into account, in particular in spacetimes without symmetry. Also, causality violations or a non-trivial topological structure of spacetime (apart from a "hole" that is meant to model a non-transparent deflector) may change the global features of light cones dramatically.

3.3 Citeria for Multiple Imaging

We are now ready to turn to the discusssion of multiple imaging situations. To find out how many images an observer at p would see of a light source with worldline γ we have to determine the intersection of γ with the past light cone of p. The following proposition shows that in past-distinguishing spacetimes the occurence of cut points is necessary for multiple imaging.

Proposition 6. *Assume that the spacetime $(\mathcal{M}, g, \mathcal{T}^+)$ is past-distinguishing at a point p and that the past lightlike cut locus of p is empty. Then for every timelike C^∞ curve $\gamma : I \longrightarrow \mathcal{M}$ there is at most one past-pointing lightlike geodesic that starts at p and terminates on γ.*

Proof. By contradiction, assume that we have two past-pointing lightlike geodesics λ_1 and λ_2 from p to γ. If they meet γ at the same point, Proposition 3 shows that the past lightlike cut locus of p cannot be empty. So let us assume that the two geodesics meet γ in two different points $\lambda_1(s_1) = q_1 \neq \lambda_2(s_2) = q_2$, with q_1 in the past of q_2 (say). Then we can join the section of γ between q_1 and q_2 to λ_2 to get a past-pointing causal curve from p to q_1 that is not a lightlike geodesic. By a well-known theorem (see Hawking and Ellis [29], Proposition 4.5.10) this implies that q_1 is in $\mathcal{I}^-(p)$. Together with the past-distinguishing assumption this makes sure that the past cut point of p along λ_1 must exist which gives the desired contradiction. □

A slightly weaker version of this result, assuming the strong causality condition rather than the past-distinguishing condition, was given in Perlick [58].

Next we give a sufficient criterion for multiple imaging. The examples studied at the end of the preceding subsection suggest that a past light cone forms several sheets after past-pointing light rays have passed through cut points or conjugate points, and that this gives rise to multiple imaging situations. The following proposition puts this general idea into precise form.

Proposition 7. *Fix, in an arbitrary spacetime $(\mathcal{M}, g, \mathcal{T}^+)$, a point p and a past-pointing lightlike geodesic $\lambda : I \longrightarrow \mathcal{M}$ with $\lambda(s_o) = p$. Assume that, for some parameter $s_1 > s_o$ in I, $\lambda(s_1)$ is a conjugate point or the past cut point (or both) of p along λ. Then, for every parameter value $s \in I$ with $s > s_1$, there is a timelike curve γ through $\lambda(s)$ that can be reached from p along at least two past-pointing lightlike geodesics.*

Proof. We first show that, for $s > s_1$, the point $\lambda(s)$ can be reached from p along a past-pointing timelike curve. If $\lambda(s_1)$ is the past cut point of p along

λ, this follows directly from the definition of cut points. If it is a conjugate point, it follows from Proposition 1. Now we take such a timelike curve and perturb it slightly near p. In this way we get a timelike curve that intersects the past light cone of p in $\lambda(s)$ and in another point close to p, so it can be reached from p along two different past-pointing lightlike geodesics. □

Together with Proposition 5 this result implies that multiple imaging takes place whenever a past light cone fails to be an embedded submanifold. This is true, in particular, whenever a past light cone forms a caustic.

Proposition 7 can be used to prove that multiple imaging occurs in large classes of spacetimes. E. g., it is well known that, under conditions which are to be considered as fairly general from a physical point of view, a lightlike geodesic must be either incomplete or contain a pair of conjugate points. Those "fairly general conditions" are, e. g., the weak energy condition and the socalled generic condition. We do not want to go into this matter here. We just mention that results of this kind have played a crucial part in the development of the Penrose-Hawking singularity theorems and we refer to the detailed discussion in Hawking and Ellis [29], in particular to Proposition 4.4.5. We also mention that the weak energy condition need not hold pointwise but that some integrated version of the weak energy condition would do, see Tipler [77], Borde [6], Roman [67] and Kánnár [34]. In view of these results it seems justified to say that the occurence of conjugate points along lightlike geodesics is the rule rather than the exception. But then Proposition 7 implies that the occurence of multiple imaging is the rule rather than the exception.

One has to keep in mind that the worldline γ in Proposition 7 must be constructed in a particular way. There is no guarantee that the real universe for which $(\mathcal{M}, g)$ is a mathematical model contains a real light source (i. e., a galaxy or a quasar) that travels on this worldline γ. Therefore it would be nice to have an analogous proposition in which both the point p and the worldline γ are to be prescribed. Such a proposition holds in globally hyperbolic spacetimes and will be proven in Sect. 4, see Proposition 17 below.

Proposition 7, which is a fairly simple corollary of standard theorems, was given in Perlick [58]. Already earlier, Padmanabhan and Subramanian [51] had shown that the existence of conjugate points along a lightlike geodesic is sufficient for multiple imaging. However, their proof is completely different from ours and it uses a lot of additional assumptions on the topological and causal structure of spacetime most of which slip in surreptitiously. On the basis of these additional assumptions, Padmanabhan and Subramanian [51] were also able to show that the existence of conjugate points along lightlike geodesics is necessary for multiple imaging. We have already emphasized that this is not true in arbitrary spacetimes, please recall Example 1 at the end of the preceding subsection. We are now going to investigate topological and causal conditions on the spacetime that allow to prove such a result.

Example 1 might suggest that in simply connected spacetimes multiple imaging situations without conjugate points cannot occur. A more careful analysis shows that it is not the topology of (4-dimensional) spacetime but rather the topology of (3-dimensional) space that matters. To make this notion precise we have to consider a timelike C^∞ vector field V (*observer field*) on $\mathcal{M}$. The existence of such an observer field is guaranteed on all of $\mathcal{M}$ owing to the time-orientability assumption (c) of Definition 1, cf., e. g., Wald [80], Lemma 8.1.1. With such a V chosen, we may call any two points of $\mathcal{M}$ *equivalent* if they lie on a common integral curve of V. The corresponding quotient space, equipped with the quotient topology, will be denoted by $\mathcal{S}_V$ and can be interpreted as the *space* with respect to the observer field V. Please note that $\mathcal{S}_V$ need not satisfy the Hausdorff axiom; as a counter-example one may consider any timelike vector field on Minkowski space with one point removed. Also, there is no guarantee that $\mathcal{S}_V$ admits a smooth manifold structure such that the natural projection $\pi_V : \mathcal{M} \longrightarrow \mathcal{S}_V$ becomes a submersion; as a counter-example one may consider a timelike vector field V with an integral curve that is almost periodic. For later purpose we state the following result.

Proposition 8. *Let $(\mathcal{M}, g, \mathcal{T}^+)$ be a spacetime that does not contain a closed timelike curve and let V be a timelike C^∞ vector field on $\mathcal{M}$. If the quotient space $\mathcal{S}_V$ satisfies the Hausdorff axiom, $\mathcal{S}_V$ admits a C^∞ structure such that the natural projection $\pi_V : \mathcal{M} \longrightarrow \mathcal{S}_V$ makes $\mathcal{M}$ into a fiber bundle over $\mathcal{S}_V$ with typical fiber diffeomorphic to $\mathbb{R}$.*

For a proof we refer to Harris [27], Theorem 2. Note that Harris' assumption of V being complete is unnecessary since every nowhere vanishing vector field on $\mathcal{M}$ can be made into a complete vector field by multiplication with an appropriate positive function. To prove this, one puts a complete Riemannian metric h on $\mathcal{M}$. This is possible since, by a famous theorem of Whitney [86] (see also, e. g., Hirsch [31], p. 55) every n-dimensional paracompact manifold can be smoothly embedded as a closed submanifold into $\mathbb{R}^{2n+1}$; pulling back the Euclidean metric gives the desired complete Riemannian metric. It is then easy to check that the vector field $h(V,V)^{-1/2}\,V$ is complete, cf., e. g., Abraham and Marsden [1], Proposition 2.1.21.

For the following consideration we only need the topological structure on $\mathcal{S}_V$. We define for any point $p \in \mathcal{M}$ the set $\mathcal{S}_V^p$, called the *space visible to p with respect to V*, in the following way. We say that a point in $\mathcal{S}_V$ is in $\mathcal{S}_V^p$ if and only if the integral curve of V which is represented by that point either passes through p or can be reached from p along a past-pointing lightlike geodesic in $\mathcal{M}$. We are now ready to formulate the desired proposition.

Proposition 9. *Choose a timelike C^∞ vector field V on an arbitrary spacetime $(\mathcal{M}, g, \mathcal{T}^+)$. Fix a point $p \in \mathcal{M}$ and assume that $\mathcal{S}_V^p$, the space visible to p with respect to V, is simply connected. If the past lightlike conjugate locus of p is empty, any integral curve of V can be reached from p along at most one past-pointing lightlike geodesic.*

Proof. As the past lightlike conjugate locus of p is empty, the map e_p^- of (10) is an immersion. Then its image, the past light cone of p, is an immersed lightlike submanifold of $\mathcal{M}$. Since a timelike vector cannot be tangent to a lightlike submanifold, each integral curve of V intersects the image of e_p^- transversely. Hence, the combination of e_p^- with the projection $\pi_V : \mathcal{M} \longrightarrow \mathcal{S}_V$ gives a homeomorphism locally around each point, i. e., it gives a covering map from $\mathcal{C}_p^-$ onto $\mathcal{S}_V^p$. As a covering map onto a simply connected space must be a (global) homeomorphism, no integral curve of V can intersect the past light cone of p more than once and the past light cone of p cannot have self-intersections or almost self-intersections. The latter implies that e_p^- is even an embedding, i. e., it is impossible that a point can be reached from p along two different past-pointing lightlike geodesics. □

The condition of $\mathcal{S}_V^p$ being simply connected prohibits, in particular, situations such as in Example 1 where a non-transparent deflector is modeled by a hole. However, it may also be violated in situations with transparent deflectors, viz., if the visible universe has a non-trivial spatial topology.

For some situations of interest, at least, Proposition 9 says that multiple imaging requires the occurence of conjugate points. This is a valuable result since the existence of conjugate points along a lightlike geodesic allows to estimate the Ricci tensor along that geodesic. If we take Einstein's field equation into account, this estimate of the Riccci tensor can be rewritten as an estimate on the energy density, see Padmanabhan and Subramanian [51]. It is this observation that makes it physically interesting to investigate whether in a multiple imaging situation conjugate points must occur.

We summarize the results found sofar in the following way. The occurence of conjugate points or of cut points along a past-pointing lightlike geodesic is always sufficient for multiple imaging. If the past-distinguishing condition is satisfied at the observer's position, the occurence of cut points is necessary as well. If the space visible to p (with respect to an observer field) has a simply connected topology, then the occurence of conjugate points is also necessary.

We end this subsection with a group of propositions characterizing the special situation that a worldline γ meets the caustic of the past light cone of p. We first show that this is an exceptional situation.

Proposition 10. *Let $\gamma : I \longrightarrow \mathcal{M}$ be a timelike C^∞ curve in an arbitrary spacetime $(\mathcal{M}, g, \mathcal{T}^+)$. Then the set of all points $p \in \mathcal{M}$ such that γ does not meet the caustic of the past light cone of p is dense in $\mathcal{M}$.*

Proof. For each p in some open subset $\mathcal{U} \subseteq \mathcal{M}$, we consider the map e_p^- of (10) and identify its domain $\mathcal{C}_p^-$ with $\mathbb{R}^3 \setminus \{0\}$. This can be done with the help of local coordinates in the tangent bundle. Then the assignment $p \longmapsto e_p^-$ gives a continuous embedding from $\mathcal{U}$ into the space $C^1(\mathbb{R}^3 \setminus \{0\}, \mathcal{M})$ of C^1 maps from $\mathbb{R}^3 \setminus \{0\}$ into $\mathcal{M}$, equipped with the weak (or compact-open) topology. For the definition of this topology we refer to Hirsch [32], p. 34. By the transversality theorem (see, e. g., Hirsch [32], Theorem 2.1), the maps which are transverse to γ form a dense subset of $C^1(\mathbb{R}^3 \setminus \{0\}, \mathcal{M})$. Thus, the points p for which e_p^- is transverse to γ are dense in $\mathcal{U}$. Please recall that, by definition, e_p^- is transverse to γ at a point $q = e_p^-(X)$ if either $q \notin \gamma$ or the image of $T_X e_p^-$ and the tangent space of γ span all of $T_q\mathcal{M}$. Clearly, if γ meets the caustic of the past light cone of p at $e_p^-(X)$, transversality cannot be satisfied since the image of $T_X e_p^-$ is at most two-dimensional. □

In other words, by a "small perturbation" of the point p we can always achieve that a given worldline γ stays away from the caustic. But then multiple imaging situations are restricted by the following result.

Proposition 11. *If, in an arbitrary spacetime $(\mathcal{M}, g, \mathcal{T}^+)$, a timelike C^∞ curve $\gamma : I \longrightarrow \mathcal{M}$ does not meet the caustic of the past light cone of a point p, then there are at most denumerably many past-pointing lightlike geodesics that start at p and terminate on γ.*

Proof. Consider the pre-image $\mathcal{A}_{p,\gamma} = (e_p^-)^{-1}(\gamma)$ of γ under the map (10). If γ does not meet the caustic, e_p^- is an immersion at each point $X \in \mathcal{A}_{p,\gamma}$, i. e., it maps a neighborhood of X in C_p^- onto a 3-dimensional submanifold which is lightlike and, thus, transverse to γ. This proves that the points in $\mathcal{A}_{p,\gamma}$ are isolated, i. e., that there are only finitely many in each compact subset of $\mathcal{C}_p^-$. Since $\mathcal{C}_p^- \simeq \mathbb{R}^3 \setminus \{0\}$ can be covered with denumerably many compact sets this completes the proof. □

These two propositions justify our earlier claim that for multiple imaging situations Case C is generic and Case D is exceptional. However, again we emphasize that these results crucially depend on our idealization of assuming a pointlike source. They are, of course, no longer true if the worldline γ is replaced with a worldsheet or a worldtube.

Proposition 11 implies that a Case D situation is possible only if γ meets the caustic of the past light cone of p. In other words, if this light cone does not develop a caustic, then it is impossible for the observer at p to see an extended image such as an arc or a ring. We end this section by proving our earlier claim that, in a Case D situation, all parts of an extended (and connected) image show the light source at the same age.

Proposition 12. *Let p be a point and $\gamma : I \longrightarrow \mathcal{M}$ a timelike C^∞ curve in a spacetime $(\mathcal{M}, g, \mathcal{T}^+)$. Let $\mathcal{A}_{p,\gamma}$ denote the pre-image of γ under the map e_p^- which was introduced in* (10). *Then each connected component of $\mathcal{A}_{p,\gamma}$ is mapped by e_p^- onto a single point.*

Proof. Assume the image is not a single point. Then, by continuity of the exponential map, it is a one-dimensional timelike submanifold, viz., a portion of γ. This contradicts the fact that the image of $T_X e_p^-$ is always lightlike, so the light cone cannot contain a timelike curve. □

3.4 Fermat's Principle

For many applications it is useful to characterize the lightlike geodesics between a point and a timelike curve in a spacetime as the solutions of a variational problem. There are several versions of such a variational problem which may be viewed as general-relativistic generalizations of the traditional *Fermat principle.* The oldest versions, which hold on static or stationary spacetimes only, date back to Weyl [84] and Levi-Civita [41]. They are also discussed in several modern text-books and review articles, see, e. g., Frankel [17] or Straumann [76] for the static case and Landau and Lifschitz [40] or Brill [8] for the stationary case. For a discussion from a mathematical point of view we refer to Masiello [44]. Here we want to present a more general version of Fermat's principle which holds on *arbitrary* spacetimes. Its formulation is due to Kovner [36] and the proof that the solution curves of this variational problem are, indeed, the lightlike geodesics was given by Perlick [55]. The same version of Fermat's principle is also discussed in Schneider, Ehlers and Falco [70]. As an aside, we mention that this version of Fermat's principle may be generalized to the case of light rays in media, see Perlick [59] for a detailed exposition, and to the case of extended (i. e., non-pointlike) observers and light sources, see Perlick and Piccione [60]. According to the framework of this article we shall not discuss these generalizations here.

It is our goal to characterize, in an arbitrary spacetime $(\mathcal{M}, g, \mathcal{T}^+)$, the lightlike geodesics from a point p to a timelike worldline γ by a variational principle. To that end we have to specify (i) the set of *trial curves*, i. e., the set of curves among which the solutions to the variational problem are to be sought, and (ii) the functional that is to be extremized. The set of trial curves, denoted by $C^\infty_{p,\gamma}$ henceforth, is defined in the following way.

Definition 4. Fix, in an arbitrary spacetime $(\mathcal{M}, g, \mathcal{T}^+)$, a point $p \in \mathcal{M}$ and an embedded timelike C^∞ curve $\gamma : I \longrightarrow \mathcal{M}$. Then $C^\infty_{p,\gamma}$ is, by definition, the set of all C^∞ immersions $\lambda : [0,1] \longrightarrow \mathcal{M}$ with the following properties.
(a) λ is lightlike, i. e., $g\big(\lambda'(s), \lambda'(s)\big) = 0$ for all $s \in [0,1]$.
(b) λ starts at p and terminates on γ, i. e., $\lambda(0) = p$ and there is a $\tau(\lambda) \in I$ such that $\lambda(1) = \gamma\big(\tau(\lambda)\big)$. Since γ is assumed to be an embedding, this defines a unique assignment $\lambda \longmapsto \tau(\lambda)$.
(c) $g\big(\lambda'(1), \gamma'(\tau(\lambda))\big) < 0$, where $\tau(\lambda) \in I$ is defined through (b).

Roughly speaking, the space of trial curves can be characterized as the set of all ways to go from p to γ at the speed of light. Our decision to define all trial curves on the interval $[0,1]$ is a matter of convenience only. Condition (c) of Definition 4 restricts to future-oriented or past-oriented curves, depending on whether γ is future-oriented or past-oriented. For applications to gravitational lensing we are interested in the case that γ is a *past-pointing* parametrization of the worldline of a light source, see Fig. 1.

Condition (b) of Definition 4 defines a map $\tau : C^\infty_{p,\gamma} \longrightarrow \mathbb{R}$. We refer to τ as to the *arrival time functional* henceforth. This will be the functional to be extremized. – Finally, we need the following definition.

Definition 5. For a curve $\lambda \in C^{\infty}_{p,\gamma}$, as defined in Definition 4, a C^{∞} *variation* of λ in $C^{\infty}_{p,\gamma}$ is a C^{∞} map $\eta :]-\varepsilon_o, \varepsilon_o[\times [0,1] \longrightarrow \mathcal{M}$, for some $\varepsilon_o > 0$, such that $\eta(0,\cdot\,) = \lambda$ and $\eta(\varepsilon,\cdot\,) \in C^{\infty}_{p,\gamma}$ for all $\varepsilon \in]-\varepsilon_o, \varepsilon_o[$. The vector field $X : [0,1] \longrightarrow T\mathcal{M}$, $s \longmapsto X(s) = \big(\eta(\,\cdot\,, s)\big)'(0)$ is called the *variational vector field* of η. X is called *non-trivial* if $X(s)$ and $\lambda'(s)$ are non-collinear for some $s \in [0,1]$.

We want to prove that among all trial curves the lightlike geodesics are the stationary points of the arrival time. To that end we shall need the following characterization of variational vector fields.

Lemma 1. *For a C^{∞} vector field $X : [0,1] \longrightarrow T\mathcal{M}$ along $\lambda \in C^{\infty}_{p,\gamma}$, the following two properties are equivalent.*
(a) *X is the variational vector field of a C^{∞} variation η of λ in $C^{\infty}_{p,\gamma}$.*
(b) *$g(\nabla_{\lambda'}X, \lambda') = 0$, $X(0) = 0$, and $X(1) \,||\, \gamma'\big(\tau(\lambda)\big)$.*

The implication "(a)⇒(b)" is obvious since the desired properties of X follow just by differentiating the defining properties of trial curves. For a proof of the converse implication, which is more cumbersome, the reader is refered to Perlick [55].

We are now ready to formulate and prove the general-relativistic Fermat principle.

Theorem 1. (Fermat's principle) *Let $(\mathcal{M}, g, T^+)$ be an arbitrary spacetime, fix a point $p \in \mathcal{M}$ and an embedded timelike C^{∞} curve $\gamma : I \longrightarrow \mathcal{M}$. Then for a trial curve $\lambda \in C^{\infty}_{p,\gamma}$ the following two properties are equivalent.*
(a) *λ is a geodesic or a reparametrization thereof.*
(b) *For all C^{∞} variations η of λ in $C^{\infty}_{p,\gamma}$, the equation $\frac{d}{d\varepsilon}\tau\big(\eta(\varepsilon,\cdot)\big)\big|_{\varepsilon=0} = 0$ holds true.*

Proof. We first observe that the definition of the arrival time τ implies the equation $\eta(\varepsilon, 1) = \gamma\big(\tau\big(\eta(\varepsilon,\cdot\,)\big)\big)$ for each C^{∞} variation η of λ in $C^{\infty}_{p,\gamma}$. Differentiating with respect to ε and setting $\varepsilon = 0$ yields

$$X(1) = \gamma'\big(\tau(\lambda)\big)\, \frac{d}{d\varepsilon}\tau\big(\eta(\varepsilon,\,\cdot\,)\big)\Big|_{\varepsilon=0} . \tag{12}$$

We now prove the implication "(a)⇒(b)". By assumption, there is a C^{∞} function $w : [0,1] \longrightarrow \mathbb{R}$ such that $\nabla_{\lambda'}\lambda' = w\,\lambda'$. Now let X be the variational vector field of a C^{∞} variation η of λ in $C^{\infty}_{p,\gamma}$. Then we find $w\, g(X, \lambda') = g(X, \nabla_{\lambda'}\lambda') = g(X, \lambda')' - g(\nabla_{\lambda'}X, \lambda')$. The last term vanishes by Lemma 1. Upon integration, we find

$$g\big(X(1), \lambda'(1)\big) = g\big(X(0), \lambda'(0)\big) \exp\Big(\int_0^1 w(s)\, ds \Big) . \tag{13}$$

Since $X(0) = 0$, this implies $g\big(X(1), \lambda'(1)\big) = 0$. But then the desired result can be read from (12) because the timelike tangent vector to γ cannot be orthogonal to the lightlike tangent vector to λ. – To prove the converse implication "(b)⇒(a)" we define a vector field $U_\lambda : [0,1] \longrightarrow T\mathcal{M}$

by parallel transporting the vector $\gamma'(\tau(\lambda))$ along λ, i. e., $\nabla_{\lambda'} U_\lambda = 0$ and $U_\lambda(1) = \gamma'(\tau(\lambda))$. Clearly, U_λ is everywhere timelike, so $g(\lambda', U_\lambda)$ has no zeros. Now let $Z : [0,1] \longrightarrow T\mathcal{M}$ be any C^∞ vector field along λ with $Z(0) = 0$ and $Z(1) = 0$. We use this Z to define a new vector field X along λ by

$$X(s) = Z(s) - \Big(\int_0^s \frac{g(\nabla_{\lambda'} Z, \lambda')}{g(U_\lambda, \lambda')}\Big|_{\tilde{s}} \, d\tilde{s} \Big) \, U_\lambda(s) \, . \tag{14}$$

With the help of Lemma 1 it is easy to verify that X is the variational vector field of a C^∞ variation η of λ in $C^\infty_{p,\gamma}$. Hence, we can read from (12) that our hypothesis implies $X(1) = 0$. As $Z(1) = 0$ and $g\big(U_\lambda(1), \lambda'(1)\big) \neq 0$, the integral in (14) must vanish for $s = 1$. Upon integration by parts, this results in

$$\int_0^1 g\Big(Z \, , \nabla_{\lambda'} \frac{\lambda'}{g(U_\lambda, \lambda')} \Big)\Big|_s \, ds = 0 \, . \tag{15}$$

Since Z was an arbitrary C^∞ vector field along λ with $Z(0) = 0$ and $Z(1) = 0$, the fundamental lemma of variational calculus implies that

$$\nabla_{\lambda'} \frac{\lambda'}{g(U_\lambda, \lambda')} = 0 \, , \tag{16}$$

i. e., that $\nabla_{\lambda'} \lambda'$ is a multiple of λ'. □

Theorem 1 can be phrased as saying that among all ways to go from p to γ at the speed of light, the light rays are characterized as the stationary points of the arrival time τ. The analogy to the traditional Fermat principle is obvious. For some applications it might be convenient to choose γ as parametrized by proper time, $g(\gamma', \gamma') = -1$. However, Theorem 1 is true for any other smooth parametrization as well. The arrival time functional $\tau : C^\infty_{p,\gamma} \longrightarrow \mathbb{R}$ changes, of course, if the parametrization is changed, but the new arrival time functional has the same stationary points as the old one.

Please note that, by Theorem 1, a light ray may be a local minimum, a local maximum or a saddle of τ. We shall see in the next subsection that, actually, local maxima do not occur.

From Theorem 1 we can easily rederive the more special versions of Fermat's principle which are given in many text-books on general relativity. To illustrate this claim we consider the special case of a *conformally static spacetime*. More precisely, we need the following assumptions.

(a) $\mathcal{M}$ is diffeomorphic to $\mathcal{S} \times \mathbb{R}$, with some 3-dimensional manifold $\mathcal{S}$.

(b) The metric takes the form

$$g = e^{2f(x,t)} \Big(h_{\mu\nu}(x) \, dx^\mu \otimes dx^\nu \;-\; dt \otimes dt \Big) \tag{17}$$

where t denotes the projection from $\mathcal{M} \simeq \mathcal{S} \times \mathbb{R}$ onto the second factor, $x = (x^1, x^2, x^3)$ are coordinates on $\mathcal{S}$ and the Einstein summation convention is used for greek indices running from 1 to 3. (Coordinates on $\mathcal{S}$ are used for notational convenience only. It will not be necessary to assume that $\mathcal{S}$ can be covered by a single coordinate system.)

(c) $t \circ \gamma$ is constant, i. e., γ is vertical with respect to the product structure of $\mathcal{M} \simeq \mathcal{S} \times \mathbb{R}$.

In this situation the spacetime geometry is time independent up to an overall factor $e^{2f(x,t)}$ and the worldline γ is at rest in the "space" $\mathcal{S}$. Fermat's principle takes its simplest form if we use for γ a parametrization adapted to t, i. e., $dt(\gamma') = 1$. Since all trial curves are lightlike, we can then read from (17) that the arrival time functional is given by

$$\tau(\lambda) = \int_0^1 \sqrt{h_{\mu\nu}\big(x(s)\big)\frac{dx^\mu(s)}{ds}\frac{dx^\nu(s)}{ds}}\, ds\,. \tag{18}$$

Here $s \longmapsto x(s)$ denotes the projection onto the first factor of $s \longmapsto \lambda(s) \in \mathcal{M} \simeq \mathcal{S} \times \mathbb{R}$. By (18), $\tau(\lambda)$ is exactly the *length* of the projected curve $s \longmapsto x(s)$, measured with the spatial metric $h = h_{\mu\nu}(x)\, dx^\mu \otimes dx^\nu$. Hence, in this special case Theorem 1 says that a trial curve is a light ray if and only if its projection to $\mathcal{S}$ traces out an h-geodesic. This result is due to Weyl [84], apart from the fact that Weyl restricted to the static case, i. e., he did not allow the function f in (17) to depend on t. There is a straightforward generalization from the (conformally) static to the (conformally) stationary case which is essentially due to Levi-Civita [41]. The solution curves are then no longer h-geodesics but modified by a kind of Coriolis force. For a detailed discussion, including several examples, the reader is refered to Perlick [56].

3.5 Morse Index Theory for Fermat's Principle

Fermat's principle admits several interesting applications to gravitational lensing. E. g., Schneider [69] has shown that Fermat's principle can be used in the derivation of the so-called lens equation of the quasi-Newtonian approximation formalism, see also Schneider, Ehlers and Falco [70]. Again in the quasi-Newtonian approximation, Blandford and Narayan [5] have used Fermat's principle to give a topological classification of images. Owing to the approximation assumptions, in these situations it suffices to consider Fermat's principle on conformally static spacetimes. An application to gravitational lensing of Theorem 1 where the conformally static or conformally stationary version would not do was worked out by Kovner [36]. He considered a gravitational wave sweeping over a gravitational lensing situation and calculated, to within certain approximations, the influence of the wave on the arrival times and on the positions of the images at the observer's sky. This line of thought was further developed by Faraoni [14] who assumed the spacetime to be a first order (but non-stationary) perturbation of Minkowski space and used a coordinate version of Fermat's principle, given in Perlick

[55], to calculate integral formulae for the arrival time and for the deflection angle.

Here we want to discuss a different application of Fermat's principle to gravitational lensing. The basic idea is to formulate a Morse theory for Fermat's principle, in analogy to the classical Morse theory of Riemannian geometry, and to investigate the significance of the Morse relations in view of gravitational lensing. As a first step towards this goal, we establish a Morse index theorem for Fermat's principle, thereby investigating whether a solution curve yields a local minimum, a local maximum or a saddle of the arrival time functional. A full Morse theory has to presuppose a globally hyperbolic spacetime and will be the subject of Sect. 4.2 below.

As a preparation, it is certainly useful to recall the classical Morse index theory of Riemannian geometry which was developed by Morse [48] in the 1930s. Let p and q be two points in a Riemannian manifold $(\mathcal{N}, h)$, i. e., in a manifold with a positive definite metric. Then, among all sufficiently regular curves $\alpha : [0,1] \longrightarrow \mathcal{N}$ with $\alpha(0) = p$ and $\alpha(1) = q$, the geodesics are characterized as the stationary points of the *energy functional* $E(\alpha) = \int_0^1 h\big(\alpha'(s), \alpha'(s)\big)\, ds$. To find out whether a geodesic α is a local minimum, a local maximum or a saddle of E one has to calculate the Hessian $\mathrm{Hess}_\alpha(E)$ of E at the point α (i. e., the "second variation") and to determine the index and the extended index of $\mathrm{Hess}_\alpha(E)$. Please recall that the *index* (or the *extended index*, respectively) of a bilinear form is the maximal dimension of a subspace on which this bilinear form is negative definite (or negative semi-definite, respectively). The classical Morse index theorem says that $\mathrm{Hess}_\alpha(E)$ is non-degenerate if and only if the endpoint $\alpha(1)$ is not conjugate to the initial point $\alpha(0)$ along the geodesic α and that the extended index of $\mathrm{Hess}_\alpha(E)$ is equal to the number of points $\alpha(s)$, $s \in\,]0,1]$, that are conjugate to $\alpha(0)$ along α. Here each conjugate point is to be counted with its multiplicity. Based on the Morse index theorem, Morse was able to establish a number of theorems, now summarized under the name of Morse theory, to the effect that the number of geodesics joining two points p and q in a *complete* Riemannian manifold is related to the topology of the space of sufficiently regular curves joining these two points. Morse proved these results by considering the energy functional on the finite-dimensional space of broken geodesics with N breakpoints between p_1 and p_2, and then letting $N \to \infty$. For a detailed review of Morse's work we refer to Milnor [46]. Later Palais and Smale [52] [53] brought forward a fresh approach to Morse theory by considering functionals on infinite-dimensional Hilbert manifolds. It was then no longer necessary to approximate the space of trial curves by N-dimensional spaces and to consider the limit $N \to \infty$ afterwards. It is this Palais-Smale version of Morse theory we want to apply to Fermat's principle.

It should be mentioned that a Morse index theory for the geodesic variational problem (i. e., extremizing the energy functional between two points) exists not only for geodesics in Riemannian manifolds but also for timelike and lightlike geodesics in Lorentzian manifolds, see Beem, Ehrlich and Easley [4] for a detailed exposition. However, this is not what we are interested in. We want to characterize lightlike geodesics between a point and a timelike curve (not between two points), and the functional we are going to extremize is the arrival time (not the Lorentzian analogue of the energy functional).

The following exposition closely follows Perlick [57]. It is our goal to establish a Morse index theorem for Fermat's principle in an infinite-dimensional Hilbert manifold setting à la Palais-Smale. To that end we have to modify Fermat's principle, as it was given in Theorem 1, a little bit. First, we observe that, according to Definition 4, all trial curves $\lambda \in C^{\infty}_{p,\gamma}$ are of class C^{∞}. It is well known that C^{∞} maps from one manifold into another do not form a Hilbert manifold but, at the very best, a Fréchet manifold. Since this is too weak for applying Morse theory, we shall replace the C^{∞} condition on the trial curves by a Sobolev H^r condition in order to get a Hilbert manifold. Second, we observe that the arrival time functional τ is invariant under reparametrization. As a consequence, its Hessian at a solution curve λ is always degenerate because it vanishes on the infinite dimensional vector space of trivial variational vector fields (please recall Definition 5). We shall solve this problem by imposing a parametrization fixing condition upon the trial curves.

To work this out we have to introduce the Hilbert manifold of Sobolev H^r curves. For background material on this subject the reader is refered to Schwartz [71]. The same book also contains a review of the Palais-Smale version of Morse theory. For $f_1, f_2 \in C^{\infty}\big([0,1], \mathbb{R}^n\big)$, we define

$$< f_1 \,|\, f_2 >_r = \sum_{i=0}^{r} \int_0^1 f_1^{(i)}(s) \cdot f_2^{(i)}(s)\, ds \tag{19}$$

where $f_1^{(i)}$ denotes the i-th derivative of f_1 and the dot denotes the standard scalar product in $\mathbb{R}^n$. It is easy to check that this scalar product makes $C^{\infty}\big([0,1], \mathbb{R}^n\big)$ into a real pre-Hilbert space. The completion of this pre-Hilbert space is, by definition, the *Sobolev space* $H^r\big([0,1], \mathbb{R}^n\big)$. For $r = 0$ this gives the real Lebesgue space $L^2\big([0,1], \mathbb{R}^n\big)$ whose complex version is known to every physicist from quantum mechanics. For integers $r \geq 1$, $H^r\big([0,1], \mathbb{R}^n\big)$ can be (and will be henceforth) identified with the space of all C^{r-1} maps from $[0,1]$ into $\mathbb{R}^n$ whose r-th derivatives exist almost everywhere and are locally square integrable.

Now we introduce the notion of H^r curves in a manifold. Let $\mathcal{M}$ be a real, finite-dimensional C^{∞} manifold whose topology satisfies the Hausdorff axiom and the second countability axiom. Then we define

$$H^r\big([0,1], \mathcal{M}\big) = \Big\{ \lambda : [0,1] \longrightarrow \mathcal{M} \,\Big|\, j \circ \lambda \in H^r\big([0,1], \mathbb{R}^n\big) \Big\} \tag{20}$$

where $j : \mathcal{M} \longrightarrow \mathbb{R}^n$ is a C^{∞} embedding. A well-known theorem of Whitney [86] (see also, e. g., Hirsch [31], p. 55) guarantees the existence of such an embedding for $n \geq 2\dim(\mathcal{M})+1$. It is easy to show that the set $H^r([0,1],\mathcal{M})$ is independent of which j has been chosen. Moreover, it is a well-known result of Palais and Smale [53] that the inclusion map $H^r([0,1],\mathcal{M}) \longrightarrow H^r([0,1],\mathbb{R}^n)$ induced by j makes $H^r([0,1],\mathcal{M})$ into a C^∞ submanifold of the Hilbert space $H^r([0,1],\mathbb{R}^n)$ and that the manifold structure thereby established on $H^r([0,1],\mathcal{M})$ is, again, independent of j. Thus, we may view $H^r([0,1],\mathcal{M})$ as an infinite dimensional real C^∞ Hilbert manifold in its own right.

To define the modified space of trial curves we restrict the original space $C^\infty_{p,\gamma}$ of Definition 4 by a parametrization fixing condition. For $\lambda \in C^\infty_{p,\gamma}$ we define a vector field $U_\lambda : [0,1] \longrightarrow T\mathcal{M}$ by parallel transporting the vector $\gamma'(\tau(\lambda))$ along λ, as in the proof of Theorem 1. Then the condition $g(U_\lambda,\lambda') = \text{const.}$ singles out exactly one parametrization for each trial curve. Please note that this condition singles out an affine parametrization along each geodesic. Now we define the modified space of trial curves in the following way.

Definition 6. Fix a point p and a timelike embedded C^∞ curve $\gamma : I \longrightarrow \mathcal{M}$. Then the space $H^2_{p,\gamma}$ is, by definition, the set of all $\lambda \in H^2([0,1],\mathcal{M})$ with the following properties.
(a) $g(\lambda',\lambda') = 0$.
(b) $\lambda(0) = p$ and there is a $\tau(\lambda) \in I$ such that $\lambda(1) = \gamma(\tau(\lambda))$.
(c) $g(U_\lambda,\lambda') = \text{const.} < 0$.

It can be shown that $H^2_{p,\gamma}$ is, indeed, an infinite dimensional C^∞ Hilbert submanifold of $H^2([0,1],\mathcal{M})$ and that the arrival time functional $\tau : H^2_{p,\gamma} \longrightarrow \mathbb{R}$ defined by (b) is a C^∞ map. For a proof of these facts we refer to Perlick [57]. This result remains true if the H^2 condition in Definition 6 is replaced with an H^r condition for $r > 2$; it is not true, however, for $r = 1$. Now Fermat's principle, i. e., Theorem 1, can be reformulated in the following way.

Theorem 2. *A curve $\lambda \in H^2_{p,\gamma}$ is a geodesic if and only if the differential of the arrival time functional $\tau : H^2_{p,\gamma} \longrightarrow \mathbb{R}$ has a zero at λ.*

The proof, which is worked out in Perlick [57], is a straightforward translation of the proof of Theorem 1 into an H^2 setting.

For the Morse index theorem we have to calculate the Hessian $\text{Hess}_\lambda(\tau)$ of τ at a geodesic λ. We find the following result.

Theorem 3. (Morse index theorem) *Let $\lambda \in H^2_{p,\gamma}$ be a geodesic. The index of* $\text{Hess}_\lambda(\tau)$ *is equal to the number of points $\lambda(s)$, $s \in \,]0,1[$, that are conjugate to $\lambda(0)$ along λ. The extended index of* $\text{Hess}_\lambda(\tau)$ *is equal to the number of points $\lambda(s)$, $s \in \,]0,1]$, that are conjugate to $\lambda(0)$ along λ. In both cases conjugate points are to be counted with their multiplicities.*

The proof of this theorem is given in Perlick [57]. The strategy of this proof is to relate the second variational formula for Fermat's principle to the second variational formula for the geodesic variational problem. For lightlike geodesics, the latter is worked out in Beem, Ehrlich and Easley [4], Chap. 10. We have already emphasized the differences between these two variational problems. Nonetheless, their second variational formulae turn out to be essentially the same. Thereby, the proof of Theorem 3 comes as a corollary of the Morse index theorem proven in Beem, Ehrlich and Easley.

Theorem 3 has the following immediate consequences.

(a) $\mathrm{Hess}_\lambda(\tau)$ is non-degenerate if and only if $\lambda(1)$ is not conjugate to $\lambda(0)$ along λ.

(b) The index and the extended index of $\mathrm{Hess}_\lambda(\tau)$ are finite for all geodesics $\lambda \in H^2_{p,\gamma}$. Hence, λ cannot be a local maximum of τ.

(c) A geodesic $\lambda \in H^2_{p,\gamma}$ is a strict local minimum of τ if and only if λ does not contain a point conjugate to $\lambda(0)$

(d) A geodesic $\lambda \in H^2_{p,\gamma}$ is a saddle of τ if it contains a point $\lambda(s)$ which is conjugate to $\lambda(0)$ for some $s \in]0,1[$.

In view of gravitational lensing, the index has the following interpretation. At each conjugate point, infinitesimally neighboring light rays "cross over" from one side of λ to the other. This is associated with a side-reversion of the image. Thus, light rays with an even index yield a mirror image of those with an odd image. This is observable if our pointlike light source (e. g., the core of a galaxy) is surrounded by some non-symmetrical structure (e. g., irregular lobes or jets).

In Sect. 4.2 below we use the Morse index theorem to develop a full Morse theory for light rays joining a point and a timelike curve in a globally hyperbolic spacetime and we discuss applications to gravitational lensing.

4 Gravitational Lensing in Globally Hyperbolic Spacetimes

We have seen in the preceding section that the geometry of multiple imaging situations is strongly influenced by the topological and causal structure of spacetime. Such global effects are usually ignored in the astronomical literature on gravitational lensing where, typically more implicitly than explicitly, the deflector is assumed to be embedded in a universe without topological or causal pathologies.

In this section we get somewhat closer to the standard astronomer's point of view by restricting to spacetimes without causal pathologies. More precisely, we are going to consider spacetimes that are globally hyperbolic according to the following definition.

Definition 7. For a spacetime $(\mathcal{M}, g, \mathcal{T}^+)$, a subset $\mathcal{S}$ of $\mathcal{M}$ is called a *Cauchy surface* if each inextendible causal curve in $\mathcal{M}$ intersects $\mathcal{S}$ in exactly one point. A spacetime is called *globally hyperbolic* if it admits a Cauchy surface.

The name "globally hyperbolic" was introduced by Leray [39] in 1952. It refers to the fact that a global existence and uniqueness theorem for hyperbolic partial differential equations can be established only on spacetimes with this property. Actually, our Definition 7 of global hyperbolicity does not coincide with Leray's original definition, but it is well known that the two definitions are equivalent, see, e. g., Wald [80], p. 209. Basic properties of globally hyperbolic spacetimes are also reviewed in Hawking and Ellis [29], in O'Neill [50] and in Beem, Ehrlich and Easley [4].

One can show that a Cauchy surface $\mathcal{S}$ is a 3-dimensional topological submanifold of $\mathcal{M}$, see, e. g., Theorem 8.1.3 and Theorem 8.3.1 in Wald [80]. Mimicking the proof of Proposition 6.3.1 in Hawking and Ellis [29], one may even show that $\mathcal{S}$ is a C^{1-} (i. e., Lipschitz) submanifold of $\mathcal{M}$. In general, a Cauchy surface will not be a C^1 submanifold of $\mathcal{M}$.

In Sect. 3.3 we have introduced, for each timelike C^∞ vector field V on $\mathcal{M}$, the quotient space $\mathcal{S}_V = \mathcal{M}/_\sim$, where two points in $\mathcal{M}$ are considered equivalent if they can be connected by an integral curve of V. We shall now use Proposition 8 to show that in a globally hyperbolic spacetime $\mathcal{S}_V$ comes not only with a topological but even with a differentiable structure. To that end, we first observe that the existence of a Cauchy surface makes sure that there are no closed timelike curves in $\mathcal{M}$. Since every integral curve of V intersects a Cauchy surface $\mathcal{S}$ exactly once, the restriction of the natural projection

$$\pi_V : \mathcal{M} \longrightarrow \mathcal{S}_V \tag{21}$$

to $\mathcal{S}$ gives a homeomorphism from $\mathcal{S}$ onto $\mathcal{S}_V$. Since $\mathcal{S}$, being a topological submanifold of a Hausdorff space, must satisfy the Hausdorff axiom, this proves that $\mathcal{S}_V$ satisfies the Hausdorff axiom. By Proposition 8, there is a C^∞ structure on $\mathcal{S}_V$ such that $\pi_V : \mathcal{M} \longrightarrow \mathcal{S}_V$ makes $\mathcal{M}$ into a fiber bundle over $\mathcal{S}_V$ with fiber diffeomorphic to $\mathbb{R}$. This argument implies that any two Cauchy surfaces in $\mathcal{M}$ must be homeomorphic and that, for any two timelike C^∞ vector fields V and V' on $\mathcal{M}$ the quotient manifolds $\mathcal{S}_V$ and $\mathcal{S}_{V'}$ must be homeomorphic. According to a famous theorem of Moise [47] any 3-dimensional topological manifold admits exactly one differentiable structure. Hence, $\mathcal{S}_V$ and $\mathcal{S}_{V'}$ must even be diffeomorphic.

Geroch [20] has established the important fact that every globally hyperbolic spacetime admits a continuous function $t : \mathcal{M} \longrightarrow \mathbb{R}$ such that the set $t^{-1}(t_o)$ is a Cauchy surface for each $t_o \in \mathbb{R}$. It is widely believed that such a *Cauchy time function* t can be chosen differentiable, employing a smoothing argument of Seifert [72]. However, the details of the proof have never been worked out completely, and several dedicated experts who tried to do so failed. In any case, every globally hyperbolic spacetime admits a continuous Cauchy time function t which can be combined with the C^∞ projection π_V determined by a timelike C^∞ vector field V to give a homeomorphism $(\pi, t) : \mathcal{M} \longrightarrow \mathcal{S}_V \times \mathbb{R}$. Hence, the topology of a globally hyperbolic spacetime is determined by the topology of any of its Cauchy surfaces. Note, however, that $\mathcal{S}_1 \times \mathbb{R}$ may be homeomorphic to $\mathcal{S}_2 \times \mathbb{R}$ without $\mathcal{S}_1$ being homeomorphic to $\mathcal{S}_2$. Therefore, the topology of a globally hyperbolic spacetime does *not* determine the topology of its Cauchy surfaces. E. g., Newman and Clarke [49] contrived a spacetime with topology $\mathbb{R}^4$ that admits a Cauchy surface which is not homeomorphic to $\mathbb{R}^3$.

It is a matter of debate whether the assumption of global hyperbolicity is to be considered as "reasonable" from a physical point of view. It is certainly true that most physicists consider the validity of a global existence and uniqueness theorem for wave equations as a requirement any "reasonable" spacetime should satisfy. Moreover, only globally hyperbolic spacetimes arise from globally solving the initial value problem of Einstein's (vacuum) field equation. From the viewpoint of global Lorentzian geometry, however, global hyperbolicity is a very strong assumption. In particular, any globally hyperbolic spacetime has to satisfy, at each point p, the strong causality condition and, thus, the distinguishing conditions and the causality condition, recall Definition 3. Also, it is easy to verify that removing a point, a worldline or a worldtube from any spacetime necessarily results in a spacetime that is *not* globally hyperbolic. This remark is important for gravitational lensing where non-transparent deflectors are modeled by excising worldlines or worldtubes from spacetime. – We summarize these observations in the following way.

> In view of gravitational lensing situations, restricting to globally hyperbolic spacetimes means restricting to the case of a transparent deflector in a spacetime without causality violation whose topology is a product of space and time.

Since a Cauchy surface may have a complicated topology, the restriction to globally hyperbolic spacetimes does not exclude universes with "handles" etc.

4.1 Criteria for Multiple Imaging in Globally Hyperbolic Spacetimes

In Sect. 3.3 we have formulated several criteria for multiple imaging in arbitrary spacetimes. These results can be considerably strengthened if we restrict to globally hyperbolic spacetimes. The reason is that light cones in globally hyperbolic spacetimes cannot be too pathological. In particular, the following technically important proposition holds true.

Proposition 13. *For a point p in a globally hyperbolic spacetime $(\mathcal{M}, g, \mathcal{T}^+)$ the past light cone of p united with $\{p\}$ gives a closed subset of $\mathcal{M}$.*

Proof. Let q_n be a sequence in the past light cone of p that converges towards a point $q \neq p$ in $\mathcal{M}$. We want to show that q is, again, in the past light cone of p. To that end we choose a Cauchy surface $\mathcal{S}$ through q. This is possible since, by the above-mentioned result of Geroch, every globally hyperbolic spacetime can be foliated into (continuously embedded) Cauchy surfaces. Convergence of the q_n implies that their pre-image under e_p^-, using the notation of (10), is contained in a compact subset of $T_p\mathcal{M}$. So, by passing to a subsequence we find a sequence X_n in $\mathcal{C}_p^-$ with $e_p^-(X_n) = q_n$ that converges towards a vector X in the closure of $\mathcal{C}_p^-$. Since $q \neq p$, X is different from zero and, thus, lightlike. We do not know yet if X is in the domain of the exponential map. Let λ_n denote the geodesic with $\lambda_n'(0) = X_n$ and λ the geodesic with $\lambda'(0) = X$. Since $\mathcal{S}$ is a Cauchy surface, λ_n must intersect $\mathcal{S}$ in a point $\tilde{q}_n$ and λ must intersect $\mathcal{S}$ in a point $\tilde{q}$. Since geodesics depend continuously on their initial conditions, the $\tilde{q}_n$ converge towards $\tilde{q}$. On the other hand, the q_n converge towards q. As each geodesic λ_n intersects $\mathcal{S}$ exactly once, this is possible only if $q = \tilde{q}$ which implies that q is in the past light cone of p. □

In addition, it can be shown that in the globally hyperbolic case the past lightlike conjugate locus and the past lightlike cut locus of p are closed subsets of $\mathcal{M}$, see Beem, Ehrlich and Easley [4], Propositions 9.27 and 9.29 in combination with our Proposition 13. We have already seen in Example 3 at the end of Sect. 3.2 that this is not true without the assumption of global hyperbolicity. – The following proposition says that for globally hyperbolic spacetimes the name "cut point" is, indeed, justified because such a point indicates an intersection of geodesics. It was already mentioned that the analogous statement for complete Riemannian manifolds is known as *Poincaré Theorem* and dates back to Poincaré [66] and Whitehead [85].

Proposition 14. (Poincaré Theorem for lightlike geodesics) *Let p and q be two points in a globally hyperbolic spacetime $(\mathcal{M}, g, \mathcal{T}^+)$. Assume that q is in the past lightlike cut locus but not in the past lightlike conjugate locus of p. Then there are at least two past-pointing lightlike geodesics from p to q. The past light cone of p has a transverse self-intersection at q.*

A proof can be found in Beem, Ehrlich and Easley [4], Theorem 9.15. The light cone must have a transverse self-intersection at q since otherwise the two lightlike geodesics would arrive with collinear tangent vectors at q. As the assumption of global hyperbolicity excludes the possibility of having a closed lightlike geodesic through p, this collinearity would imply that the two geodesics are the same.

If q is in the cut locus and in the conjugate locus, then it may be impossible to reach it from p along a second geodesic, even in a globally hyperbolic spacetime. This is exemplified by the point q in Fig. 5. However, since a conjugate point indicates an intersection with an "infinitesimally neighboring geodesic" the name "cut point" might be viewed as justified in this case as well.

Proposition 14 has the consequence that in globally hyperbolic spacetimes Proposition 5 admits the following converse.

Proposition 15. *Fix a point p in a globally hyperbolic spacetime $(\mathcal{M}, g, \mathcal{T}^+)$ and assume that the map e_p^- of* (10) *is a C^∞ embedding, i. e., that the past* *light cone of p is an embedded submanifold of $\mathcal{M}$. Then the past lightlike cut locus of p is empty.*

Proof. By contradiction, assume that q is the past cut point of p along some lightlike geodesic. If q is also conjugate to p we are done since this proves that e_p^- is not even an immersion. If q is not conjugate to p, Proposition 14 implies that the past light cone of p has a self-intersection at q, so e_p^- cannot be an embedding. □

Together with Propositions 5, 6 and 7 this result implies that in a globally hyperbolic spacetime there is a multiple imaging situation for an observer at p if and only if the past light cone of p fails to be an embedded submanifold of $\mathcal{M}$.

Propositions 14 and 15 are not true without the assumption of global hyperbolicity. To see this consider the light cone of Fig. 4. Divide a spherical section of this cone which is close to p into two hemispheres and excise one of them, together with its boundary, from spacetime. Then, if the division has been chosen appropriately, the light cone becomes an embedded submanifold since one half of the light rays is cut off before the light cone forms a self-intersection. However, the cut locus remains unchanged since the remaining light rays can be reached by the same timelike curves from p as before.

Another important feature of globally hyperbolic spacetimes is the following existence result for lightlike geodesics.

Proposition 16. *Let p be a point and γ an inextendible timelike curve in a globally hyperbolic spacetime $(\mathcal{M}, g, \mathcal{T}^+)$ such that $\gamma \cap \mathcal{I}^-(p) \neq \emptyset$. Then there is a past-pointing lightlike geodesic λ from p to γ that is completely contained in the boundary of $\mathcal{I}^-(p)$. This geodesic does not pass through the past cut point of p or through a point conjugate to p before it reaches γ.*

Proof. Global hyperbolicity implies that γ cannot be completely contained in $\mathcal{I}^-(p)$ because it must reach every Cauchy surface in the future of p. Therefore our assumptions imply that γ intersects $\partial\mathcal{I}^-(p)$ in some point q. It is well known (see, e. g., Wald [80], Theorem 8.1.6) that every point in $\partial\mathcal{I}^-(p)$ can be reached from p along a past-pointing lightlike geodesic. This geodesic cannot pass through the past cut point of p before it reaches q since after passing through the cut point a lightlike geodesic stays inside the open set $\mathcal{I}^-(p)$. By Proposition 2 this implies that the geodesic cannot pass through a point conjugate to p before reaching q. □

This proposition gives, in particular, sufficient conditions for the existence of a past-pointing lightlike geodesic from p to γ that does not contain a point conjugate to p. A similar result was proven by Uhlenbeck [78], Corollary 4.8, with the help of Morse theory. In Sect. 4.2 we shall comment on her work in more detail.

As a corollary of Proposition 16 we immediately get the following sufficiency criterion for multiple imaging. As an illustration the reader may use Fig. 5 with an appropriately placed worldline γ.

Proposition 17. *Let p be a point and $\gamma : I \longrightarrow \mathcal{M}$ an inextendible timelike curve in a globally hyperbolic spacetime $(\mathcal{M}, g, \mathcal{T}^+)$. Assume that there is a past-pointing lightlike geodesic λ from p to γ that passes through a point conjugate to p or through the past cut point of p (or both) before it reaches γ. Then there are at least two past-pointing lightlike geodesics from p to γ.*

Proof. The existence of the lightlike geodesic λ implies that γ intersects the closure of $\mathcal{I}^-(p)$. Since γ is inextendible, this means that it must intersect $\mathcal{I}^-(p)$. But then Proposition 16 gives a past-pointing lightlike geodesic from p to γ which is different from λ because the latter contains a conjugate point or the past cut point of p. □

This proposition says that, under certain assumptions, the existence of conjugate points or cut points along a lightlike geodesic is sufficient for multiple imaging. Such a sufficiency criterion was already proven in Proposition 7 for arbitrary spacetimes. The new feature of Proposition 17 is that in a globally hyperbolic spacetime the worldline γ can be freely prescribed (except for the condition of being inextendible, i. e., "sufficiently long"). In view of applications to gravitational lensing, this is a great advantage since a real light source such as a galaxy or a quasar cannot be expected to travel along a worldline that is constructed as in the proof of Proposition 7.

4.2 Morse Theory in Globally Hyperbolic Spacetimes

In Sect. 3.5 we have established a Morse index theory for Fermat's principle. Now we want to discuss the possibility of developing a full-fledged Morse theory, relating the number of solution curves to the topology of the space of trial curves. Whereas the Morse index theory works perfectly well on an arbitrary spacetime, the full Morse theory requires a globally hyperbolic spacetime. This is in analogy to the case of Riemannian geodesics where the Morse index theory works on arbitrary Riemannian manifolds but the full Morse theory has to presuppose a complete Riemannian manifold.

For the geodesic problem on complete Riemannian manifolds, Morse theory exists in two versions. The first version, invented by Morse [48] in the 1930s and nicely reviewed by Milnor [46], considers for the space of trial curves the finite dimensional manifold of broken geodesics between two points with N breakpoints. For the final results one has to consider the limit $N \to \infty$. The second version, brought forward by Palais and Smale [52] [53] in the 1960s, considers for the space of trial curves the infinite dimensional Hilbert manifold of H^1 curves between two points. (For the notion of H^r curves please recall Sect. 3.5.) Both versions have been carried over to general relativity. For Lorentzian manifolds, the most natural analogue of the geodesic problem in complete Riemannian manifolds is the timelike geodesic problem in globally hyperbolic spacetimes. A Morse theory for this situation was developed independently by Uhlenbeck [78] and Woodhouse [87] who both used finite dimensional approximation techniques in the spirit of Morse and Milnor. An attempt to find an infinite dimensional Hilbert manifold version, in the spirit of Palais and Smale, was brought forward by Everson and Talbot [13] but, unfortunately, turned out to be fatally flawed, see Erratum to [13]. In any case, this timelike geodesic problem has no relevance to gravitational lensing where we are interested in lightlike geodesics between a point and a timelike curve, not in timelike geodesics between two points.

For our purpose, the relevant variational problem is Fermat's principle. A Morse theory based on a version of Fermat's principle in globally hyperbolic spacetimes was invented by Uhlenbeck [78] who, in analogy to her treatment of the timelike geodesic problem in the same paper, used finite dimensional approximation techniques. Her results were used by McKenzie [45] to formulate conditions under which in a gravitational lensing situation the number of images must be odd. As to an infinite dimensional version of Morse theory for Fermat's principle, the most natural starting point seems to be the formalism established in Sect. 3.5. Unfortunately, the parametrization fixing condition on the trial curves, i. e., condition (c) of Definition 6, together with the fact that the trial curves are of type H^2 rather than of type H^1, leads to many technical problems. For that reason Giannoni, Masiello and Piccione [22] [23] used a slightly different Hilbert manifold setting as the starting point. This approach led, indeed, to a full Morse theory for lightlike geodesics between a point and a timelike curve in a globally hyperbolic spacetime. In the rest of this subsection we shall review their main result and discuss some of its implications. Since the mathematical details are highly technical we have to refer to the original articles [22] and [23] for the proofs.

It was already mentioned that the general setting for treating variational problems in terms of infinite dimensional Hilbert manifolds is due to Palais and Smale [52] [53]. In this setting one considers differentiable functions $F : \mathcal{X} \longrightarrow \mathbb{R}$ on a real Hilbert manifold $\mathcal{X}$. In applications to variational problems, $\mathcal{X}$ is the space of trial curves (or, more generally, trial maps) which is typically infinite dimensional, F is the functional to be extremized, and the critical points of F (i. e., the points where the differential of F has a zero) are the solutions of the variational problem. It is the goal to relate the number of critical points of F to the topology of $\mathcal{X}$. More precisely, one wants to relate the number N_k of critical points where the Hessian of F has index k to the k-th *Betti number* B_k of $\mathcal{X}$. Formally, B_k is defined for each topological space $\mathcal{X}$ in terms of the k-th singular homology space $H_k(\mathcal{X})$ with coefficients in a field $\mathbb{F}$ (The results of Morse theory are true for any choice of $\mathbb{F}$). For the definition of singular homology spaces the reader is refered, e. g., to Dold [11], p. 32, or to Spanier [75], p. 173. $H_k(\mathcal{X})$ is a vector space over $\mathbb{F}$ and B_k is, by definition, the dimension of this vector space. Geometrically, B_o is the number of connected components of $\mathcal{X}$ and, for $k \geq 1$, B_k can be interpreted as the number of those "holes" in $\mathcal{X}$ that prevent a k-cycle with coefficients in $\mathbb{F}$ from being a boundary. In particular, if $\mathcal{M}$ is contractible to a point, then

$B_k = 0$ for all $k \geq 1$. Palais and Smale were able to establish the following result, see Corollary (3), p. 338, in Palais [52].

Assume that $F : \mathcal{X} \longrightarrow \mathbb{R}$ is of class C^3 at least and satisfies the following conditions.

(1) F is a *Morse function*, i. e., at each critical point of F the Hessian of F is non-degenerate.

(2) F is bounded from below.

(3) F satisfies the socalled *Condition* C, also known as *Palais-Smale condition*: There is a complete Riemannian metric h on $\mathcal{X}$ such that the following holds. If $\mathcal{S}$ is any subset of $\mathcal{X}$ on which F is bounded and $||dF||$ is not bounded away from zero, then there is a critical point of F adherent to $\mathcal{S}$. Here $||\cdot||$ denotes the norm induced by the metric h.

Then the Morse inequalities

$$N_k \geq B_k \,, \quad k \geq 0 \tag{22}$$

and the relation

$$\sum_{k=0}^{\infty}(-1)^k N_k = \sum_{k=0}^{\infty}(-1)^k B_k \tag{23}$$

hold true. The right-hand side of (23) is, by definition, the *Euler characteristic* χ of $\mathcal{X}$. If one introduces the notation $N_+ = \sum_{i=0}^{\infty} N_{2i}$ and $N_- = \sum_{i=0}^{\infty} N_{2i+1}$, then (23) takes the form

$$N_+ - N_- = \chi \,. \tag{24}$$

Please note that the N_k and B_k need not be finite.

The geometric idea behind this result is the following. It turns out that the topology of the sublevel set $\mathcal{X}_t = \{x \in \mathcal{X} \,|\, F(x) \leq t\}$ remains unchanged if t varies over an interval which does not contain a critical value of F (i. e., a value taken by F at some critical point). On intervals containing a critical value, the topology of the sublevel set changes by "attaching a handle" for each critical point, with the special type of the handle determined by the index of the Hessian of F at the critical point. This result was first proven by Morse for functions on compact (and thus finite dimensional) manifolds, see Milnor [46]. In that case Condition C is automatically satisfied. As a matter of fact, Condition C was introduced as a sufficient condition for proving the same "handle-body theorem" without the compactness assumption.

If we want to apply this general result to our variational problem, we have to check if the assumptions on F are satisfied by the arrival time functional $\tau : H^2_{p,\gamma} \longrightarrow \mathbb{R}$ discussed in Sect. 3.5. The Morse index theorem tells us that τ is a Morse function if and only if γ does not intersect the caustic of the past light cone of p. By Proposition 10, this is the case for almost all p once

γ has been chosen. Moreover, the Morse index theorem tells us that N_k is the number of past-ponting lightlike geodesics from p to γ that pass through k conjugate points before arriving at γ, counting each conjugate point with multiplicity. The second condition of F being bounded from below is easily checked to be true on globally hyperbolic spacetimes. The main problem comes with the third condition, i. e., with Condition C. Giannoni, Masiello and Piccione [22] [23] found it necessary to modify the whole setting a little bit before they were able to verify Condition C. First they considered H^1 trial curves, rather than H^2 trial curves, i. e., curves which are differentiable almost everywhere and whose derivative is locally square integrable. Unfortunately, the equation $g(\lambda', \lambda') = 0$ (almost everywhere) does not define a submanifold of $H^1\big([0,1], \mathcal{M}\big)$, contrary to the H^2 case. Therefore Giannoni, Masiello and Piccione replaced this with the equation $g(\lambda', \lambda') = -\varepsilon^2$ which, for fixed $\varepsilon > 0$, defines a submanifold and considered the limit $\varepsilon \to 0$ afterwards. Second, they dropped the parametrization condition (c) of Definition 6. This has the effect that every critical point of the arrival time functional now comes together with all its (H^1) reparametrizations, i. e., τ cannot be a Morse function on this modified space of trial curves. Therefore Giannoni, Masiello and Piccione switched to a new functional Q that is related to the arrival time functional in a similar fashion as the energy functional $E(\alpha) = \int_0^1 h\big(\alpha'(s), \alpha'(s)\big)\, ds$ to the length functional $\ell(\alpha) = \int_0^1 \sqrt{h\big(\alpha'(s), \alpha'(s)\big)}\, ds$ in Riemannian geometry. In this modified setting Giannoni, Masiello and Piccione were, indeed, able to verify Condition C, thereby establishing a full Morse theory for the variational problem at hand. Their main result can be phrased in the following way.

Theorem 4. (Morse relations for lightlike geodesics) *Let $(\mathcal{M}, g, \mathcal{T}^+)$ be a globally hyperbolic spacetime, fix a point $p \in \mathcal{M}$ and a past-pointing timelike C^∞ curve $\gamma : I \longrightarrow \mathcal{M}$ from an open interval I into $\mathcal{M}$ such that $p \notin \gamma$. Assume that γ is closed in $\mathcal{M}$ and does not intersect the caustic of the past light cone of p. Let $H^1_{p,\gamma}$ denote the topological subspace of $H^1\big([0,1], \mathcal{M}\big)$ consisting of all $\lambda \in H^1\big([0,1], \mathcal{M}\big)$ with $\lambda(0) = p$, $\lambda(1) \in \gamma$, $g(\lambda', \lambda') = 0$ and λ' past-pointing almost everywhere. Let B_k be the k-th Betti number of $H^1_{p,\gamma}$ and N_k the number of past-pointing lightlike geodesics from p to γ that pass through k conjugate points before reaching γ, counting each conjugate point with multiplicity. Then the Morse relations* (22) *and* (23) *hold true.*

This result is implied by Theorem 1.7 of Giannoni, Masiello and Piccione [23]. Actually, they prove a slightly more general result since they consider curves confined to a subset Λ of $\mathcal{M}$ with certain properties. Our Theorem 4 is the version for $\Lambda = \mathcal{M}$ in which case the assumptions placed by Giannoni, Masiello and Piccione upon the functional are satisfied, for each pair (p, γ), if and only if the spacetime is globally hyperbolic.

The Morse relations have several interesting implications. Information on the N_k, i. e., on the number of images in gravitational lensing situations,

place restrictions upon the Betti numbers and the other way round. If we want to make full use of such results we need, of course, some methods of determining the topology of the curve space $H^1_{p,\gamma}$ which is a difficult task in general. Before commenting on this problem we list some consequences of the Morse relations. Under the assumptions of Theorem 4, the following is true.
(a) The Morse inequality for $k = 0$, i. e. $N_o \geq B_o$, implies that there are at least B_o past-pointing lightlike geodesics from p to γ which are free of conjugate points, where B_o is the number of connected components of $H^1_{p,\gamma}$. This strengthens Proposition 16.
(b) If $H^1_{p,\gamma}$ is non-empty (i. e., $\gamma \cap \mathcal{I}^-(p) \neq \emptyset$) and not contractible, we have $B_o \geq 1$ and $B_k \geq 1$ for some $k \geq 1$. Then there is a multiple imaging situation, $N_o + N_k \geq 2$, and at least one past-pointing lightlike geodesic from p to γ must contain a conjugate point.
(c) If the number of past-ponting lightlike geodesics from p to γ is finite, then all the Betti numbers B_k must be finite.
(d) If we write the Morse relation (23) in the form of (24), we find $N_+ + N_- = 2N_- + \chi$. Thus, in a gravitational lensing situation with finitely many images the total number of images is odd if and only if the Euler characteristic χ is odd.

It is an interesting problem to determine all globally hyperbolic spacetimes in which gravitational lensing always leads to an odd number of images. (Please recall that the assumption of global hyperbolicity implicitly restricts to transparent deflectors.) With the help of Morse theory we have reduced this to the problem of determining the Euler characteristic of the curve space $H^1_{p,\gamma}$, for each pair (p, γ) that satisfies the assumptions of Theorem 4. In some special cases, this can be achieved in the following way, please cf. McKenzie [45] for a similar investigation.

Let us assume that the assumptions of Theorem 4 are satisfied and choose a timelike C^∞ vector field V on $\mathcal{M}$ such that V is tangent to γ. This is possible, see Proposition 5.1 in Giannoni, Masiello and Piccione [22]. Then the projection (21) defines a map $\lambda \longmapsto \hat{\lambda} = \pi_V \circ \lambda$ from $H^1_{p,\gamma}$ to the space $\hat{H}^1_{p,\gamma} = \{\hat{\lambda} \in H^1([0,1], \mathcal{S}_V) \mid \hat{\lambda}(0) = \pi_V(p)\,,\ \hat{\lambda}(1) = \pi_V(\gamma)\,\}$. This map is obviously continuous. Moreover, it is injective since a past-pointing lightlike curve is uniquely determined by its initial point and by its spatial projection. In general, however, it need not be surjective because for some curves in the target space the lightlike lift may terminate (at the "boundary" of $\mathcal{M}$) before γ has been reached. This certainly happens whenever there is a *particle horizon* in the sense that for some point $q \in \mathcal{I}^-(p)$ there is no past-pointing causal curve from q to γ. As a simple example where particle horizons occur one may consider Minkowski space restricted to the region $t > 0$. Let us say that the *lightlike lifting property* is satisfied for the pair (p, γ) if the map $\lambda \longmapsto \hat{\lambda} = \pi_V \circ \lambda$ gives a homeomorphism from $H^1_{p,\gamma}$ onto $\hat{H}^1_{p,\gamma}$. As C^∞ curves are dense in the set of H^1 curves and the lifting procedure is obviously H^1-continuous, the lightlike lifting property is satisfied if every C^∞

curve in $\hat{H}^1_{p,\gamma}$ is the projection of a curve λ in $H^1_{p,\gamma}$. An analytical condition that guarantees the lightlike lifting property is the socalled *metric growth condition* of Uhlenbeck [78] which was employed by McKenzie [45]. We can now prove the following result.

Theorem 5. (Odd number theorem) *Assume that all the assumptions of Theorem 4 are satisfied and let V be a timelike C^∞ vector field on $\mathcal{M}$ such that V is tangent to γ. Moreover, assume that the lightlike lifting property is satisfied for (p, γ) and that the space $\mathcal{S}_V$ is contractible. Then the number of past-pointing lightlike geodesics from p to γ is (infinite or) odd.*

Proof. If $\mathcal{S}_V$ is contractible, the curve space $\hat{H}^1_{p,\gamma}$ is contractible. To prove this one fixes a particular C^∞ curve $\hat{\lambda}_o \in \hat{H}^1_{p,\gamma}$ and considers a differentiable map $\phi : [0,1] \times [0,1] \times \mathcal{S}_V \longrightarrow \mathcal{S}_V$ such that $\phi(s,0,x) = x$ and $\phi(s,1,x) = \hat{\lambda}_o(s)$ for all $s \in [0,1]$ and $x \in \mathcal{S}_V$. The existence of such a map is guaranteed since $\mathcal{S}_V$ is contractible. (It is true that contractibility is defined in terms of homotopies of *continuous* maps. However, according to a well-known theorem every continuous map between two manifolds is homotopic to a C^∞ map, see, e. g., Bott and Tu [7], Proposition 17.8, p. 213.) Now the desired contraction $\Phi : [0,1] \times \hat{H}^1_{p,\gamma} \longrightarrow \hat{H}^1_{p,\gamma}$ is defined by $\Phi(t, \hat{\lambda})(s) = \phi(s,t,\hat{\lambda}(s))$. Since the lightlike lifting property is satisfied, this implies that $H^1_{p,\gamma}$ is contractible, i. e., the Morse relations hold with $B_o = 1$ and $B_k = 0$ for $k > 0$ which implies $\chi = 1$. If we write the Morse relation (23) in the form of (24), we find $N_+ + N_- = 2N_- + 1$, i. e., $N_+ + N_-$ is (infinite or) odd. □

This theorem can be phrased as saying that a transparent deflector produces an odd number of images provided that there are no particle horizons and the spatial topology is trivial. It is, of course, true that particle horizons do occur in many cosmological models which are of physical interest. So, in a sense, the lightlike lifting property may be considered as a reasonable assumption only in gravitational lensing situations where cosmological aspects can be ignored.

An obvious example where the lightlike lifting property is satisfied is the case that V is a complete and hypersurface-orthogonal conformal Killing vector field. In that case Fermat's principle reduces to the geodesic problem for a Riemannian metric h on $\mathcal{S}_V$, as outlined at the end of Sect. 3.4, and global hyperbolicity is easily checked to be equivalent to completeness of the Riemannian manifold $(\mathcal{S}_V, h)$. Hence, the Morse theory for Fermat's principle reduces to the standard Morse theory for Riemannian geodesics.

It is an open problem to determine the Euler characteristic of the curve space $H^1_{p,\gamma}$ in cases where particle horizons do occur, i. e., where the lightlike lifting property is violated. Apparently no results in this direction exist so-far. The socalled "chronological homotopy theory" employed by Woodhouse [87] might be of help in this connection. The latter is closely related to the Lorentzian fundamental groups of Smith [74] and to the notion of "future one-connectedness" of Flaherty [15] [16] which is also discussed in Beem, Ehrlich and Easley [4].

The formalism presented in this subsection has a (much simpler) analogue in the quasi-Newtonian approximation of gravitational lensing. A Morse theory for the latter situation was developed by Petters [61] [63]. In that case the space of trial curves is genuinely finite dimensional, so one can apply the techniques of Morse without having to consider a limit $N \to \infty$. In particular, Petters [61] used this formalism to prove an odd number theorem. Already earlier, it was shown by Burke [9] that in the quasi-Newtonian approximation every transparent deflector produces an odd number of images. Burke used a fairly simple argument from differential topology, rather than Morse theory, cf. Schneider, Ehlers and Falco [70], p. 172, and Lombardi [42]. A similar argument will be used in the next section to prove an odd number theorem, without invoking Morse theory, for asymptotically simple and empty spacetimes, see Theorem 6 below.

It should be mentioned that, actually, there are several gravitational lens candidates where an even number of images is observed. Usually astronomers are not troubled by this fact because they found good reasons to assume that in those cases one of the images is too faint to be seen. Also, it might be possible that one image is hidden behind the deflector, or that two images are so close together that they are mistaken for being just one image.

5 Gravitational Lensing in Asymptotically Simple and Empty Spacetimes

In elementary optics one often considers "light sources at infinity" which are characterized by the fact that all light rays emitted from such a source are parallel to each other. In this section we want to introduce the notion of "light sources at infinity" for general-relativistic spacetimes. To that end we have to restrict to a special class of spacetimes called "asymptotically simple and empty". Roughly speaking, an asymptotically simple spacetime is a spacetime for which the notion of "(future- or past-pointing) light rays going out to infinity" makes sense. The following definition, which is essentially due to Penrose [54], puts this vague idea into precise form, cf., e. g. Hawking and Ellis [29], p. 222.

Definition 8. A spacetime $(\mathcal{M}, g, \mathcal{T}^+)$ is called *asymptotically simple* if there is a strongly causal spacetime $(\tilde{\mathcal{M}}, \tilde{g}, \tilde{\mathcal{T}}^+)$ with the following properties.
(a) $\mathcal{M}$ is an open submanifold of $\tilde{\mathcal{M}}$ with a non-empty boundary $\partial\mathcal{M}$.
(b) There is a C^∞ function $\Omega : \tilde{\mathcal{M}} \longrightarrow \mathbb{R}$ such that $\mathcal{M} = \{p \in \tilde{\mathcal{M}} | \Omega(p) > 0\}$, $\partial\mathcal{M} = \{p \in \tilde{\mathcal{M}} | \Omega(p) = 0\}$, and the equation $\tilde{g} = \Omega^2 g$ holds on $\mathcal{M}$.
(c) Every inextendible lightlike geodesic in $\mathcal{M}$ has two endpoints on $\partial\mathcal{M}$.

$(\mathcal{M}, g, \mathcal{T}^+)$ is called *asymptotically simple and empty* if, in addition,
(d) there is a neighborhood $\mathcal{U}$ of $\partial\mathcal{M}$ in $\tilde{\mathcal{M}}$ such that the Ricci tensor of g vanishes on $\mathcal{U} \cap \mathcal{M}$.

Asymptotically simple and empty spacetimes are good models for isolated gravitating bodies. Condition (d) of Definition 8 is a way of saying that, sufficiently far away from the gravitating body under consideration, Einstein's vacuum field equation is satisfied. This is a reasonable model for a deflector producing gravitational lensing as long as cosmological aspects can be ignored.

Conditions (b) and (c) of Definition 8 imply that in an asymptotically simple spacetime all lightlike geodesics are complete. Indeed, since on $\mathcal{M}$ the equation $\tilde{g} = \Omega^2 g$ is supposed to hold, every lightlike g-geodesic becomes a lightlike $\tilde{g}$-geodesic by changing the affine parameter according to $ds/d\tilde{s} = \Omega^{-2}$. As Ω is zero on $\partial\mathcal{M}$, this implies that $s \to \pm\infty$ if the geodesic approaches $\partial\mathcal{M}$. Hence, it is justified to interpret the elements of $\partial\mathcal{M}$ as points at infinity or, more precisely, as those points at infinity which can be reached along light rays. Thus, our plan to consider "light sources at infinity" naturally leads to considering "worldlines" contained in $\partial\mathcal{M}$.

In view of gravitational lensing, the observation that in an asymptotically simple and empty spacetime all lightlike geodesics are complete has the following interesting consequence. By a well-known theorem (see Hawking and Ellis [29], Proposition 4.4.5) a complete lightlike geodesic must contain a pair of conjugate points if the weak energy condition and the socalled "generic condition" are satisfied along this geodesic. By Proposition 7, the occurence of conjugate points gives rise to multiple imaging. This result may be interpreted as saying that in almost all physically reasonable spacetimes which are asymptotically simple and empty multiple imaging takes place.

Before turning our attention to "light sources at infinity" we have to recall some basic facts about asymptotically simple and empty spacetimes. First we use part (b) of Definition 8 to define a vector field Z on $\tilde{\mathcal{M}}$ by the equation $d\Omega = \tilde{g}(Z, \cdot)$. It is well-known that condition (d) of Definition 8 implies that Z is non-vanishing and $\tilde{g}$-lightlike at each point of $\partial\mathcal{M}$. For a proof we refer to Hawking and Ellis [29], p. 222. (Please note that Hawking and Ellis include the assumption of $d\Omega$ having no zeros on $\partial\mathcal{M}$ into the definition of asymptotically simple spacetimes. However, it is a well-known result of Penrose [54] that, with the additional assumption (d) of asymptotical emptiness, this property must be automatically satisfied.) As a consequence, $\partial\mathcal{M}$ is a $\tilde{g}$-lightlike hypersurface of $\tilde{\mathcal{M}}$, ruled by the integral curves of Z which are $\tilde{g}$-lightlike geodesics (up to parametrization). Those lightlike geodesics are called the *generators* of $\partial\mathcal{M}$.

In combination with assumption (c) of Definition 8, the property of $\partial\mathcal{M}$ being a $\tilde{g}$-lightlike hypersurface implies that $\partial\mathcal{M}$ has two connected components: $\mathcal{J}^+$ (pronounced "scri plus") where future-pointing g-lightlike geodesics terminate and $\mathcal{J}^-$ (pronounced "scri minus") where past-pointing g-lightlike geodesics terminate. – We now state an important proposition, essentially due to Geroch [21], which determines the global structure of asymptotically simple and empty spacetimes.

Proposition 18. *Let $(\mathcal{M}, g, \mathcal{T}^+)$ be an asymptotically simple and empty spacetime. Then $(\mathcal{M}, g, \mathcal{T}^+)$ is globally hyperbolic and every Cauchy surface is homeomorphic to $\mathbb{R}^3$. Either component $\mathcal{J}^\pm$ of $\partial\mathcal{M}$ can be diffeomorphically mapped onto $S^2 \times \mathbb{R}$ in such a way that each generator of $\mathcal{J}^\pm$ is mapped onto an $\mathbb{R}$-line. Here S^2 denotes the 2-dimensional sphere.*

If the reader wants to verify the proof of this proposition he or she should consult Newman and Clarke [49] who clarified a subtlety overlooked in the original work of Geroch [21] and in Hawking and Ellis [29], Proposition 6.4.9..

After these preparations we are now ready to discuss gravitational lensing situations with light sources at infinity. To that end we consider, in an asymptotically simple and empty spacetime, a sequence of timelike C^∞ curves $\gamma_n : I \longrightarrow \mathcal{M}$ that approach, for $n \to \infty$, a curve $\gamma : I \longrightarrow \mathcal{J}^-$. We want to assume that γ is an immersed curve of class C^1 at least, and that the limit is in the C^1 sense, i. e., that not only $\lim_{n\to\infty} \gamma_n(s) = \gamma(s)$ in $\tilde{\mathcal{M}}$ but also $\lim_{n\to\infty} \gamma_n'(s) = \gamma'(s)$ in $T\tilde{\mathcal{M}}$. Since $\gamma_n'(s)$ is g-timelike and thus $\tilde{g}$-timelike, $\gamma'(s)$ is either $\tilde{g}$-timelike or $\tilde{g}$-lightlike. The first case is impossible, since $\mathcal{J}^-$ is a lightlike hypersurface with respect to $\tilde{g}$, and the second case is possible only if $\gamma'(s)$ is tangent to a generator of $\mathcal{J}^-$. We are thus led to the following conclusion. In an asymptotically simple and empty spacetime, the worldline of a light source at infinity is to be identified with (a section of) a generator of $\mathcal{J}^-$.

Please note that this does *not* mean that light sources at infinity move at the speed of light. The (physical) metric g is not defined on $\partial\mathcal{M}$, i. e., it does not make sense to speak of the causal character of a curve $\gamma : I \longrightarrow \mathcal{J}^-$ with respect to g. The (unphysical) metric $\tilde{g}$ is but a formal device to introduce a geometric structure on the set of points at infinity. The causal character of curves in $\partial\mathcal{M}$ with respect to $\tilde{g}$ has no direct physical interpretation.

Henceforth we restrict to light sources at infinity with inextendible worldlines, i. e., to (maximal) generators of $\mathcal{J}^-$. From Proposition 18 we know that the set of generators of $\mathcal{J}^-$ is a manifold diffeomorphic to the 2-sphere S^2. Hence, the set of all light sources at infinity is in one-to-one correspondence with the points of S^2. On the other hand, we can consider for any $p \in \mathcal{M}$ the set of all one-dimensional g-lightlike subspaces of $T_p\mathcal{M}$. This, again, gives a manifold diffeomorphic to S^2 which may be called the *sky* at p. Clearly, each point of this manifold determines a g-lightlike past-pointing geodesic through p uniquely up to parametrization (i. e., it determines a light ray arriving at p), and vice versa. Hence, the points of this manifold can, indeed, be identified with the points at the celestial sphere of an observer at p. This construction defines for each $p \in \mathcal{M}$ a C^∞ map

$$f_p : S^2 \longrightarrow S^2 \tag{25}$$

by assigning to each point x of the sky at p a light source $f_p(x)$ at infinity by extending the lightlike geodesic tangent to x until it reaches $\mathcal{J}^- \simeq S^2 \times \mathbb{R}$

and projecting onto the first factor afterwards. Henceforth we refer to this map f_p as to the *lens map* at p for light sources at infinity. The lens map can be written in an obvious way with the help of the exponential map. Based on the same idea, one may try to establish a similar lens map in arbitrary spacetimes. The problem is that in the general situation there is no natural "source sphere", i. e., no analogue of the sphere at infinity. Nonetheless, a kind of general lens map can be established, as was recently demonstrated by Frittelli and Newman [19]. Their formalism, which is based on the Hamilton-Jacobi equation for families of light rays, will not be used in this article.

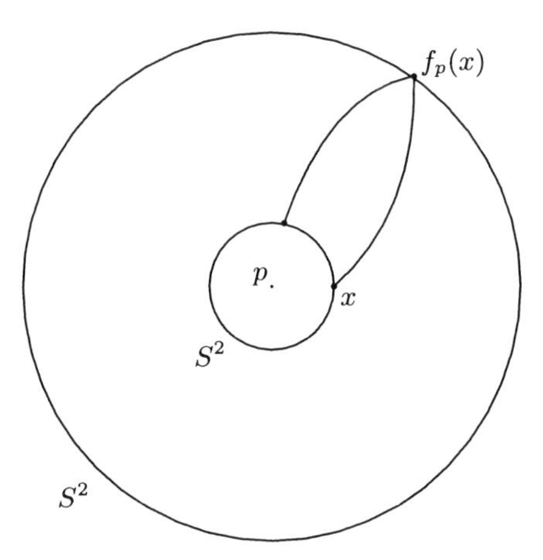

Fig. 6. In this picture the smaller sphere is meant to represent the sky at p and the bigger sphere is meant to represent light sources at infinity. The lens map (25) assigns to each point x of the sky at p a light source $f_p(x)$ at infinity. If f_p is not one-to-one, then there is multiple imaging for light sources at infinity.

The lens map (25) obviously gives all informations on how light sources at infinity are seen by an observer at p. For each light source at infinity, represented by a point $y \in S^2$, the set $f_p^{-1}(y)$ gives all points at the sky of p where this light source is seen, see Fig. 6. If $f_p^{-1}(y)$ consists of more than one point, then there is multiple imaging. Please recall that $y \in S^2$ is called a *regular value* of f_p if for all $y \in S^2$ with $f_p(x) = y$ the differential $T_x f_p : T_x S^2 \longrightarrow T_y S^2$ has maximal rank (i. e., is surjective). It is easy to check that y is a regular value of the lens map if and only if the generator represented by y does not insersect the caustic of the past light cone of p. Here and in the rest of this subsection the term "light cone" always refers to the light cone in $\tilde{\mathcal{M}}$ with respect to the metric $\tilde{g}$ since we need the extension of the "physical" light cone to $\mathcal{J}^-$. Owing to the well-known Theorem of Sard (see, e. g., Guillemin and Pollack [26], p. 39, or Hirsch [31], p. 69) almost all points $y \in S^2$ are regular values of the lens map. This is in agreement with

Proposition 10 according to which the situation that a light source passes through a caustic point is to be viewed as "exceptional". By compactness of S^2, $f_p^{-1}(y)$ is finite for any regular value y. Hence, the observer at p sees finitely many images of each light source at infinity that does not pass through the caustic of the past light cone of p.

We shall now establish the remarkable result that, for each light source at infinity which does not pass through the caustic of the observer's past light cone, the number of images is odd. Based on this observation we shall then prove that the same result is true for a light source moving inside $\mathcal{M}$ provided that its worldline is inextendible and does not approach $\mathcal{J}^-$. This *odd number theorem* for light sources in asymptotically simple and empty spacetimes will emerge as an application of elementary differential topology; in particular, it will not be necessary to invoke the considerable technical apparatus of Morse theory that was discussed in Sect. 4.2 above.

The proof of our odd number theorem is based on a simple idea. However, the details of the proof look a little bit involved since it is necessary to introduce some cumbersome notation. For that reason we first give the general idea upon which the proof is based. This general idea of an odd-number argument is popular with astronomers who usually present it on the understanding that space and time can be described in a Newtonian fashion. Then the argument goes like this (cf. Schneider, Ehlers and Falco [70], p. 176). Consider all light rays that reach at a particular instant an observer at the point p. Parametrize these light rays with time, in a past-pointing way such that they have $t = 0$ on their arrival at p. Then for all $t > 0$ we can consider the *wavefront* $\mathcal{W}_t$, defined as the set of all points in space that are crossed by at least one of the considered light rays at the time t. For small t, $\mathcal{W}_t$ is a sphere around p. For larger t, $\mathcal{W}_t$ will develop self-intersections because the light rays are influenced by gravitating masses in the universe. Whenever the wavefront crosses a light source, given by a point moving in space in dependence of time, this gives rise to an image of the light source seen at p. As the wavefronts develop as continuous deformations of a sphere, the following definition makes sense. We say that the light source is *outside* of the wavefront $\mathcal{W}_t$ if the position of the light source at t can be connected to p by a curve that intersects $\mathcal{W}_t$ an odd number of times and *inside* otherwise. Here we have to restrict to curves that intersect the wavefront transversely and only at points where the wavefront is an immersed submanifold, i. e., not at caustic points. Now we assume that the light source itself never touches a wavefront with a tangent velocity vector and that it stays away from caustic points. Then, whenever the light source meets a wavefront, it changes from outside to inside or vice versa. For small t the light source is outside. For large t it is inside, provided that all light rays go out to infinity and the light source does not go out to infinity. This implies that, in total, the wavefront crosses the light source an odd number of times, i. e., that there is an odd number of images.

Instead of the somewhat vague notion of being inside or outside one may use he socalled "mapping degree" from differential topology. Using this terminology the above argument was written down, apparently for the first time, in the introduction of McKenzie [45]. Gottlieb [25] tried to translate this general idea into a Lorentzian manifold setting and came to the conclusion that the argument does not work without quite special and unrealistic assumptions. However, this conclusion is largely based on the fact that Gottlieb implicitly restricts to multiple imaging situations without time delay. What we want to show in the following is that, with the help of the mapping degree, the above-mentioned odd number argument can be made into a precise theorem in asymptotically simple and empty spacetimes. The crucial point is, of course, that in this case it is guaranteed that all light rays go out to infinity. As an aside, we mention that a slightly different mapping degree argument was used by Lombardi [42] to prove an odd number theorem in the quasi-Newtonian approximation formalism. This is closely related to Burke's [9] original proof of an odd number theorem, again in the quasi-Newtonian approximation, using the index of vector fields. The latter is also discussed in Schneider Ehlers and Falco [70].

The essential tool for our proof is the *mapping degree* of a C^1 map $F : \mathcal{M}_1 \longrightarrow \mathcal{M}_2$ between oriented manifolds of the same dimension. In the following we briefly summarize the basic facts about this notion. For more background material the reader is refered to Westenholz [83], Guillemin and Pollack [26] and Dold [11] who, in this order, treat the subject with an increasing amount of abstract mathematics. First we recall that, by the Sard theorem already mentioned above, almost all points $y \in \mathcal{M}_2$ are regular values of F. If, for such a regular value y, the pre-image $F^{-1}(y)$ is contained in a compact set and thus finite, the *local degree of* F *at* y can be defined by the equation

$$\deg_y(F) = \sum_{x \in F^{-1}(y)} \operatorname{sgn}(x) \, . \tag{26}$$

where $\operatorname{sgn}(x)$ is, by definition, equal to $+1$ if the tangent map $T_x F$ preserves orientation and equal to -1 if $T_x F$ reverses orientation. One can then establish the following facts.

(a) If $\mathcal{M}_1$ and $\mathcal{M}_2$ are both compact without boundary, $\deg_y(F)$ is the same for all regular values y. In this case one calls $\deg(F) = \deg_y(F)$ simply the *degree of* F. The degree is a homotopic invariant in the sense that a second C^1 map $\tilde{F} : \mathcal{M}_1 \longrightarrow \mathcal{M}_2$ has the same degree as F if and only if F can be continuously deformed into $\tilde{F}$.

(b) If $\mathcal{M}_1$ and $\mathcal{M}_2$ are compact with smooth boundaries, and if F restricts to a map $\partial F : \partial\mathcal{M}_1 \longrightarrow \partial\mathcal{M}_2$, then

$$\deg_y(F) = \deg(\partial F) \tag{27}$$

for all regular values $y \in \mathcal{M}_2 \setminus \partial\mathcal{M}_2$.

We are now ready to state and prove the desired odd number theorem.

Theorem 6. *For any point p in an asymptotically simple and empty space-time $(\mathcal{M}, g, \mathcal{T}^+)$, the following holds true.*
(a) *The lens map (25) has degree one,* $\deg(f_p) = 1$. *As a consequence, each generator of $\mathcal{J}^-$ that does not pass through the caustic of the past light cone of p can be reached from p along an odd number of lightlike geodesics. (We have already seen above that this number is finite.)*
(b) *Let I be an open interval and $\gamma : I \longrightarrow \mathcal{M}$ a timelike embedded C^∞ curve such that γ is closed in $\mathcal{M}$ and has no endpoint on $\mathcal{J}^-$. (This is a way of saying that γ is inextendible in $\mathcal{M}$ and, in the past direction, does not go out to infinity approaching the velocity of light.) If γ does not pass through the caustic of the past light cone of p, then the number of past-pointing lightlike geodesics from p to γ is finite and odd.*

Proof. Fix a curve γ that satisfies the assumptions of (b). We want to construct a timelike C^∞ vector field V on $\mathcal{M}$ that is tangent to γ and smoothly extends to the vector field Z on $\mathcal{J}^-$ given by $\tilde{g}(Z, \cdot) = d\Omega$. First we choose a future-pointing timelike C^∞ vector field V_1 on some neighborhood $\mathcal{U}_1$ of γ in $\mathcal{M}$ such that V_1 is tangent to γ. This is possible since γ is an embedding and γ is closed in $\mathcal{M}$; the proof can be patterned after the proof of Proposition 5.1 in Giannoni, Masiello and Piccione [22]. Then we choose a future-pointing timelike C^∞ vector field V_2 on an open subset $\mathcal{U}_2$ of $\mathcal{M}$ whose closure in $\tilde{\mathcal{M}}$ covers $\mathcal{J}^-$ such that V_2 smoothly extends to Z on $\mathcal{J}^-$. The existence of such a vector field V_2 follows from the fact that $\mathcal{J}^-$ is a closed embedded $\tilde{g}$-lightlike submanifold of $\tilde{\mathcal{M}}$. Since γ does not approach $\mathcal{J}^-$, the domains of V_1 and V_2 can be chosen disjoint. Finally, we choose a future-pointing timelike C^∞ vector field V_3 on an open subset $\mathcal{U}_3$ of $\mathcal{M}$ such that $\mathcal{U}_1$, $\mathcal{U}_2$ and $\mathcal{U}_3$ cover $\mathcal{M}$ and the closure of $\mathcal{U}_3$ in $\tilde{\mathcal{M}}$ has void intersection with γ and with $\mathcal{J}^-$. Then we get the desired vector field V by combining V_1, V_2 and V_3 with a partition of unity. For this vector field V, we consider the projection (21) onto the 3-manifold of integral curves of V. By Proposition 18, $\mathcal{S}_V$ is homeomorphic to $\mathbb{R}^3$. Since, by the theorem of Moise [47] already mentioned above, any 3-dimensional topological manifold admits a unique differentiable strcuture, $\mathcal{S}_V$ must even be diffeomorphic to $\mathbb{R}^3$ or, what is the same, to the open unit ball $B = \{x \in \mathbb{R}^3 \,|\, |x| < 1\}$. Since V extends to the vector field Z on $\mathcal{J}^-$ that is tangent to the generators, π_V extends to a C^∞ map

$$\overline{\pi}_V : \mathcal{M} \cup \mathcal{J}^- \longrightarrow \overline{B} \tag{28}$$

between manifolds with boundaries. – Now the vector field V defines a timelike vector V_p at the point p. We choose three spacelike tangent vectors E_1, E_2, E_3 at p with $\tilde{g}(E_\mu, V_p) = 0$ and $\tilde{g}(E_\mu, E_\nu) = -\tilde{g}(V_p, V_p)\,\delta_{\mu\nu}$. Then each $x \in \mathbb{R}^3$ defines a past-pointing lightlike $\tilde{g}$-geodesic λ_x by initial conditions $\lambda_x(0) = p$ and $\lambda_x'(0) = x^1 E_1 + x^2 E_2 + x^3 E_3 - |x|\, V_p$. Since we use the $\tilde{g}$-affine parametrization, rather than the g-affine parametrization, λ_x arrives at $\mathcal{J}^-$ at a finite parameter value $v(x) \in \mathbb{R}$. Having established the projection (28), the map $x \longmapsto \lambda_x$ and the map $x \longmapsto v(x)$, we are now ready to prove part (a) of the proposition. For each $x \in S^2$, the curve $\overline{\pi}_V \circ \lambda_x$ in $\overline{B}$ starts at the origin and reaches the boundary at some parameter value $v(x)$. Hence, for each $r \in \left]0,1\right]$, there is a unique parameter value $u(r,x) \leq v(x)$ such that $\left|\overline{\pi}_V\big(\lambda_x(s)\big)\right|$ is smaller than r for $0 < s < u(r,x)$ and equal to r for $s = u(r,x)$. Then the assignment $x \longmapsto \Phi_r(x) = \frac{1}{r}\overline{\pi}_V\big(\lambda_x\big(u(r,x)\big)\big)$ gives a C^∞ map $\Phi_r : S^2 \longrightarrow S^2$. For $r = 1$ we get the lens map (25), $\Phi_1 = f_p$. For r sufficiently small, Φ_r is an orientation preserving diffeomorphism, hence $\deg(\Phi_r) = 1$. This follows from the fact that the light cone looks like the Minkowski light cone if we restrict to sufficiently short light rays. Since the degree is a homotopic invariant, letting r vary from some small value up to the value 1 shows that the lens map has degree one. Now each generator of $\mathcal{J}^-$ is the pre-image of a point $y \in S^2 = \partial\overline{B}$ under the map $\overline{\pi}_V$. This point y is a regular value of the lens map if and only if the generator does not meet the caustic of the past light cone of p. Under this condition $f_p^{-1}(y)$ is finite, by compactness of S^2. Let us denote by $n_\pm$ the number of points $x \in f_p^{-1}(y)$ such that $\mathrm{sgn}(x) = \pm 1$. Then the definition of the degree implies $n_+ - n_- = \deg(f_p)$. Since $\deg(f_p) = 1$, this gives $n_+ + n_- = 2n_- + 1$, i. e., the number of points in $f_p^{-1}(y)$ is odd. – Now we prove part (b). To that end we consider the map $F : \overline{B} \longrightarrow \overline{B}$ defined by $F(x) = \overline{\pi}_V\big(\lambda_x(v(x)\,|x|)\big)$. Clearly, the restriction of this map to the boundary gives the lens map $\partial F = f_p : \partial\overline{B} = S^2 \longrightarrow \partial\overline{B} = S^2$. If our curve γ does not meet the caustic of the past light cone of p, the point $y \in B$ with $\overline{\pi}_V^{-1}(y) = \gamma$ is a regular value of F. Hence, by (27), $\deg_y(F) = \deg(f_p) = 1$. By compactness of $\overline{B}$, the set $F^{-1}(y)$ is finite. As in the proof of part (a), we get $n_+ + n_- = 2n_- + 1$, where $n_\pm$ denotes the number of elements $x \in F^{-1}(y)$ with $\mathrm{sgn}(x) = \pm 1$. Thus, the number of elements in $F^{-1}(y)$ is odd. □

The fact that the lens map has degree one implies, in particular, that the lens map is surjective which was not obvious from the start. The reader should also consult Kozameh, Lamberti and Reula [38] who state in Lemma 1 a result which is essentially equivalent to the fact that, in our terminology, the lens map has degree one. This paper [38] belongs to a long series of articles by Ted Newman, Carlos Kozameh and various coauthors on studying general relativity in terms of the Hamilton-Jacobi equation for families of lightlike geodesics. For reviews on this topic we refer to Chap. 7 of Joshi [33] and to Kozameh [37]. In particular, these authors have found interesting results on the geometry of "light cone cuts at infinity", i. e., of intersections of light cones with $\mathcal{J}^+$ or $\mathcal{J}^-$ in an asymptotically simple and empty spacetime, which are of relevance in view of gravitational lensing.

As stressed already at the beginning of this subsection, an asymptotically simple and empty spacetime is a good model for an *isolated* gravitating body, i. e., if cosmological aspects are ignored. Rudiments of cosmology can be introduced by modifying condition (d) of Definition 8. E. g., one could require

$\mathrm{Ric} = \Lambda g$ near $\partial\mathcal{M}$ with a positive or negative cosmological constant Λ, rather than the vacuum field equation $\mathrm{Ric} = 0$. The resulting spacetimes are called *asymptotically deSitter* for $\Lambda > 0$ and *asymptotically anti-deSitter* for $\Lambda < 0$. It was verified already by Penrose [54] that then $\partial\mathcal{M}$ is no longer $\tilde{g}$-lightlike but rather $\tilde{g}$-spacelike for $\Lambda > 0$ and $\tilde{g}$-timelike for $\Lambda < 0$. In the latter case we can consider immersed worldlines in $\partial\mathcal{M}$ which are $\tilde{g}$-timelike. For such "light sources at infinity" in an asymptotically anti-deSitter spacetime we have Fermat's principle in the version of Theorem 1, viewed in the spacetime $(\tilde{\mathcal{M}}, \tilde{g}, \tilde{T}^+)$, at our disposal. This observation was used by Woolgar [88] to prove a positive energy theorem for asymptotically anti-deSitter spacetimes.

The class of asymptotically simple spacetimes seems particularly appropriate for discussing gravitational lensing in a Lorentzian geometry setting. As only a few results in this direction have been worked out so far, all dedicated experts are invited to join the work in this interesting field.

References

1. Abraham, R., Marsden, J. (1978) Foundations of mechanics. Benjamin-Cummings, Reading, Massachusetts
2. Arnold, V. (1990) Singularities of Caustics and Wave Fronts. Kluver, Dordrecht
3. Arnold, V., Gusein-Zade, S., Varchenko, A. (1985) Singularities of Differentiable Maps . I . Birkhäuser, Boston
4. Beem, J., Ehrlich, P., Easley, K. (1996) Global Lorentzian Geometry. Dekker, New York
5. Blandford, R., Narayan, R. (1986) Fermat's principle, caustics, and the classification of gravitational lens images. Astrophys. J. **310**, 568–582
6. Borde, A. (1987) Geodesic focusing, energy conditions and singularities. Class. Quantum Grav. **4**, 343–356
7. Bott, R., Tu, L. W. (1982) Differential forms in algebraic topology. Springer, New York
8. Brill, D. (1973) Observational contacts of general relativity. In Israel, W. (Ed.) Relativity, Astrophysics and Cosmology, Proceedings of Banff Summer School 1972. Reidel, Dordrecht, 127–152
9. Burke, W. L. (1981) Multiple gravitational imaging by distributed masses. Astrophys. J. **244**, L1
10. Chwolson, O. (1924) Über eine mögliche Form fiktiver Doppelsterne. Astronomische Nachrichten **221**, 329
11. Dold, A. (1980) Lectures on algebraic topology. Springer, Berlin
12. Eddington, A. S. (1920) Space, time, and gravitation. Cambridge Univ. Press, Cambridge
13. Everson, J., Talbot, C. (1976) Morse theory on timelike and causal curves. Gen. Rel. Grav. **7**, 609–622. Erratum (1978) **9**, 1047
14. Faraoni, V. (1992) Nonstationary gravitational lenses and the Fermat principle. Astrophys. J. **398**, 425–428

15. Flaherty, F. (1975) Lorentzian manifolds of non-positive curvature. I. Proc. Symp. Pure Math. **27**, No. 2, 395–399
16. Flaherty, F. (1975) Lorentzian manifolds of non-positive curvature. II. Proc. Amer. Math. Soc. **48**, 199–202
17. Frankel, T. (1979) Gravitational curvature. Freeman, San Francisco
18. Friedrich, H., Stewart, J. (1983) Characteristic initial data and wavefront singularities in general relativity. Proc. Roy. Soc. London **A 385**, 345–371
19. Frittelli, S., Newman, Ezra T. (1998) An exact universal gravitational lensing equation. preprint gr-qc/9810017
20. Geroch, R. (1970) Domain of dependence. J. Math. Phys. **11**, 417–449
21. Geroch, R. (1971) Space-time structure from a global viewpoint. In Sachs, R. K. (Ed.) General relativity and cosmology, Enrico Fermi School, Course XLVII. Academic Press, New York, 71–103
22. Giannoni, F., Masiello, A., Piccione, P. (1997) A variational theory for light rays in stably causal Lorentzian manifolds: Regularity and multiplicity results. Commun. Math. Phys. **187**, 375–415
23. Giannoni, F., Masiello, A., Piccione, P. (1998) A Morse theory for light rays on stably causal Lorentzian manifolds. Ann. Inst. H. Poincaré, Physique Theoretique **69**, 359–412
24. Gott, J. R. (1985) Gravitational lensing effects of a vacuum string: exact solutions. Astrophys. J. **288**, 422–427
25. Gottlieb, D. (1994) A gravitational lens need not produce an odd number of images. J. Math. Phys. **35**, 5507–5510
26. Guillemin, V., Pollack, A. (1974) Differential topology. Prentice-Hall, Englewood Cliffs, NJ
27. Harris, S. (1992) Conformally stationary spacetimes. Class. Quantum Grav. **9**, 1823–1827
28. Hasse, W., Kriele, M., Perlick, V. (1996) Caustics of wavefronts in general relativity. Class. Quantum Grav. **13**, 1161–1182
29. Hawking, S. W., Ellis, G. F. R. (1973) The large scale structure of space-time. Cambridge Univ. Press, Cambridge
30. Helfer, A. D. (1994) Conjugate points on spacelike geodesics or pseudo-self-adjoint Morse-Sturm-Liouville systems. Pacific J. Math. **164**, 321–340
31. Hirsch, M. W. (1976) Differential topology. Springer, New York
32. Hiscock, W. (1985) Exact gravitational field of a string. Phys. Rev. D **31**, 3288–3290
33. Joshi, P. (1993) Global aspects in gravitation and cosmology. Clarendon Press, Oxford
34. Kánnár, J. (1991) A note on the existence of conjugate points. Class. Quantum Grav. **8**, L179–L184
35. Klingenberg, W. (1982) Riemannian geometry. De Gruyter, Berlin
36. Kovner, I. (1990) Fermat principle in gravitational fields. Astrophys. J. **351**, 114–120
37. Kozameh, C. (1998) Dynamics of null surfaces in general relativity. In Dadhich, N., Narlikar, J. (Eds.) Gravitation and relativity: At the turn of the millenium, Proceedings of GR 15 Conference, 1997. IUCAA, Pune, 139–152
38. Kozameh, C., Lamberti, P. W., Reula, O. (1991) Global aspects of light cone cuts. J. Math. Phys. **32**, 3423–3426
39. Leray, J. (1952) Hyperbolic differential equations. Institute for Advanced Study, Princeton

40. Landau, L.D., Lifshits, E.M. (1959) Course of theoretical physics. II: Theory of fields. Addison–Wesley, Reading, Massachusetts and Pergamon, London
41. Levi-Civita, T. (1917) Statica Einsteiniana. Atti della reale accademia dei lincei, Seria quinta, Rendiconti, Classe di scienze fisiche, matematiche e naturali **26**, 458–470
42. Lombardi, M. (1998) An application of the topological degree to gravitational lenses. Modern Phys. Lett. *A 13*, 83–86
43. Low, R. (1998) Stable singularities of wave-fronts in general relativity. J. Math. Phys. **39**, 3332–3335
44. Masiello, A. (1994) Variational methods in Lorentzian geometry. Pitman Research Notes in Mathematics Series 309. Longman Scientific & Technical, Essex
45. McKenzie, R. H. (1985) A gravitational lens produces an odd number of images. J. Math. Phys. **26**, 1592–1596
46. Milnor, J. (1963) Morse theory. Ann. Math. Studies No. 51, Princeton
47. Moise, E (1956) Affine structures in 3-manifolds. V. Ann. Math. **56**, 96–114
48. Morse, M. (1934) The calculus of variations in the large. Am. Math. Soc. Colloquium Publications XVIII, Providence, Rhode Island
49. Newman, R. P. C., Clarke, C. J. S. (1987) An $\mathbb{R}^4$ spacetime with a Cauchy surface which is not $\mathbb{R}^3$. Class. Quantum Grav. **4**, 53–60
50. O'Neill, B. (1983) Semi-Riemannian geometry. Academic Press, New York
51. Padmanabhan, T., Subramanian, K. (1988) The focusing equation, caustics and the condition of multiple imaging by thick gravitational lenses. Mon. Not. Roy. Astron. Soc. **233**, 265–284
52. Palais, R. (1963) Morse theory on Hilbert manifolds. Topology **2**, 299–340
53. Palais, R., Smale, S. (1964) A generalized Morse theory. Bull. Amer. Math. Soc. **70**, 165–172
54. Penrose, R. (1964) Conformal treatment of infinity. In deWitt, C. M., deWitt, B. (Eds.) Relativity, groups and topology, Les Houches Summer School 1963. Gordon and Breach, New York, 565–587
55. Perlick, V. (1990) On Fermat's principle in general relativity. I. The general case. Class. Quantum Grav. **7**, 1319–1331
56. Perlick, V. (1990) On Fermat's principle in general relativity. II. The conformally stationary case. Classs. Quantum Grav. **7**, 1849–1867
57. Perlick, V. (1995) Infinite dimensional Morse theory and Fermat's principle in general relativity. I. J. Math. Phys. **36**, 6915–6928
58. Perlick, V. (1996) Criteria for multiple imaging in Lorentzian manifolds. Class. Quantum Grav. **13**, 529–537
59. Perlick, V. (1999) Ray optics, Fermat's principle and applications to general relativity. To appear in Lecture Notes of Physics, Series m, Springer, Heidelberg
60. Perlick, V., Piccione, P. (1998) A general-relativistic Fermat principle for extended light sources and extended receivers. Gen. Rel. Grav. **30**, 1461–1476
61. Petters, A. (1992) Morse theory and gravitational microlensing. J. Math. Phys. **33**, 1915–1931
62. Petters, A. (1993) Arnold's singularity theory and gravitational lensing. J. Math. Phys. **34**, 3555–3581
63. Petters, A. (1995) Multiplane gravitational lensing. I. Morse theory of image counting. J. Math. Phys. **36**, 4263–4275

64. Petters, A. (1995) Multiplane gravitational lensing. II. Global geometry of caustics. J. Math. Phys. **36**, 4276–4295
65. Petters, A., Levine, H., Wambsganss, J. (1999) Singularity theory and gravitational lensing. To appear with Birkhäuser, Boston
66. Poincaré, H. (1905) Sur les lignes géodésiques des surfaces convexes. Trans. Amer. Math. Soc. **6**, 237–274
67. Roman, T. A. (1988) On the "averaged weak energy condition" and Penrose's singularity theorem. Phys. Rev. **D 37**, 546–548
68. Sasaki, M. (1993) Cosmological gravitational lens equation. Its validity and limitation. Prog. Theor. Phys. **83**, 467–491
69. Schneider, P. (1984) A new formulation of gravitational lens theory, time-delay and Fermat's principle. Astron. Astrophys. **143**, 413–420
70. Schneider, P., Ehlers, J., Falco, E. (1992) Gravitational lenses. Springer, Heidelberg
71. Schwartz, J. T. (1969) Nonlinear functional analysis. Gordon and Breach, New York
72. Seifert, H.-J. (1967) Kausale Lorentzräume. Doctoral Thesis, Hamburg University
73. Seitz, S., Schneider, P., Ehlers, J. (1994) Light propagation in arbitrary spacetimes and the gravitational lens approximation. Class. Quantum Grav. **11**, 2345–2373
74. Smith, J. W. (1960) Fundamental groups on a Lorentz manifold. Amer. J. Math. **82**, 873–890
75. Spanier, E. (1966) Algebraic topology. McGraw-Hill, New York
76. Straumann, N. (1984) General relativity and relativistic astrophysics. Springer, Berlin
77. Tipler, F. J. (1978) General relativity and conjugate ordinary differential equations. J. Diff. Equat. **30**, 165–174
78. Uhlenbeck, K. (1975) A Morse theory for geodesics on a Lorentz manifold. Topology **14**, 69–90
79. Vilenkin, A. (1981) Gravitational field of vacuum domain walls and strings. Phys. Rev. D **23** , 852–857
80. Wald, R. (1984) General relativity. University of Chicago Press, Chicago
81. Walsh, D., Carlswell, R., Weyman, R. (1979) 0957 +561 A,B: twin quasistellar objects or gravitational lens? Nature **279**, 381–384
82. Wambsganss, J. (1998) Gravitational lensing in astronomy. http://www.livingreviews.org/Articles/Volume1/1998-12wamb/
83. Westenholz, C. v. (1978) Differential forms in mathematical physics. North-Holland, Amsterdam
84. Weyl, H. (1917) Zur Gravitationstheorie. Annalen der Physik **54**, 117–145
85. Whitehead, J. H. C. (1935) On the covering of a complete space by the geodesics through a point. Ann. Math. **36**, 679–704
86. Whitney, H. (1936) Differentiable manifolds. Ann. Math. **37**, 645–680
87. Woodhouse, N. (1976) An application of Morse theory to space-time geometry. Commun. Math. Phys. **46**, 135–152
88. Woolgar, E. (1994) The positivity of energy for asymptotically anti-deSitter spacetimes. Class. Quantum. Grav. **11**, 1881–1900

Jürgen Ehlers – Bibliography

1. Exakte Lösungen der Einstein–Maxwellschen Feldgleichungen für statische Felder, *Z. Physik* 140, 394–408 (1955).
2. Beiträge zur Theorie der statischen Vakuumfelder in der klassischen und der erweiterten relativistischen Gravitationstheorie, *Z. Physik* 143, 239–248 (1955).
3. Elektrostatische Felder in der erweiterten Gravitationstheorie, *Z. Physik* 146, 515–526 (1956).
4. Die Evolution der Physik, Enzyklopädisches Stichwort zum gleichnamigen Buch von A. Einstein und L. Infeld, 196–202, Rowohlt, 1956.
5. Konstruktionen und Charakterisierungen von Lösungen der Einsteinschen Gravitationsfeldgleichungen, Dissertation Universität Hamburg, Hamburg, 1958.
6. (with R.K. Sachs) Erhaltungssätze für die Wirkung in elektromagnetischen und gravischen Strahlungsfeldern, *Z. Physik* 155, 498–506 (1959).
7. Transformations of Static Exterior Solutions of Einstein's Gravitational Field Equations into Different Solutions by Means of Conformal Mappings, Les Théories Relativistes De La Gravitation, CNRS, Paris, 1959, p. 275.
8. Exterior Solutions of Einstein's Gravitational Field Equations Admitting a Two-Dimensional Abelian Group of Isometric Correspondences, Col. sur la theorie de la relativite, 49–57, Gauthier-Villar, Paris, 1960.
9. (with P. Jordan and W. Kundt) Strenge Lösungen der Feldgleichungen der allgemeinen Relativitätstheorie, Akad. Wiss. Mainz Abh., Math.-Nat. Kl., Jahrgang 1960, Nr. 2.
10. (with P. Jordan and R.K. Sachs) Beiträge zur Theorie der reinen Gravitationsstrahlung, Akad. Wiss. Mainz Abh., Math.-Nat. Kl., Jahrgang 1961, Nr. 1, p. 3–62.
11. Beiträge zur relativistischen Mechanik Kontinuierlicher Medien, Akad. Wiss. Mainz Abh., Math.-Nat. Kl., Jahrgang 1961, Nr. 11, p. 763–837.
12. Relativistic Hydrodynamics and Its Relation to Interior Solutions of the Gravitational Field Equations, Recent Developments in General Relativity, Pergamon Press, Warzawa, 1962, p. 201.
13. (with W. Kundt) Exact Solutions of the Gravitational Field Equations, in *The Theory of Gravitation*, ed. by L. Witten, Wilay, New York, 1962.
14. (with P. Jordan and W. Kundt) Quantitatives zur Diracschen Schwerkrafthypothese, *Z. Physik* 178, 501–518 (1964).
15. Gravitational Waves, Chapter in *Relativita Generale*, ed. by C. Cattaneo, Rome: Centro Internationale Matematico Estivo Publications, 1965.
16. (with R. Penrose and W. Rindler) Energy Conversation as the Basis of Relativistic Mechanics, *Am. J. Phys.* 33, 995–997 (1965).

17. Neuere Entwicklungen in der Allgemeinen Relativitätstheorie, Physikertagung 1965 Frankfurt, 286–303, Teubner, Stuttgart, 1965.
18. Generalized Electromagnetic Null Fields and Geometrical Optics, in *Perspectives in Geometry and Relativity*, ed. by B. Hoffmann, p. 127–133, Indiana University Press, Bloomington and London, 1966.
19. (with I. Robinson and W. Rindler) Quaterions, Bivectors, and the Lorentz Group, in *Perspectives in Geometry and Relativity*, ed. by B. Hoffmann, p. 134–149, Indiana University Press, Bloomington and London, 1966.
20. Exact Solutions, in *International Conference on Relativistic Theories of Gravitation*, Vol. II, ed. by H. Bondi et al., University of London, London, 1966.
21. Zum Übergang von der Wellenoptik zur geometrischen Optik in der allgemeinen Relativitätstheorie, *Zeitschrift f. Naturforschung* 22a, 1328–1332 (1967).
22. (with E.L. Schücking) Zur Hönl-Dehnenschen Formulierung des Machschen Prinzips, *Zeitschr. f. Physik* 206, 438–491 (1967).
23. (with P. Geren and R.K. Sachs) Isotropic Solutions of the Einstein–Liouville Equations, *Journal of Mathematical Physics* 99, 1344–1349 (1968).
24. (with W. Rienstra) Isotropic Solutions of the Liouville and Poisson Equations, *Astrophysical Journal* 155, 105–116 (1969).
25. Probleme und Ergebnisse der modernen Kosmologie, Mitteilungen der Astronomischen Gesellschaft 27, 73 (1969).
26. Relativistic Kinetic Theory of Charged Gases, in *La Magneothydrodynamique Classique et Relativiste*, Editions du Centre National De La Recherche Scientifique, 221–229, Paris, 1970.
27. (with W. Rindler) A Gravitationally Induced (Machian) Magnetic Field, *Physics Letters* 32A, 257 (1970).
28. General Relativity and Kinetic Theory, in *General Relativity and Cosmology*, ed. by R.K. Sachs, 1–70, Academic Press, New York and London, 1971.
29. General-relativistic kinetic theory of gases in *Relativistic Fluid Dynamics*, ed. by C. Cattaneo, Centro Internazionale Matematico Estivo, Edizioni, Cremonese, 301–388, Rome, 1971.
30. Kinetic Theory of Gases in General Relativity, in *Relativity and Gravitation*, ed. by C.J. Kuper and A. Peres, 145–154, Gordon and Breach, New York, 1971.
31. (with R.K. Sachs) Kinetic Theory and Cosmology, in *Astrophysics and General Relativity*, vol. 2, ed. by Chretien, S. Deser, J. Goldstein, 331–383, Gordon and Breach, New York, 1971.
32. (with W. Rindler) An Electromagnetic Thirring Problem, *Physics Rev.* D 4, 3543–3552 (1971).
33. (with F.A.E. Pirani and A. Schild) The Geometry of Light Propagation and Free Fall, in *General Relativity*, ed. by L. O'Raifeartaigh, 63–84, Clarendon Press, Oxford, 1972.
34. (with A. Schild) Geometry in a Manifold with Projective Structure, *Commun. Math. Phys.* 32, 119–146 (1973).
35. The Nature and Structure of Spacetime, in *The Physicist's Conception of Nature*, J. Mehra (ed.), 293–313, Reidel, Dordrecht, 1973.
36. Survey of General Relativity Theory, in *Relativity, Astrophysics and Cosmology*, W. Israel (ed.), 1–125, Reidel, Dordrecht, 1973.

37. Kinetic Theory of Gases in General Relativity Theory, in *Lectures in Statistical Physics*, W.C. Schive and J.S. Turner (eds.), *Lecture Notes in Physics* Vol. 28, 78–105, Springer, Berlin Heidelberg, 1974.
38. The Geometry of the (Modified) GHP–Formalism, *Commun. Math. Phys.* 37, 327–329 (1974).
39. Gravitation, Raumzeitstruktur und Astrophysik, im Jahrbuch der Max-Planck-Gesellschaft 1974, 47–60, München, 1974.
40. Progress in Relativistic Mechanics, Thermodynamics and Continuum Mechanics, in *General Relativity and Gravitation*, G. Shaviv and J. Rosen (eds.), 213–232, Wiley, New York, 1975.
41. (with G. Börner and E. Rudolph) Relativistic Spin Precession in Two-Body Systems, *Astronomy and Astrophysics* 44, 417–420 (1975).
42. (with G. Börner and E. Rudolph) Relativistic Effects in the Binary Pulsar PSR 1913–16; *Mitt. Astr. Ges.* 38, 117–119 (1976).
43. Kosmologische Beobachtungen und ihre Beziehungen zum Standard-Weltmodell, *Mitt. Astr. Ges.* 38, 41–54 (1976).
44. (with A. Rosenblum, J.N. Goldberg and P. Havas) Comments on Gravitational Radiation and Damping and Energy Loss in Binary Systems, *Ap. J.* 208, L77–L81 (1976).
45. (with E. Rudolph) Dynamics of extended bodies in general relativity, *J. Gen. Rel. Grav.* 8, 197–217 (1977).
46. (with E. Köhler) Path Structures on Manifolds, *J. Math. Phys.* 18, 2014–2018 (1977).
47. Weak-Field Approximations and Equations of Motion in General Relativity, in Proceedings of the Intern. School of Relativistic Astrophysics (3rd course), Erice/Sizilien. MPI-PAE/Astro 138 (1977).
48. Leben und Werk Albert Einsteins, *Umschau in Wissenschaft und Technik*, 79. Jahrg., Heft 1, 7–10 (1979).
49. Albert Einstein, *The Unesco Courier*, May Issue, 4–8 (1979).
50. Einstein's Theory of Gravitation, in *Einstein Symposium Berlin*, Nelkowski, Hermann, Poser, Schrader, Seiler (Herausg.), Lecture Notes in Physics 100, 10–35, Springer, Berlin,1979.
51. Introduction. Survey of Problems, in *Isolated Gravitating Systems in General Relativity*, J. Ehlers (ed.), North–Holland, Publishing Co., Amsterdam 1979.
52. (with R. Breuer) Propagation of high-frequency electromagnetic waves through a magnetized plasma in curved spacetime, I, Proc. R. Soc. Lond. A370, 389–406 (1980).
53. Isolated Systems in General Relativity, In Proceedings of the Ninth Texas Symposium on Relativistic Astrophysics, J. Ehlers, J.J. Perry, and M. Walker (eds.), Ann. N.Y. Acad. Sci. Vol. 336, 279–294 (1980).
54. Christoffel's Work on the Equivalence Problem for Riemannian Spaces and its Importance for Modern Field Theories of Physics, in: *E.B. Christoffel: The Influence of his work on Mathematics and the Physical Sciences*, P.L. Butzner, F. Fehér (eds.), 526–542, Birkhäuser, Basel, 1981.
55. (with R. Breuer) Propagation of high-frequency electromagnetic waves through a magnetized plasma in curved spacetime, II, Proc. R. Soc. Lond. A 374, 65–86 (1981).
56. (with R. Breuer) Propagation of Electromagnetic Waves through Magnetized Plasmas in Arbitrary Gravitational Fields, Astron. Astrophys. 96, 293–295 (1981).

57. Über den Newtonschen Grenzwert der Einsteinschen Gravitationstheorie, in *Grundlagenprobleme der modernen Physik*, J. Nitsch, J. Pfarr, E.W. Stachow (eds.), 65–84, Bibliogr. Institut, Mannheim, 1981.
58. Some advances and problems in classical general relativity, in *Proceedings of the sixth international conference on mathematical physics*, R. Schrader and R. Seiler (eds.), 411–416, Springer, Berlin,1982.
59. Albert Einstein und die Gravitationstheorie, in *Meyers Grosses Universallexikon*, Bibliogr. Institut, Mannheim, 1982.
60. Relations between the Galilei–invariant and the Lorentz–invariant theories of collisions, in *Space, Time and Mechanics*, D. Mayr and G. Sámann (eds), 21–37, Reidel, Dordrecht, 1983.
61. Problems of relativistic celestial mechanics, in *Aspetti Matematici Della Teorica Della Relativit* , 85–99, Atti Dei Convegni Lincei 57, Roma, 1983.
62. (with Martin Walker) Gravitational radiation and the quadrupole formula, in *General Relativity and Gravitation*, 10th Internat. Conf. on General Relativity and Gravitation, Padua, B. Bertotti, F. de Felice, A. Pascolini (eds.), 125–137, Reidel, Dordrecht, 1984.
63. On limit relations between, and approximate explanations of, physical theories, in *Logic, Methodology and Philosophy of Science*, Vol. VII, B. Marcus et al. (eds.), 387–403, Elsevier Science Publ., Amsterdam, 1986.
64. (with P. Schneider) Self-consistent probabilities for gravitational lensing in inhomogeneous universes, Astron. Astrophys. 168, 57–61 (1986).
65. Summary, in *Gravitational Collapse and Relativity*, Proc. of Yamada Conference XIV, Sato, H. and Nakamura, T. (eds.), 4–10, World Scientific, Singapore, 1986.
66. (with W. Rindler) How far can observable relations determine a Robertson–Walker metric? Astron. Astrophys. 174, 1–4 (1987).
67. (with A.R. Prasanna and R.A. Breuer) Propagation of gravitational waves through pressureless matter, *Class. Quant. Grav.* 4, 253–264 (1987).
68. Folklore in relativity and what is really known, in *General Relativity and Gravitation*, Proc. 11th Intern. Conf. on General Relativity and Gravitation, Stockholm, Mc Callum, M.A.H. (ed.), 61–81, Cambridge University Press, Cambridge, 1987.
69. Raum, Zeit, Materie im Großen: Kosmologie, in *Der Raum*, Schubert, V. (Hrsg.), 151–182, Eos-Verlag, St. Ottilien, 1987.
70. Hermann Weyl's Contributions to the General Theory of Relativity, in *Exakte Wissenschaften und ihre philosophische Grundlegung*, Deppert, W. et al. (eds.), 83–105, Verlag Peter Lang, Frankfurt, 1988.
71. Grundfragen der Relativistischen Physik und Quantenphysik (Abschlussbericht), Mitt. Deutsche Akad. d. Naturf. Leopoldina 32, 107–111 (1988).
72. (with G. Börner) Was there a big bang? Astron. Astrophys. 204, 1–2 (1988).
73. Einführung in die Raum-Zeit-Struktur mittels Lichtstrahlen und Teilchen, in *Philosophie und Physik der Raum-Zeit*, Andretsch, J. et al. (eds.), 145–162, BI-Wissenschaftsverlag, Mannheim, 1988.
74. Summary and Overview, in *Highlights in gravitation and cosmology* (Proc. Intern. Conf. on Gravitation and Cosmology, Goa, India), Iyer, B.R. et al. (eds.), 431–441, Cambridge University Press, Cambridge, 1988.
75. (with W. Rindler) A phase space representation of Friedmann-Lemaître universes containing both dust and radiation and the inevitability of the big bang, *Mon. Not. R. Astr. Soc.* 238, 503–521 (1989).

76. (with W. Rindler) A Novel Phase Space Representation & Classification of the Friedmann-Lemaître Universes, in *Proc. of the 14th Texas Symposium on Relativistic Astrophysics*, Fenyves, E.J. (ed.), *Annals of the New York Academy of Sciences* 571, 62–67 , New York 1989.
77. Summary, in *General Relativity and Gravitation*, Ashby, N. et al. (eds.), 491–502, Cambridge, 1990.
78. (with G. Schäfer), 75 Jahre Allgemeine Relativitätstheorie, *Phys. Bl.* 46, 481–484 (1990).
79. The Newtonian limit of general relativity, in *Classical Mechanics And Relativity: Relationship And Consistency*, Ferrarese, G. (ed.), 95–106, Napoli-Bibliopolis, Napoli, 1991.
80. (with P. Hübner) Inflation in Curved Model Universes with Noncritical Density, *Class. Quant. Grav.* 8, 333–346 (1991).
81. Stochastische Herleitung relativistischer Aussagen?, *Phu D 1.* 70–71 (1991).
82. Remarks on the Relation Between Machian Ideas and General Relativity, in *Ernst Mach and the Development of Physics*, Prosser, V. and Folta, J. (eds.), 83–89, Universitas Carolina Pragensis, Prague, 1991.
83. Ist die Raumzeit krumm?, in *Jahrbuch der Bayer. Akad. Wissensch.* 1991, 1–12, C.H. Beck Verlag, München, 1992.
84. Bemerkungen über die relativistische Himmelsmechanik, in *Naturgesetzlichkeit und Kosmologie in der Geschichte* (Festschrift für Ulrich Grigull), Bialas, V. (ed.), 108–115, Fr. Steiner Verlag, Stuttgart, 1992.
85. (with P. Schneider and E.E. Falco) *Gravitational Lenses*, Springer, Berlin,1992.
86. (with P. Schneider) Gravitational lensing, in *Relativistic Gravity Research*, Lecture Notes in Physics 410, Ehlers, J. and Schäfer, G. (eds.), 1–45, Springer, Berlin,1992.
87. Epilog, in *Vom Urknall zum komplexen Universum*, Börner, G., Ehlers, J. and Meier, H. (eds.), 195–203, Piper-Verlag, München, 1993.
88. Ist die Raumzeit krumm?, *Naturwiss. Rundschau* 46. Jg., Nr. 1, 1–6 (1993); Reprint of Nr. 83.
89. (with S. Kind) Initial-boundary value problem for the spherically symmetric Einstein equations for a perfect fluid, *Class. Quant. Grav.* 10, 2123–2136 (1993).
90. (with S. Kind and B.G. Schmidt) Relativistic stellar oscillations treated as an initial value problem, *Class. Quant. Grav.* 10, 2137–2152 (1993).
91. Schwerkraft und Weltall – Erkenntnisse und offene Fragen über den Aufbau und die Entwicklung der Welt im Großen, in *Horizonte. Wie weit reicht unsere Erkenntnis heute?*, Wilke, G. et al. (eds.), 135–142, Verhdl. Ges. Deutscher Naturf. und Ärzte, 117. Versammlg., Wissensch. Verlagsges., Stuttgart, 1993.
92. (with P. Schneider) Gravitational Lensing, in *General Relativity and Gravitation* 1992, Proceed. of GR13, Gleiser, R.J. et al (eds.), 21–40, Instit. of Physics Publ., Bristol, 1993.
93. (with R. Rindler) Successive Schwarzschild spheres and other rigidity frontiers in spherically symmetric dust-plus-vacuum spacetimes, *Physics Letters* A 180, 197–202 (1993).
94. Contributions to the Relativistic Mechanics of Continuous Media, GRG 25, 1225–1266 (1993). English translation of Nr. 11.

95. (with T. Buchert) Lagrangian theory of gravitational instability of Friedmann-Lemaître cosmologies – second-order approach: an improved model for non-linear clustering, *Mon. Not. R. Astron. Soc.* 264, 375–387 (1993).
96. (with M. Bartelmann and P. Schneider) Timescales of isotropic and anisotropic cluster collapse, *Astron. Astrophys.* 280, 351–359 (1993).
97. Experimentierkunst und Scharfsinn in der Theorie, zu H. Hertz' 100. Todestag, Berliner Tagesspiegel 31.12.93, p. 15.
98. Der Nobelpreis für Physik 1993, *Phu D* 94.1, 64–68 (1994).
99. Ein Universalgenie der Wissenschaft, zum hundertsten Todestag des Naturforschers Hermann von Helmholtz, Berliner Tagesspiegel 7.9.94, p. 27.
100. (with S. Seitz and P. Schneider) Light propagation in arbitrary spacetimes and the gravitational lens approximation, *Class. Quantum Grav.* 11, 2345–2373 (1994).
101. (with H.-J. Fahr) Urknall oder Ewigkeit, Bild der Wissenschaft 6, 84–86 (1994).
102. (with H. Friedrich, ed.) Canonical Gravity: From Classical to Quantum, Lecture Notes in Physics 434, Springer, Heidelberg, 1994.
103. Machian Ideas and General Relativity, in *Mach's Principle*, J. Barbour and H. Pfister (eds.), 458–469, Birkhäuser, Basel, 1995.
104. Spacetime Structures, in *Physik, Philosophie und die Einheit der Wissenschaften*, L. Krüger und B. Falkenburg (eds.), 165–176, Spektrum Akad.-Verlag, Heidelberg, 1995.
105. Gedenkansprache anlä sslich der Enthüllung der Gedenktafel für Albert Einstein, Berlin-Brandenburgische Akademie der Wissenschaften, Berichte und Abh., Bd. 1, p. 304–308.
106. (with G. Börner, eds.) Gravitation, *Spektrum der Wissenschaft*, Heidelberg, 2nd edn., 1996.
107. (with A.R. Prasanna) A WBK formalism for multicomponent fields and its application to gravitational and sound waves in perfect fluids, *Class. Quant. Grav.* 13, 2231–2240 (1996).
108. (with W. Rindler) Local and Global Light Bending in Einstein's and Other Gravitational Theories , *GRG* 29, 519–529 (1997).
109. Concepts of time in Classical Physics, in *Time, Temporality, Now*, H. Atmanspacher and E. Ruhnau (eds.), 191–200, Springer, Berlin,1997.
110. Examples of Newtonian Limits of Relativistic Spacetimes, *Class. Quant. Grav.* 14, A119–A126 (1997).
111. (with Th. Buchert) Newtonian Cosmology in Lagrangian Formulation: Foundations and Perturbation Theory, *GRG* 29, 733–764 (1997).
112. (with Th. Buchert) Averaging in Newtonian Cosmology, *Astronomy & Astrophysics*, 320, 1–7 (1997).
113. 80 Years of General Relativity, in *Reviews in Modern Astronomy* 10, R.E. Schielicke (ed.), 91–100, Astron. Gesellschaft, Hamburg, 1997.
114. (with G.F.R. Ellis, G. Börner, Th. Buchert, C.J. Hogan, R.P. Kirshner, W.H. Press, G. Raffelt, F.-K. Thielemann, S. van den Bergh), What Do We Know About Global Properties of the Universe, in *The Evolution of the Universe*, S. Gottlöber and G. Börner (eds.), 51–78, Wiley, New York, 1997.
115. Der Kosmos als Objekt der Naturforschung, *Nova Acta Leopoldina*, NF76, Nr. 303, 139–147, 1997.

116. General Relativity as a Tool for Astrophysics, *Relativistic Astrophysics* H. Riffert, H. Ruder, H.-P. Nollert and W. Hehl (eds.), 1–15, Vieweg, Braunschweig, 1998.
117. Reprint of 115 in *Der Sternenbote* 41. Jg., Heft 5, Wien, 1998.
118. The Newtonian Limit of General Relativity, in *Understanding Physics*, A.K. Richter (ed.), 1–13, Copernicus Gesellschaft e.V., Katlenburg, Lindau, 1998.
119. Some Developments in Newtonian Cosmology, in *On Einstein's Path*, A Harvey (ed.), 189–202, Springer, Berlin,1999.
120. "Gravitationslinsen – Lichtablenkung in Schwerefeldern und ihre Anwendungen". Book series of the Carl Friedrich von Siemens Stiftung, Vol. 69, 1999.

Printing: Druckhaus Beltz, Hemsbach
Binding: Buchbinderei Schäffer, Grünstadt